Sustainable Solutions for Food Security

Atanu Sarkar • Suman Ranjan Sensarma
Gary W. vanLoon
Editors

Sustainable Solutions for Food Security

Combating Climate Change by Adaptation

Editors
Atanu Sarkar
Division of Community Health and
Humanities
Faculty of Medicine, Memorial University
St Johns, NL, Canada

Suman Ranjan Sensarma
KPMG
New Delhi, India

Gary W. vanLoon
School of Environmental Studies
Queen's University
Kingston, ON, Canada

ISBN 978-3-319-77877-8   ISBN 978-3-319-77878-5   (eBook)
https://doi.org/10.1007/978-3-319-77878-5

Library of Congress Control Number: 2018964086

© Springer Nature Switzerland AG 2019
This work is subject to copyright. All rights are reserved by the Publisher, whether the whole or part of the material is concerned, specifically the rights of translation, reprinting, reuse of illustrations, recitation, broadcasting, reproduction on microfilms or in any other physical way, and transmission or information storage and retrieval, electronic adaptation, computer software, or by similar or dissimilar methodology now known or hereafter developed.
The use of general descriptive names, registered names, trademarks, service marks, etc. in this publication does not imply, even in the absence of a specific statement, that such names are exempt from the relevant protective laws and regulations and therefore free for general use.
The publisher, the authors, and the editors are safe to assume that the advice and information in this book are believed to be true and accurate at the date of publication. Neither the publisher nor the authors or the editors give a warranty, express or implied, with respect to the material contained herein or for any errors or omissions that may have been made. The publisher remains neutral with regard to jurisdictional claims in published maps and institutional affiliations.

This Springer imprint is published by the registered company Springer Nature Switzerland AG
The registered company address is: Gewerbestrasse 11, 6330 Cham, Switzerland

Foreword

Climate change is no longer a myth or speculation; rather it is a clear and present threat to the entire biosphere and all of humanity. We have already started witnessing the onslaughts of climate change in various avatars, such as extreme weather patterns (drought, heavy rain, floods, cyclones, heat waves, and forest fires), sea level rise, growing numbers of climate refugees, and the spread of various illnesses. It is worth emphasizing that people from low-income countries are the ones who will bear the major brunt of climate change.

Indeed, climate change is a threat to all four dimensions of food security, i.e. its availability, accessibility, utilization, and systems stability. The inhabitants of vulnerable and food insecure areas are likely to be the first victims. People living on the coasts and floodplains and in mountains, drylands, and the Arctic are the most at risk. Agriculture-based livelihood systems that are already vulnerable to food insecurity face the risk of increased crop failure, new patterns of pests and diseases, lack of good seeds and planting materials, and loss of livestock and fisheries. Low-income people are at risk of food insecurity owing to loss of assets and lack of an adequate social safety net. Food systems are also facing looming threats of internal and international migration, resource-based conflicts, and civil unrest triggered by climate change and its consequences.

Considering the global trend in using fossil fuels, rising energy demands of the emerging economies, and lack of political consensus, there has been very limited progress in curbing greenhouse emissions following the Kyoto Protocol and subsequent international agreements including the recent Paris Climate Agreement. Even, for argument sake, if we stop all greenhouse gas emitting anthropogenic activities, it will take several decades to reach any noticeable improvement. Therefore, it is necessary to strengthen the resilience of the vulnerable people and to help them cope with this additional threat to food security. To avert food insecurity, several international agencies and experts strongly advocate climate change adaptation along with mitigation. Sustainable adaptations are believed to ensure food security without affecting the ecosystem and at the same time maintain economic growth. Scientists are actively engaged in the development of various sustainable solutions to overcome the adverse effects of climate change on food security, and many of these ingenuities are displaying promising results. On the other hand, there is a need for appropriate social policy and pro-people regulatory systems in order to bring the benefits of climate smart strategies.

Despite the fact that some sustainable technologies are already recognized as the viable solutions for adapting climate change to warrant food security, there are no proper centralized information sources of those proven (and also waiting-for-approval) solutions for researchers and policy-makers. While there are numerous publications describing impacts of climate change on food security at a global or regional scale, there has not been any book documenting and describing the potential contribution to climate adaptation of tested and replicable, sustainable solutions.

The present volume *Sustainable Solutions for Food Security: Combating Climate Change by Adaptation* has filled the long-standing gap. The editors, led by Dr. Atanu Sarkar, make a significant effort to illustrate the right perspectives, and have made a significant contribution to disseminate the knowledge in a layman's language. It is a challenge to provide an interdisciplinary perspective in the complex discourses of food insecurity, climate change, and adaptation. The editors are to be commended for recognizing the value of diversity in research and practice. They have brought together researchers from all continents, and the shared research is relevant in local and global contexts. While the volume, as a whole, represents an excellent cross section of the extremely complicated global problem, the individual chapters contribute to various specific lines of inquiry, for example, latest research and application of climate smart crops, fisheries, natural resource management, and capacity building for farming and disadvantaged communities.

I deeply appreciate the efforts of the editorial team for their hard work and hope that the volume will be very helpful for policy-makers, development practitioners, researchers, and general readers who are interested in the subject.

M. S. Swaminathan
M. S. Swaminathan Research Foundation,
Chennai, India

Foreword

The probability that climate change would harm food security was well recognized by some in the early 1990s, though it was, as far as I know, not until 2003 that the Food and Agriculture Organization (FAO) of the United Nations first highlighted climate change at the 29th session of the FAO Committee on World Food Security. In the 1990s there was a strong, even dominant view that although climate change would create winners and losers, food production overall would hold its own, even considering population growth and growing affluence for many.[1] Furthermore, and perhaps not coincidentally, the losers were mostly predicted to be farmers and poor consumers in low-income regions, such as the tropics, while winners would be farmers at high latitudes (soil quality permitting) in Russia, Canada, and Scandinavia. There has also long been a view that adaptation by farmers and others involved would make an important compensatory contribution.

That relaxed period is over, as revealed many times by the over 60 contributors of this authoritative and optimistic book, written by authors from many countries, including from several of the world's leading agricultural research institutions. Although the book has a particular emphasis on agricultural adaptation to climate change, some chapters also focus on issues such as equity and natural resource management. The authors collectively illustrate the large and diverse challenges that climate change poses to agriculture. These include rising temperature, more extremes, intensified droughts, and, in some settings, more flooding and rising salinity. There are also risks of increased pests and falling nutritional properties, even if yields can be preserved. Added to this is the growing realization that other resources needed for agriculture are limited (at least at affordable prices), such as fossil fuels, currently essential for fertilizer, for pesticides, and, in most places, for transport and powering machinery.

Suggested techniques to contribute to the solutions that are needed (and that are discussed) include crop diversification, drip irrigation, and different crop management strategies. Also important are crop breeding and genomic modification,

---

[1] Butler, C.D. (2009) C.D. Food security in the Asia-Pacific: Malthus, limits and environmental challenges *Asia Pacific Journal of Clinical Nutrition* **2009**, *18*, 577–584

including the possibility that it may be able to make rice more like maize, by converting strains of it to a C4 plant, thus reducing its need for water and nitrogen fertilizer.

Although the book has chapters on adapting the staples of wheat, maize, and rice to climate change, including in Africa, the need for greater diversification is recognized and reflected by chapters on crops such as Bambara groundnuts. Several chapters stress the additional vulnerability of agriculture (and farmers) in rain-fed areas, which often have the additional disadvantage of being underdeveloped. As one chapter points out, "the most vulnerable farmers and communities, namely, smallholder subsistence and market oriented farmers in marginal environments, are those who face the most extreme climate change-related challenges at the same time as being the least able to adapt."

The methods and policies described in this book are vital to improve food security, but so too are climate change mitigation, slower population growth (especially in countries with high fertility), and a fairer global system of economic opportunity and reward. It is my understanding, especially, of these challenges, shared with Associate Professor Atanu Sarkar who, together with his colleagues Prof. Emeritus Gary W. vanLoon and Dr. Suman Ranjan Sensarma, has edited this book, which motivated me to write this foreword, which I am delighted to do.

Colin D. Butler
Health Research Institute
University of Canberra
Canberra, ACT, Australia

College of Arts, Humanities & Social Sciences
Flinders University
Bedford Park, SA, Australia

National Centre for Epidemiology and Population Health
Australian National University
Canberra, ACT, Australia

Contents

Part I Introduction

1 Introduction 3
Atanu Sarkar, Gary W. vanLoon, and Dave Watson

Part II Climate-Smart Crops, Adaptive Breeding and Genomics

**2 Adaptation of Crops to Warmer Climates:
Morphological and Physiological Mechanisms** 27
Ullah Najeeb, Daniel K. Y. Tan, Muhammad Sarwar,
and Shafaqat Ali

3 Climate Smart Crops: Flood and Drought-Tolerant Crops 51
Camila Pegoraro, Carlos Busanello, Luciano Carlos da Maia,
and Antonio Costa de Oliveira

**4 Adaption to Climate Change: Climate Adaptive
Breeding of Maize, Wheat and Rice** 67
Dave Watson

5 Using Genomics to Adapt Crops to Climate Change 91
Yuxuan Yuan, Armin Scheben, Jacqueline Batley,
and David Edwards

6 Climate-Resilient Future Crop: Development of C_4 Rice 111
Hsiang Chun Lin, Robert A. Coe, W. Paul Quick,
and Anindya Bandyopadhyay

**7 Crop Diversification Through a Wider Use
of Underutilised Crops: A Strategy to Ensure Food
and Nutrition Security in the Face of Climate Change** 125
M. A. Mustafa, S. Mayes, and F. Massawe

Contents

8 Bambara Groundnut is a Climate-Resilient Crop: How Could a Drought-Tolerant and Nutritious Legume Improve Community Resilience in the Face of Climate Change? 151
Aryo Feldman, Wai Kuan Ho, Festo Massawe, and Sean Mayes

Part III Natural Resource and Landscape Management

9 The Challenge of Feeding the World While Preserving Natural Resources: Findings of a Global Bioeconomic Model 171
Timothy S. Thomas and Shahnila Dunston

10 Adaptation to Climate Change Through Adaptive Crop Management 191
Dave Watson

11 Water Management Technology for Adaptation to Climate Change in Rice Production: Evidence of Smart-Valley Approach in West Africa. 211
Aminou Arouna and Aristide K. A. Akpa

12 An Agroecological Strategy for Adapting to Climate Change: The System of Rice Intensification (SRI). 229
Norman Uphoff and Amod K. Thakur

13 Efficient Desalinated Water Pricing in Wetlands. 255
Oscar Alfranca

14 Drip Irrigation Technology: Analysis of Adoption and Diffusion Processes 269
Francisco Alcon, Nuria Navarro, María Dolores de-Miguel, and Andrea L. Balbo

15 Combating Climate Change Impacts for Shrimp Aquaculture Through Adaptations: Sri Lankan Perspective 287
J. M. P. K. Jayasinghe, D. G. N. D. Gamage, and J. M. H. A. Jayasinghe

16 Application of the Terroir Concept on Traditional Tea Cultivation in Uji Area 311
Fitrio Ashardiono

17 The Opportunity of Rural Space with Urban Relationships: Urban Agriculture as Contemporary Cultural Landscape for Resilience by Design 331
Luis Maldonado Rius

Contents   xi

Part IV   Strategies for Access to Food, Technology, Knowledge and Equity, and Risk Governance

18 Stakeholders' Perceptions on Effective Community Participation in Climate Change Adaptation 355
Subhajyoti Samaddar, Akudugu Jonas Ayaribilla, Martin Oteng-Ababio, Frederick Dayour, and Muneta Yokomatsu

19 Disadvantaged Communities in Indonesian Semi-Arid Regions: An Investigation of Food Security Issues in Selected Subsistence Communities in West Timor 381
Yenny Tjoe, Paulus Adrianus Ratumakin, Moazzem Hossain, and Peter Davey

20 Enhancing Food Security in Subarctic Canada in the Context of Climate Change: The Harmonization of Indigenous Harvesting Pursuits and Agroforestry Activities to Form a Sustainable Import-Substitution Strategy 409
Leonard J. S. Tsuji, Meaghan Wilton, Nicole F. Spiegelaar, Maren Oelbermann, Christine D. Barbeau, Andrew Solomon, Christopher J. D. Tsuji, Eric N. Liberda, Richard Meldrum, and Jim D. Karagatzides

21 Vulnerability Amidst Plenty? Food Security and Climate Change in Australia 437
Ruth Beilin, Michael Santhanam-Martin, and Tamara Sysak

22 Agricultural Decision Support Tools: A Comparative Perspective on These Climate Services 459
Jonathan Lambert, Nagothu Udaya Sekhar, Allison Chatrchyan, and Art DeGaetano

23 Multilevel Governance for Climate Change Adaptation in Food Supply Chains 479
Ari Paloviita and Marja Järvelä

24 Climate Change and Food Security in India: Adaptation Strategies and Major Challenges 497
Atanu Sarkar, Arindam Dasgupta, and Suman Ranjan Sensarma

Part V   Conclusion

25 Conclusion 523
Atanu Sarkar and Gary W. vanLoon

Index 531

About the Editors and Authors

Editors

Atanu Sarkar, PhD, MES, MBBS Atanu Sarkar's major research interests are (a) climate change, food security, and climate change and health; (b) agriculture and health; (c) water quality and contamination; (d) endocrine disrupting chemicals and ecosystem health; and (e) employment, occupation, and health. In recent past, he has completed several research projects on food security, water and environmental contamination, and climate change. His future research interests are (a) adaptation to climate change and sustainable food production in India and subarctic Canada and (b) water and food contamination and low-cost remediation. Environmental Health and Occupational Health, Division of Community Health and Humanities, Faculty of Medicine, Memorial University, St. John's, NL, Canada

Suman Ranjan Sensarma, DEng, MPlan Suman Ranjan Sensarma's research interests include urban governance, smart city, climate change and sustainability, and integrated disaster risk management. He worked as a country investigator for COMPON (COMparing Climate Change Policy Networks, Research Project on Climate Change), funded by the US National Science Foundation from 2009 to 2012. He delivered a number of guest lecturers/expert talks and seminars in many institutions across the country like the School of Planning and Architecture, New Delhi; Indian Institute of Public Administration, New Delhi; and IIT, Roorkee and international forums. Government Advisory, Infrastructure and Government Services (IGS), KPMG, New Delhi, India

Gary W. vanLoon, PhD Gary vanLoon was originally trained as an analytical chemist but for the past 45 years has carried out research in areas of environmental chemistry. Much of this has been directed towards understanding the behaviour of human-source chemicals in soil and water. More recently, the focus of his work has shifted towards the big-picture subject of sustainability in its broadest sense, and he is particularly interested in how one can assess (measure) it. Along with several colleagues, he has proposed and developed a methodology that can be used for this purpose, and they have applied it in a number of situations, particularly agricultural situations. Lately, they have tried to bring health issues more closely into the assessment procedures. Much of their testing has been done in low-income countries, particularly in India and more recently in Bangladesh. Department of Chemistry, Queen's University, Kingston, ON, Canada

School of Environmental Studies, Queen's University, Kingston, ON, Canada

Authors

Jonas Ayaribilla Akudugu, PhD Jonas Akudugu's research interests include local economic development promotion, decentralization, and community resource management and community development. University for Development Studies, Wa, Ghana

Aristide K. A. Akpa Biography Africa Rice Center (AfricaRice), Cotonou, Bénin

Francisco Alcón, PhD Dr. Francisco Alcón (male) is an agricultural engineer specialized on agricultural economics and policies. He obtained his PhD in Economics of Technology Adoption in 2007. He is currently working as associate professor in the Business Economics Department (Agricultural Economics Area) at Universidad

Politécnica de Cartagena. His primary research interests are in the fields of agricultural technology adoption, non-market valuation, agricultural resource management, resource management policy and planning, cost-benefit analysis, and discrete choice models. He has been awarded by the European Association of Agricultural Economists with the Best Paper of the Year 2010 Award and by the Australian Agricultural and Resource Economics Society (AARES) with the Quality of Research Discovery Award for 2012. Universidad Politécnica de Cartagena, Cartagena, Murcia, Spain

Dr. Oscar Alfranca, PhD Oscar Alfranca is a visiting professor specialized on environmental economics at the Iowa State University and the Laboratoire d'Economie Appliquée de l'INRA (France) whose research is about public policies in agri-food policies. UPC, Barcelona, Spain

Iowa State University, Ames, IA, USA

Laboratoire d'Economie Appliquée de l'INRA, Grenoble, France

Shafaqat Ali Biography Department of Environmental Sciences and Engineering, Government College University, Faisalabad, Pakistan

Aminou Arouna, PhD Aminou Arouna is an associate principal scientist and impact assessment economist at AfricaRice. He specialized in environmental and resources economics (water and land) and impact assessment of technological and institutional innovations on food security and livelihood in rural developing countries using quantitative methods. He has published in many peer-reviewed journals including *Water Management* and *Global Food Policy*. Aminou obtained his Master's degree in 2002 from the University of Abomey-Calavi (Benin) with distinction and was the best student. In 2006, Aminou was awarded a DAAD scholarship for the PhD studies at the University of Hohenheim (Germany) where he obtained the PhD degree with distinction (Magna cum laude) in 2009. Africa Rice Center (AfricaRice), Cotonou, Benin

Fitrio Ashardiono, PhD Currently, the main research focus is to develop climate change adaptation method in the agriculture industries, by utilizing terroir concept as the base framework for ensuring the environmental and socio-economic sustainability of the affected region. At the moment, his research is continuing on the tea cultivation in Uji Area in Japan. In addition to his research, he is also researching the tea cultivation in the Nuwara Eliya Region, Sri Lanka, and the Sukabumi Region in Indonesia. Based on the research from the tea cultivation, the developed terroir framework would apply to other agriculture industries which possess similar traits such as coffee cultivation and other agriculture cultivation. Asia-Japan Research Institute, Ritsumeikan Asia-Japan Research Organization, Ritsumeikan University, Kyoto, Japan

Andrea Luca Balbo, PhD As an archaeologist, her research focus is on human adaptation to, and human transformation of, different environmental contexts. Her work revolves around the Holocene epoch, with a broad geographical spectrum, from high latitudes (Northern Norway) to desert environments (Libyan Sahara, Syria). Her current research is on climate-related and water-related challenges in Mediterranean agricultural landscapes, whose origin can be tracked back to pre-industrial times. With her current and future work, she wishes to further contribute to the integration of the archaeological and anthropological perspectives and theory in sustainable development discourses. Research Group Climate Change and Security (CLISEC), Center for Earth System Research and Sustainability (CEN), University of Hamburg, Hamburg, Germany

Anindya Bandyopadhyay, PhD With a combined experience of more than 15 years in both academic and seed industry, Dr. Anindya's academic pursuit as an ardent researcher in the area of plant biotechnology as well as professional experience as a manager/group leader in agriculture/plant biotechnology sector has equipped him with handling large research team, national and international research consortiums, management and facilitation of research projects, research strategy development, and operation. Conducting plant biotechnology studies pertaining to development of high-yielding crops for the developing world towards a sustainable agriculture system is always the core of Dr. Anindya's past and present activities in the professional arena. His research experience and interest also span in the area of application of modern technologies in the crop science research such as CRISPR system-mediated genome editing, high-throughput crop transformation systems, gene discovery, plant molecular biology, and automated phenomics studies. After a successful stint at the International Rice Research Institute as a leader of Molecular Biology and deputy coordinator of International C4 Rice Consortium, Dr. Anindya presently leads genome editing programme at Syngenta Innovation Center with a future goal of applying genome editing for the discovery of novel breeding technologies and introduction of beneficial traits in the crop plants towards a climate-resilient sustainable agriculture. Syngenta, Beijing, People's Republic of China

Christine D. Barbeau, BSc, MES, PhD Christine Barbeau's research interests include climate change adaptation, agroforestry, web-based decision-support tools, and decision-support apps. School of Environment, Resources, and Sustainability, University of Waterloo, Waterloo, ON, Canada

Jacqueline Batley, PhD, MSc, BSc (Hons) Professor Jacqueline Batley is an ARC Future Fellow at the University of Western Australia. She moved to Australia in 2002 as a senior research scientist at DPI-Victoria and then led a research group at the University of Queensland as an ARC QEII Research Fellow, from 2007 to 2014. She has received several awards for her research including a University of Queensland Foundation Research Excellence Award, an ARC QEII Fellowship, and an ARC Future Fellowship. Jacqueline has expertise in the field of plant molecular biology, genetics, and genomics, gained from working in both industry and academia. Her areas of interest include genetic and genomic analysis and, specifically, genome sequence analysis, pan genomics, SNP analysis, and the role of structural variation for applications such as genetic diversity, genetic mapping, LD, GWAS, evolutionary, population, and comparative genomic studies, as well as the molecular characterization of agronomic traits. She is currently focussing on blackleg resistance in *Brassicas*. School of Biological Sciences, University of Western Australia, Crawley, WA, Australia

Ruth Beilin, PhD Professor Ruth Beilin is a landscape and environmental sociologist. Her research is in the area of land and water management and focuses on how people live in the landscape and what their practices are in shaping our social ecological interfaces. She is a former Director of the Office for Environmental Programs, a Scientific Advisor for the New Zealand Government's National Land and Water Challenge, and an author and co-author of over 100 publications. Environmental Social Science, School of Ecosystem and Forest Sciences, Faculty of Science, University of Melbourne, Parkville, VIC, Australia

Carlos Busanello, PhD Carlos Busanello's research interests include functional genomics in rice, bioinformatics, and abiotic stress response in rice. He improved the proposal and the manuscript and edited the manuscript. Plant Genomics and Breeding Center, School of Agronomy Eliseu Maciel, Federal University of Pelotas, Pelotas, RS, Brazil

Allison Morrill Chatrchyan, PhD, MA, BA As a social scientist, Dr. Chatrchyan's work focuses on climate change and the interactions between social, environmental, and agricultural systems. She facilitates interdisciplinary research and extension teams and helps develop resources and tools for climate change adaptation and mitigation. Her research focus is on assessing views and actions on climate change, multilevel climate change governance mechanisms, and climate change policies and institutions. She is codeveloping a plan for a new Climate Master Extension Volunteer Program for Climate Smart Communities with NIFA funds. Dr. Chatrchyan helped establish the Cornell Climate Smart Farming programme, the University Climate Change Seminar series, and has led the Cornell Delegations to COP21, COP22, and COP23. Recent publications have been published by Routledge; Springer International; *Renewable Agriculture and Food Systems*; *Journal of Extension*; *Weather, Climate, and Society*; and *WIREs Climate Change*. Department of Development Sociology, Cornell University, Ithaca, NY, USA

Department of Earth and Atmospheric Sciences, Cornell University, Ithaca, NY, USA

Cornell Institute for Climate Smart Solutions, Cornell University, Ithaca, NY, USA

Robert A. Coe Biography C4 Rice Centre, International Rice Research Institute (IRRI), Los Baños, Philippines

Arindam Dasgupta, PhD, MA, MPhil Arindam Dasgupta's research interests include urban sprawl and urban land use, urban governance and basic services, smart city, development projects, and socio-economic change.

At present he is doing a project funded by the University Grants Commission (UGC) on the status of urban basic services and role of urban governance (E-governance) on some selected municipalities and municipal corporation in West Bengal, India. He wants to study on urban food security for urban poor in the backdrop of climate change in the future. He has worked as research associate in the Indian Institute of Advanced Study, Rashtrapati Niwas, Shimla, for 3 years (2003–2005) and delivered expert lectures there. Besides he has also delivered invited lectures and presented papers in different international conferences, seminar, and workshops. Department of Geography, Chandernagore College, Hooghly, West Bengal, India

Peter John Davey, PhD, MEnvComHlth, BBus Dr. Peter Davey has research interests in better understanding and improving policy and professional practice in the disciplines of environmental protection and pollution, coastal planning, and health quarantine, with full consideration of the United Nations-SDG and UNISDR-DRR Frameworks in developing countries. Peter conducts academic research designed to guide more resilient and sustainable livelihoods, with a focus on the complexity of internal migration as a result of climate disruption and increasing disaster events using qualitative approaches, community needs assessment tools and process, and impact and outcome evaluation. Peter has skills in forming partnerships and coalitions, memorandum of understandings at the local level, and the formulation and implementation of strategic community planning, monitoring the success factors and challenges of government and non-government policy initiatives. Research has been focused in several Asia and Pacific countries/ministries/agencies including Australia, Indonesia, Malaysia, and Bangladesh and more recently in Brazil. Peter is a UNISDR-DRR "Train the Trainer" and in this capacity conducts regular intensive environmental health and disaster management training courses identifying problems and research options for future risk reduction and dissemination of information in communities. Centre of Excellence for Sustainable Development for Indonesia, Griffith University, Brisbane, QLD, Australia

Environmental Protection in School of Environment and Natural Science (ESC), Griffith University, Brisbane, QLD, Australia

Frederick Dayour, MPhil Frederick Dayour is a faculty at the UDS, Ghana. He is now a tourism PhD research student at the University of Surrey, UK. His PhD is fully sponsored by the University of Surrey for 3 years. His research interests comprise ICTs in tourism, backpacker tourist behaviours, and tourism and terrorism and climate change. His works have been published in the following journals: *Annals of Tourism Research, Tourism Planning and Development, Tourism Management Perspectives, Journal of Quality Assurance in Hospitality and Tourism*, and *Risk, Hazards, and Crises in Public Policy*. Department of Community Development, University for Development Studies (UDS), Wa, Ghana

Arthur DeGaetano, PhD, MS, BS Art DeGaetano is a professor in the Department of Earth and Atmospheric Sciences and Director of the Northeast Regional Climate Center (NRCC). The mission of the NRCC is to enhance the use and dissemination of climate information to a wide variety of sectors in the Northeast. Art serves as an editor for the American Meteorological Society *Journal of Applied Meteorology and Climatology*. His research focuses on applied climatology which involves the development of methods and data sets that provide climatological information to decision-makers in a variety of fields. Four more specific research areas make up this general focus: modelling climate influences on man-made and biological systems, documenting observed variations in the climate system, improving climate data quality, and assessing climate impacts. Recent publications have been published in journals including *Theoretical and Applied Climatology*, the *International Journal of Climatology*, the *Journal of Applied Meteorology and Climatology*, and the *Bulletin of the American Meteorological Society*. Department of Earth and Atmospheric Sciences, Northeast Regional Climate Center (NRCC), Cornell University, Ithaca, NY, USA

Shahnila Dunston, MS Shahnila Dunston's current research focuses on climate change adaptation and modelling of agricultural crops and technologies going out into the future. She has contributed to the development of the International Food Policy Research Institute's (IFPRI) multi-market partial equilibrium model of agriculture: the International Model for Policy Analysis of Agricultural Commodities and Trade (IMPACT). As a part of her research, she has developed and analysed future scenarios for projects for USAID, the Bill and Melinda Gates Foundation, FAO, etc. Previously, Shahnila has also worked on topics related to biofuels and water resource management. International Food Policy Research Institute (IFPRI), Washington, DC, USA

David Edwards, PhD David Edwards' research interests include the structure and expression of plant genomes, the discovery and application of genome variation, and applied bioinformatics, with a focus on crop plants and accelerating crop improvement in the face of climate change. School of Biological Sciences, University of Western Australia, Perth, WA, Australia

Aryo Benjamin Feldman, PhD Dr. Aryo Feldman joined Crops for the Future, Malaysia, in 2014 as a coordinator for field evaluation trials of global Bambara groundnut (*Vigna subterranea* (L.) Verdc.) germplasm with partners from Southeast Asia and sub-Saharan Africa. He specializes in plant physiology and has considerable training in agronomy, plant breeding, and genetics. Dr. Feldman received his Bachelor's, Master's, and PhD degrees from the University of Nottingham, the latter of which was funded by the Monsanto Beachell-Borlaug International Scholars Programme. His passion is in international agriculture, particularly within the scope of small-scale, resource-poor farmers in low-income countries. He is concerned about how it can contribute to relieving global hunger, poverty, and environmental degradation while remaining critical of development issues. Crops for the Future, Semenyih, Selangor, Malaysia

D. G. N. D. Gamage Biography Faculty of Livestock, Fisheries and Nutrition, Department of Aquaculture & Fisheries, Wayamba University of Sri Lanka, Gonawila, Sri Lanka

Moazzem Hossain, PhD Moazzem Hossain's current research interests are in the areas of global warming-induced climate change and growth in Asia and sustainability. He has been working with climate change issues since 2007. Over the last 10 years, he has published in three outlets covering these areas: books, journal articles, and book chapters. Twelve books, journal articles, and book chapters have been published on climate change and related areas over the last 10 years. In addition he has contributed numerous oped articles in print media and has participated in a few live TV talk shows with the invitation of Channel NewsAsia based in Singapore. One of these was on Indonesia's climate adaptation and mitigation during the visit by the former president Barak Obama to Indonesia few years back. Department of International Business and Asian Studies, Griffith Business School, Griffith University, Brisbane, QLD, Australia

Wai Kuan Ho, PhD Wai Kuan's research area includes genetic improvement in crops, particularly on molecular mechanisms underlying genetic control of agronomic traits. Having experience in biomarker development in oil palm to predict embryogenesis potential, her current interest is currently focusing on functional genomic and transcriptomic studies to assist in crop breeding programmes. Crops for the Future, University of Nottingham Malaysia Campus, Semenyih, Selangor, Malaysia

Marja Järvelä, PhD Over the years Järvelä has completed many research projects related to local sustainable development and adaptation to climate change in Finland, Europe, Russia, and West Africa. With the background in social and public policy studies, she has focused on topics of social resilience in community, ways of life, and participation. Since 2011 she is member and vice-chair of the Finnish Climate Change Panel, and her main activities in research are connected accordingly to transdisciplinary investigations related to national and regional climate policy. Department of Social Sciences and Philosophy, University of Jyväskylä, Jyväskylä, Finland

Jayasinghe Mudiyanselage Pushpakumara Jayasinghe, PhD Jayasinghe Mudiyanselage Pushpakumara Jayasinghe is an emeritus professor at present and affiliated to Wayamba University whose research interests include climate change impacts and adaptations in aquaculture, brackish water environment assessment and management, coastal zone management, and coastal sediment management. Faculty of Livestock Fisheries and Nutrition, Wayamba University of Sri Lanka, Gonawila, Sri Lanka

J. M. H. A. Jayasinghe Biography   Faculty of Engineering, University of Moratuwa, Moratuwa, Sri Lanka

Jim D. Karagatzides, BA, MSc, PhD   Jim Karagatzides's research interests include agroforestry, plant-environment interactions, plant nutrient allocation, nutrient cycling, ecosystems ecology, human toxicology, and environmental education.   Engineering and Environmental Technologies, Georgian College, Barrie, ON, Canada

Jonathan E. Lambert, MA, BS, BS   Jonathan Lambert has been involved in research in climate and environmental science for over 8 years and has specialized in fields such as paleotempestology (study of ancient hurricanes), climatology, ocean acidification, climate change and agriculture, and oceanography. At Cornell, he was involved in modifying and creating online agricultural decision tools that empower farmers—via data-driven user-friendly platforms—to make more science-based decisions in the face of climate variability and change. He currently seeks to understand past changes in climate using oceanic paleoclimate proxies such as foraminifera in the Pacific Ocean and is attempting to determine the role of specific Western Pacific air-sea interactions in the regional to global climate system. He has been awarded the NOAA Ernest F. Hollings Scholarship and the Udall Scholarship for undergraduate work, a Dean's Fellowship and Columbia Climate and Society Scholarship for graduate work, and awards for posters at regional conferences and symposia.   Department of Earth and Environmental Sciences, Lamont-Doherty Earth Observatory, Columbia University, New York, NY, USA

Eric N. Liberda, BES, MES, MASc, PhD   Eric Liberda's research interests include environment and health, toxicology, inhalational toxicology, exposure assessment, and indigenous health.   School of Occupational and Public Health, Ryerson University, Toronto, ON, Canada

Hsiang Chun Lin Biography   C4 Rice Centre, International Rice Research Institute (IRRI), Los Baños, Philippines

Luciano Carlos da Maia, PhD   Luciano Carlos da Maia's research interests include functional genomics in rice, bioinformatics, abiotic stress response in rice, and maize breeding. He improved the proposal and the manuscript and edited the manuscript.   Plant Genomics and Breeding Center, School of Agronomy Eliseu Maciel, Federal University of Pelotas, Pelotas, RS, Brazil

Festo Massawe, PhD   Professor Festo Massawe is professor of Crop Science and head of the School of Biosciences at the University of Nottingham, Malaysia Campus. He is also codirector of the CFF-UNMC Doctoral Training Partnership. His research interests range from crop physiology to breeding and the use of biotechnological approaches for genetic improvement of crop plants. His research focuses on sustainable crops and crop diversification using underutilized crops. He has published in various scientific journals and co-authored book chapters mainly on research and development of underutilized crops. As codirector of the CFF-UNMC DTP, he has worked with the DTP team to support over 49 postgraduate students working across disciplinary boundaries to develop solutions for global challenges including food insecurity.   University of Nottingham Malaysia Campus, Semenyih, Selangor, Malaysia

Sean Mayes, PhD Dr. Sean Mayes is an associate professor in Crop Genetics in the University of Nottingham and acting theme leader with Crops for the Future in Malaysia. He joined the University in 2004 and was made an associate professor in 2008. During a leave of absence between 2012 and 2015, he was based in Malaysia while helping to establish Crops for the Future as a programme and theme leader. He still runs research groups in both the UK and Malaysia and is codirector of the CFF-UNMC Doctoral Training Partnership.

He has developed and applied genetic markers to a wide range of animal and especially crop species, with a focus on the application of marker-assisted selection, and has just coedited a comprehensive book on oil palm breeding, genetics, and genomics (eds. Aik Chin Soh, Sean Mayes and Jeremy Roberts). To date, he has published over 100 peer-reviewed journal articles.   South Labs, Plant and Crop Sciences, Biosciences, University of Nottingham, Loughborough, Leicestershire, UK

Crops for the Future, Semenyih, Selangor, Malaysia

Richard Meldrum, BSc, MPH, PhD Richard Meldrum's research interests include public health, food safety, and microbiology.   School of Occupational and Public Health, Ryerson University, Toronto, ON, Canada

María Dolores de Miguel Gómez Dr. María Dolores de Miguel Gómez is an agronomical engineer who graduated at the ETSI Agronomists of Cordoba. Her speciality is agricultural economics. She has acquired the most important management positions at the UPCT as the Director of the Department of Business Studies, Director of the ETSIA, and Commissioner for equal opportunities. Also, she was General Director of Research at the Regional Ministry for Education and Culture in the Region of Murcia. She participated in 12 research projects about competition. She is co-author of books and chapters, published by national publishing companies. Furthermore, she published her work in national and international scientific journals with high impact factor. She assisted in national and international scientific congresses. Her work is especially concentrated in socioeconomic analysis of agricultural firms, as well as studies on cost-effectiveness, efficiency, adoption, and diffusion of technology in different agricultural sectors of national relevance, especially in the sectors of viticulture, horticulture, and citrus products. Furthermore, she worked in the area of water management and equal opportunities.   Departamento de Economía de la Empresa, Universidad Politécnica de Cartagena (UPCT), Cartagena, Murcia, Spain

About the Editors and Authors

Maysoun Abdelmoniem Mustafa, PhD Maysoun manages CFF-UNMC DTP—a Doctoral Training Partnership between Crops for the Future and the University of Nottingham Malaysia Campus—where she oversees the professional and academic development of postgraduate students. CFF-UNMC DTP addresses food security in a changing climate by encouraging research on agricultural diversification. Having earned a BSc (Hons) in Plant Biotechnology, and a PhD in Biosciences from the University of Nottingham, she is a postharvest biologist with a chief interest in reducing food losses and enhancing nutritional value. Maysoun is actively engaged in education and outreach, aspiring to boost sustainable agricultural practices through proactive and inclusive research. She joined Southeast Asian Council for Food Security and Fair Trade (SEACON), a regional NGO that facilitates collaboration across a network of ASEAN NGOs in 2015. At SEACON she conducted projects for enhancing smallholders' livelihoods, such as stewardship programmes for best farming practices and promoting knowledge-sharing through ICT platforms. Doctoral Training Partnership, Crops for the Future/University of Nottingham, Semenyih, Selangor, Malaysia

Udaya Sekhar Nagothu, PhD, MSc, MSc Sekhar has more than 25 years of research and development experience in natural resource management and environment-related areas. His focus is mainly on socio-economics, stakeholder, institutional, and policy issues in sectors including agriculture, forestry, water, and aquaculture. He has coordinated several large interdisciplinary projects in Norway, India, Vietnam, Sri Lanka, Cambodia, Philippines, the Balkans (Montenegro, Latvia, Macedonia, Albania), and Africa (Kenya, Tanzania, Ethiopia, Rwanda, Malawi, South Africa). He has published articles in several international journals focusing on interdisciplinary research. He has edited and written five books on climate change, water, sustainable agriculture, and food security. Centre for International Development, Norwegian Institute of Bioeconomy Research, Aas, Norway

Maren Oelbermann, BSc, MSc, PhD Maren Oelbermann's research interests include agroforestry and intercropping, carbon sequestration, greenhouse gas emissions, climate change resilience, soil health, sustainable agriculture, riparian land-use systems, and biochar. School of Environment, Resources, and Sustainability, University of Waterloo, Waterloo, ON, Canada

Dr. Antonio Costa de Oliveira, PhD Dr. Antonio Costa de Oliveira's research interest includes cereal genomics, with a focus on abiotic stress tolerance. He has contributed to the International Rice Genome Sequencing Project (IRGSP), Brachypodium Genome Initiative, and International Oryza Alignment Sequencing Project (IOMAP) (sequencing of O. glumaepatula). Dr. Costa de Oliveira has published more than 272 peer-reviewed papers. He was awarded CNPq Scientific Initiation Fellowship, 1985, and CNPq Graduate Fellowship, 1986–1988. He graduated with distinction and honour: Citation from the City Council of Santa Maria for efforts in Biotechnology, 1991; CNPq Biotechnology Visiting Fellow at UFRGS,

1990; CNPq Graduate Fellowship, 1992–1996; Nomination for the F. Umbarger Outstanding Graduate Student in Research, 1996; CNPq Fellowship for Productivity in Research, 1998–2000 and 2003–present, highest level (1A) since 2018; Citation from the Press of the City of Pelotas, Personality of 2000 in Agricultural Research; Member of the Brazilian Society of Genetics, 1987–present; Member of the Crop Science Society of America, 1995–present; Member of the Genetics Society of America, 1998–2004; Member (co-founder) of the Brazilian Society of Plant Breeding, 2000–present; Member of the American Association for the Advancement of Science, 2002–present; Ad hoc Consultant for CNPq and CAPES (Brazilian Agencies for Research and Graduate Programs); Ad hoc Consultant for EMBRAPA (Brazilian Institute of Agricultural Research); 2003 World Technology Award in Biotechnology (given to IRGSP); 2005 Agronomy Society of Pelotas, Outstanding Research Award; and 2008, Honour Citation by the Press of the City of Pelotas. Plant Genomics and Breeding Center, Federal University of Pelotas, Capão do Leão, RS, Brazil

Martin Oteng-Ababio, PhD Dr. Martin Oteng-Ababio is an associate professor in the Department of Geography and Resource Development, University of Ghana, and coordinates the Disaster Risk Reduction Programme. He received his PhD from the same University in 2007. His research interests include urban environmental management, socio-economic measures for disaster risks, preparedness for response, and informality and the built environment and urban infrastructural services. He has published over 60 peer-reviewed articles and project managed several major international collaborative research to successful completion. In April 2015, he was the proud recipient of the 2013/2014 Best Researcher Award from the School of Social Sciences, University of Ghana, having won the Second Prize of the Japanese Award for Outstanding Research on Development in 2011 and the Best Paper Award—*Southern Africa Society for Disaster Reduction (SASDiR) Conference*—2012. He is a fellow of the American Association of Geographers, Ghana Geographers Association, and Global Alliance for Disaster Research Institutes (GADRI). Department of Geography and Resource Development, University of Ghana, Accra, Ghana

Ari Paloviita, DSc Ari Paloviita specializes on sustainable food systems and works in Food System Studies Research Group at the University of Jyväskylä, Finland. His past research interests include climate change adaptation, food security, and food system vulnerability. Currently (2017–2020), his research focuses on sustainable protein systems as part of the ScenoProt project, funded by the Strategic Research Council of the Academy of Finland. He is also an adjunct professor in Environmental Management at the University of Eastern Finland. He has coedited a book with Professor (emerita) Marja Järvelä, titled *Climate Change Adaptation and Food Supply Chain Management* (Routledge, 2015). Department of Social Sciences and Philosophy, University of Jyväskylä, Jyväskylä, Finland

About the Editors and Authors

Nuria Navarro Pay, BA Nuria Navarro Pay has an interest on social behaviour and cultural studies. Centro Integrado de Formación y Experiencias Agrarias (CIFEA), Molina de Segura, Murcia, Spain

Camila Pegoraro, PhD Camila Pegoraro's research interests include functional genomics in rice, rice biofortification, and abiotic stress response in rice. She improved the proposal and the manuscript and edited the manuscript. Plant Genomics and Breeding Center, School of Agronomy Eliseu Maciel, Federal University of Pelotas, Pelotas, RS, Brazil

W. Paul Quick Biography C4 Rice Centre, International Rice Research Institute (IRRI), Los Baños, Philippines

Paulus Adrianus K. L. Ratumakin, MPA Paulus's major research interests are philosophy, public policy, rural development, and community-based management of common pool resources. His present research includes social justice for fisherman: a policy analysis of protection of fishermen, fish raisers, and salt farmers. His past research included (a) fisherman and farmer in reading the climate and season, an analysis of the knowledge of fisherman and farmer on the climatic and seasonal information; (b) implementation of public housing policy for low-income households in Kupang City, East Nusa Tenggara, Indonesia; and (c) local knowledge in sustainability water management. His future plan will be to continue to focus on rural development, particularly enterprise management by government at the village level as a more sustainable approach to public services to improve the wellbeing of the rural people. He (a) assisted in the Hauhena Water Project and developed a community-based water management system in a remote hamlet in Timor Tengah Selatan (TTS) Regency of West Timor and (b) assisted and monitored the development of a community-based forest management system in Semau Island, East Nusa Tenggara Province, Indonesia. Catholic University of Widya Mandira, Kupang, Nusa Tenggara Timur, Indonesia

PIKUL (Non-Governmental Organisation), Kupang, Nusa Tenggara Timur, Indonesia

Luis Maldonado Rius Luis Maldonado Rius is a lecturer in the Department of Agri-Food Engineering and Biotechnology (EAB) with a Bachelor's degree in Environmental and Landscape Engineering, and a Master's degree in Landscape Architecture (MBLandArch). He is UPC's coordinator of EMiLA (European Master in Landscape Architecture). His thesis "Paper Gardens" shows the reappearance of the garden at the end of the last century as an indicator of the research and design interests of the current landscape architecture. His research and professional practice focused on the planning and design of open space linking agriculture and forests with landscape processes in urban systems and new forms of nature. His ongoing research analyses bottom-up approaches in consumption and food production looking for models to design landscape adaptation to new food systems and global challenges in highly urbanized areas. Escola Superior d'Agricultura de Barcelona (ESAB), Universitat Politècnica de Catalunya, UPC, Castelldefels, Barcelona, Spain

Subhajyoti Samaddar, PhD Subhajyoti Samaddar is an associate professor in Disaster Prevention Research Institute (DPRI), Kyoto University, Japan. His research interests are household preparedness behaviour, disaster risk communication, community participation, and implementation science. He has an interdisciplinary academic background including PhD in Disaster Management from Kyoto University and Master of Planning from the School of Planning and Architecture (SPA), New Delhi, and MA in Social Anthropology. He has been involved in different global research projects on disaster risk management and climate change adaptation. Dr. Samaddar has conducted in-depth field studies in different countries including India, Bangladesh, Japan, and Ghana. He is the recipient of international award "Hazards 2000" in 2016 and "Young Scientist Award" by DPRI, Kyoto University, in 2010. He is the secretariat member of GADRI (Global Alliance of Disaster Research Institute). He has published widely in premier international journals on disaster risk studies. Disaster Prevention Research Institute, Kyoto University, Kyoto, Japan

Michael Santhanam-Martin, PhD, GDSS, BA, BE (Hons) Dr. Michael Santhanam-Martin is an applied social scientist and a member of the Rural Innovation Research Group, in the Faculty of Veterinary and Agricultural Sciences, at the University of Melbourne. His research focuses on the sustainability of agricultural systems, with emphases on the social dimensions of sustainability and on governance for sustainability. His PhD research examined how industry governance processes in the Australian dairy industry account for social sustainability at community scale. Other interests include family farm succession processes and how they can be supported, farm employment relationships and agricultural workforce development, and farmers' climate change adaptation decision-making. Agricultural Production Systems, Faculty of Veterinary and Agricultural Sciences, University of Melbourne, Parkville, VIC, Australia

Muhammad Sarwar, PhD During his PhD, he conducted research on the effect of heat stress in cotton crop and the mitigation techniques. In his last 6-month stay at Macquarie University, he studied the regulation of ethylene and carbon metabolisms in heat-stressed cotton (paper published in Frontiers by Najeeb et al.). He also developed protocols for microscopic study of developing cotton pollen. He found that endogenous ethylene plays limited role in heat-induced yield injury to cotton. Currently, he is conducting research on the effect of heat stress on wheat, and he is applying silver nitrate as foliar spray for inducing thermotolerance. I am also conducting research on maize crop by growing it under different locations to see the effect of heat stress on heat-tolerant and heat-susceptible hybrids. In the future, he wants to learn proteomic techniques to see the expression of different proteins under stressful and normal conditions. He got honorarium for 2016–2017. Ayub Agricultural Research Institute, Faisalabad, Pakistan

Armin Scheben Biography School of Biological Sciences and Institute of Agriculture, University of Western Australia, Perth, WA, Australia

Andrew Solomon, DLE, DA Andrew Solomon's research interests include treaty rights, indigenous health, and wellbeing. Fort Albany First Nation, Fort Albany, ON, Canada

Nicole F. Spiegelaar, BSc, MES Nicole F. Spiegelaar's research interests include environmental psychology, mental health, indigenous food systems, and traditional healing. Department of Physical and Environmental Sciences, University of Toronto Scarborough, Toronto, ON, Canada

Tamara Sysak, PhD Tamara Sysak has over 15 years environmental research and policy experience in corporate, government, and academic capacities. Tamara completed her PhD at the University of Melbourne where she researched drought and farmer's adaptation strategies in rural and regional settings for her doctoral thesis. She has published internationally from this and prior research on rural and regional biosecurity and risk. Tamara received scholarships from both the Future Farming Industries Cooperative Research Centre and the Primary Industries Adaptation Research Network. She is currently a social researcher at the University of the Sunshine Coast. University of the Sunshine Coast, Sippy Downs, QLD, Australia

Daniel Kean Yuen Tan, PhD, BAppSc (Hort Tech) Hons1 Daniel has developed a broccoli development model for vegetable growers in southeast Queensland. His research is in the area of crop agronomy, specializing in crop abiotic stress and farming systems. His specific interests within crop abiotic stress are in physiology, especially high temperature tolerance. Research outcomes (e.g. stress detection methods and markers) are used by plant breeders in their development of stress-tolerant crops. His ongoing work on abiotic stress and farming systems research in rice, wheat, chickpea, and cotton has been supported by the Cotton Research and Development Corporation, the Grains Research and Development Corporation, and the Department of Foreign Affairs and Trade and the Australian Centre for International Agricultural Research. He currently has collaborative research links at CSIRO Plant Industry (Narrabri); the University of Oxford; NSW Department of Agriculture; Applied Horticultural Research (Sydney); Texas A&M University (USA); the United States Department of Agriculture (Lubbock, Texas, USA); ICRISAT, India; and the Cocoa and Coconut Institute (Papua New Guinea). Sydney Institute of Agriculture, School of Life and Environmental Sciences, Faculty of Science, The University of Sydney, Sydney, NSW, Australia

Amod Kumar Thakur, PhD, MS He is a Fellow of the Indian Society for Plant Physiology, and in 2015 Thakur was awarded its J. J. Chinoy Gold Medal for lifetime contributions in 2015. In 2012–2013, he was a Fulbright-Nehru Senior Research Fellow for 8 months at Cornell University in the USA. Before that, he spent 3 months at Cornell in 2011 on a USDA Norman Borlaug International

Agricultural Science and Technology Fellowship. Plant Physiology, ICAR-Indian Institute of Water Management, Bhubaneswar, Odisha, India

Timothy Scott Thomas, PhD Dr. Thomas began his career studying determinants of tropical deforestation. While he continues to have a research interest in the area of land-use change, over the last decade, his main emphasis has been on studying climate change and its impact on agricultural productivity. Such research includes the use of biophysical and bioeconomic models. He has co-authored several books and journal articles in both areas of interest. International Food Policy Research Institute (IFPRI), Washington, DC, USA

Yenny Tjoe, PhD Dr. Tjoe's research interest is to examine the intersection between poverty, economic development, and sustainability, particularly concerning with the vulnerable tribal communities. Her earlier research included assessing the impacts of fiscal decentralization on poverty reduction in East Nusa Tenggara Province and Indonesia. Her current research investigates how tribal communities in West Timor sustain their livelihoods facing pressure arising from monetization and modernization. Her future plan will be to continue to focus on how development can be beneficial to communities in Indonesia, focusing on sustainable water management and domestic waste management. Her significant contribution as the Hauhena Water Project leader is supplying clean water directly to 30 households in a remote hamlet in Timor Tengah Selatan (TTS) Regency of West Timor. She received 2013 Griffith Excellent Teacher Award under the DVC-A Letters of Commendation. Griffith University, Archerfield, QLD, Australia

Christopher J. D. Tsuji, BA, MA Christopher Tsuji's research interests include indigenous health and wellbeing and anthropology. Osgoode Hall Law School, York University, Toronto, ON, Canada

Leonard J. S. Tsuji, BSc, DDS, PhD Leonard Tsuji's research focuses on environment and health. Department of Physical and Environmental Sciences, University of Toronto Scarborough, Toronto, ON, Canada

Najeeb Ullah, PhD For the last 12 years, Dr. Najeeb Ullah has been studying abiotic stress tolerance mechanisms in plants. He has studied the effect of waterlogging, shade and elevated CO_2, and heavy metal on various crop species. The research encompasses various aspects of abiotic stress tolerance from leaf physiology to carbon metabolism and reproductive biology and devising strategies to mitigate stress-induced injury in plants. His current research focuses on exploring the opportunities to improve wheat performance to future warmer climates. The project aims to understand the role of leaf stay-green trait in heat stress tolerance in wheat and will lead to the development of reliable high-throughput phenotyping techniques for screening of heat-tolerant wheat lines. Queensland Alliance for Agriculture and Food Innovation, Centre for Crop Science, The University of Queensland, Toowoomba, QLD, Australia

About the Editors and Authors

Norman Uphoff, PhD Norman Uphoff spent three decades on applied social science research on agricultural and rural development, including local institutions, participatory approaches, and irrigation and natural resource management, and then the past two decades doing research and application for agroecologically based sustainable development, with a focus on the System of Rice Intensification (SRI). He is a recipient of the first Olam International Award for Innovation in Food Security, selected by the panel of the Agropolis Fondation, Montpelier, France, in 2015. He is chief editor for the volume on *Biological Approaches to Sustainable Soil Systems*, CRC Press, 2006, with 102 contributors from 28 countries. He is also Director of the Cornell International Institute for Food, Agriculture and Development (CIIFAD) and engaged in multidisciplinary, collaborative work in more countries for sustainable agricultural and rural development, 1990–2005. Government, Cornell University, Ithaca, NY, USA

International Agriculture, Cornell University, Ithaca, NY, USA

SRI International Network and Resources Center (SRI-Rice), International Programs, College of Agriculture and Life Sciences, Cornell University, Ithaca, NY, USA

David James Watson, PhD, MSc, BSc David has four principal areas of expertise: (1) guiding and managing demand-driven agricultural research and agricultural/rural development interventions; (2) the transformation and commercialization of agrarian systems (both developed countries and developing countries), where in a developed country context, analysis has focused on food and farming systems, while in a developing country context, analysis has focused on the development of pro-poor value-chains, understanding roles and opportunities for small-scale producers and economically marginalized groups (women and youths); (3) sustainable agriculture, agri-environmental and rural development education, training, and extension; and (4) rural, agricultural, and agri-environmental policy development. He has three principal modes of developing and applying this expertise: (1) understanding agrarian development through both active research and research management, (2) influencing agrarian development, and (3) elucidating agrarian development through sustainable agriculture and agri-environmental education and academic discourse, namely, the publication of sustainable agro-food system literature (i.e. books, papers, and educational curricula). Currently David is managing the CGIAR Research Program (CRP) on maize. International Maize and Wheat Improvement Centre (CIMMYT), Texcoco, Mexico D.F., Mexico

Meaghan J. Wilton, BSc, MSc Meaghan Wilton's research interests include sustainable agriculture, agroforestry, intercropping, food security, sustainable intensification, plant nutrition, stable carbon isotopes in soils, greenhouse gas emissions from soils, and climate adaptation and mitigation. School of Environment, Resources, and Sustainability, University of Waterloo, Waterloo, ON, Canada

Department of Physical and Environmental Sciences, University of Toronto Scarborough, Toronto, ON, Canada

Muneta Yokomatsu, PhD Muneta Yokomatsu is working in a field of economic analysis of disaster risk management, where he has developed the methods of cost-benefit analysis of disaster prevention and infrastructure management. He was awarded "Outstanding Civil Engineering Achievement Award" by Japan Society of Civil Engineers twice for his results in "Catastrophe Risks and Economic Valuation of Disaster Prevention Investment" in 2000 and "Local Assets as Artifacts and a Resident's Identity Formation: Category-Selection-Model Approach" in 2015. He was also given "Aniello Amendola Distinguished Service Award" by the International Society of Integrated Disaster Risk Management (IDRiM) in 2015. He is now intensively working on simulation models of macroeconomic dynamics under disaster risk and disaster mitigation investment. Disaster Prevention Research Institute, Kyoto University, Kyoto, Japan

Yuxuan Yuan Biography School of Biological Sciences and Institute of Agriculture, University of Western Australia, Perth, WA, Australia

Part I
Introduction

Chapter 1
Introduction

Atanu Sarkar, Gary W. vanLoon, and Dave Watson

More than any other human activity, agriculture is fundamental to the survival and well-being of the human population. In 2016, a total of 2.73 Gt of food grains were produced worldwide. This fundamental food source is alone enough to supply sufficient nutritional kilocalories for the entire global population. And nutrition is supplemented by the many other crops and livestock that are part of the overall food system (FAO 2016). Yet, in the same year, it is estimated that around 815 million people, some 11% of the world's population were chronically hungry. Moreover, the number was higher (by 38 million) than in the previous year and this rise can largely be attributed to conflict combined with climate effects such as more frequent droughts or floods (FAO 2017). These two issues are also connected. Exacerbated by climate-related shocks, the number of conflicts is on the rise augmenting the challenges of maintaining food security. Indeed, existing household level poverty and, at the macro level, the slowing down of national/regional economies has drained foreign exchange and fiscal revenues, eventually affecting both food availability through reduced import capacity and food access through reduced fiscal space to protect poor households against rising domestic food prices have worsened food security. The global population is expected to increase from 7.6 billion in 2017 to 8.5 billion by 2030, and, perhaps, over 9.5 billion by 2050 (UN). In some regions,

---

A. Sarkar (✉)
Division of Community Health and Humanities, Faculty of Medicine, Memorial University, St Johns, NL, Canada
e-mail: atanu.sarkar@med.mun.ca

G. W. vanLoon
Department of Chemistry, Queen's University, Kingston, ON, Canada

School of Environmental Studies, Queen's University, Kingston, ON, Canada
e-mail: gary.vanloon@chem.queensu.ca

D. Watson
CGIAR Research Program on Maize, International Maize and Wheat Improvement Center (CIMMYT), Texcoco, México

© Springer Nature Switzerland AG 2019
A. Sarkar et al. (eds.), *Sustainable Solutions for Food Security*,
https://doi.org/10.1007/978-3-319-77878-5_1

such as sub-Saharan Africa, the population is likely to double by 2050 (PRB 2013). Whilst uncontrolled population growth is posing a major challenge to continuing effort in improving global food security; climate change has begun to pose a formidable threat to our surrounding agro-ecosystem. Based on the Intergovernmental Panel on Climate Change (IPCC) 2014 report, the following schematic diagram shows the cascading effects of climate change on food insecurity (IPCC 2014) (see Fig. 1.1, page 17).

Various models for production of major crops show that climate change will fundamentally alter global food production patterns, particularly the yields of four crop groups (maize, wheat, rice and oil seeds, covering almost 70% of global crop

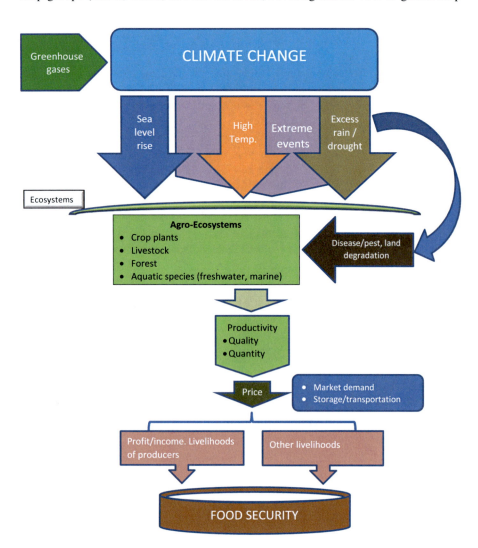

Fig. 1.1 Climate change and food security

harvested area) (Frieler et al. 2015; Rosenzweig et al. 2014; Nelson et al. 2014). Several wheat production zones in the European Union, and maize production zones in the USA, will witness a significant decline due to reduced water availability during the growing season and heat damage during flowering (Müller and Elliott 2015). However, these high-income regions are also believed to have greater flexibility for adaptation due to financial and technological capacity. Yields of maize, sorghum and millet, occupying the highest crop areas within all of Africa will be directly impacted by climate change (Thomas and Rosegrant 2015). Most of the climate models agree that production systems in low-income countries in semi-tropical and tropical parts of the globe are likely to be adversely affected by increased temperatures (Lobell et al. 2013; Deryng et al. 2014; Cooper et al. 2014; Horton et al. 2015). In South Asia, by 2100 yields of maize, wheat and rice could decrease by up to 30%. In Bangladesh, rice production could fall by 8% and wheat production by 32% as early as 2050 (Hossain and Teixeira da Silva 2013).

Whilst increased mean temperatures are important, it is predicted that it is temperature extremes that will have a more significant impact on agricultural production (Deryng et al. 2014). Major crops such as maize, wheat and rice are currently grown over a wide range of temperatures, which vary in situ from season to season. Cereal crops have consistently displayed the capacity to cope well under the predicted mean average temperatures; at least up to the average temperatures predicted by 2050. Indeed, Fischer et al. (2014) suggest that in recent (hot) years, numbering more than 25% of years, cropping systems have already experienced predicted average temperatures and that cropping system performance in these years can be reliably used as predictors of performance under future climates and serve as early tests of adaptation strategies. What is worrying to scientists is the fact that the expected greater frequency of temperature extremes (IPCC 2007) will expose these crops to temperatures so far not experienced; especially in strategic agricultural production areas in developing countries (Deryng et al. 2014). For example, Africa is warming faster than the global average (Collier et al. 2008) and by the end of this century, growing season temperatures are predicted to exceed the most extreme seasonal temperatures recorded in the past century (Battisti and Naylor 2009). Indeed, by 2050 maximum temperatures within maize-growing areas of sub-Saharan Africa could increase by 2.1–3.6 °C (Cairns et al. 2012). Climate change will, therefore, severely test farmers' resourcefulness and adaptive capacity (Adger et al. 2007; Cairns et al. 2013). In May 2016, India recorded its highest temperature ever when a town in the western state of Rajasthan reached 51 °C. The searing heat across south Asia has already critically damaged crops and destabilized food security in the region (CNN 2016). Temperature extremes (heat waves) at vulnerable stages of crop reproduction cycles have been shown to have devastating effects, often rendering whole fields barren (Cooper et al. 2014; Horton et al. 2015; Gong et al. 2015). According to Deryng et al. (2014), by 2080, extreme heat stress at anthesis (flowering period of a plant) could account for up to 45% global maize yield losses and reduce projected gains in spring wheat yields by as much as 52%. In the case of maize, the American corn-belt, the Middle East, west and south Asia, and northeast

China are increasingly at risk from yield losses due to heat stress, most especially during the reproductive stage of their life cycle.

There are also conflicting predictions regarding crop production in Central Eurasia (Russian Federation, Ukraine and Kazakhstan), Scandinavia (Norway) Some studies suggest increasing production due to warmer temperatures, longer growing seasons, decrease of frosts and positive impact of higher atmospheric concentrations of CO_2 on crops. However, other modeling experiments show decline of agricultural potential due to increasing frequency of droughts and possibly more weeds, pests and diseases, particularly in the most productive semi-arid zone (Lioubimtseva et al. 2015; Uleberg et al. 2014).

In line with extreme temperature events, changes in rainfall distribution are also hard to predict; differing widely between geographic regions (Fischer et al. 2014) with many models not even agreeing on the direction of change (IPCC 2007). For example, whilst there seems to be agreement regarding increased precipitation in East Africa and decreased precipitation in Southern Africa (IPCC 2007), there is considerable disagreement regarding predicted changes in precipitation in West Africa (IPCC 2007; Cooper et al. 2008). However, there is a general consensus that, together with extreme temperatures, drought will be more frequent in the near future and will severely limit crop production (Cooper et al. 2014; Horton et al. 2015); especially in highly vulnerable areas of sub-Saharan Africa, Asia and Latin America, with sub-Saharan Africa and South Asia (Masih et al. 2014; Nyasimi et al. 2014). Brown and Funk (2008) suggest that smallholder farmers in the tropics and subtropics will be much less able to cope with climate change because they have far fewer adaptation options open to them compared to farmers in developed countries. The predicted changes in temperature and precipitation, especially in SSA and Asia, will further accentuate the intensity and frequency of drought, further increasing the vulnerability of smallholder farmers producing under rain-fed conditions with limited inputs (Cairns et al. 2012, 2013; Masih et al. 2014; Shiferaw et al. 2014). Combined drought and heat extreme effects are predicted to hit hardest in SSA, which is particularly vulnerable due to the range of projected impacts, multiple stresses and low adaptive capacity (IPCC 2007). Low yields in this region are largely associated with drought stress, declining soil fertility (Sanchez 2002; Kumwenda et al. 1998) and severe soil degradation (Nkonya et al. 2015), weeds, pests, diseases, low input availability, low input use and inappropriate seeds (Recha et al. 2012).

Although there are few studies on high nutritional value products (roots and tubers, pulses, vegetables, fruits and other horticultural products), production of apples, cherries, grapevines, nuts and other horticulture products in USA will decline (HLPE 2012; Porter et al. 2014). As far as coffee is concerned, Brazil is expected to get initial benefits of rising temperature, but coffee yields in Costa Rica, Nicaragua and El Salvador are expected to be reduced (Porter et al. 2014).

Climate change may increase the negative impact of pests and weeds by expanding geographic ranges, invading new suitable zones and allowing for their establishment in these new areas. For instance, "desert locust," the most dangerous of all migratory pests are expected to wreak havoc in Africa due to potential changes in temperature, rainfall and wind patterns (Cressman 2013). Climate change is also expected to increase plants' susceptibility to pests and other diseases due to weakened immune systems (Pautasso et al. 2012).

In addition to increasing temperatures and increasingly erratic rainfall, global warming is contributing to sea level rise. Whilst sea level rise is not predicted to affect maize and wheat production in any meaningful way, it is expected to impact on rice production in mega-deltas and coastal zones through uncontrolled flooding and soil salinization (Wassmann et al. 2009).

Livestock production is affected by climate change in several ways including decreasing productivity due to low yields of forages and feed crops, more animal illnesses (such as heat stress, lack of drinking water, increasing winter survival of vectors and pathogens) (Crescio et al. 2010; Porter et al. 2014).

Heat stress, drought stress, pathogen attacks and pest outbreaks increase tree mortality and declining productivity of forests. Moreover, the dead trees favour greater fire disturbance (Settele et al. 2014; Ciais et al. 2008). Tropical forests are more vulnerable to drought and fire-induced mortality during extreme dry periods, since moisture is the key to their survival. There are ample evidences that frequency and severity of forest fire are increasing, due to a combination of land-use change, low moisture and drought (Miles et al. 2006).

Aquatic animals are mostly cold-blooded; their metabolic rates, growth, reproductive cycles (reaching sexual maturity) are significantly affected by rising temperature (Perry et al. 2005; Pörtner 2008). In addition, mass poleward migration of various fish species is resulting in the rapid "tropicalization" of mid- and high-latitude systems. Changing climate, habitat, salinity (and freshwater) content, acidification and phytoplankton, is resulting in a large-scale redistribution of global marine fish catch potential, such as increase in high-latitude regions and decline in the tropics (Cheung et al. 2010). Thus small-scale fisheries in less developed and economically poor tropical regions are particularly vulnerable to climate change impacts (IPCC 2014). Furthermore, rising sea levels and floods displace brackish and freshwater in river deltas affecting aquaculture practices, eliminating local species and destroying wetlands (IPCC 2013). Extreme events such as cyclones or hurricanes, can affect the ability of marine and aquatic ecosystems such as coral reefs and mangroves to provide services crucial for livelihoods and food security of aquatic species. The bulk of the fish raised or caught in inland waters is produced in least developed or developing countries and consumed locally (World Bank et al. 2010). The inland fishery production is also threatened by changes in precipitation, increases in temperature and sea level, water management and the frequency and intensity of extreme climate events (Brander 2007)

Genetic resources for food and agriculture can play a crucial role in meeting the challenges of food insecurity due to climate change and also to sustain agricultural productivity and livelihoods (FAO 2015; Asfaw and Lipper 2011). Climate change is one of the key factors for erosion of the existing pool of genetic resources, such as the variety and variability of animals, plants and micro-organisms used by farmers, pastoralists, forest dwellers and fishers. These resources are essential for improving the quality and output of food production. As a part of adaptation strategies, novel crop varieties, animal breeds or fish and forest species populations that will be required for survival in the changing climate conditions will have to come from the existing pool of genetic resources. The new varieties of species are expected to have high tolerance to heat, drought, flooding, frost, salinity, pest, and diseases and thus can maintain adequate yields to fulfill our needs. Thus, we have reached at the stage of crisis where the nature is on the verge of losing important genetic resources for ever.

When productivity (quality and quantity) declines, its economic and social consequences at various scales are inevitable. The consequences manifest in lowering agricultural incomes, making food markets and prices unstable and the patterns of trade and investment patterns unpredictable. Eventually, the consequences of instability take their toll by ruining the farming communities, particularly the small producers in developing countries. It has been predicted that there will be a substantial 12% decline in the food self-sufficiency ratio of developing countries by 2050 (HLPE 2011). Farmers will increasingly be compelled to sell productive capital, for instance cattle and lands in order to absorb income shocks. This will reduce the capacity to invest and ultimately limit their capacity to face other essential expenditures, such as health of family members and education for children. At the macro-economic level low productivity can prompt an increase in commodities' prices (such as food and feed) impacting socio-economic conditions of the people, particularly the low-income households, who spend a significant proportion of their household budgets on food. In fact, combined analysis of multi-models projected the mean price rise for maize, rice and wheat by 2050 are 87%, 31% and 44% compared with 2010 levels for an optimistic scenario of low population and high-income growth and using mean results from four climate scenarios (Nelson et al. 2010).

The introduction of modern agricultural practices (also known as the Green Revolution) in the last century transformed several nations from being net food importers to exporters and pulled them out of perennial crises of food insecurity. The Green Revolution was essentially driven by extensive input practices, including the introduction of selective high yielding seed varieties, intensive use of agrochemicals and irrigation (river and groundwater). For example in India, government supports, and plant breeding programs, stressed three major cereals (rice, wheat, maize), but neglected several traditional, nutritionally superior crops including sorghum, millets and pulses (Sarkar and vanLoon 2015). The traditional multi-functionality of agriculture has been largely ignored by the Green Revolution and the ecosystem functions that mitigate the environmental impacts are not given adequate attention (van Loon et al. 2005). Hence, emerging consequences of climate change have been disrupting the progress of food production in the erstwhile highly successful regions of the Green Revolution. For example, in the semi-arid regions of southern India (Fig. 1.2), the traditional drought-resistant crops such as sorghum, millets and pulses were replaced by a climate-sensitive crop like rice. Initial success of high production of rice due to institutional and infrastructural support of credit, seeds, agrochemicals, intensive network of large dam irrigation encouraged the farmers to embrace the Green Revolution (Fig. 1.3). However, aging irrigation infrastructure (Fig. 1.4) and lack of regular government support have brought uncertainty to the current prospect of Green Revolution. High demand for water for

1 Introduction

Fig. 1.2 Semi-arid landscape of north Karnataka (southern Indian state) ideal of local coarse cereals like sorghum, millets and pulses

rice cultivation amid frequent droughts (Figs. 1.4 and 1.5) resulted in crop failure pushing the poor and marginalized farmers into debt traps where poverty is perpetuated, in the absence of any climate adaptation strategy (Sarkar et al. 2011).

Climatic risks, with its impending uncertainty can discourage investments in agricultural development. The predicted and expected consequences of climate change on agricultural production are generally most adverse for the areas inhabited by poor and food-insecure smallholders and for agro-economy based countries, mostly located in South Asia and sub-Saharan Africa (Ericksen et al. 2011). One important economic consequence of climate change may be to alter investment patterns in such a way as to reduce long-term productivity and resilience of agricultural systems at household and national levels. Greater uncertainty can reduce incentives to invest in agricultural production, and this trend will particularly hit poor family farmers and smallholders with limited or no access to credit for farm inputs, technologies, and insurance. Even novel climate-smart technologies like climate resilient crops, natural resource management (soil, water, nutrients), risk prediction and forecasting tools may not accessible to the most vulnerable farming communities, unless any radical changes occur at the public policy level.

Lastly, the net impact of a climatic change on food security depends on the intensity of the climate shock and the vulnerability of the food system of specific context (country/region) (IPCC 2012). In other words, vulnerability of food insecurity due

Fig. 1.3 Green revolution has changed the agriculture practice—rice has become a common crop

Fig. 1.4 Intensive canal system and water reservoir of river dam are the legacy of the Green Revolution. Current dilapidated irrigation infrastructure and insufficient water in the reservoir due to droughts raise question on sustainability of the Green Revolution

1 Introduction

Fig. 1.5 Traditional rainwater-fed pond for irrigation is no longer reliable due to erratic precipitation

to climate change means the propensity of the food system to be incapable of delivering food security outcomes under climate change. In this regard, some characteristics of a food system make it more or less vulnerable to a set of climate change risks. For example, a farm relying on a single crop is particularly vulnerable to a pest infestation or price volatility. On the other hand, a highly diversified system is less vulnerable to both pests and price fluctuations that affect specifically one type of production. Any rain-fed system is more vulnerable to a drought than an irrigated one. In rain-fed areas, households that are completely dependent on agriculture are more vulnerable, from an economic point of view, to drought than the households having other alternative sources of income or any form of social protection.

Due to non-existent social protection in many developing countries, poor households are faced with hard choices between consumption and asset smoothing in response to a climate shock (Gitz and Meybeck 2012). Therefore, poor resilience makes these farmers more vulnerable in subsequent climate shocks due to the complex, cascading, multidimensional and multiscale nature of their vulnerabilities. Thus, the drought affected farmers become more vulnerable to the next drought by losing productive capital, increasing financial debt and reducing household assets. Eventually, from one level of crisis to another, vulnerabilities can add themselves or amplify each other. It is noteworthy that food insecurity vulnerability affects educational attainment of the next generation and indebtedness due to out of pocket expenditure on healthcare leading to economic impediments, with long-term effects (Hoddinot 2006).

Despite the fact that gender roles are rapidly changing in food system, gender perspective of climate change and food security has not yet become a priority discourse. This is unfortunate as feminization of the agriculture workforce is a global phenomenon (FAO 2011). However, women face constraints in terms of accessing credit, decision making, land tenure, social norms (restriction) preventing women from seizing off-farm opportunities, access to weather/climate information and information of appropriate technologies (FAO 2011; Pattnaik et al. 2018; Lambrou and Nelson 2010). As a result, women in rural areas may experience the effects of climate change more acutely than men.

Addressing climate change and food insecurity cannot be dealt with by "business as usual.". Rather, it requires a climate-sensitive approach that aligns actions for immediate humanitarian assistance for the starving population and long-term multi-sector development. According to the Food and Agriculture Organization (FAO), climate change is a threat to all four dimensions of food security i.e. its availability, accessibility (economically and physically), utilization (the way it is used and assimilated by the human body) and systems stability (of these three dimensions) (FAO 2017). Climate change is now beginning to affect every aspect of life on Earth, sometimes in unpredictable ways and it is certain that the changes will continue to grow in nature and magnitude. Agriculture is, of course, one component of human life that suffers major impacts when local climates change. The impacts mean that local food supplies everywhere will be affected, and the sum of these local effects is that we can expect increasing uncertainties about global food security. It is important to note that whilst agriculture responds negatively or positively to climate changes, at the same time agriculture itself impacts the nature of the global climate. Being a system that is managed by humans, it then becomes possible and necessary to develop practices that can mitigate the stresses caused by changing climate and also that can contribute to ways of reducing possible further changes in climate (OECD 2015).

In response to malnutrition around the world, in 2015 the global community pledged to eliminate hunger by 2030 as one of its sustainable development goals (SDGs). SDG-2 (*end hunger, achieve food security and improved nutrition, and promote sustainable agriculture*) clearly focuses on strengthening capacity for adaptation to adverse outcomes of agriculture practices due to climate change, such as … ensuring *sustainable food production systems and implement resilient agricultural practices that increase productivity and production, that help maintain ecosystems … progressively improve land and soil quality* (UN 2014). SDG-2 has also addressed issues pertaining to the complex mechanisms of food insecurity as the fallout of climate change, such as; (a) economic stability, equity of vulnerable populations (small scale farmers, women, indigenous peoples, family farmers, pastoralists and fishers), knowledge dissemination, additional employment, (b) maintaining genetic diversity of seeds, cultivated plants, farmed and domesticated animals and their related wild species, preservation of gene pools and traditional knowledge, (c) increasing investment in agriculture including enhanced cooperation in rural infrastructure, agricultural research and extension services, technology development, gene banks in developing countries, (d) increase market access by preventing trade restrictions, timely access to market information, proper functioning of food commodity markets and to help limit extreme food price volatility. SDG-13 specifically addresses taking urgent action to combat climate change and its impacts by strengthening resilience and adaptive capacity to climate-related hazards and raising awareness, and human and institutional capacity on climate change mitigation, adaptation, impact reduction and early warning (UN 2014).

1.1 Adaptation and Mitigation

Actions required against food insecurity due to climate change can be categorized as either adaptation or mitigation. According to the FAO, "adaptation" to climate change involves deliberate adjustments in natural or human systems and behaviours to reduce the risks to people's lives and livelihoods (FAO 2008). Adaptations build on scientific understanding of the nature of the climate changes occurring as well as of the natural ecosystem where responsive actions are required. Adaptations require comprehensive and holistic scientific research that establishes new ways of doing agriculture that will achieve goals of sustained and increased food production without adding to resource and environmental stresses. Adaptations also require knowledge of the socio-economic environment—the human resources available in the region, and therefore the strengths and limitations of the population in its ability to modify agricultural practices. There is clearly a "capacity building" component to any adaptation strategy.

Mitigation of climate change involves actions to reduce greenhouse gas emissions and sequester or store carbon in the short term, and development choices that will lead to low emissions in the long term. Mitigation refers to changes in practice that are able to reduce agriculture's own negative impact on the environment. At best, mitigations will bring about positive impacts on the surroundings so that local and even global stresses are moderated. Mitigations therefore point to long term improvements in the environmental situation. Like adaptation, mitigation requires both scientific/technical innovation as well as development of human capacity to activate the innovation.

Adaptation and mitigation include overlapping and mutually supportive ideas. Facing the challenges of climate change means taking sustained action in both areas. In this book, we focus on research initiatives that are essentially directed toward innovations in adaptation.

Adaptations take into account the changing realities of local climates. Recognizing present and future weather patterns, new agricultural management practices are developed. Importantly, these adaptations may or may not have a complementary effect on the overall agroecosystem. The ability to adapt should, at the very least, not be an impediment to efforts directed at stabilizing climate systems and other aspects of the environment. Initiatives directed in areas that can be considered adaptation include:

- Good water management so that irrigation is available more widely, whilst ensuring limited waste of water through evaporation or excessive drainage
- Providing strong support for prevention of catastrophic damage to agriculture—support, for example, through the building of dams to ensure that excessive rainfall is saved and is then available for controlled release
- Development of crops that have the ability to flourish in the new environment—crops that are heat resistant, drought or flood resistant, resistant to newly encountered pests
- Reduced and climate smart inputs—fertilizer and pesticide inputs that are more natural where possible, or that have a low embodied carbon content. The use of precision agriculture either in its sophisticated GPS-controlled system or through accurate small scale mapping and careful manual application of chemicals in low-income countries
- Providing ongoing accurate information—better forecasting and early warning systems to predict weather and environmental threats. Accompanying this is support for local communities and farmers to access the information in a timely way
- Reducing deforestation, and in some cases integrating forestry and agriculture in a mutually supportive environment that may reduce stresses like erosion, evapotranspiration and pest problems
- Control on urbanization so that it does not continue to occupy some of the best and most valuable land in countries around the world. Where urbanization has occurred, developing sustainable methods of urban food production
- Enhancing the overall environment by encouraging biodiversity both in the natural surroundings and in the diversified cropping patterns over both time and space

(Smit and Skinner 2002; Backlund et al. 2008.)

1.1.1 Scale of Adaptation Practices

Recommending and choosing what could be best practices for adaptation in a given situation require careful analysis of the biophysical, socio-economic and political environment. Also required is a clear view of the goal of promoting a particular innovation. A number of issues must be taken into consideration.

The spatial scale at which the particular activity functions must be considered. Some innovations can be made by individual farmers even in the absence of higher level policy initiatives or financial support. Adoption of precision farming methods, for example, is clearly a choice for individual farmers as would be the installation of an efficient drip irrigation system. And there are many other examples. Nevertheless, almost all of these farmer-based management choices can also be encouraged by some measure of broader support often through policies that provide financial incentives. Other initiatives are clearly placed in the broader regional or national sphere—things like large dams to control water flows or carbon markets. Many other issues, such as plant breeding to achieve particular goals, involve government or private sector research and extensive carefully planned trials.

The temporal scale also must be considered. Adaptations such as changing rice management from traditional production by transplanting seedlings into flooded fields to water-saving direct seeded rice or using the system of rice intensification can provide immediate benefits in terms of reduced water and energy consumption and reduced release of methane into the atmosphere. Other adaptations encompass a longer-term view. Carbon markets to support technologies that promote enhanced storage of carbon-containing residues in the soil can be expected to show positive results only over a scale of decades or more.

In general, there will be many stakeholders who are affected by adaptation decisions. And the role of all stakeholders has to be accounted for (FAO 2016; Wreford et al. 2010).

1.1.2 Bringing New Practices to the Agricultural Community Everywhere: Implementation

Implementation of innovative strategies focused on adaptation is and will continue to be an ongoing challenge. All changes in management of agriculture and livestock rearing require resources—financial, technical and human—to effect the new way of operating. How to pay for improved agricultural practices is one of the issues that must be considered. In many countries financial support for agriculture has been an on-going practice. Traditionally, the most common goals of financial subsidies have been to maintain the welfare of a healthy farming community and to ensure good productivity that can supply nutrition to growing populations.

Whilst financial support for farmers has been and continues to be a widespread practice, in the climate change context, it is essential that the current subsidies and other forms of financial support do not operate against the principles of adaptation and mitigation such as by supporting highly intensive resource-consuming practices. Beyond this, as a response to changing climate, it is also important to introduce subsidies that will promote explicit sustainable adaptation and mitigation practices. Policies of this type are required to be effective at all levels—for the individual farmer, for particular sectors and at the national or international levels.

Adaption practices can emanate from within the farming community itself, be developed and implemented through the private sector or via government support.

1.1.3 Bringing Laboratory, Policy Makers and Farmers Together

As technological innovations that address the goals of climate-smart agriculture to promote sustainable intensification are developed, at the same time the innovations must be accompanied by initiatives to promote dissemination of the new knowledge and capacity-building within the farming community. Again, this is a particularly critical problem in low-income countries. In many situations, the private sector can facilitate knowledge transfer and capacity building. But it is important that the profit imperatives of the private sector do not put costs out of reach of farmers. Furthermore, national and international bodies must ensure that sound, sustainability-focused advice is promoted and readily available to farmers in every part of the globe.

Around the world, scientists are working to develop new varieties of seeds and other modern technologies to combat the adverse effects of climate change on food security. However, their endeavors will not yield any tangible benefits, unless appropriate commercialization models are appropriated. The ideal commercialization should be based on scientific validity and equity. For example, proper seed commercialization ensures delivering high-quality seed to the marketplace, so that farmers get the best possible chance of consistently producing high yielding and valuable crops. Furthermore, to address the equity issue, farmers' access to seed distribution systems that facilitate dissemination of improved varieties and functioning markets for produce are critical for the benefits of the smallholders. Public–private–community partnerships offer opportunities to catalyze new approaches and investment whilst accelerating integrated research and development and commercial supply chain-based solutions (FAO 2016; Kurukulasuriya and Rosenthal 2003; OECD 2013).

Since Intellectual Property Protection allows the corporate sector to control commercialization of their products, it creates difficulties for farming communities and researchers to access patented technologies when developing crops for humanitarian purposes in the resource constraint countries. Groups such as the *Public Intellectual Property Resource for Agriculture* may serve as viable alternative in promoting licensing practices for developing new crops and technologies (Meridian Institute 2017).

1.1.4 Innovative and Effective Communication for Knowledge Transfer Related to Adaptation and Risk Prediction

Effective communication is one of the top priorities for risk reduction; importantly, it requires collaborative and participatory approaches. This is essential as a long-term strategy. Risk communication can be defined as "both a one-way transfer of hazard and risk-related information and their management, and as a two-way exchange of related information, knowledge, attitudes, and/or values". Risk communication and training that goes along with it helps farmers to make sound choices related to adaptation and mitigation and to invest in preventive measures that will reduce adverse impacts on people and social systems. Currently and increasingly in the future, the internet, satellite communication, and social media can play significant roles in early warning, risk assessment, and follow-up actions (Wreford et al. 2010).

1.1.5 Crop Insurance

To ensure the sustainability of agriculture, insurance is needed against risks of weather (in the short term) and climate (in the long term) as well as against market risks that depend on other factors. Crop insurance is widely available in medium- and high-income countries, but is less common in poorer countries. Whilst insurance itself is not an adaptation strategy, it can provide the financial stability that enables farmers to establish adaptive technologies in preparation for future risk stresses. However, it can become a financial cushion that implicitly encourages farmers to continue risk-prone practices; it therefore must be accompanied by strategies that encourage movement toward more sustainable management methods (Smit and Skinner 2002).

1.1.6 Water Management

Around the globe in various locations, there are serious issues of water oversupply or undersupply. Especially critical are the drought-prone regions of low-income countries where it is becoming more and more difficult for farmers to sustain a viable agricultural operation. Climate change is exacerbating the problem, in part because of the high level of unpredictable variability that is accompanying the changes. Besides water scarcity, there are many instances of overuse leading to declining water tables and also growing instances of salinity of soils in formerly high productivity areas. Together all these factors point to a growing critical need to manage water resources carefully. Given these issues, as a long term measure, adaptation requires enhancing water use efficiency and management with a strong emphasis on equity of supply. This will then require strategies like altering crops and irrigation systems to take account of changing rainfall patterns, repairing water delivery systems, ensuring proper drainage, and the conjunctive use of groundwater. These are generally large-scale and long-term initiatives and generally will require government and private investments on a regional level.

There are new advances in efficient water delivery systems, such as the development of improved sprinklers and underground drip irrigation systems. These are, however, costly to implement, and it is imperative that research be done on equivalent systems appropriate for poor farmers. Some such technologies have been proposed and applied, but most are suitable only for very small holdings and require considerable manual labour.

Finally, there are initiatives that can be taken to improve green water management. Building of bunds, trenches, large and small pits for rainwater collection and retention can all be effective. Windbreaks, mulching with plant residues or plastics, steps to increase the organic matter content of soils can all serve to conserve soil moisture. Many of these things can be done by communities or by individual farmers at a relatively small cost.

Finally, the concept of virtual water can be increasingly employed to make decisions about crop selection in regions or countries. Coupled with rationale trade decisions, this can ensure that water resources are used appropriately and simultaneously ensure adequate food supplies everywhere (Rosegrant et al. 2008).

In the book we have focused on various adaptations to climate change in order sustain food security and the chapters are broadly divided into three categories: (a) climate smart crops, adaptive breeding and genomics, (b) natural resource management, and (c) strategies for access to food, technology, knowledge and equity, and risk governance. We acknowledge that adaptation is a very complex, dynamic and context specific approach and therefore we have emphasized on local level success stories with in depth analysis. We have also encouraged some chapters on global perspectives to let the readers update on general overview of adaptation.

References

Adger, W. N., Agrawala, S., Mirza, M. M. Q., Conde, C., O'Brien, K., Pulhin, J., et al. (2007). *Assessment of adaptation practices, management options, constraints and capacity. Climate change 2007: Impacts, adaptation and vulnerability. Contribution of Working Group II to the Fourth Assessment Report of the Intergovernmental Panel on Climate Change.* Cambridge; New York, NY: Cambridge University Press. Retrieved December 22, 2017, from https://www.ipcc.ch/pdf/assessment-report/ar4/wg2/ar4-wg2-chapter17.pdf.

Asfaw, S., & Lipper, L. (2011). *Economics of PGRFA management for adaptation to climate change: A review of selected literature. Commission on Genetic Resources for Food and Agriculture. Background Study Paper No. 60.* Rome: FAO. Retrieved December 22, 2017, from http://www.fao.org/docrep/meeting/023/mb695e.pdf.

Backlund, P., Janetos, A., & Schimel, D. (2008). *The effects of climate change on agriculture, land resources, water resources, and biodiversity in the United States. A Report by the U.S. Climate Change Science Program and the Subcommittee on Global Change Research.* Washington, DC: U.S. Department of Agriculture. 362.

Battisti, D. S., & Naylor, R. L. (2009). Historical warnings of future food insecurity with unprecedented seasonal heat. *Science, 323,* 240–244.

Brander, K. M. (2007). Global fish production and climate change. *Proceedings of the National Academy of Sciences of the United States of America, 104*(50), 19709–19714. https://doi.org/10.1073/pnas.0702059104.

Brown, M., & Funk, C. C. (2008). Food security under climate change. *Science, 319,* 580–581.

Cairns, J. E., Sonder, K., Zaidi, P. H., Verhulst, N., Mahuku, G., Babu, R., et al. (2012). Maize production in a changing climate: Impacts, adaptation, and mitigation strategies. *Advances in Agronomy, 114,* 1–57.

Cairns, J. E., Hellin, J., Sonder, K., Araus, J. L., MacRobert, J. F., Thierfelder, C., et al. (2013). Adapting maize production to climate change in sub-Saharan Africa. *Food Security, 5,* 345–360. https://doi.org/10.1007/s12571-013-0256-x.

Cheung, W. W. L., Lam, V. W. Y., Sarmiento, J. L., Kearney, K., Watson, R., Zeller, D., et al. (2010). Large-scale redistribution of maximum fisheries catch potential in the global ocean under climate change. *Global Change Biology, 16*, 24–35.

Ciais, P., Schelhaas, M. J., Zaehle, S., Piao, L., Cescatti, A., Liski, J., et al. (2008). Carbon accumulation in European forests. *Nature Geoscience, 1*(7), 425–429.

CNN. (2016). *Mercury rising: India records its highest temperature ever.* May 23. Retrieved December 22, 2017, from http://edition.cnn.com/2016/05/20/asia/india-record-temperature/.

Collier, P., Conway, G., & Venables, T. (2008). Climate change and Africa. *Oxford Review of Economic Policy, 24*, 337–353.

Cooper, P. J. M., Dimes, J., Rao, K. P. C., Shapiro, B., Shiferaw, B., & Twomlow, S. (2008). Coping better with current climatic variability in the rain-fed farming systems of sub-Saharan Africa: An essential first step in adapting to future climate change? *Agriculture, Ecosystems and Environment, 126*, 24–35.

Cooper, M., Messina, C. D., Podlich, D., Radu Totir, L., Baumgarten, A., & Hausmann, N. J. (2014). Predicting the future of plant breeding: Complementing empirical evaluation with genetic prediction. *Crop & Pasture Science, 65*, 311–336. https://doi.org/10.1071/CP14007.

Crescio, M. I., Forastiere, F., Maurella, C., Ingravalle, F., & Ru, G. (2010). Heat-related mortality in dairy cattle: A case crossover study. *Preventive Veterinary Medicine, 97*(3), 191–197.

Cressman, K. (2013). Climate change and locusts in the WANA Region. In M. V. K. Sivakumar, R. Lal, R. Selvaraju, & I. Hamdan (Eds.), *Climate change and food security in West Asia and North Africa* (pp. 131–143). Dordrecht: Springer. https://doi.org/10.1007/978-94-007-6751-5_7.

Deryng, D., Conway, D., Ramankutty, N., Price, J., & Warren, R. (2014). Global crop yield response to extreme heat stress under multiple climate change futures. *Environmental Research Letters, 9*, 034011. Retrieved December 22, 2017, from http://iopscience.iop.org/article/10.1088/1748-9326/9/3/034011/pdf.

Ericksen, P., Thornton, P., Notenbaert, A., Cramer, L., Jones, P., & Herrero, M. (2011). *Mapping hotspots of climate change and food insecurity in the global tropics. CCAFS Report 5.* Copenhagen: CGIAR Research Program on Climate Change, Agriculture and Food Security (CCAFS). Retrieved December 22, 2017, from www.ccafs.cgiar.org.

FAO. (2008). *Climate change and food security: A framework document.* Rome: Food Agriculture Organization of the United Nations.

FAO. (2011). *The State of food and agriculture 2010–2011. Women in agriculture: Closing the gender gap for development.* Rome: Food Agriculture Organization of the United Nations. Retrieved December 22, 2017, from http://www.fao.org/docrep/013/i2050e/i2050e.pdf.

FAO. (2015). *Coping with climate change – The roles of genetic resources for food and agriculture.* Rome: Food Agriculture Organization of the United Nations. Retrieved December 22, 2017, from http://www.fao.org/3/a-i3866e.pdf.

FAO. (2016). *The state of food and agriculture: Climate change, agriculture and food security.* Rome: Food and Agriculture Organization of the United Nations. Retrieved from http://www.fao.org/publications/sofa/2016/en/.

FAO, IFAD, UNICEF, WFP, WHO. (2017). *The State of food security and nutrition in the world 2017. Building resilience for peace and food security.* Rome: Food Agriculture Organization of the United Nations. Retrieved from http://www.fao.org/3/a-I7695e.pdf.

Fischer, R. A., Byerlee, D., & Edmeades, G. O. (2014). *Crop yields and global food security: Will yield increase continue to feed the world?* Canberra, ACT: Australian Centre for International Agricultural Research. Retrieved December 22, 2017, from http://aciar.gov.au/publication/mn158.

Frieler, K., Levermann, A., Elliott, J., Heinke, J., Arneth, A., Bierkens, M. F. P., et al. (2015). A framework for the cross-sectoral integration of multi-model impact projections: Land use decisions under climate impacts uncertainties. *Earth System Dynamics, 6*, 447–460. https://doi.org/10.5194/esd-6-447-2015.

Gitz, V., & Meybeck, A. (2012). Risks, vulnerabilities and resilience in a context of climate change. In A. Meybeck, J. Lankoski, S. Redfern, N. Azzu, & V. Gitz (Eds.), *Building resilience for adaptation to climate change in the agriculture sector, Proceedings of a joint FAO/OECD Workshop* (pp. 19–36). Rome: FAO. Retrieved December 22, 2017, from http://www.fao.org/docrep/017/i3084e/i3084e.pdf.

Gong, F. P., Wu, X., Zhang, H., Chen, Y., & Wang, W. (2015). Making better maize plants for sustainable grain production in a changing climate. *Frontiers in Plant Science, 6*, 835. https://doi.org/10.3389/fpls.2015.00835.

HLPE. (2011). *Price volatility and food security*. A report by the High Level Panel of Experts on Food Security and Nutrition of the Committee on World Food Security (report 1). Rome. Retrieved December 22, 2017, from http://www.fao.org/fileadmin/user_upload/hlpe/hlpe_documents/HLPE-price-volatility-and-food-security-report-July-2011.pdf.

HLPE. (2012). *Food security and climate change*. A report by the High Level Panel of Experts on Food Security and Nutrition of the Committee on World Food Security (report 3). Rome. Retrieved December 22, 2017, from http://www.fao.org/fileadmin/user_upload/hlpe/hlpe_documents/HLPE_Reports/HLPE-Report-3-Food_security_and_climate_change-June_2012.pdf.

Hoddinot, J. (2006). Shocks and their consequences across and within households in rural Zambia. *Journal of Development Studies, 42*(2), 301–321.

Horton, D. E., Johnson, N. C., Singh, D., Swain, D. L., Rajaratnam, B., & Diffenbaugh, N. S. (2015). Contribution of changes in atmospheric circulation patterns to extreme temperature trends. *Nature, 522*, 465–469. https://doi.org/10.1038/nature14550.

Hossain, A., & Teixeira da Silva, J. A. (2013). Wheat production in Bangladesh: Its future in the light of global warming. *AoB Plants, 5*, pls042. https://doi.org/10.1093/aobpla/pls042.

IPCC. (2007). *Fourth assessment report: Synthesis*. Retrieved December 22, 2017, from http://www.ipcc.ch/pdf/assessment-report/ar4/syr/ar4_syr.pdf.

IPCC. (2012). Managing the risks of extreme events and disasters to advance climate change adaptation - Special Report of the Intergovernmental Panel on Climate Change. In C. B. Field, C. Barros, T. F. Stocker, D. Qin, D. J. Dokken, K. L. Ebi, et al. (Eds.). Cambridge; New York, NY: Cambridge University Press. 582 pp. Retrieved December 22, 2017, from https://www.ipcc.ch/pdf/special-reports/srex/SREX_Full_Report.pdf.

IPCC. (2013). Climate change 2013: The physical science basis. In T. F. Stocker, D. Qin, G. K. Plattner, M. Tignor, S. K. Allen, et al. (Eds.), *Working Group I Contribution to the Fifth Assessment Report of the Intergovernmental Panel on Climate Change. Summary for policymakers*. Cambridge; New York, NY: Cambridge University Press. 1535 p. Retrieved December 22, 2017, from https://www.ipcc.ch/pdf/assessment-report/ar5/wg1/WGIAR5_SPM_brochure_en.pdf.

IPCC. (2014). Climate change 2014: Synthesis report. In Core Writing Team, R. K. Pachauri, & L. A. Meyer (Eds.), *Contribution of Working Groups I, II and III to the Fifth Assessment Report of the Intergovernmental Panel on Climate Change*. Geneva: IPCC. 151 pp. Retrieved December 22, 2017, from http://www.ipcc.ch/pdf/assessment-report/ar5/syr/SYR_AR5_FINAL_full_wcover.pdf.

Kumwenda, J. D. T., Waddington, S. R., Snapp, S. S., Jones, R. B., & Blackie, M. J. (1998). Soil fertility management in Southern Africa. In D. Byerlee & C. K. Eicher (Eds.), *Africa's emerging maize revolution* (p. 305). Boulder, CO: Lynne Rienner Publishers.

Kurukulasuriya, P., & Rosenthal, S. (2003). *Climate change and agriculture: A review of impacts and adaptations, Climate Change Series, Paper 91*. Washington, DC: The World Bank. Retrieved December 22, 2017, from http://documents.worldbank.org/curated/en/757601468332407727/pdf/787390WP0Clima0ure0377348B00PUBLIC0.pdf.

Lambrou, Y., & Nelson, S. (2010). *Farmers in a changing climate – Does gender matter? Food security in Andhra Pradesh, India*. Rome: FAO. Retrieved December 22, 2017, from http://www.fao.org/docrep/013/i1721e/i1721e00.pdf.

Lioubimtseva, L., Dronin, N., & Kirilenko, A. (2015). Grain production trends in the Russian Federation, Ukraine and Kazakhstan in the context of climate change and international trade.

In A. Elbehri (Ed.), *Climate change and food systems: Global assessments and implications for food security and trade* (pp. 211–244). Rome: Food Agriculture Organization of the United Nations (FAO).

Lobell, D. B., Hammer, G. L., McLean, G., Messina, C., Roberts, M. J., & Schlenker, W. (2013). The critical role of extreme heat for maize production in the United States. *Nature Climate Change, 3*, 497–501.

Masih, I., Maskey, S., Mussá, F. E. F., & Trambauer, P. (2014). A review of droughts on the African continent: A geospatial and long-term perspective. *Hydrology and Earth System Sciences, 18*(9), 3635–3649.

Miles, L., Newton, A. C., DeFries, R. S., Ravilious, C., May, I., Blyth, S., et al. (2006). A global overview of the conservation status of tropical dry forests. *Journal of Biogeography, 33*(3), 491–505.

Meridian Institute. (2017). *Public intellectual property resource for agriculture.* Retrieved December 22, 2017, from http://merid.org/Content/Projects/Public_Intellectual_Property_Resource_for_Agriculture.aspx.

Müller, C., & Elliott, J. (2015). The global gridded crop model intercomparison: Approaches, insights and caveats for modelling climate change impacts on agriculture at the global scale. In A. Elbehri (Ed.), *Climate change and food systems: Global assessments and implications for food security and trade* (pp. 28–59). Rome: Food Agriculture Organization of the United Nations (FAO).

Nelson, G. C., Rosegrant, M. W., Palazzo, A., Gray, I., Ingersoll, C., Robertson, R., et al. (2010). *Food security, farming, and climate change to 2050: Scenarios, results, policy options.* Washington, DC: International Food Policy Research Institute (IFPRI).

Nelson, G. C., Valin, H., Sands, R. D., Havlik, P., Ahammad, H., Deryng, D., et al. (2014). Climate change effects on agriculture: Economic responses to biophysical shocks. *Proceedings of the National Academy of Sciences of the United States of America, 111*(9), 3274–3279.

Nkonya, E., Mirzabaev, A., & Von Braun, J. (2015). *Economics of land degradation and improvement: A global assessment for sustainable development.* New York, NY: Springer.

Nyasimi, M., Amwata, D., Hove, L., Kinyangi, J., & Wamukoya, G. (2014). *Evidence of impact: Climate-smart agriculture in Africa. CCAFS Working Paper No. 86.* Copenhagen: CGIAR Research Program on Climate Change, Agriculture and Food Security (CCAFS).. Retrieved December 22, 2017, from https://ccafs.cgiar.org/sites/default/files/research/attachments/climate_smart_farming_successes_Africa.pdf.

OECD. (2013). *Policy instruments to support green growth in agriculture. OECD green growth studies.* Paris: OECD. Retrieved February 06, 2018, from http://www.oecd.org/environment/policy-instruments-to-support-green-growth-in-agriculture-9789264203525-en.htm.

OECD. (2015). *Agriculture and climate change, trade and agriculture directorate.* Paris: Organization of Economic Cooperation and Development. September. Retrieved February 06, 2018, from https://www.oecd.org/tad/sustainable-agriculture/agriculture-climate-change-september-2015.pdf.

Pattnaik, I., Lahiri-Dutt, K., Lockie, S., & Pritchard, B. (2018). The feminization of agriculture or the feminization of agrarian distress? Tracking the trajectory of women in agriculture in India. *Journal of the Asia Pacific Economy, 23*, 138. https://doi.org/10.1080/13547860.2017.1394569.

Pautasso, M., Döring, T. F., Garbelotto, M., Pellis, L., & Jeger, M. J. (2012). Impacts of climate change on plant diseases – Opinions and trends. *European Journal of Plant Pathology, 133*(1), 295–313.

Perry, A. L., Low, P. J., Ellis, J. R., & Reynolds, J. D. (2005). Climate change and distribution shifts in marine fishes. *Science, 308*(5730), 1912–1915.

PRB. (2013). World population data sheet 2013. Population reference bureau. . Retrieved December 22, 2017, from www.prb.org/Publications/Datasheets/2013/2013-world-population-data-sheet.aspx.

Porter, J. R., Xie, L., Challinor, A. J., Cochrane, K., Howden, S. M., Iqbal, M. M., et al. (2014). Food security and food production systems. In C. B. Field, V. R. Barros, D. J. Dokken, K. J. Mach, M. D. Mastrandrea, T. E. Bilir, et al. (Eds.), *Climate change 2014: Impacts, adaptation, and vulnerability. Part A: Global and sectoral aspects. Contribution of Working Group II to the Fifth Assessment Report of the Intergovernmental Panel on Climate Change* (pp. 485–533). Cambridge; New York, NY: Cambridge University Press. Retrieved December 22, 2017, from http://www.ipcc.ch/pdf/assessment-report/ar5/wg2/WGIIAR5-Chap7_FINAL.pdf.

Pörtner, H. O. (2008). Ecosystem effects of ocean acidification in times of ocean warming: A physiologist's view. *Marine Ecology Progress Series, 373,* 203–217.

Rosenzweig, C., Elliott, J., Deryng, D., Ruane, A. C., Müller, C., Arneth, A., et al. (2014). Assessing agricultural risks of climate change in the 21st century in a global gridded crop model intercomparison. *Proceedings of the National Academy of Sciences of the United States of America, 111*(9), 3268–3273. https://doi.org/10.1073/pnas.1222463110.

Rosegrant, M. W., Ewing, M., Yohe, G., Burton, I., Huq, S., & Valmonte-Santos, R. (2008). *Climate change and agriculture: Threats and opportunities.* Bonn: Deutsche Gesellschaft für Technische Zusammenarbeit (GTZ) GmbH. Climate Protection Programme for Developing Countries.

Recha, C. W., Makokha, G. L., Traore, P. S., Shisanya, C., Lodoun, T., & Sako, A. (2012). Determination of seasonal rainfall variability, onset and cessation in semi-arid Tharaka District, Kenya. *Theoretical and Applied Climatology, 108,* 479–494.

Sarkar, A., Patil, S., Hugar, L. B., & van Loon, G. (2011). Sustainability of current agriculture practices, community perception, and implications for ecosystem health: An Indian study. *EcoHealth, 8*(4), 418–431. https://doi.org/10.1007/s10393-011-0723-9.

Sarkar, A., & vanLoon, G. W. (2015). Modern agriculture and food and nutrition insecurity: Paradox in India. *Public Health, 129*(9), 1291–1293.

Settele, J., Scholes, R., Betts, R., Bunn, S., Leadley, P., Nepstad, D., et al. (2014). Terrestrial and inland water systems. In C. B. Field, V. R. Barros, D. J. Dokken, K. J. Mach, M. D. Mastrandrea, T. E. Bilir, et al. (Eds.), *Climate change 2014: Impacts, adaptation, and vulnerability. Part A: Global and sectoral aspects. Contribution of Working Group II to the Fifth Assessment Report of the Intergovernmental Panel on Climate Change* (pp. 271–359). Cambridge; New York, NY: Cambridge University Press. Retrieved December 22, 2017, from https://www.ipcc.ch/pdf/assessment-report/ar5/wg2/WGIIAR5-Chap4_FINAL.pdf.

Smit, B., & Skinner, M. W. (2002). Adaptation options in agriculture to climate change: A typology. *Mitigation and Adaptation Strategies for Global Change, 7,* 85–114.

Sanchez, P. (2002). Soil fertility and hunger in Africa. *Science, 295,* 2019–2020.

Shiferaw, B., Tesfaye, K., Kassie, M., Abate, T., Prasanna, B. M., & Menkir, A. (2014). Managing vulnerability to drought and enhancing livelihood resilience in sub-Saharan Africa: Technological, institutional and policy options. *Weather and Climate Extremes, 3,* 67–79.

Thomas, T., & Rosegrant, M. (2015). Climate change impact on key crops in Africa: Using crop models and general equilibrium models to bound the predictions. In A. Elbehri (Ed.), *Climate change and food systems: Global assessments and implications for food security and trade.* Rome: FAO.

Uleberg, E., Hanssen-Bauer, I., van Oort, B., & Dalmannsdottir, S. (2014). Impact of climate change on agriculture in Northern Norway and potential strategies for adaptation. *Climatic Change, 122,* 27–39.

UN. (2014). *Goal 2. Sustainable development knowledge platform.* United Nations. December 1. Retrieved December 22, 2017, from https://sustainabledevelopment.un.org/?page=view&nr=164&type=230 and https://sustainabledevelopment.un.org/sdg13.

van Loon, G. W., Patil, S. G., & Hugar, L. B. (2005). *Agricultural sustainability—Strategies for assessment.* New Delhi: Sage.

Wassmann, R., Jagadish, S. V. K., Heuer, S., Ismail, A., Redoña, E., Serraj, R., et al. (2009). Climate change affecting rice production: The physiological and agronomic basis for possible adaptation strategies. In L. Sparks Donald (Ed.), *Advances in agronomy* (Vol. 101, pp. 59–122). Burlington, MA: Academic Press.

World Bank/FAO/WorldFish Center. (2010). *The hidden harvests: The global contribution of capture fisheries. Agriculture and Rural Development Department, Sustainable Development Network. June.* Washington, DC: World Bank. Retrieved December 22, 2017, from http://siteresources.worldbank.org/EXTARD/Resources/336681-1224775570533/TheHiddenHarvestsConferenceEdition.pdf.

Wreford, A., Moran, D., & Adger, N. (2010). *Climate change and agriculture: Impacts, adaptation and mitigation.* Paris: OECD. Retrieved December 22, 2017, from https://www.cabdirect.org/cabdirect/abstract/20103233880.

Part II
Climate-Smart Crops, Adaptive Breeding and Genomics

Chapter 2
Adaptation of Crops to Warmer Climates: Morphological and Physiological Mechanisms

Ullah Najeeb, Daniel K. Y. Tan, Muhammad Sarwar, and Shafaqat Ali

Abstract Increased surface temperature is one of the major reasons for reduced crop productivity in many parts of the world. Response to elevated temperature varies among crop species—a certain threshold temperature has been determined for each crop above which it suffers yield losses. Thus, some crop species, e.g. summer crops (cotton, rice, sorghum), are considered relatively more tolerant to high temperature than winter crops (wheat, barley, chickpeas, faba bean). Heat-induced yield penalties in crops are the result of inhibited vegetative growth or impaired reproductive development. High temperature can cause cellular injury, leading to catastrophic collapse of cellular organization and functioning and ultimately, growth inhibition. Similarly, reproductive structures, especially pollen are highly sensitive to elevated temperatures and a heat shock event at reproductive phase impairs fertilisation and consequently increases fruit or seed abortion. Tolerance to high temperature is associated with a range of physiological and morphological adaptations in plants. For example, plants can tolerate heat-induced damage through foliar orientation, stomatal regulation and stimulation of antioxidative defence systems. These adaptive mechanisms are regulated by stress responsive genes, encoding for specific proteins, e.g. heat shock proteins, which enable plants to survive under extreme environments. This chapter discusses various adaptive, avoidance and acclimation

---

U. Najeeb (✉)
Queensland Alliance for Agriculture and Food Innovation, Centre for Plant Science, The University of Queensland, Toowoomba, QLD, Australia

The University of Sydney, Plant Breeding Institute, Sydney Institute of Agriculture, School of Life and Environmental Sciences, Faculty of Science, Sydney, NSW 2006, Australia
e-mail: n.ullah@uq.edu.au

D. K. Y. Tan
The University of Sydney, Plant Breeding Institute, Sydney Institute of Agriculture, School of Life and Environmental Sciences, Faculty of Science, Sydney, NSW 2006, Australia

M. Sarwar
Agronomic Research Institute, Ayub Agricultural Research Institute, Faisalabad, Pakistan

S. Ali
Department of Environmental Sciences and Engineering, Government College University, Faisalabad, Pakistan

© Springer Nature Switzerland AG 2019
A. Sarkar et al. (eds.), *Sustainable Solutions for Food Security*,
https://doi.org/10.1007/978-3-319-77878-5_2

mechanisms of heat tolerance in plants. It also highlights the breeding and management techniques used for inducing heat stress tolerance in crop plants.

2.1 Introduction

Since the Industrial Revolution, global air temperature has been rising at a considerable rate. Climate models projected 1–4 °C increase in the mean global temperature by 2100 (IPCC 2007). In fact, the Northern Hemisphere has recently experienced the highest recorded temperatures in the past 1400 years and this trend may continue in future (Hansen et al. 2015). In addition, high frequency of extreme weather events such as heat waves, long-term drought and flooding are also expected to increase (IPCC 2013). Despite some positive effects of global warming, e.g. longer growing season (McCarthy 2001) and expansion of agricultural production in the higher latitudes (Lotze-Campen et al. 2009; Olesen and Bindi 2002), this trend will adversely affect the crop production in many parts of the world (IPCC 2007). Maximum damage is expected to the productive agricultural regions of Africa and Asia (Fischer et al. 2002; Ortiz et al. 2008), e.g. 15–35% yield losses in crops for Asia and the Middle East has been estimated under an average increase of 3–4 °C air temperature (FAO 2009). These climatic changes will more likely cause a serious impact to the people living in developing countries where approximately 50% of the population is directly or indirectly associated with the agriculture industry.

High temperature may damage plant development directly by increasing membrane fluidity and denaturing proteins or indirectly impairing various plant functions through inactivation of enzymes, production of toxic compounds and reactive oxygen species (ROS). Plants require an optimum temperature for continuing normal growth and development, and a temperature which irreversibly damages the developmental processes of a crop results in heat stress for that specific crop. Threshold temperatures for different crop species have been established but generally a transient increase of 10–15 °C above ambient temperature can cause heat shock or stress. Any temperature above threshold can significantly impair important physiological processes influencing plant growth and yield. For example, a short-term exposure to extreme temperature can cause cellular injury, leading to a catastrophic collapse of cellular organization.

Out of more than 200,000 identified plant species on the planet, only 20–25 species contribute to major dietary intake for human (Füleky 2009). For example, wheat, rice and maize alone contribute to 75% of the total grain production in the world (Lobell and Gourdji 2012). Further, to meet the demand of nine billion people by 2050, a 70% increase in food production is required (FAO 2009), which means 38% more yield increase than those achieved historically (Tester and Langridge 2010). In past three decades, a significant increase in the yield of food crops has been achieved through breeding and agronomic improvements (Ainsworth and Ort 2010; Teixeira et al. 2013). However, this continuous increase in food production may be restricted by environmental variability and increased global surface temperature. For example, the data suggest a significant decline in the yields of major food crops under a scenario of 2 °C of warming without adaptation (Challinor et al. 2014). This review discusses the effect of heat stress on major crops, plant survival mechanisms and potential sources of heat tolerance.

2.2 Heat Sensitivity in Crops

Optimum temperature ranges for different crops may considerably differ for different developmental stages and yield components, e.g. plant architecture, photosynthetic capacity, resource partitioning and reproductive success in most of the crops are strongly influenced by high temperature (Table 2.1). In general, the crops grown under tropical regions tend to be relatively more heat tolerant than the crops grown in temperate regions (Wahid et al. 2007). Similarly, summer crops such as cowpea and rice can sustain higher temperatures of 37 °C and 40 °C, respectively, at emergence (Akman 2009), compared with winter crops such as chickpea, lentils and lettuce (Covell et al. 1986).

2.2.1 Winter Crops

Winter crops including cereals such as wheat and barley, and pulses such as chickpeas and faba bean are regarded as heat sensitive. High temperatures negatively influence various aspects of growth such as phenological development, tillering, reproductive structures, grain formation and grain filling in these species. In addition, elevated temperature may alter source–sink relationships either through inhibited carbon assimilation and respiration and/or through impaired carbohydrate partitioning and redistribution (Reynolds et al. 2004; Calderini et al. 2006). Changes in the C and N availability affect starch and protein metabolism in leaves, leading to poor yield and crop quality (Yang et al. 2011).

Wheat, one of the most widely grown food crops and major contributors to our dietary intake, is highly sensitive to high temperatures. In wheat crops, a 3–5% grain yield reduction for each degree rise in the average temperature above 15 °C has been recorded under controlled environment (Gibson and Paulsen 1999). Similarly, field data suggested that yield losses can be in the order of 190 kg ha^{-1} for each degree rise in average temperature, and in some cases, it can cause even greater damage to yield than water availability (Kuchel et al. 2007; Bennett et al. 2012). Another summer cereal, barley (*Hordeum vulgare* L.), which is considered more adaptable to a wide range of environmental conditions, had reduced grain yield at air temperatures above 25 °C (Hossain et al. 2012).

Similar to cereals, winter legumes are also sensitive to high temperatures. For example, optimum temperature for chickpea vegetative and reproductive growth is

Table 2.1 Heat sensitivity of crops (ranges of optimum and critical temperature, yield losses)

Crop species	Optimum temperature	Critical temperature	Growth inhibition	Yield losses
Cotton	20–32 °C (Burke et al., 1988; Reddy et al., 1991).	>35 °C (Bibi et al., 2008)	50% decline in shoot biomass at 40/30 °C (Reddy et al., 1991).	110 kg ha^{-1} loss of seed cotton yield for each °C rise in temperature above maximum (Singh et al. 2007). 10% yield reduction for each °C rise in temperature above average ambient (Pettigrew, 2008).
Rice	30–35 °C for seed germination (Krishnan et al., 2011). 22–31 °C for growth (Yoshida, 1981; Krishnan et al., 2011).	30 °C (Krishnan et al., 2011).	10% reduction in tillering by increasing temperature from 29/21 to 37/29 °C (Krishnan et al., 2011).	7–8% yield loss for each °C rise in temperature from 28/21 to 34/27 °C (Krishnan et al., 2011).
Wheat	15–20 °C for vegetative growth (Palta et al., 1994). 12–22 °C for reproductive growth (Farooq et al. 2011).	25 °C for grain filling (Gate et al., 2010).	15% and 25% reduction in shoot dry weight and plant growth rate, respectively, by increasing temperature from 22/16 to 40/20 °C (Peck and McDonald, 2010)	3% yield reduction for each 1 °C rise in temperature above 15 °C (Wardlaw et al. 1989). Up to 10% grain yield reduction by post-anthesis exposure to 40/20 °C (Peck and McDonald, 2010). 10% grain yield reduction by each °C rise in night temperature above optimum (Prasad et al. 2008).

(continued)

2 Heat Resistant Crops   31

Table 2.1 (continued)

Crop species	Optimum temperature	Critical temperature	Growth inhibition	Yield losses
Chickpea	20–26 °C (Wang et al., 2006)	30–35 °C for flowering (Basu et al., 2009).	Up to >30 °C reduction in crop duration period by increasing temperature above optimum (Summerfield et al., 1984)	Up to 308 kg ha^{-1} reduction in yield by each °C rise in temperature above mean season temperature (Kalra et al., 2008).
Faba bean	15–23 °C (Patrick and Stoddard, 2010).	28 °C (Bishop et al. 2016)	A short exposure to 32/27 °C severely inhibits reproductive development (Gross and Kigel, 1994).	6.9 g per plant yield reduction by increasing temperature from 18 to 34 °C (Bishop et al. 2016)
Date Palm (*Phoenix dactylifera* L.)	30 °C for pollen germination (Maryam et al., 2015).	50 °C for growth (Chao and Robert, 2007)		
Cactus (*Cactaceae*)	35–40 °C (Gerwick et al., 1977)	55 °C (Downton et al., 1984).		

20–25 °C and temperatures ≥30 °C can cause detrimental effects on seed yield. High temperatures during flowering influence chickpea grain yield by reducing pod formation and seed set (Devasirvatham et al. 2012). A significant reduction in growth, seed and pod set has also been recorded in common bean (*Phaseolous vulgaris* L.) at 32 °C (Suzuki et al. 2001), in pea (*Pisum sativum* L.) at 30 °C (McDonald and Paulsen 1997) and cowpea (*Vigna unguiculata* L.) at 33 °C (Ahmed et al. 1992), while at temperatures above 28 °C were detrimental for growth and yield of faba bean crops (Bishop et al. 2016).

2.2.2 Summer Crops

Summer crops such as cotton, rice and sorghum are relatively tolerant to high temperature but they may experience a degree of yield losses when exposed to an extremely high temperature. For example, cotton crop is widely grown under hot, semi-arid regions throughout the world. The optimum range of temperature for cotton growth and development is 28–30 °C but the air temperature exceeds this limit in most cotton growing regions. In southwest parts of USA, cotton crop is cultivated under arid environments, where air temperatures above 40 °C are common throughout the growing season (Radin 1992). Similarly, cotton is grown in the subcontinent regions, where air temperature often reaches to 45 °C. Despite continuing to grow, the cotton crop is sensitive to high temperatures and a strong negative correlation has been established between high temperature and lint yields, with a yield reduction of 110 kg ha^{-1} for each 1 °C increase in maximum day temperature (Singh et al. 2007)

Rice, another major summer cereal crop, is also facing a serious challenge from global warming, and the crop has experienced a steady decline in grain yield from 1979 to 2003 due to increased night temperatures. It has been estimated that each 1 °C temperature rise in the mean temperature during the growing season may cause up to 15% reduction in rice grain yield (Peng et al. 2004). Other summer crops such as corn and soybean have also shown 17% reduction in yield for each degree increase in average growing season temperature above the optimum (Lobell and Gregory 2003).

2.2.3 Extremely Heat-Resistant Crops

Unlike animals, plants are sessile and cannot escape from environmental stresses directly by moving away from the stressed zone. Instead, they have developed sophisticated mechanisms to protect their tissues from overheating or heat-induced cellular injury (heat tolerance). This thermo-tolerance mechanism in plants primarily depends on the extent of high temperature to which a tissue is exposed and the ability of a plant to acclimate to stress (hardening or acquired thermo-tolerance). Desert plant species, for example, display a high level of acclimation to the temperatures unsuitable for growth of most arable crop species. Many of the desert succulents, such as the prickly pear (*Opuntia* spp.) can tolerate a short exposure to heat shock of up to 70 °C (Nobel 1988)—a temperature which is lethal for most cultivated crops. Another example of heat-tolerant plants is members of *Agavaceae* family, which are well adapted to arid and semiarid regions where air temperature often rises to 55 °C (Lujan et al. 2009). Similarly cultivated cacti such as *Nopalea cochenillifera, Opuntia robusta, Selenicereus megalanthus* and *Hylocereus undatus* have shown tolerance to tissue temperature up to 45 °C (Nobel and De la Barrera 2002; Zutta et al. 2011). Succulents such as *Agave deserti and Opuntia ficus-indica* have the ability to acclimate heat shock within few days of heat stress (Nobel 1988). These plant species achieve high levels of heat tolerance through various adaptive mechanisms. For example, agaves and many other succulent plants adapt to adverse environmental conditions through crassulacean acid metabolism (CAM), which allows them to conserve water during the hot daylight hours by regulating stomatal conductance (Larcher 1995). In addition, presence of thick cuticles on epidermis may protect plants from heat shock by reducing absorbance to short-wave radiation (Nobel and Smith 1983). More recently Lujan et al. (2009) linked this extreme tolerance high temperature in agave species (*A. tequilana*) with higher stomatal density on leaves and their ability to open the stomata under high temperature.

2.3 Morphological Adaptations

The capability of a crop species to survive and produce good yield under high temperature is termed as heat stress tolerance (Wahid et al. 2007). Vignjevic et al. (2015) proposed that a crop can achieve heat tolerance by (1) maintaining photosynthesis and assimilate supply to fruits during the heat period (2) sustaining assimilation of carbohydrates and nutrients following heat episode and (3) re-mobilising water soluble carbohydrates to supplement assimilate supply.

Plant responses to heat stress are mediated by an intrinsic capacity of basal thermo-tolerance and the ability to gain thermo-tolerance after acclimation. Plant species may also sustain heat-induced injury by maintaining the internal body temperature to a level that is lower than the lethal stress level (heat avoidance). This is achieved through structural modifications such as leaf orientation, reflection of solar radiation, leaf shading and/or through stomatal regulation and transpiration cooling. The ability of a crop genotype to maintain physiological functioning despite an increased in tissues temperatures is regarded as heat tolerance. High temperature tolerant genotypes display superior photosynthetic capacity, carbon allocation and nutrient uptake than the sensitive genotypes under high temperatures (Cui et al. 2006). Thus, plant breeders may seek genotypes with superior physiological traits under high temperature to achieve additive gene action for yield enhancement (Reynolds and Langridge 2016). However, development of genotypes tolerant to extreme temperatures is still challenging due to the fact that traits associated with high yield such as crop architecture, carbon metabolism, resource uptake and distribution and reproductive accomplishment are variably influenced across the developmental stages (Driedonks et al. 2016). Selection on the basis of better physiology combined with superior agronomic traits may result in high performing genotypes (Fig. 2.1).

2.3.1 Leaf Structural Modifications

Plants have developed mechanisms to survive under high temperature either through long term evolutionary phenological/morphological adaptations or through short-term avoidance/acclimation. Leaf structural modifications have commonly been observed as a step towards adaption to high temperature. Some plant species can avoid heat-induced injury by reducing the absorption of solar radiation by leaf morphological changes such as development of small hairs (tomentose), leaf rolling or leaf shedding (Wahid et al. 2007). Glaucous or waxy leaves can protect leaves by regulating stomatal behaviour and reflecting excessive radiation (Mondal 2012).

Some plant species have the capacity to avoid direct sunlight by changing the angle of leaf blades into a position parallel to sun rays (paraheliotropism), this protects leaves from heat-induced burning. Development of thinner leaves with low specific leaf weight and prostrate growth pattern may also play a role in heat

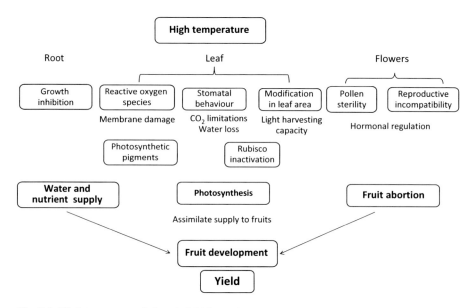

Fig. 2.1 High temperature-induced yield losses in crops

tolerance by increasing radiation use efficiency and moisture conservation (Richards 1996). Similarly, leaf hyponasty (elevation) has also been suggested as a mechanism to minimise heat-induced damage by reducing direct exposure to solar radiation in plants (Van Zanten et al. 2009).

Selection of genotypes based on these superior leaf characteristics for heat-stress tolerance can be a step towards developing heat-tolerant genotypes. This can be achieved by exploiting existing crop germplasm, wild relatives and land races. For example, Bennett et al. (2012), identified a quantitative trait loci (QTL) for glaucousness in wheat genotypes that contributes 50% of the genetic variance under high temperature. They suggested that glaucous leaves can be an ideal trait for breeding heat and drought tolerant genotypes. Similarly, genetic variation for better growth rates and early ground cover has been observed in wheat genotypes, although its association with high temperature tolerance still needs exploration (Cossani and Reynolds 2012). Crawford et al. (2012) observed a significant variation in the shoot architecture and water use of *Arabidopsis* under different temperatures, and suggested that these traits may be useful for selecting heat-tolerant genotypes.

2.3.2 Canopy Temperature

Greater transpiration and high stomatal conductance imply cooler leaves with improved carbon assimilation and crop production rates (Jones 2013). Cooler canopies coupled with deep rooting traits are often associated with superior drought and heat stress tolerance. These traits have been used for selecting genotypes with better radiation use efficiency, light interception and adaptation to stressed environment (Devasirvatham et al. 2016).

Canopy temperature permits rapid phenotyping of traits such as transpirational flux and root depth (Lopes and Reynolds 2010). A cooler canopy trait is associated with a number of advantages such as better root system with greater water uptake ability, high intrinsic radiation use efficiency and photo protective mechanism that helps maintain high radiation use efficiency (Reynolds and Trethowan 2007). Chauhan et al. (2009) reported that wheat genotypes with an ability to maintain cooler canopies under high temperature were more tolerant to heat stress than the genotypes with warmer canopies. Similarly, tetraploid wheat showed an ability to maintain cooler leaves and thus better growth and yield compared with the bread wheat under high temperatures (Bahar et al. 2008).

2.3.3 Root System Architecture

The root system plays a key role in water uptake from the soil profile and a better understanding of root anatomy and functioning may help in selecting genotypes with superior water use efficiency for high temperature (Pinto and Reynolds 2015). Studies on root length density showed that plant growth under variable environment greatly depends on root performance. For example, increased root length density promoted photosynthetic efficiency of heat-stressed plants by increasing phosphorus uptake (Manske et al. 2000).

Most of crop breeding programs have mainly been focused on studying the changes in the aboveground plant traits in response to heat stress. The major challenges in studying root level responses are the difficulties in in situ evaluation of root traits (Uga et al. 2015), root phenotypic plasticity to changing soil conditions and lack of cost-effective high-throughput screening techniques. Protocols for exploring root system responses to high temperatures under field conditions have been established but comprehensive studies are required to understand the role of plant belowground parts in yield stability under variable climatic conditions.

2.3.4 Reproductive Development

High temperature stress changes the physiological mechanism and growth pattern, however, plant response to these changes vary from one physiological stage to another. For instance, at an early crop growth stage, high temperature affects seed germination, seedling vigour and establishment, and photosynthetic rates at tillering (Schapendonk et al. 2007). During late vegetative phase and pre-anthesis, high temperature can change plant phenology (speed up developmental rates), while during the reproductive phase, heat stress can cause flower abscission through pollen sterility or inhibit carbohydrate remobilisation (Devasirvatham et al. 2016).

Reproductive structures, particularly, pollen development, is very sensitive heat-induced damage and therefore pollen germination tests are often used for screening thermo-tolerant crop species. High temperature-induced damage to pollen development has been recorded in many plants. For example, pollen sterility has been observed in response to 35 °C in rice and chickpea (Endo et al. 2009; Devasirvatham et al. 2010), 40 °C in tomato (Giorno et al. 2013), 30 °C in wheat and barley (Saini et al. 1984; Oshino et al. 2007). Further early stages of pollen development were more sensitive to heat-induced injury. Song et al. (2015) indicated that sporogenous cell to tetrad formation phases (16–24 days prior to anthesis) are the most sensitive stages of pollen development to high temperature. Pollen in heat-resistant genotypes, in contrast, have shown a greater ability to produce heat shock proteins (HSPs), and increasing the expression of HSP has been found effective in increasing pollen stability and consequently, thermo-tolerance in cotton and tobacco (Burke and Chen 2015).

Maternal tissues of pistil and female gametophyte are considered relatively more tolerant to high temperature; however, heat-induced injuries to ovaries and megasporogenesis have also been recorded in wheat (Saini et al. 1983), apricot (Rodrigo and Herrero 2002), cherry, peach and plum (Endo et al. 2009; Hedhly 2011). Apart from damage to reproductive structures and reduced fertilization efficiency, high temperature may influence post-pollination processes such as pollen tube growth, fertilization, endosperm formation and embryo development, leading to poor fruit/seed set (Barnabs et al. 2008). Thus, heat-induced damage to reproductive phase is more complicated and more research is required to address this issue.

2.4 Physiological Response

Understanding of physiological mechanisms, regulating the processes of yield enhancement and stability is fundamental for population improvement in target environments (Fischer 2007). Advancement in phenotyping approaches has strengthened physiological breeding, enabling breeders to quantify the impact of physiological factors on crop production under field conditions. This will certainly provide opportunities to modulate physiological mechanism of plants to cope with future climatic hazards.

2.4.1 Leaf Chlorophyll

Chlorophyll degradation is one of the earliest noticeable responses of plants under high temperatures. This damage may lead to leaf senescence, which is particularly common during reproductive phase of crops, and may reduce assimilates supply to developing fruits. Therefore, high leaf chlorophyll and the plant's capacity to maintain green pigment (termed stay-green) under high temperature are often linked with heat tolerance (Reynolds et al. 2007). Thermo-tolerance potential of a stay-green genotype is determined through its total leaf chlorophyll content at anthesis along with the rate and duration of leaf senescence under high temperature (Harris et al. 2007). Stay-green genotypes can regulate heat-induced leaf area degradation through various mechanisms such as increased stomatal conductance as in wheat (Chauhan et al. 2009) or through increased nitrogen supply to shoots as in sorghum (Borrell et al. 2001). A range of crop species has been investigated for the stay-green trait (Harris et al. 2007; Kumari et al. 2007) but breeding for this trait has been limited mainly to wheat and sorghum.

Field-based screening methods have been developed for stay-green genotypes; for example, spectral reflectance normalised difference vegetation index (NDVI) has shown a significant association of leaf green colour with yield under heat stress (Lopes and Reynolds 2012). Other approaches such as cellular metabolic process study can be beneficial to identify heat-tolerant genotypes. Khanna-Chopra and Chauhan (2015) established that a higher level of total antioxidant capacity during grain filling was associated with delayed senescence, which in turn helps to maintain photosynthesis and assimilate supply to developing fruits under heat stress. However, more work in the exploration of specific indices for stay-green may be useful to identify heat-tolerant genotypes.

2.4.2 Leaf Water Status

Leaf relative water content, leaf water potential, stomatal conductance and rate of transpiration are strongly influenced by changes in leaf/canopy temperatures. Under dry environments, elevated temperature increases vapour pressure deficits leading to reduced leaf water content and water potential. Plants may avoid overheating of canopy by increasing transpiration, although most plant species tend to conserve water over temperature regulation, particularly when the temperatures exceed critical limits. Elevated temperature increases the hydraulic conductivity of membranes and plant tissues due to increased aquaporin activity, membrane fluidity and permeability. For example, increasing day temperature from 15 to 40 °C can reduce transpiration and growth in wheat and sorghum (Machado and Paulsen 2001), while the heat-sensitive genotypes have greater reduction in leaf water potential than tolerant genotypes (Almeselmani et al. 2009). Since water is required for various reproductive processes such as pollen ripening, rapid extension of stamen filaments and fruit development, restricted water supply may lead to poor fertilisation and yield losses.

2.4.3 Photosynthesis/Photorespiration

Photosynthesis, a key physiological process, is highly sensitive to high temperature stress. Heat-induced photosynthesis reduction is often attributed to an increased rate of photorespiration. This is primarily because of changes in CO_2 and O_2 affinity and kinetics of rubisco, which are regulated by high temperatures (Jordan and Ogren 1984; Sharkey et al. 2005). Photosynthesis inhibition in response to high temperature may also be attributed to non-photorespiratory processes such as heat-induced damage to photosystem I and II and subsequent reduction in electron transport and Calvin cycle activity (Prasad et al. 2008). Similarly, high temperature may reduce photophosphorylation efficiency by disrupting the structures of membrane proteins, increasing membrane permeability and loss of the electrolytes (Wahid et al. 2007). Photosynthetic reduction has been observed in crops in response to a wide range of temperatures, e.g. 28 °C in chickpea (Devasirvatham et al. 2010), 32 °C in tomato (Islam 2011), 40 °C in cotton and sorghum (Kim and Portis 2005; Djanaguiraman et al. 2010) and 45 °C in fingered citron (Chen et al. 2012).

2.5 Oxidative Damage and Plant Defence Mechanism

High temperature accelerates the generation reactive oxygen species (ROS) such as superoxide radicals (O^{-2}), hydroxyl radicals (OH^{-1}) and hydrogen peroxide (H_2O_2) in plant tissues, which damage cellular organelles such as chloroplast and mitochondria (Grigorova et al. 2012). For example, a short term (20 min) exposure to heat (40–44 °C) can increase lipid peroxidation and production of oxidative species, and subsequently inhibit carbon assimilation in wheat (Kreslavski et al. 2009). Similarly, heat stress can increase lipid peroxidation and H_2O_2 production in a range of agricultural crops including chickpea (Kaushal et al. 2011), wheat (Hameed et al. 2012), maize (Kumar et al. 2012a), rice (Mohammed and Tarpley 2011) and tobacco (Tan et al. 2011).

2.5.1 Antioxidant and Hormones

In response to ROS-induced injury, the plants have evolved an antioxidant defence system, consisting of enzymatic and non-enzymatic components such as ascorbate peroxidase, dehydroascorbate reductase, glutathione S-transferase, superoxide dismutase, catalase, guaiacol peroxidase, glutathione reductase, glutathione, ascorbate and tocopherols (Noctor and Foyer 1998). These antioxidants can detoxify the harmful ROS and protect plants from heat-induced cellular injury. Balla et al. (2009) linked the high temperature tolerance in wheat genotypes with the increased activities of enzymatic antioxidants. Similarly, relatively higher activity of antioxidants such as superoxide dismutase (SOD), catalase (CAT), ascorbate peroxidase (APX), and glutathione reductase (GR) was observed in heat tolerant compared with sensitive genotypes of *Brassica juncea* at 45 °C temperature (Rani et al. 2013). These studies imply that heat tolerance in crops may be increased by modulating the antioxidant defence mechanism of plants either through breeding or management techniques.

The capacity of plants to adapt and survive under high temperatures may be increase through hormonal regulation. This defence mechanism is activated by modulating the internal hormonal levels, e.g. abscisic acid, which regulates stomatal behaviour and mediates dehydration signalling pathways in heat stressed plants (Kumar et al. 2012b) or through external application of hormones. For example, exogenous application of hormones such as auxin and salicylic acid can effectively protect plants from heat-induced injury (Sakata et al. 2010; Fahad et al. 2016). In contrast, gibberellins induce high-temperature sensitivity, thus tolerance to high temperature can be increased by blocking gibberellic acid synthesis (Vettakkorumakankav et al. 1999).

2.5.2 Heat Shock Proteins

To cope with environmental stress, plants upregulate the production of various heat shock proteins (HSPs) and transcription factors (HSF) (Wahid et al. 2007). Five major HSP groups have been identified in plant species, namely HSP100, HSP90, HSP70, HSP60 and HSP20 (Swindell et al. 2007). These HSPs may increase heat tolerance in crops by stabilising membranes and chromatin structures. For example, relatively greater accumulation of HSP18, HSP26 and HSP100 have been observed in heat tolerant compared with susceptible genotypes under high temperature (Sharma-Natu et al. 2010; Sumesh et al. 2008; Chauhan et al. 2012). Lujan et al. (2009) linked higher thermo-tolerance of *Agave tequilana* leaves with an increased expression of transcripts for small HSPs such as HSP-CI and HSP-CII. Similarly, a significant increase in heat tolerance has been achieved in cotton and tobacco plants by increasing the expression of HSP101 through genetic transformation (Burke and Chen 2015). In addition to HSF and HSPs, plants may accumulate a range of low molecular mass organic compounds (osmolytes) such as sugars, polyols, proline and glycine betaine (Wahid et al. 2007). These compounds can help maintaining cell turgor pressure and water absorption in stressed plants by reducing cell osmotic potential, although the exact mechanism through which osmolytes induce heat-stress tolerance in plants is not completely understood (Kumar et al. 2013).

2.6 Manipulating Heat Stress Tolerance in Crops

2.6.1 Breeding for Heat-Stress Tolerance

Development of genotypes with superior yield and adaptation to climatic conditions has been one of the primary goals of crop breeders/scientists. The rising atmospheric temperatures and episodes of high temperature during the reproductive phase have increased the concern for improving physiological, morphological and molecular traits of plants associated with heat tolerance. Expansion of genetic variability in the crop gene pool can be an important strategy for breeding programs aimed at developing heat-tolerant genotypes. Although existence of genetic diversity for heat tolerance has been reported in conventional varieties, efforts have been made to explore new sources of genetic diversity for heat tolerance in cultivated crops. For example, cultivated genotypes are often cross-bred with their wild relative, as in the case of breeding of *Triticum aestivum* with its key ancestors such as *Aegilops tauchii* and *Triticum durum* to select genotypes with superior performance under high temperatures.

Selection of genotypes based on yield performance under elevated temperature is one of the most effective strategies; however, it may not be feasible for large scale screening due to high costs associated with conducting yield trials. Therefore, marker traits have been used for selecting and screening of heat tolerance genotypes. Edmeades et al. (1999) proposed that traits used for selecting stress tolerance should be easily measurable, highly heritable and exhibit a strong association with crop yield. Lipid membrane thermo-stability is often used an indicator of high temperature tolerance in crops and it is measured through a simple electrolyte leakage method (Blum and Ebercon 1981). A strong correlation has been established between cell membrane thermo-stability and grain yield for many crops including wheat (Blum et al. 2001), cotton (Cottee et al. 2012) and rice (Agarie et al. 1998). Similarly, changes in tetrazolium triphenyl chloride levels in mitochondria, which correspond to cell viability has been suggested as an indicator of heat tolerance in plants (Towill and Mazur 1975). Mass screening of wheat genotypes for heat tolerance has also been done using the stay-green character, a visual rating method for easy and quick screening (Cossani and Reynolds 2012). Several other physiological indicators of heat stress such as canopy temperature, stomatal conductance and photosynthetic rate, which are highly correlated with field performance and yield have also been used for screening crops for high temperature tolerance (Reynolds and Langridge 2016).

In addition to morphological and physiological markers, several molecular markers have also been used for selecting heat stress tolerant crops. Since heat stress tolerance in plants is regulated by multiple genes, where each gene has a minor effect individually but significant effect when acting together, thus QTL mapping and subsequent marker-assisted selection may be used for selecting thermo-tolerant genotypes. Various QTLs were identified for wheat traits associated with heat stress tolerance including stay-green, leaf senescence and canopy temperature (Farooq et al. 2011). Mason et al. (2010) identified 15 QTLs associated with yield and yield component in wheat, suggesting that for heat tolerance, main spikes could be used for the identification of QTLs genomic regions. Similarly, QTLs associated with heat tolerance have been identified for *Arabidopsis*, azuki bean, barley, *Brassica*, cowpea, maize, potato, rice, sorghum and tomato (Jha et al. 2014).

2.6.2 Genetic Manipulation

Modern biotechnology and genetic engineering tools offer novel solution to resolve issues in plant production systems. Several genes associated with physiological traits, conferring heat tolerance have been incorporated into candidate crops, although, compared with other abiotic stresses such as drought, salt or cold, the application of genetic engineering techniques for heat tolerance has been quite limited. Genes associated with the key stress regulatory pathways such as high glycine betaine production, rubisco activase stability, modified fatty acid composition of lipid membrane have been successfully used for improving heat tolerance in plants (Yang et al. 2005). In addition, constitutive expression of specific HSPs have been shown to enhance heat tolerance. For example, increased heat tolerance has been achieved by altering the transcription factor and thus expression level of HSP in *Arabidopsis* (Lee et al. 1995) and carrot (Malik et al. 1999). Genetic improvement of proteins associated with osmotic adjustment, detoxification of ROS, photosynthetic reactions, production of polyamines and protein biosynthesis process are other potential targets for increasing high temperature tolerance in plants through genetic engineering.

2.6.3 Crop Management

In a series of glasshouse and field experiments, Sarwar et al. (2017) studied the effect of high temperatures on cotton crops. Exposure to 38/24 °C and 45/30 °C (controlled environments) can reduce seed cotton yield up to 22% and 63%, respectively. A similar reduction in seed cotton yield was recorded when reproductive phase of the crops was exposed to elevated temperature by staggering sowing dates under field conditions. This reduction in cotton yield was strongly associated with damage to lipid membrane, causing abscission of fruiting structures. Controlling heat-induced cellular membrane damage via exogenous application of various growth substances such as hydrogen peroxide and ascorbic acid may also lead to better thermo-tolerance in cotton. Field studies by Rahman et al. (2005) and Rauf et al. (2007) established that high temperature regimes can influence seed physical traits such as seed volume, seed density and seed surface area. Variation in response to high temperature was linked with membrane stability in a wide range of cotton cultivars grown under field condition (Rahman et al. 2004). Similar data were obtained by Azhar et al. (2009) who found that high temperature significantly reduced seed cotton yield and fibre quality in the heat-sensitive varieties compared with heat-tolerant varieties. This reduction in lint yield was strongly associated with membrane stability of cotton varieties. In China, high temperature spells during July and August resulted into lower boll set and subsequent lint yield reduction. Liu et al. (2006) suggested that increasing temperature from 10 to 50 °C reduces pollen germination and pollen tube growth rate, which are strongly associated with boll retention. Thus, high temperature at flowering can induce fruit shedding through pollen abortion. Late planted wheat crop, on the other hand, displayed a degree of tolerance in lipid membrane under high temperature through an increased level of antioxidants (Almeselmani et al. 2006). However, exposure to higher temperature (35–40 °C) at pollination and at milking stage significantly reduced membrane stability index and grain yield. Kumar et al. (2012) linked this heat-induced damage in wheat to increased leaf proline decreased H_2O_2 contents. Foliar spray of salicylic acid can induce thermo-tolerance by activating many underlying mechanisms such as heat hock proteins and antioxidant enzymes (Kaur et al. 2009). Exposure to high temperature at pre- and post-anthesis in field-grown wheat crop increased membrane peroxidation leading to photosynthesis reduction. Heat acclimation process pre- and post-anthesis reduced ROS-induced higher antioxidants and maintained the normal photosynthetic rate (Wang et al. 2011).

Various crop management techniques such as no tillage/stubble retention (soil-water conservation), timely fertilization (critical growth stages) and modification in sowing time are commonly practised for protecting crops from heat-induced injury. Increased water supply during the heat stress period has been effective in improving grain-filling rate, duration and size of wheat (Dupont et al. 2006). In some parts of the world, crop sowing time is often adjusted to protect the critical growth phases of crop from high temperatures. For example, grain filling period of most of the grain crops is very sensitive to heat stress and thus early planting has been recommended to avoid terminal heat stress.

Tolerance to moderate heat stress in wheat crops can also be increased by appropriate fertilizer application, although this may not be an effective strategy under extreme temperatures such as 37 °C (Dupont et al. 2006). Timely application of nutrients to coincide with the key developmental stages, e.g. nitrogen supplementation at 1 cm spike length, has already been in practice for improving heat tolerance in wheat. Apart from fertilizer application, various other chemicals such as osmo-protectants, phytohormones, signalling molecules, polyamines and trace elements have been found effective in increasing heat stress tolerance in crops under controlled environments although these have limited application in broadacre agriculture.

2.7 Conclusions

We comprehensively reviewed the overall effects of warmer climates on important crops. Negative effects of warmer climates on various morphological, biochemical and physiological behaviour of plants have been recorded for both controlled and field environments. Warmer climate can change the overall biochemistry of plants by accelerating ROS generation in plant cells, which subsequently induce cellular and injury and impair normal functioning. Some plant species have an intrinsic capability to overcome this heat-induced cellular damage though activation of antioxidant defence mechanism, regulation of heat shock proteins and morphological adaptation, however, most of the commonly grown crops suffer severe damages. Management techniques such as optimal sowing time, cropping density, fertilisation and use of growth regulators/osmoprotectants may improve performance of crops under heat-stressed environments. Similarly, selection of heat-resistant varieties and genetic engineering play a major role in breeding crops for the future warmer climates.

References

Agarie, S., Hanaoka, N., Ueno, O., Miyazaki, A., Kubota, F., Agata, W., & Kaufman, P. B. (1998). Effects of silicon on tolerance to water deficit and heat stress in rice plants (*Oryza sativa* L.), monitored by electrolyte leakage. *Plant Production Science, 1*, 96–103.

Ahmed, F. E., Hall, A. E., & DeMason, D. A. (1992). Heat Injury during floral development in Cowpea (*Vigna unguiculata, Fabaceae*). *American Journal of Botany, 79*, 784–791.

Ainsworth, E. A., & Ort, D. R. (2010). How do we improve crop production in a warming world? *Plant Physiology, 154*, 526–530.

Akman, Z. (2009). Comparison high temperature stress in maize, rice and sorghum by plant growth regulators. *Journal of Animal and Veterinary Advances, 8*, 358–361.

Almeselmani, M., Deshmukh, P. S., & Sairam, R. K. (2009). High temperature stress tolerance in wheat genotypes: Role of antioxidant defence enzymes. *Acta Agronomica Hungarica, 57*, 1–14.

Almeselmani, M., Deshmukh, P. S., Sairam, R. K., Kushwaha, S., & Singh, T. (2006). Protective role of antioxidant enzymes under high temperature stress. *Plant Science, 171*, 382–388.

Azhar, F., Ali, Z., Akhtar, M., Khan, A., & Trethowan, R. (2009). Genetic variability of heat tolerance, and its effect on yield and fibre quality traits in upland cotton (*Gossypium hirsutum* L.). *Plant Breeding, 128*, 356–362.

Bahar, B., Yildirim, M., Barutcular, C., & Ibrahim, G. (2008). Effect of canopy temperature depression on grain yield and yield components in bread and durum wheat. *Notulae Botanicae Horti Agrobotanici Cluj-Napoca, 36*, 34–37.

Balla, K., Bencze, S., Janda, T., & Veisz, O. (2009). Analysis of heat stress tolerance in winter wheat. *Acta Agronomica Hungarica, 57*, 437–444.

Barnabs, B., Jäger, K., & Fehér, A. (2008). The effect of drought and heat stress on reproductive processes in cereals. *Plant, Cell and Environment, 31*, 11–38.

Basu, P. S., Masood, A., & Chaturvedi, S. K. (2009). Terminal heat stress adversely affects chickpea productivity in Northern India–strategies to improve thermotolerance in the crop under climate change." W3 Workshop Proceedings: Impact of Climate Change on Agriculture.

Bennett, D., Izanloo, A., Reynolds, M., Kuchel, H., Langridge, P., & Schnurbusch, T. (2012). Genetic dissection of grain yield and physical grain quality in bread wheat (*Triticum aestivum* L.) under water-limited environments. *Theoretical and Applied Genetics, 125*, 255–271.

Bibi, A., Oosterhuis, D., & Gonias, E. (2008). Photosynthesis, quantum yield of photosystem II and membrane leakage as affected by high temperatures in cotton genotypes. *Journal of Cotton Science, 12*, 150–159.

Bishop, J., Potts, S. G., & Jones, H. E. (2016). Susceptibility of Faba Bean (*Vicia faba* L.) to heat stress during floral development and anthesis. *Journal of Agronomy and Crop Science, 202*, 508–517.

Blum, A., & Ebercon, A. (1981). Cell membrane stability as a measure of drought and heat tolerance in wheat. *Crop Science, 21*, 43–47.

Blum, A., Klueva, N., & Nguyen, H. T. (2001). Wheat cellular thermotolerance is related to yield under heat stress. *Euphytica, 117*, 117–123.

Borrell, A., Hammer, G., & Oosterom, E. (2001). Stay-green: A consequence of the balance between supply and demand for nitrogen during grain filling? *Annals of Applied Biology, 138*, 91–95.

Burke, J. J., Mahan, J. R., & Hatfield, J. L. (1988). Crop-specific thermal kinetic windows in relation to wheat and cotton biomass production. *Agronomy Journal, 80*, 553–556.

Burke, J. J., & Chen, J. (2015). Enhancement of reproductive heat tolerance in plants. *PLoS One, 10*, e0122933.

Calderini, D. F., Reynolds, M. P., & Slafer, G. A. (2006). Source–sink effects on grain weight of bread wheat, durum wheat, and triticale at different locations. *Crop & Pasture Science, 57*, 227–233.

Challinor, A. J., Watson, J., Lobell, D. B., Howden, S. M., Smith, D. R., & Chhetri, N. (2014). A meta-analysis of crop yield under climate change and adaptation. *Nature Climate Change, 27*, 1–5.

Chauhan, H., Khurana, N., Nijhavan, A., Khurana, J. P., & Khurana, P. (2012). The wheat chloroplastic small heat shock protein (sHSP26) is involved in seed maturation and germination and imparts tolerance to heat stress. *Plant, Cell and Environment, 35*, 1912–1931.

Chauhan, S., Srivalli, S., Nautiyal, A., & Khanna-Chopra, R. (2009). Wheat cultivars differing in heat tolerance show a differential response to monocarpic senescence under high-temperature stress and the involvement of serine proteases. *Photosynthetica, 47*, 536–547.

Chao, C. C. T., & Robert, R. K. (2007). The date palm (Phoenix dactylifera L.): overview of biology, uses, and cultivation. *Hortscience, 42*(5), 1077–1082.

Chen, W. R., Zheng, J. S., Li, Y. Q., & Guo, W. D. (2012). Effects of high temperature on photosynthesis, chlorophyll fluorescence, chloroplast ultrastructure, and antioxidant activities in fingered citron. *Russian Journal of Plant Physiology, 59*, 732–740.

Cossani, C. M., & Reynolds, M. P. (2012). Physiological traits for improving heat tolerance in wheat. *Plant Physiology, 160*, 1710–1718.

Cottee, N. S., Bange, M. P., Wilson, I. W., & Tan, D. K. (2012). Developing controlled environment screening for high-temperature tolerance in cotton that accurately reflects performance in the field. *Functional Plant Biology, 39*, 670–678.

Covell, S., Ellis, R. H., Roberts, E. H., & Summerfield, R. J. (1986). The influence of temperature on seed germination rate in grain legumes: A comparison of chickpea, lentil, soybean, and cowpea at constant temperatures. *Journal of Experimental Botany, 37*, 705–715.

Crawford, A. J., McLachlan, D. H., Hetherington, A. M., & Franklin, K. A. (2012). High temperature exposure increases plant cooling capacity. *Current Biology, 22*, 396–397.

Cui, L., Cao, R., Li, J., Zhang, L., & Wang, J. (2006). High temperature effects on ammonium assimilation in leaves of two Festuca arundinacea cultivars with different heat susceptibility. *Plant Growth Regulation, 49*, 127–136.

Devasirvatham, V., DKY, T., Trethowan, R. M., Gaur, P. M., & Mallikarjuna, N. (2010). *Food security from sustainable agriculture. Proceedings of the 15th Australian Society of Agronomy Conference: Impact of high temperature on the reproductive stage of chickpea* (pp. 15–18). Lincoln: Australian Society of Agronomy.

Devasirvatham, V., Tan, D. K., & Trethowan, R. M. (2016). Breeding strategies for enhanced plant tolerance to heat stress. In J. M. Al-Khayri, S. M. Jain, & D. V. Johnson (Eds.), *Advances in plant breeding strategies: Agronomic, abiotic and biotic stress traits* (pp. 447–469). Cham: Springer International Publishing.

Devasirvatham, V., Tan, D. K. Y., Gaur, P. M., Raju, T. N., & Trethowan, R. M. (2012). High temperature tolerance in chickpea and its implications for plant improvement. *Crop & Pasture Science, 63*, 419–428.

Djanaguiraman, M., Prasad, P. V. V., & Seppanen, M. (2010). Selenium protects sorghum leaves from oxidative damage under high temperature stress by enhancing antioxidant defense system. *Plant Physiology and Biochemistry, 48*, 999–1007.

Downton, W., John, S., Joseph, A. B., & Jeffrey, R. S. (1984). Tolerance of photosynthesis to high temperature in desert plants. *Plant Physiology, 74*, 786–790.

Driedonks, N., Rieu, I., & Vriezen, W. H. (2016). Breeding for plant heat tolerance at vegetative and reproductive stages. *Plant Reproduction, 29*, 67.

Dupont, F. M., Hurkman, W. J., Vensel, W. H., Tanaka, C., Kothari, K. M., Chung, O. K., & Altenbach, S. B. (2006). Protein accumulation and composition in wheat grains: Effects of mineral nutrients and high temperature. *European Journal of Agronomy, 25*, 96–107.

Edmeades, G. O., Chapman, S. C., & Lafitte, H. R. (1999). Selection improves drought tolerance in tropical maize populations: I. Gains in biomass, grain yield, and harvest index. *Crop Science, 39*, 1306–1315.

Endo, M., Tohru, T., Kazuki, H., Shingo, K., Kentaro, Y., Masahiro, O., Atsushi, H., Masao, W., & Makiko, K. K. (2009). High temperatures cause male sterility in rice plants with transcriptional alterations during pollen development. *Plant and Cell Physiology, 50*, 1911–1922.

Fahad, S., Hussain, S., Saud, S., Hassan, S., Ihsan, Z., Shah, A. N., & Alghabari, F. (2016). Exogenously applied plant growth regulators enhance the morpho-physiological growth and yield of rice under high temperature. *Frontiers in Plant Science, 7*, 1250.

Food and Agriculture Organisation (FAO) of the United Nations. (2009). *Declaration of the World Summit on Food Security*, Rome, November 16–18, 2009.

Farooq, M., Bramley, H., Palta, J. A., & Siddique, K. H. (2011). Heat stress in wheat during reproductive and grain-filling phases. *Critical Reviews in Plant Sciences, 30*, 491–507.

Fischer, G., Shah, M., & Van, V. H. (2002). *Climate change and agricultural vulnerability, world summit on sustainable development*. Laxenburg: IIASA.

Fischer, R. (2007). Understanding the physiological basis of yield potential in wheat. *Journal of Agricultural Science-Cambridge, 145*, 99.

Füleky, G. (2009). *Cultivated plants, primarily as food resources. Encyclopedia of life supper systems (EOLSS)* (Vol. I). Paris: UNESCO.

Gate, P., & Brisson, N. (2010). Advancement of phenological stages and shortening of phases. In N. Brisson & F. Levrault (Eds.), *Climate change, agriculture and forests in France: simulations of the impacts on the main species. The Green Book of the CLIMATOR project (2007–2010)* (pp. 65–78). Angers, France: ADEME.

Gerwick, B. C., George, J. W., & Ernest, G. U. (1977). Effects of temperature on the Hill reaction and photophosphorylation in isolated cactus chloroplasts. *Plant Physiology, 60*, 430–432.

Gibson, L. R., & Paulsen, G. M. (1999). Yield components of wheat grown under high temperature stress during reproductive growth. *Crop Science, 39*, 1841–1846.

Giorno, F., Wolters-Arts, M., Mariani, C., & Rieu, I. (2013). Ensuring reproduction at high temperatures: The heat stress response during anther and pollen development. *Plants, 2*, 489–506.

Gross, Y., & Kigel, J. (1994). Differential sensitivity to high temperature of stages in the reproductive development of common bean (Phaseolus vulgaris L.). *Field Crops Research, 36*, 201–212.

Grigorova, B., Vassileva, V., Klimchuk, D., Vaseva, I., Demirevska, K., & Feller, U. (2012). Drought, high temperature, and their combination affect ultrastructure of chloroplasts and mitochondria in wheat (*Triticum aestivum* L.) leaves. *Journal of Plant Interactions, 7*, 204–213.

Hameed, A., Goher, M., & Iqbal, N. (2012). Heat stress-induced cell death, changes in antioxidants, lipid peroxidation, and protease activity in wheat leaves. *Journal of Plant Growth Regulation, 31*, 283–291.

Hansen, J., Sato, M., Hearty, P., Ruedy, R., Kelley, M., Masson-Delmotte, V., Russell, G., Tselioudis, G., Cao, J., & Rignot, E. (2015). Ice melt, sea level rise and superstorms: Evidence from paleoclimate data, climate modeling, and modern observations that 2°C global warming is highly dangerous. *Atmospheric Chemistry and Physics, 15*, 20059–20179.

Harris, K., Subudhi, P., Borrell, A., Jordan, D., Rosenow, D., Nguyen, H., Klein, P., Klcin, R., & Mullet, J. (2007). Sorghum stay-green QTL individually reduce post-flowering drought-induced leaf senescence. *Journal of Experimental Botany, 58*, 327–338.

Hedhly, A. (2011). Sensitivity of flowering plant gametophytes to temperature fluctuations. *Environmental and Experimental Botany, 74*, 9–16.

Hossain, A., Teixeira da Silva, J. A., Lozovskaya, M. V., & Zvolinsky, V. P. (2012). High temperature combined with drought affect rainfed spring wheat and barley in South-Eastern Russia: I. Phenology and growth. *Saudi Journal of Biological Sciences, 19*, 473–487.

IPCC. (2007). In S. Solomon, D. Qin, M. Manning, R. B. Alley, T. Berntsen, N. L. Bindoff, & Z. C. Chen (Eds.), *Climate change 2007: The physical science basis. Contribution of Working Group I to the Fourth Assessment Report of the Intergovernmental Panel on Climate Change.* Cambridge: Cambridge University Press.

IPCC. (2013). In T. F. Stocker Qin, D. Plattner, G. K. Tignor, M. Allen, S. K. Boschung, & J. Naue (Eds.), *Summary for policymakers climate change 2013: The physical science basis Contribution of Working Group I to the Fifth Assessment Report of the Intergovernmental Panel on Climate Change.* Cambridge: Cambridge University Press.

Islam, M. T. (2011). Effect of temperature on photosynthesis, yield attributes and yield of tomato genotypes. *International Journal of Experimental Agriculture, 2*, 8–11.

Jha, U. C., Bohra, A., & Singh, N. P. (2014). Heat stress in crop plants: Its nature, impacts and integrated breeding strategies to improve heat tolerance. *Plant Breeding, 133*, 679–701.

Jones, H. G. (2013). *Plants and microclimate: A quantitative approach to environmental plant physiology.* Cambridge: Cambridge University Press.

Jordan, D. B., & Ogren, W. L. (1984). The CO_2/O_2 specificity of ribulose 1, 5-bisphosphate carboxylase/oxygenase. *Planta, 161*, 308–313.

Kalra, N., Chakraborty, D., Sharma, A., Rai, H. K., Jolly, M., Chander, S., Kumar, P. R., Bhadraray, S., Barman, D., Mittal, R. B., & Lal, M. (2008). Effect of increasing temperature on yield of some winter crops in northwest India. *Current Science*, 82–88.

Kaur, P., Ghai, N., & Sangha, M. K. (2009). Induction of thermotolerance through heat acclimation and salicylic acid in brassica species. *African Journal of Biotechnology, 8*, 619.

Kaushal, N., Gupta, K., Bhandhari, K., Kumar, S., Thakur, P., & Nayyar, H. (2011). Proline induces heat tolerance in chickpea (Cicer arietinum L.) plants by protecting vital enzymes of carbon and antioxidative metabolism. *Physiology and Molecular Biology of Plants, 17*, 203.

Khanna-Chopra, R., & Chauhan, S. (2015). Wheat cultivars differing in heat tolerance show a differential response to oxidative stress during monocarpic senescence under high temperature stress. *Protoplasma, 1*, 11.

Kim, K., & Portis, A. R. (2005). Temperature dependence of photosynthesis in Arabidopsis plants with modifications in Rubisco activase and membrane fluidity. *Plant and Cell Physiology, 46*, 522–530.

Kreslavski, V. D., Lyubimov, V. Y., Shabnova, N. I., Balakhnina, T. I., & Kosobryukhov, A. A. (2009). Heat-induced impairments and recovery of photosynthetic machinery in wheat seedlings. Role of light and prooxidant-antioxidant balance. *Physiology and Molecular Biology of Plants, 15*, 115–122.

Krishnan, P., Ramakrishnan, B., Reddy, K. R., & Reddy, V. R. (2011). High-temperature effects on rice growth, yield, and grain quality. In *Advances in agronomy* (Vol. 111, pp. 87–206). Academic Press.

Kuchel, H., Williams, K., Langridge, P., Eagles, H. A., & Jefferies, S. P. (2007). Genetic dissection of grain yield in bread wheat. II. QTL-by-environment interaction. *Theoretical and Applied Genetics, 115*, 1015–1027.

Kumar, R. R., Goswami, S., Sharma, S. K., Singh, K., Gadpayle, K. A., Singh, S. D., & Rai, R. D. (2013). Differential expression of heat shock protein and alteration in osmolyte accumulation under heat stress in wheat. *Journal of Plant Biochemistry and Biotechnology, 22*, 16–26.

Kumar, R. R., Goswami, S., Sharma, S. K., Singh, K., Gadpayle, K. A., Kumar, N., Rai, G. K., Singh, M., & Rai, R. D. (2012). Protection against heat stress in wheat involves change in cell membrane stability, antioxidant enzymes, osmolyte, H_2O_2 and transcript of heat shock protein. *International Journal of Plant Physiology and Biochemistry, 4*, 83–91.

Kumar, S., Gupta, D., & Nayyar, H. (2012a). Comparative response of maize and rice genotypes to heat stress: Status of oxidative stress and antioxidants. *Acta Physiologiae Plantarum, 34*, 75–86.

Kumar, S., Kaushal, N., Nayyar, H., & Gaur, P. (2012b). Abscisic acid induces heat tolerance in chickpea (*Cicer arietinum* L.) seedlings by facilitated accumulation of osmoprotectants. *Acta Physiologiae Plantarum, 34*, 1651–1658.

Kumari, M., Singh, V. P., Tripathi, R., & Joshi, A. K. (2007). Variation for staygreen trait and its association with canopy temperature depression and yield traits under terminal heat stress in wheat. In H. T. Buck, J. R. Nisi, & N. Salomon (Eds.), *Wheat production in stressed environments* (pp. 357–363). Dordrecht: Springer.

Larcher, W. (1995). *Physiological plant ecology*. New York, NY: Springer.

Lee, J. H., Hubel, A., & Schoffl, F. (1995). Derepression of the activity of genetically engineered heat shock factor causes constitutive synthesis of heat shock proteins and increased thermotolerance in transgenic Arabidopsis. *The Plant Journal, 8*, 603–612.

Liu, Z., Yuan, Y. L., Liu, S. Q., Yu, X. N., & Rao, L. Q. (2006). Screening for high-temperature tolerant cotton cultivars by testing in vitro pollen germination, pollen tube growth and boll retention. *Journal of Integrative Plant Biology, 48*, 706–714.

Lobell, D. B., & Gregory, P. A. (2003). Climate and management contributions to recent trends in US agricultural yields. *Science, 299*, 1032–1032.

Lobell, D., & Gourdji, S. (2012). The influence of climate change on global crop productivity. *Plant Physiology, 160*, 1686–1697.

Lopes, M. S., & Reynolds, M. P. (2010). Partitioning of assimilates to deeper roots is associated with cooler canopies and increased yield under drought in wheat. *Functional Plant Biology, 37*, 147–156.

Lopes, M. S., & Reynolds, M. P. (2012). Stay-green in spring wheat can be determined by spectral reflectance measurements (normalized difference vegetation index) independently from phenology. *Journal of Experimental Botany, 63*, 3789–3798.

Lotze-Campen, H., Schellnhuber, H. J., et al. (2009). Climate impacts and adaptation options in agriculture: What we know and what we don't know. *Journal für Verbraucherschutz und Lebensmittelsicherheit, 4*, 145–150.

Lujan, R., Fernando, L., Luz, M., Rita, B., Gladys, I. C., & Nieto-sotelo, J. (2009). Small heat-shock proteins and leaf cooling capacity account for the unusual heat tolerance of the central spike leaves in *Agave tequilana* var. Weber. *Plant, Cell and Environment, 32*, 1791–1803.

Machado, S., & Paulsen, G. M. (2001). Combined effects of drought and high temperature on water relations of wheat and sorghum. *Plant and Soil, 233*, 179–187.

Malik, M. K., Slovin, J. P., Hwang, C. H., & Zimmerman, J. L. (1999). Modified expression of a carrot small heat shock protein gene, Hsp17.7, results in increased or decreased thermotolerance. *The Plant Journal, 20*, 89–99.

Manske, G. G. B., Ortiz-Monasterio, J. I., Van Ginkel, M., Gonzalez, R. M., Rajaram, S., Molina, E., & Vlek, P. L. G. (2000). Traits associated with improved P-uptake efficiency in CIMMYT's semi dwarf spring bread wheat grown on an acid Andisol in Mexico. *Plant and Soil, 221*, 189–204.

Maryam, M., Fatima, B., Haider, M. S., Abbas, S., Naqvi, M., Ahmad, R., & Khan, I. A. (2015). Evaluation of pollen viability in date palm cultivars under different storage temperatures. *Pakistan Journal of Botany, 47*, 377–381.

Mason, R. E., Mondal, S., Beecher, F. W., Pacheco, A., Jampala, B., Ibrahim, A. M., & Hays, D. B. (2010). QTL associated with heat susceptibility index in wheat (*Triticum aestivum* L.) under short-term reproductive stage heat stress. *Euphytica, 174*, 423–436.

McCarthy, J. J. (2001). *Climate change: Impacts, adaptation, and vulnerability: Contribution of Working Group II to the third assessment report of the Intergovernmental Panel on Climate Change*. Cambridge: Cambridge University Press.

McDonald, G. K., & Paulsen, G. M. (1997). High temperature effects on photosynthesis and water relations of grain legumes. *Plant and Soil, 196*, 47–58.

Mohammed, A. R., & Tarpley, L. (2011). High night temperature and plant growth regulator effects on spikelet sterility, grain characteristics and yield of rice (*Oryza sativa* L.) plants. *Canadian Journal of Plant Science, 91*, 283–291.

Mondal, S. (2012). *Defining the molecular and physiological role of leaf cuticular waxes in reproductive stage heat tolerane in wheat.* Doctoral dissertation, Texas A and M University, College Station.

Nobel, P. S. (1988). *Environmental biology of agaves and cacti.* New York, NY: Cambridge University Press.

Nobel, P. S., & Smith, S. D. (1983). High and low temperature tolerances and their relationships to distribution of agaves. *Plant, Cell and Environment, 6*, 711–719.

Nobel, P. S., & De la Barrera, E. (2002). High temperatures and net CO_2 uptake, growth, and stem damage for the hemiepiphytic cactus *Hylocereus undatus. Biotropica, 34*, 225–231.

Noctor, G., & Foyer, C. H. (1998). Ascorbate and glutathione: Keeping active oxygen under control. *Annual Review of Plant Biology, 49*, 249–279.

Olesen, J. E., & Bindi, M. (2002). Consequences of climate change for European agricultural productivity, land use and policy. *European Journal of Agronomy, 16*, 239–262.

Ortiz, R., Braun, H. J., Crossa, J., Crouch, J. H., Davenport, G., Dixon, J., Dreisigacker, S., Duveiller, E., He, Z., & Huerta, J. (2008). Wheat genetic resources enhancement by the International Maize and Wheat Improvement Center (CIMMYT). *Genetic Resources and Crop Evolution, 55*, 1095–1140.

Oshino, T., Abiko, M., Saito, R., Ichiishi, E., Endo, M., Kawagishi-Kobayashi, M., & Higashitani, A. (2007). Premature progression of anther early developmental programs accompanied by comprehensive alterations in transcription during high-temperature injury in barley plants. *Molecular Genetics and Genomics, 278*, 31–42.

Palta, J. A., Kobata, T., Turner, N. C., & Fillery, I. R. (1994). Remobilisation of carbon and nitrogen in wheat as influenced by postanthesis water deficits. *Crop Science, 34*, 118–124.

Patrick, J. W., & Stoddard, F. L. (2010). Physiology of flowering and grain filling in faba bean. *Field Crops Research, 115*, 234–242.

Peck, A. W., & McDonald, G. K. (2010). Adequate zinc nutrition alleviates the adverse effects of heat stress in bread wheat. *Plant and Soil, 337*, 355–374.

Peng, S., Huang, J., Sheehy, J. E., Laza, R. C., Visperas, R. M., Zhong, X., Centeno, G. S., Khush, G. S., & Cassman, K. G. (2004). Rice yields decline with higher night temperature from global warming. *Proceedings of the National Academy of Sciences of the United States of America, 101*, 9971–9975.

Pettigrew, W. T. (2008). The effect of higher temperatures on cotton lint yield production and fiber quality. *Crop Science, 48*, 278–285.

Pinto, R. S., & Reynolds, M. P. (2015). Common genetic basis for canopy temperature depression under heat and drought stress associated with optimized root distribution in bread wheat. *Theoretical and Applied Genetics, 128*, 575–585.

Prasad, P., Pisipati, S., Ristic, Z., Bukovnik, U., & Fritz, A. (2008). Impact of nighttime temperature on physiology and growth of spring wheat. *Crop Science, 48*, 2372–2380.

Radin, J. W. (1992). Reconciling water-use efficiency of cotton in field and laboratory. *Crop Science, 32*, 13–18.

Rahman, H., Malik, S. A., & Saleem, M. (2004). Heat tolerance of upland cotton during the fruiting stage evaluated using cellular membrane thermostability. *Field Crops Research, 85*, 149–158.

Rahman, H., Malik, S. A., & Saleem, M. (2005). Inheritance of seed physical traits in upland cotton under different temperature regimes. *Spanish Journal of Agricultural Research, 3*, 225–231.

Rani, B., Dhawan, K., Jain, V., Chhabra, M. L., & Singh, D. (2013). *High temperature induced changes in antioxidative enzymes in Brassica juncea (L) Czern and Coss.* M. Sc. Dissertations, CCS HAU Hisar, India.

Rauf, S., Khan, T. M., Naveed, A., & Munir, H. (2007). Modified path to high lint yield in upland cotton (*Gossypium hirsutum* L.) under two temperature regimes. *Turkish Journal of Biology, 31*, 119–126.

Reddy, V., Baker, D., & Hodges, H. (1991). Temperature effects on cotton canopy growth, photosynthesis, and respiration. *Agronomy Journal, 83*, 699–704.

Reynolds, M., & Langridge, P. (2016). Physiological breeding. *Current Opinion in Plant Biology, 31*, 162–171.

Reynolds, M., & Trethowan, R. (2007). Physiological interventions in breeding for adaptation to abiotic stress. *Frontis, 21*, 127–144.

Reynolds, M. P., Pierre, C. S., Saad, A. S., Vargas, M., & Condon, A. G. (2007). Evaluating potential genetic gains in wheat associated with stress-adaptive trait expression in elite genetic resources under drought and heat stress. *Crop Science, 47*(Suppl 3), S-172–S-189.

Reynolds, M. P., Trethowan, R., Crossa, J., Vargas, M., & Sayre, K. D. (2004). Erratum to physiological factors associated with genotype by environment interaction in wheat. *Field Crops Research, 85*, 253–274.

Richards, R. (1996). Defining selection criteria to improve yield under drought. *Plant Growth Regulation, 20*, 157–166.

Rodrigo, J., & Herrero, M. (2002). Effects of pre-blossom temperatures on flower development and fruit set in apricot. *Scientia Horticulturae, 92*, 125–135.

Saini, H. S., Sedgley, M., & Aspinall, D. (1983). Effect of heat-stress during floral development on pollen tube growth and ovary anatomy in wheat (*Triticum aestivum* L.). *Functional Plant Biology, 10*, 137–144.

Saini, H. S., Sedgley, M., & Aspinall, D. (1984). Development anatomy in wheat of male sterility induced by heat stress, water deficit or abscisic acid. *Functional Plant Biology, 11*, 243–253.

Sakata, T., Oshino, T., Miura, S., Tomabechi, M., Tsunaga, Y., & Higashitani, N. (2010). Auxins reverse plant male sterility caused by high temperatures. *Proceedings of the National Academy of Sciences USA, 107*, 8569–8574.

Sarwar, M., Saleem, M., Najeeb, U., Shakeel, A., Ali, S., & Bilal, M. (2017). Hydrogen peroxide reduces heat-induced yield losses in cotton (*Gossypium hirsutum* L.) by protecting cellular membrane damage. *Journal of Agronomy and Crop Science, 203*, 429–441.

Schapendonk, A., Xu, H., Van, D. P. P., & Spiertz, J. (2007). Heat-shock effects on photosynthesis and sink-source dynamics in wheat (*Triticum aestivum* L.). *Journal of Life Sciences, 55*, 37–54.

Sharkey, T. D., et al. (2005). Effects of moderate heat stress on photosynthesis: Importance of thylakoid reactions, rubisco deactivation, reactive oxygen species, and thermotolerance provided by isoprene. *Plant, Cell & Environment, 28*, 269–277.

Sharma-Natu, P., Sumesh, K. V., & Ghildiyal, M. C. (2010). Heat shock protein in developing grains in relation to thermotolerance for grain growth in wheat. *Journal of Agronomy and Crop Science, 196*, 76–80.

Singh, R. P., Prasad, P. V. V., Sunita, K., Giri, S. N., & Reddy, K. R. (2007). Influence of high temperature and breeding for heat tolerance in cotton. *Advances in Agronomy, 93*, 313–385.

Song, G., Wang, M., Zeng, B., Zhang, J., Jiang, C., Hu, Q., Geng, G., & Tang, C. (2015). Anther response to high-temperature stress during development and pollen thermotolerance heterosis as revealed by pollen tube growth and in vitro pollen vigor analysis in upland cotton. *Planta, 241*, 1271–1285.

Summerfield, R. J., Hadley, P., Roberts, E. H., Minchin, F. R., & Rawsthorne, S. (1984). Sensitivity of Chickpeas (Cicer arietinum) to hot temperatures during the reproductive period. *Experimental Agriculture, 20*, 77–93.

Sumesh, K. V., Sharma-Nat, P., & Ghildiyal, M. C. (2008). Starch synthase activity and heat shock protein in relation to thermal tolerance of developing wheat grains. *Biologia Plantarum, 52*, 749–753.

Suzuki, K., Tsukaguchi, T., Takeda, H., & Egawa, Y. (2001). Decrease of pollen stainability of green bean at high temperatures and relationship to heat tolerance. *Journal of the American Society for Horticultural Science, 126*, 571–574.

Swindell, W. R., Huebner, M., & Weber, A. P. (2007). Transcriptional profiling of Arabidopsis heat shock proteins and transcription factors reveals extensive overlap between heat and non-heat stress response pathways. *BMC Genomics, 8*, 125.

Tan, W., Meng, Q. W., Brestic, M., Olsovska, K., & Yang, X. (2011). Photosynthesis is improved by exogenous calcium in heat-stressed tobacco plants. *Journal of Plant Physiology, 168*, 2063–2071.

Teixeira, E. I., Fischer, G., Van Velthuizen, H., Walter, C., & Ewert, F. (2013). Global hot-spots of heat stress on agricultural crops due to climate change. *Agricultural and Forest Meteorology, 170*, 206–215.

Tester, M., & Langridge, P. (2010). Breeding technologies to increase crop production in a changing world. *Science, 327*, 818–822.

Towill, L. E., & Mazur, P. (1975). Studies on the reduction of 2, 3, 5-triphenyltetrazolium chloride as a viability assay for plant tissue cultures. *Canadian Journal of Botany, 53*, 1097–1102.

Uga, Y., Kitomi, Y., Ishikawa, S., & Yano, M. (2015). Genetic improvement for root growth angle to enhance crop production. *Breeding Science, 65*, 111.

Van Zanten, M., Voesenek, L. A. C. J., Peeters, A. J. M., & Millenaar, F. F. (2009). Hormone- and light-mediated regulation of heat-induced differential petiole growth in Arabidopsis. *Plant Physiology, 151*, 1446–1458.

Vettakkorumakankav, N. N., Falk, D., Saxena, P., & Fletcher, R. A. (1999). A crucial role for gibberellins in stress protection of plants. *Plant and Cell Physiology, 40*, 542–548.

Vignjevic, M., Xiao, W., Jørgen, E. O., & Bernd, W. (2015). Traits in spring wheat cultivars associated with yield loss caused by a heat stress episode after anthesis. *Journal of Agronomy and Crop Science, 1*, 32–48.

Wahid, A., Gelani, S., Ashraf, M., & Foolad, M. (2007). Heat tolerance in plants: An overview. *Environmental and Experimental Botany, 61*, 199–223.

Wang, J., Gan, Y. T., Clarke, F., & McDonald, C. L. (2006). Response of chickpea yield to high temperature stress during reproductive development. *Crop Science, 46*, 2171–2178.

Wang, X., Cai, J., Jiang, D., Liu, F., Dai, T., & Cao, W. (2011). Pre-anthesis high-temperature acclimation alleviates damage to the flag leaf caused by post-anthesis heat stress in wheat. *Journal of Plant Physiology, 168*, 585–593.

Wardlaw, I. F., Dawson, I. A., & Munibi, P. (1989). The tolerance of wheat to high temperatures during reproductive growth. 2. Grain development. *Crop & Pasture Science, 40*, 15–24.

Yang, F., Jørgensen, A. D., Li, H., Sondergaard, I., Finnie, C., Svensson, B., Jiang, D., Wollenweber, B., & Jacobsen, S. (2011). Implications of high-temperature events and water deficits on protein profiles in wheat (*Triticum aestivum* L. cv. Vinjett) grain. *Proteomics, 11*, 1684–1695.

Yang, X., Liang, Z., & Lu, C. (2005). Genetic engineering of the biosynthesis of glycine betaine enhances photosynthesis against high temperature stress in transgenic tobacco plants. *Plant Physiology, 138*, 2299–2309.

Yoshida, S. (1981). Fundamentals of rice crop science. *Int. Rice Res. Inst.*

Zutta, B. R., Nobel, P. S., Aramians, A. M., & Sahaghian, A. (2011). Low-and high-temperature tolerance and acclimation for chlorenchyma versus meristem of the cultivated cacti *Nopalea cochenillifera, Opuntia robusta, and Selenicereus megalanthus. Journal of Botany, 2011*, 347168. https://doi.org/10.1155/2011/347168.

Chapter 3
Climate Smart Crops: Flood and Drought-Tolerant Crops

Camila Pegoraro, Carlos Busanello, Luciano Carlos da Maia, and Antonio Costa de Oliveira

Abstract The increasing demands of world's population make it mandatory to increase food production. Global climate changes currently threaten this goal, through many adverse conditions, such as water stress by excess (waterlogging or flooding) or scarcity (drought). One alternative to overcome this challenge is the development of stress resilient crops, since an increase in the number of affected areas and the intensity of stress occurring in those, is happening as a consequence of climate changes. Molecular, physiological and epigenetic changes may be key to achieve tolerance to these stresses and are discussed in this chapter.

Keywords Genetic variability · Transgenic · Gene expression · Epigenetic

3.1 Flood Tolerant Crops

The current world population is estimated at 7.2×10^9 individuals and will reach by 2050, a figure around 9.2×10^9 (Fita et al. 2015). Therefore, to increase food production is mandatory in order to satisfy the predicted demand. However, the agricultural yield is extremely dependent of environmental conditions, which are influenced by global climate changes that currently threaten our planet. Taking this into account, the current and future challenge is to ensure food security facing a constant increase in population and global climate change impact, which are associated with a decrease in yield potential. One alternative for plant scientists to overcome this challenge is the development of stress resilient crops, i.e. cultivars tolerant to drought, heat, cold, salinity, flooding and/or new pests and diseases, since an increase in the number of affected areas and the intensity of stress occurring in those, is happening as a consequence of climate changes. Tolerant cultivars can be obtained from classical (hybridization followed by selection) breeding methods or modern tools of genetic engineering (genetic transformation, genome editing).

C. Pegoraro · C. Busanello · L. C. da Maia · A. Costa de Oliveira (✉)
Plant Genomics and Breeding Center, School of Agronomy Eliseu Maciel, Federal University of Pelotas, Pelotas, RS, Brazil

© Springer Nature Switzerland AG 2019
A. Sarkar et al. (eds.), *Sustainable Solutions for Food Security*,
https://doi.org/10.1007/978-3-319-77878-5_3

Predictions indicate an increase in extreme hydrological events due to climatic changes (Delgado et al. 2014), which will lead to an increase in flooded areas. Flooding is an environmental condition common to many areas of the globe. However, the majority of plant species of economical importance is susceptible to this stress. Water excess can negatively affect crop production due to seeding delays, changes in plant cycle, increase in disease susceptibility and, depending on the duration of the stress, lead to plant death.

Under flooding conditions, a restriction in atmospheric O_2 and CO_2 occurs, ethylene (C_2H_4) diffusion is difficulted, electrochemical changes in the soil result in higher concentrations of toxic elements such as manganese (Mn^{2+}), iron (Fe^{2+}) and sulphide (H_2S, HS^- and S^{2-}), and reduction of light availability. This condition induces an energetic crisis in plants, caused by limiting oxidative phosphorylation and low photosynthetic rates. These events characterize flooding as a complex stress (Voesenek and Bailey-serres 2015).

Two types of flooding stress can be observed: waterlogging, in which the roots and part of the shoot are submerged; and complete submergence, in which the whole plant remains submerged (Ahmed et al. 2013). Rice (*Oryza sativa*), an important crop, presents an ability to grow and develop in waterlogged environments without any harm. Rice tolerates waterlogged soils, because it can develop aerenchyma, facilitating O_2 diffusion from shoot to submerged parts, and to avoid O_2 loss as a result of radial diffusion to the rhyzosphere (radial oxygen loss—ROL) through the induction of a barrier against ROL (Shiono et al. 2011). However, the majority of rice genotypes does not survive under complete submergence, except the genotype FR13A (*Oryza sativa* ssp. *indica*), which is highly tolerant and survives for 2 weeks in this condition (Xu et al. 2006).

C_2H_4 accumulation under flooding conditions occur due to the induction of enzymes 1-aminocyclopropane-1-carboxylic acid (ACC) synthase and ACC oxydase. This hormone has an important role in the response of adaptation to flooding stress, because it contributes to coordinate cell death for aerenchyma formation (Yukiyoshi and Karahara 2014). Besides, in complete submergence C_2H_4 is involved in internode elongation through an increase in giberellic acid (GA) and the sensibility of cells to this hormone, which occurs due to a reduction in abscisic acid (ABA) levels (Hattori et al. 2009; Kende et al. 1998; Nishiuchi et al. 2012). This mechanism is called escape strategy (Bailey-Serres and Voesenek 2008), since it places shoot above water levels, allowing the plant to restart aerobic metabolic activity.

In rice, the escape strategy is dependent on a polygenic locus that codes for two ethylene responsive transcript factors (Ethylene Responsive Factor—ERF), Snorkel1 (SK1) and SK2 (Hattori et al. 2009). However, the fast shoot elongation is only favourable if the energy cost is compensated by benefits, such as improved O_2 availability and restoration of photosynthesis. An imbalance between cost and benefit is probable if complete submergence is too deep to allow the renovated contact with air, and under these circumstances, the majority of irrigated rice cultivars die within a week. Besides, even if the plant reaches the surface, the energy cost is too high, and it cannot recover when the water recedes (Nishiuchi et al.

2012). On the other hand, rice genotypes that can survive under longer complete submergence conditions present slow growth and energy and carbohydrate conservation during submergence, restarting development after stress. This is called quiescence strategy (Kudahettige et al. 2011). Rice varieties such as FR13A, which exploit this strategy, have higher survival rates during long submergence periods than if they had an escape strategy (Setter and Laureles 1996).

In the *SUB1* (*Submergence 1*) locus, on rice chromosome 9, three ERF coding genes are present: *SUB1A*, *SUB1B* and *SUB1C*. Two of these genes, *SUB1B* and *SUB1C* are present in all rice accessions analysed so far. However, *SUB1A* presence varies among genotypes (Xu et al. 2006). In indica varieties, the *SUB1A* is present, and presents two alleles, *SUB1A-1* and *SUB1A-2*. *SUB1A-1* allele is present on the cultivar FR13A, tolerant to complete submergence, and the allele *SUB1A-2* in the remaining genotypes sensitive to submergence (Xu et al. 2006). All genotypes that are tolerant to complete submergence present the *SUB1A-1* allele (Singh et al. 2010).

ERFs are transcription factors (TFs), and the identification of those TFs associated with plant tolerance is a very important information to reveal defence mechanism puzzles facing abiotic stress conditions. When a plant is under abiotic stress condition, there is an activation of signaling routes, which involve the perception and signal transduction, expression of abiotic responsive genes and physiological and metabolic response activations (Wang et al. 2016). Transcription factors take part in this signaling route. These regulatory elements bind to cis elements present in the promoter regions of stress response genes, leading to gene activation or inactivation (Fig. 3.1).

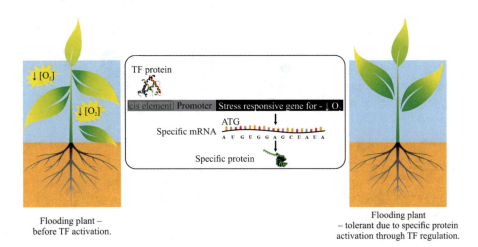

Fig. 3.1 Basic representation of transcriptional regulation via transcription factors (TF). TFs belonging to different families are induced in conditions of biotic and abiotic stresses (the scheme is displaying flooding stress) and bind to the promoter region of genes whose products are associated to stress response mechanisms, activating the transcription of these genes

The participation of TFs from the families AP2/EREBP (AP2/ethylene responsive element binding factor—ERF—DNA-binding), MYB, WRKY, NAC (NAM, ATAF and CUC transcription factors), bZIP (basic leucine zipper), bHLH (basic helix-loop-helix), HD-ZIP (Homeodomain Leucine Zipper) and HSF (Heat shock factor) in the response to abiotic stresses in plants has been widely studied (Belamkar et al. 2014; Chen et al. 2012; Guo et al. 2016; Ji et al. 2015; Roy 2015; Shao et al. 2015; Sornaraj et al. 2015). The major TFs involved in the defense of plants under flooding conditions are AP2/EREBP, where one can find ERFs SUB1 and SK (Hattori et al. 2009; Xu et al. 2006), described previously. Also, other TFs such as MYB, WRKY, NAC (Valliyodan et al. 2014) and HSF (Banti et al. 2010) have important roles. Based on these findings, it becomes evident that TFs have a high potential to be used in the development of stress tolerant crops via genetic transformation. This potential relies on the fact that one TF can regulate a large number of genes associated with stress responses.

Transcript regulation via TFs (Fig. 3.2) is dependent of epigenetic landmarks, i.e. the TF accessibility to promoters is influenced and controlled by chromatin condensation state and DNA methylation (Locatelli et al. 2009; Medvedeva et al. 2014). The presence of methylated DNA and condensed chromatin in the gene coding region also influences its expression, since transcription elongation is inhibited (Hohn et al. 1996). Abiotic stresses establish epigenetic modifications,

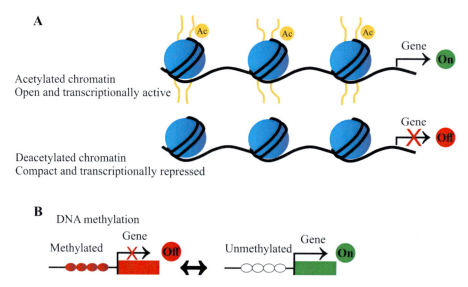

Fig. 3.2 Basic representation of epigenetic regulation. Histones are proteins that associate to compact DNA. (**a**) When acetylated, histones promote a lower degree of compaction, allowing the access of the transcription machinery to DNA, when deacetylated, they promote a higher degree of compaction, making the access to DNA cumbersome. Thus, acetylated and deacetylated promoter regions activate or block transcription, respectively. (**b**) The methylation of the DNA sequence that corresponds to the promoter region, blocks the access of the transcription machinery, not allowing gene activation. Adapted from Verdin and Ott 2015

leading to changes in gene expression. Some epigenetic changes induced by stress are erased when the stress is removed, while other are maintained and act as stress memory marks, and can be inherited through mitotic and meiotic divisions. These changes that act as stress memory marks do help the plant to better respond to subsequent stresses (Chinnusamy and Zhu 2009), and can be related to plant tolerance to abiotic stresses.

In rice, there was an increase in the expression of *ADH1* (alcohol dehydrogenase 1) and *PDC1* (pyruvate decarboxylase 1) under submergence conditions (Tsuji et al. 2006). This increase was due to the tri-methylation of H3 histone present in the region corresponding to the *ADH1* and *PDC1* genes, in the submergence period, and to H3 acetylation in the final period. During re-aeration, methylation and acetylation returned to initial levels, indicating clearly that these epigenetic modifications occurred as a function of O_2 deficiency stress caused by the submergence.

The gene *VIN-3* (vernalization insensitive 3), that codes for a chromatin remodeling protein, was induced in response to hypoxia stress in *Arabidopsis thaliana* (Bond et al. 2009). Furthermore, the authors observed that mutant plants lacking VIN-3 are more sensitive to hypoxia, being this gene associated to root and shoot survival. Another relevant information, is that VIN-3 does not regulate genes associated to O_2 deficiency in plants, such as *ADH1*, *PCD1*, *PDC2* and *GLB1* (plant haemoglobin class 1).

Epigenetic changes in response to stress present a potential application in classic plant breeding. This potential value is justified by the influence of epigenetic changes in agronomic characters of high economical importance, such as flowering time, plant stature, pathogen (Peng and Zhang 2009) and abiotic stress (Garg et al. 2015) tolerance. There are variations in the plant DNA methylation patterns, and in one form, called type I, methylated alleles are inherited in a stable form for many generations (Kakutani 2002). In rice, for example, tolerance to *Xanthomonas oryzae* pv. *oryzae*, caused by artificially induced epigenetic changes, was maintained through nine generations (Akimoto et al. 2007), time which corresponds to the commercial lifespan of a new rice variety.

Gene expression can be regulated post-transcriptionally by the action of microRNAs (miRNAs). miRNAs (Fig. 3.3) recognize target mRNAs as a function of sequence complementarity, and suppress target gene expression through forwarding either for degradation or translation inhibition (Li et al. 2013). miRNAs are involved in plant response to abiotic stresses. However, miRNAs expression is stress, species, genotype and tissue dependent (Zhang 2015). Interestingly, the major targets of stress-responsive miRNAs are some TFs (Fang et al. 2014a, b). Thus, only one miRNA is capable of regulating the expression of a large number of genes, which would be target of the TF silenced by the miRNA.

Maize (*Zea mays*) plants under submergence were reported to have 39 differentially expressed miRNAs (Zhang et al. 2008), miRNAs that were expressed in the initial submergence period have as target TFs (HD-ZIP, auxin response factor— ARF, SCL, WRKY) and miRNAs induced 24 h after submergence had as target genes involved in carbohydrate and energy metabolism and on reactive oxygen and

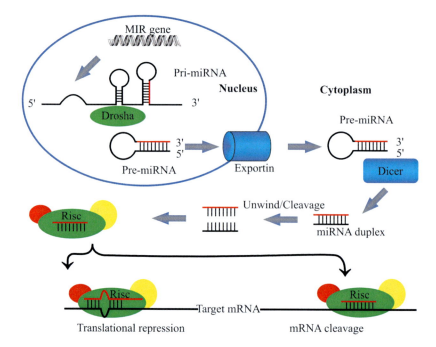

Fig. 3.3 Basic representation of miRNA synthesis and activity. miRNAs are transcribed from MIR genes, composed by one or more groups of sequences internally complementary that fold to form hairpin-like structures (pri-miRNAs). The pri-miRNAs are cleaved by the nuclear endonuclease Drosha, producing short hairpins, the pre-miRNAs. In the cytoplasm, pre-miRNAs are cleaved by the Dicer endonuclease generating miRNAs still paired to the complementary sequence. The miRNA is open, and the unnecessary strand is discarded by the action of a ribonuclease. The remaining strand is complementary to a specific mRNA, and is taken to this particular mRNA by the RNA induced silencing complex (RISC). After duplex formation, there is induction of specific mRNA degradation or blocking of translation of this specific mRNA. Adapted from Calore and Fabbri 2012

acetaldehyde species elimination. These results evidence the participation of miRNAs in the mechanisms of plant adaptation facing submergence condition.

Arabidopsis thaliana plants under hipoxia conditions present changes in the accumulation of 46 miRNA species, from 19 families (Moldovan et al. 2010). It is evident that under hypoxic conditions, miRNAs are majorly involved in gene repression. These evidences indicate that miRNAs present an important role in plant response to many abiotic stresses, therefore, these elements have become a new target for plant breeding aiming at tolerance to different abiotic stresses (Zhang and Wang 2015), such as flooding. Thus, miRNAs can be inserted or silenced, aiming the silencing or induction of target genes, respectively, associated to flooding tolerance.

Another strategy for obtaining flooding tolerant plants, as important as the ones mentioned before, is the use of the gene pools present in germplasm banks. Genes associated to the tolerance to adverse environments can be absent in modern genotypes due to the exclusion during domestication and breeding processes, as the case in the *SUB1A-1* allele. This condition can be reversed through the introgression of loci from wild into elite genotypes. The potential of wild species as donors of abiotic stress tolerant alleles, has been reported in maize, soybean, potato, wheat and other species (Mickelbart et al. 2015). An example of introgression is observed in *Zea mays* (milho) using Gaspé Flint which is a variety belonging to the Northern Flint as a segment donor and the reference elite line B73 as a recurrent parent, resulting from this cross a library of contrasting lines for many traits (Salvi et al. 2011).

Recently, the introgression of *Amblyopyrum muticum* segments into *Triticum aestivum* (wheat) through backcrosses in an attempt to develop superior varieties, as can be seen on Fig. 3.4 (King et al. 2017).

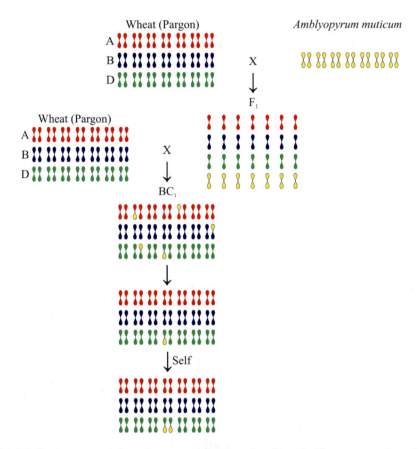

Fig. 3.4 Basic representation of segmental introgression from *Amblyopyrum muticum* into *Triticum aestivum*. Initially, a hybridization between *Triticum aestivum* and *Amblyopyrum muticum* is performed. The hybrid is backcrossed with *Triticum aestivum*. At the end, a self is made to fix the introgressed segment. Adapted from King et al. 2017

The development of flooding stress tolerant crops is a hard goal to achieve and will only have success if efforts can be joined using biotechnology tools, breeding methods and the diversity present in germplasm banks. The release of *sub1* varieties by IRRI suggests that this goal was achieved for rice and could now be targeted to other important cereals.

3.2 Drought-Tolerant Crops

Climate changes caused by natural and anthropogenic factors can result in a higher intensity of abiotic stresses and their occurrence in novel areas, leading to crop yield losses. Among these stresses is drought, which is an extreme climate event, with high impact on agriculture (Boutraa et al. 2010).

Drought stress occurs when the water available in the soil is reduced and climate conditions cause continuous water losses through transpiration or evapotranspiration. This condition results from the lack of rain during a period of time (Jaleel et al. 2009). Considering that water is vital for plant growth, development and yield, water deficit limits plant yield more than any other environmental factor (Shao et al. 2009).

Two major strategies are suggested for the mitigation of stress effects on crop yield, the development of tolerant plants and the use of irrigation. Taking into account the fact that climate changes influence directly water availability (Misra 2014), the alternative that becomes more viable is the development of tolerant cultivars and/or the efficient use of water. However, in order to make this strategy possible, it is necessary to understand the response mechanisms to this condition.

Many molecular networks, including signal transduction (Osakabe et al. 2014), epigenetic regulation (Kim et al. 2015; Zheng et al. 2013), transposable element activity (Mao et al. 2015), alternative splicing (Thatcher et al. 2016), and miRNA action (Ferdous et al. 2015) are involved in the response to drought stress. This response is controlled by abscisic acid (ABA) signaling, ion transport, and stomata opening and closing mechanisms (Osakabe et al. 2014).

Drought condition leads to ABA accumulation, which induces the expression of a large number of genes, which present the cis-element ABRE (ABA-responsive element) in their promoter regions. AREB/ABFs (ABRE-binding protein/ABRE-binding factor) are TFs that regulate ABA dependent gene expression (Nakashima et al. 2014) through the binding of ABRE elements. The superexpression of *AREB1* improved drought tolerance in rice and soybean (Barbosa et al. 2013; Oh et al. 2005). Besides ABA responsive genes, independent genes are also involved in the response and tolerance to drought. In the promoter region of these genes the cis-element DRE/CTR (dehydration responsive element/C-repeat) is found. DREB/CBF (DRE-binding protein/CRT binding factor) are TFs that bind to DRE/CRT and control the expression of a large number of stress responsive genes (Nakashima et al. 2014). The superexpression of the transcription factor *DREB1/CBF* improved drought, salinity and freezing tolerance in Arabidopsis (Kasuga et al. 1999), demonstrating the importance of these TFs for plant tolerance to drought. These studies suggest that genetic transformation with TFs can be used as a powerful tool for breeding aiming to develop drought-tolerant plants. However, new TFs must be mined.

The positive regulation of gene transcription under drought conditions has been correlated with histone modification and nucleosome density, and these epigenetic modifications depend on the stress intensity (To and Kim 2014). Furthermore, it has been demonstrated that drought response is memorized via histone modification in many genes induced under this condition (Ding et al. 2012). It was also verified that, under water stress conditions, the acetylation levels of histones was increased in genes associated with this stress (Kim et al. 2008). In rice, drought stress significantly induced four *HAT* (histone acetyltransferases) genes and increased the acetylation of four histones (Fang et al. 2014a, b), indicating that drought response is also dependent on epigenetic regulation.

The comparison of two rice varieties with contrasting levels of drought tolerance, cultivated under water stress for six generations demonstrated that the stress has a cumulative effect on DNA methylation patterns on both varieties, however with a different profile. In the sensitive variety, 27.6% of differentially methylated loci (DML) were induced by drought and 3.2% of total DML were stably transmitted to the next generations. In the tolerant variety, 48.8% of DML were induced by drought stress, and 29.8% were transmitted to the next generation (Zheng et al. 2013). The inheritance of drought-induced epigenetic changes can be a new strategy for the development of tolerant varieties via classic breeding.

Another mechanism associated with the natural variation for drought tolerance is the transposable element (TE) activity (Fig. 3.5). In maize, the presence of miniature inverted-repeat transposable elements (MITEs) in the promoter region of the *NAC* gene (NAM, ATAF, and CUC transcriptions factors) represses the expression of this gene and leads to drought susceptibility. The superexpression of the gene *ZmNAC111* in maize increases drought tolerance (Mao et al. 2015). Similar studies can be developed seeking to identify key genes inactivated by TEs, which would be associated to drought tolerance. With this understanding, one can project new genetic transformation events aiming to develop varieties tolerant to this stress.

Many reports have shown the role of alternative splicing (AS) in stress response (Fig. 3.6). This mechanism can adjust the expression of regulating (TFs) or structural genes through the production of non-functional transcripts under normal conditions, which are replaced by active variants under stress conditions. Also, an active form can be deactivated under stress conditions (Guerra et al. 2015). The AS mechanism that occurs in the gene coding for the signaling protein DROUGHT RESPONSIVE ANKYRIN1 (DRA1) in Arabidopsis is an example of intron retention involved in flooding tolerance (Sakamoto et al. 2013). The expression of *DREB2B* is regulated by AS in response to stress in rice, wheat, barley and maize (Guerra et al. 2015). Another event involving AS has been reported for the TF MYB60, which is involved in the regulation of stomata and root growth under drought stress in Arabidopsis (Oh et al. 2005). Other genes related to drought stress are also regulated by AS (Remy et al. 2013).

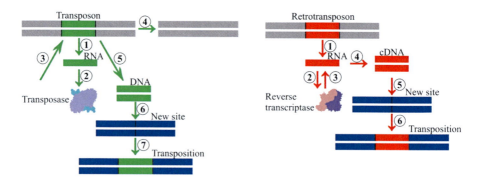

Fig. 3.5 Basic representation of transposition mechanism. Some DNA segments present the ability to move within the genome. This change can result in gene duplication or inactivation (fragment inserted in the coding region). (**a**) Transposons move via DNA. The element is excised from its site and transferred to a new, through the action of a transposase (coded by the transposon), in a process described as cut and paste. (1) RNA transcription that codes information for the transposase; (2) translation of transposase enzyme; (3) transposon cut by the action of the transposase; (4) DNA without the transposon; (5) cut transposon; (6) new site for transposon insertion; (7) transposon insertion in the new site. (**b**) Retrotransposons move through an RNA intermediate. The transposable elements are transcribed and RNA serves as a template for reverse transcription (coded by retrotransposon). The DNA formed is inserted in the genome by the action of an integrase (also coded by the retrotransposon), resulting in a new copy of the transposable element. Process called copy and paste. (1) transcription of the retrotransposon RNA (also presents information for reverse transcriptase RNA); (2) translation of the reverse transcriptase enzyme; (3) cDNA synthesis based on the retrotransposon RNA by the action of the reverse transcriptase; (4) retrotransposon cDNA; (5) new site for retrotransposon insertion; (6) retrotransposon insertion in the new site

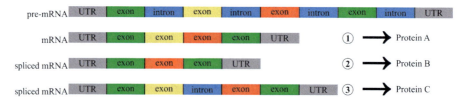

Fig. 3.6 Basic representation of alternative splicing mechanism. Some genes present the ability of generating different proteins due to alternative splicing, which occurs while the introns are being excised. There are different types of alternative splicing (exon skipping, alternative 3′ SS selection, alternative 5′ SS selection, intron retention, mutually exclusive exons, alternative promoters, alternative poly (A), each one generating a different protein). In the figure, constitutive splicing (1), exon skipping (2) and intron retention (3) are displayed

This understanding is highly relevant to plant breeding, since it indicates that part of the genetic variability present within a species is influenced by sequences involved in the regulation of AS events. Thus, AS becomes a new tool for the development of new varieties. The study of genetic variability occurs as a function of alternative splicing and the identification of genomic regions involved in the regulation of this event in crops are proposed as strategies for the selection of genotypes with a superior performance under adverse environmental conditions (Mastrangelo et al. 2012), such as drought.

Reprogramming of gene expression via miRNAs is another plant defence mechanism to deal with drought stress condition. miRNAs expression is changed in response to drought in different species such as Arabidopsis, rice, soybean, tobacco and beans (Ferdous et al. 2015). miRNAs induced or repressed by drought are potentially relevant for genetic engineering aiming the development of plants tolerant to this stress, since miRNA targets include genes that contribute negatively or positively for drought tolerance. Induced or repressed by drought are potentially relevant for genetic engineering aiming the development of drought-tolerant plants, since miRNAs targets include genes that contribute negatively or positively for drought tolerance. The induction of miRNAs means that their targets are repressed in the same conditions and vice versa (Ferdous et al. 2015). The increase in expression of genes associated with drought tolerance can be obtained through the superexpression of this gene or by silencing of the corresponding miRNA, and this tool can be used in order to obtain tolerant plants.

Therefore, one must highlight the need for amplifying variability for tolerance to drought stress. Specific genetic determinants to adaptation to abiotic stresses present in wild germplasm may have evolved in nature, or may have been lost during the breeding process, as mentioned earlier (Mickelbart et al. 2015). Pre-breeding through the introgression of desirable genes is an opportunity for the amplification of crop gene pools aiming to increase the variability for drought stress.

Abiotic stress tolerance such as drought is controlled in its majority by quantitative trait loci (QTLs), making the use of transgenic or gene edition technologies more cumbersome, since a large number of genes have to be manipulated. In this case, traditional breeding through hybridization and the accumulation of quantitative traits mediated by classic breeding methods have great importance in the development of tolerant genotypes and has contributed to yield increases and to overcome yield limits due to drought stresses. The major DNA marker applications, such as marker assisted selection (MAS), QTL mapping and gene pyramidization, are the most applied in the development of new genotypes (Nogoy et al. 2016). In rice, pyramidization associated to MAS using three QTLs enabled an increase in yield of more than 1000 kg ha^{-1} when compared to the recurrent parent under drought stress conditions at the reproductive phase (Shamsudin et al. 2016). The pyramidization of two drought tolerance genes in transgenic maize containing TsVP and BetA (encoding choline dehydrogenase from *Escherichia coli* and TsVP encoding V-H$^+$-PPase from *Thellungiella halophila*) through directed crosses produced plants that outgrew their parents regarding vigour, less growth retardation, lower anthesis–silking interval and higher yields. This suggests that the co-suppression of multiple genes is efficient and in transgenic plants could effectively increase the resistance to abiotic stress and enable a viable strategy to obtain drought-tolerant plants (Wei et al. 2011). New technologies reduced substantially the crossing cycles and increased the precision and efficiency in the development of new cultivars in plant breeding programmes (Nogoy et al. 2016).

It is evident that maintaining yield and food security are only possible if genotypes tolerant to environmental stresses are developed, such as drought stress tolerant genotypes. However, considering the complexity that is involved in plant response facing this condition, different strategies must be pursued, including those that aim to identify genes and specific mechanisms, germplasm characterization and amplification of genetic variability.

Acknowledgements The authors would like to thank Coordenação de Aperfeiçoamento de Pessoal de Nível Superior (CAPES) and Conselho Nacional de Desenvolvimento Científico e Tecnológico (CNPq) and Fundação de Amparo a Pesquisa do Rio Grande do Sul (FAPERGS) for research funding and scholarships.

References

Ahmed, F., Rafii, M. Y., Ismail, M. R., Juraimi, A. S., Rahim, H. A., Asfaliza, R., & Latif, M. A. (2013). Waterlogging tolerance of crops: Breeding, mechanism of tolerance, molecular approaches, and future prospects. *BioMed Research International, 2013*, 963525. https://doi.org/10.1155/2013/963525.

Akimoto, K., Katakami, H., Kim, H.-J., Ogawa, E., Sano, C. M., Wada, Y., & Sano, H. (2007). Epigenetic inheritance in rice plants. *Annals of Botany, 100*, 205–217. https://doi.org/10.1093/aob/mcm110.

Bailey-Serres, J., & Voesenek, L. A. (2008). Flooding stress: Acclimations and genetic diversity. *Annual Review of Plant Biology, 59*, 313–339. https://doi.org/10.1146/annurev.arplant.59.032607.092752.

Banti, V., Mafessoni, F., Loreti, E., Alpi, A., & Perata, P. (2010). The heat-inducible transcription factor HsfA2 enhances anoxia tolerance in arabidopsis. *Plant Physiology, 152*, 1471–1483. https://doi.org/10.1104/pp.109.149815.

Barbosa, E. G. G., Leite, J. P., Marin, S. R. R., Marinho, J. P., Carvalho, J. F. C., Fuganti-Pagliarini, R., et al. (2013). Overexpression of the ABA-dependent AREB1 transcription factor from *Arabidopsis thaliana* improves soybean tolerance to water deficit. *Plant Molecular Biology Reporter, 31*, 719–730. https://doi.org/10.1007/s11105-012-0541-4.

Belamkar, V., Weeks, N. T., Bharti, A. K., Farmer, A. D., Graham, M. A., & Cannon, S. B. (2014). Comprehensive characterization and RNA-Seq profiling of the HD-Zip transcription factor family in soybean (*Glycine max*) during dehydration and salt stress. *BMC Genomics, 15*, 950. https://doi.org/10.1186/1471-2164-15-950.

Bond, D. M., Wilson, I. W., Dennis, E. S., Pogson, B. J., & Jean Finnegan, E. (2009). Vernalization insensitive 3 (VIN3) is required for the response of *Arabidopsis thaliana* seedlings exposed to low oxygen conditions. *The Plant Journal, 59*, 576–587. https://doi.org/10.1111/j.1365-313X.2009.03891.x.

Boutraa, T., Akhkha, A., Al-Shoaibi, A. A., & Alhejeli, A. M. (2010). Effect of water stress on growth and water use efficiency (WUE) of some wheat cultivars (*Triticum durum*) grown in Saudi Arabia. *Journal of Taibah University for Science, 3*, 39–48. https://doi.org/10.1016/S1658-3655(12)60019-3.

Calore, F., & Fabbri, M. (2012). MicroRNAs and cancer. *The Atlas of Genetics and Cytogenetics in Oncology and Haematology, 16*, 51–59. https://doi.org/10.4267/2042/47272.

Chen, L., Song, Y., Li, S., Zhang, L., Zou, C., & Yu, D. (2012). The role of WRKY transcription factors in plant abiotic stresses. *Biochimica et Biophysica Acta, 1819*, 120–128. https://doi.org/10.1016/j.bbagrm.2011.09.002.

Chinnusamy, V., & Zhu, J.-K. (2009). Epigenetic regulation of stress responses in plants. *Current Opinion in Plant Biology, 12*, 133–139. https://doi.org/10.1016/j.pbi.2008.12.006.

Delgado, J. M., Merz, B., & Apel, H. (2014). Projecting flood hazard under climate change: An alternative approach to model chains. *Natural Hazards and Earth System Sciences, 14*, 1579–1589. https://doi.org/10.5194/nhess-14-1579-2014.

Ding, Y., Fromm, M., & Avramova, Z. (2012). Multiple exposures to drought "train" transcriptional responses in Arabidopsis. *Nature Communications, 3*, 740. https://doi.org/10.1038/ncomms1732.

Fang, H., Liu, X., Thorn, G., Duan, J., & Tian, L. (2014a). Expression analysis of histone acetyltransferases in rice under drought stress. *Biochemical and Biophysical Research Communications, 443*, 400–405. https://doi.org/10.1016/j.bbrc.2013.11.102.

Fang, Y., Xie, K., & Xiong, L. (2014b). Conserved miR164-targeted NAC genes negatively regulate drought resistance in rice. *Journal of Experimental Botany, 65*, 2119–2135. https://doi.org/10.1093/jxb/eru072.

Ferdous, J., Hussain, S. S., & Shi, B. J. (2015). Role of microRNAs in plant drought tolerance. *Plant Biotechnology Journal, 13*, 293–305. https://doi.org/10.1111/pbi.12318.

Fita, A., Rodríguez-Burruezo, A., Boscaiu, M., Prohens, J., & Vicente, O. (2015). Breeding and domesticating crops adapted to drought and salinity: A new paradigm for increasing food production. *Frontiers in Plant Science, 6*, 978. https://doi.org/10.3389/fpls.2015.00978.

Garg, R., Narayana Chevala, V., Shankar, R., & Jain, M. (2015). Divergent DNA methylation patterns associated with gene expression in rice cultivars with contrasting drought and salinity stress response. *Scientific Reports, 5*, 14922. https://doi.org/10.1038/srep14922.

Guerra, D., Crosatti, C., Khoshro, H. H., Mastrangelo, A. M., Mica, E., & Mazzucotelli, E. (2015). Post-transcriptional and post-translational regulations of drought and heat response in plants: A spider's web of mechanisms. *Frontiers in Plant Science, 6*, 57. https://doi.org/10.3389/fpls.2015.00057.

Guo, M., Liu, J.-H., Ma, X., Luo, D.-X., Gong, Z.-H., & Lu, M.-H. (2016). The plant heat stress transcription factors (HSFs): Structure, regulation, and function in response to abiotic stresses. *Frontiers in Plant Science, 7*, 114. https://doi.org/10.3389/fpls.2016.00114.

Hattori, Y., Nagai, K., Furukawa, S., Song, X.-J., Kawano, R., Sakakibara, H., Wu, J., Matsumoto, T., Yoshimura, A., Kitano, H., Matsuoka, M., Mori, H., & Ashikari, M. (2009). The ethylene response factors SNORKEL1 and SNORKEL2 allow rice to adapt to deep water. *Nature, 460*, 1026–1030. https://doi.org/10.1038/nature08258.

Hohn, T., Corsten, S., Rieke, S., Müller, M., & Rothnie, H. (1996). Methylation of coding region alone inhibits gene expression in plant protoplasts. *Proceedings of the National Academy of Sciences of the United States of America, 93*, 8334–8339.

Jaleel, C. A., Manivannan, P., Wahid, A., Farooq, M., Al-Juburi, H. J., Somasundaram, R., & Panneerselvam, R. (2009). Drought stress in plants: A review on morphological characteristics and pigments composition. *International Journal of Agriculture and Biology, 11*, 100–105.

Ji, X., Nie, X., Liu, Y., Zheng, L., Zhao, H., Zhang, B., Huo, L., & Wang, Y. (2015). A bHLH gene from *Tamarix hispida* improves abiotic stress tolerance by enhancing osmotic potential and decreasing reactive oxygen species accumulation. *Tree Physiology, 36*, 193–207. https://doi.org/10.1093/treephys/tpv139.

Kakutani, T. (2002). Epi-alleles in plants: Inheritance of epigenetic information over generations. *Plant and Cell Physiology, 43*, 1106–1111. https://doi.org/10.1093/pcp/pcf131.

Kasuga, M. I., Liu, Q., Miura, S., Yamaguchi-Shinozaki, K., & Shinozaki, K. (1999). Improving plant drought, salt and freezing tolerance by gene transfer of a single stress-inducible transcription factor. *Nature Biotechnology, 17*, 287–291. https://doi.org/10.1038/7036.

Kende, H., Van Der Knaap, E., & Cho, H. (1998). Deepwater rice: A model plant to study stem elongation. *Plant Physiology, 118*, 1105–1110. https://doi.org/10.1104/pp.118.4.1105.

Kim, J. M., Sasaki, T., Ueda, M., Sako, K., & Seki, M. (2015). Chromatin changes in response to drought, salinity, heat, and cold stresses in plants. *Frontiers in Plant Science, 6*, 114. https://doi.org/10.3389/fpls.2015.00114.

Kim, J. M., To, T. K., Ishida, J., Morosawa, T., Kawashima, M., Matsui, A., Toyoda, T., Kimura, H., Shinozaki, K., & Seki, M. (2008). Alterations of lysine modifications on the histone H3 N-tail under drought stress conditions in *Arabidopsis thaliana*. *Plant Cell Physiology, 49*, 1580–1588. https://doi.org/10.1093/pcp/pcn133.

King, J., Grewal, S., Yang, C., Hubbart, S., Scholefield, D., Ashling, S., Edwards, K. J., Allen, A. M., Burridge, A., Bloor, C., Davassi, A., da Silva, G. J., Chalmers, K., & King, I. P. (2017). A step change in the transfer of interspecific variation into wheat from Amblyopyrum muticum. *Plant Biotechnology, 15*, 217–226. https://doi.org/10.1111/pbi.12606.

Kudahettige, N. P., Pucciariello, C., Parlanti, S., Alpi, A., & Perata, P. (2011). Regulatory interplay of the Sub1A and CIPK15 pathways in the regulation of α-amylase production in flooded rice plants. *Plant Biology, 13*, 611–619. https://doi.org/10.1111/j.1438-8677.2010.00415.x.

Li, S., Liu, L., Zhuang, X., Yu, Y., Liu, X., Cui, X., Ji, L., Pan, Z., Cao, X., Mo, B., Zhang, F., & Raikhel, N. (2013). MicroRNAs inhibit the translation of target mRNAs on the endoplasmic reticulum in Arabidopsis. *Cell, 153*, 562–574. https://doi.org/10.1016/j.cell.2013.04.005.

Locatelli, S., Piatti, P., Motto, M., & Rossi, V. (2009). Chromatin and DNA modifications in the Opaque2-mediated regulation of gene transcription during maize endosperm development. *The Plant Cell, 21*, 1410–1427. https://doi.org/10.1105/tpc.109.067256.

Mao, H., Wang, H., Liu, S., Li, Z., Yang, X., Yan, J., Li, J., Tran, L. S., & Qin, F. (2015). A transposable element in a NAC gene is associated with drought tolerance in maize seedlings. *Nature Communications, 6*, 8326. https://doi.org/10.1038/ncomms9326.

Mastrangelo, A. M., Marone, D., Laidò, G., De Leonardis, A. M., & De Vita, P. (2012). Alternative splicing: Enhancing ability to cope with stress via transcriptome plasticity. *Plant Science, 185-186*, 40–49. https://doi.org/10.1016/j.plantsci.2011.09.006.

Medvedeva, Y. A., Khamis, A. M., Kulakovskiy, I. V., Ba-Alawi, W., Bhuyan, M. S. I., Kawaji, H., Lassmann, T., Harbers, M., Forrest, A. R. R., & Bajic, V. B. (2014). Effects of cytosine methylation on transcription factor binding sites. *BMC Genomics, 15*, 119. https://doi.org/10.1186/1471-2164-15-119.

Mickelbart, M. V., Hasegawa, P. M., & Bailey-Serres, J. (2015). Genetic mechanisms of abiotic stress tolerance that translate to crop yield stability. *Nature, 16*, 237–251. https://doi.org/10.1038/nrg3901.

Misra, A. K. (2014). Climate change and challenges of water and food security. I. *International Journal of Sustainable Built Environment, 3*, 153–165. https://doi.org/10.1016/j.ijsbe.2014.04.006.

Moldovan, D., Spriggs, A., Yang, J., Pogson, B. J., Dennis, E. S., & Wilson, I. W. (2010). Hypoxia-responsive microRNAs and trans-acting small interfering RNAs in Arabidopsis. *Journal of Experimental Botany, 61*, 165–177. https://doi.org/10.1093/jxb/erp296.

Nakashima, K., Yamaguchi-Shinozaki, K., & Shinozaki, K. (2014). The transcriptional regulatory network in the drought response and its crosstalk in abiotic stress responses including drought, cold, and heat. *Frontiers in Plant Science, 5*, 170. https://doi.org/10.3389/fpls.2014.00170.

Nishiuchi, S., Yamauchi, T., Takahashi, H., Kotula, L., & Nakazono, M. (2012). Mechanisms for coping with submergence and waterlogging in rice. *Rice, 5*, 2. https://doi.org/10.1186/1939-8433-5-2.

Nogoy, F. M., Song, J. Y., Ouk, S., Rahimi, S., Kwon, S. W., Kang, K. K., & Cho, Y. G. (2016). Current applicable DNA markers for marker assisted breeding in abiotic and biotic stress tolerance in rice (Oryza sativa L.). *Plant Breeding and Biotechnology, 4*, 271–284. https://doi.org/10.9787/PBB.2016.4.3.271.

Oh, S., Song, S. I., Kim, Y. S., Jang, H., Kim, S. Y., Kim, M., Kim, Y., Nahm, B. H., Kim, J., Bioscience, D., & KS, O. (2005). Arabidopsis CBF3/DREB1A and ABF3 in transgenic rice increased tolerance to abiotic stress without stunting growth. *Plant Physiology, 138*, 341–351. https://doi.org/10.1104/pp.104.059147.

Osakabe, Y., Osakabe, K., Shinozaki, K., & Tran, L.-S. P. (2014). Response of plants to water stress. *Frontiers in Plant Science, 5*, 86. https://doi.org/10.3389/fpls.2014.00086.

Peng, H., & Zhang, J. (2009). Plant genomic DNA methylation in response to stresses: Potential applications and challenges in plant breeding. *Progress in Natural Science, 19*, 1037–1045. https://doi.org/10.1016/j.pnsc.2008.10.014.

Remy, E., Cabrito, T. R., Baster, P., Batista, R. A., Teixeira, M. C., Friml, J., Sá-Correia, I., & Duque, P. (2013). A major facilitator superfamily transporter plays a dual role in polar auxin transport and drought stress tolerance in Arabidopsis. *Plant Cell, 25*, 901–926. https://doi.org/10.1105/tpc.113.110353.

Roy, S. (2015). Function of MYB domain transcription factors in abiotic stress and epigenetic control of stress response in plant genome. *Plant Signaling & Behavior, 11*, e1117723. https://doi.org/10.1080/15592324.2015.

Sakamoto, H., Nakagawara, Y., & Oguri, S. (2013). The expression of a novel gene encoding an ankyrin-repeat protein, DRA1, is regulated by drought-responsive alternative splicing. *International Journal of Biological, Biomolecular, Agricultural, Food and Biotechnological Engineering, 7*, 1120–1123.

Setter, T. L., & Laureles, E. V. (1996). The beneficial effect of reduced elongation growth on submergence tolerance of rice. *Journal of Experimental Botany, 47*, 1551–1559. https://doi.org/10.1093/jxb/47.10.1551.

Shamsudin, N. A. A., Swamy, B. P. M., Ratnam, W., Cruz, M. T. S., Sandhu, N., Raman, A. K., & Kumar, A. (2016). Pyramiding of drought yield QTLs into a high quality Malaysian rice cultivar MRQ74 improves yield under reproductive stage drought. *Rice, 9*, 21. https://doi.org/10.1186/s12284-016-0093-6.

Shao, H., Wang, H., & Tang, X. (2015). NAC transcription factors in plant multiple abiotic stress responses: Progress and prospects. *Frontiers in Plant Science, 6*, 902. https://doi.org/10.3389/fpls.2015.00902.

Shao, H.-B., Chu, L.-Y., Jaleel, C. A., Manivannan, P., Panneerselvam, R., & Shao, M.-A. (2009). Understanding water deficit stress-induced changes in the basic metabolism of higher plants - Biotechnologically and sustainably improving agriculture and the ecoenvironment in arid regions of the globe. *Critical Reviews in Biotechnology, 29*, 131–151. https://doi.org/10.1080/07388550902869792.

Shiono, K., Ogawa, S., Yamazaki, S., Isoda, H., Fujimura, T., Nakazono, M., & Colmer, T. D. (2011). Contrasting dynamics of radial O_2-loss barrier induction and aerenchyma formation in rice roots of two lengths. *Annals of Botany, 107*, 89–99. https://doi.org/10.1093/aob/mcq221.

Salvi, S., Corneti, S., Bellotti, M., Carraro, N., Sanguineti, M. C., Castelletti, S., & Tuberosa, R. (2011). Genetic dissection of maize phenology using an intraspecific introgression library. *BMC Plant Biology, 11*, 4. https://doi.org/10.1186/1471-2229-11-4.

Singh, N., Dang, T. T. M., Vergara, G. V., Mani, D., Sanchez, D., Endang, C. N. N., Merlyn, M. S., Mae, E., Ismail, T. A. M., Mackill, D. J., & Heuer, S. (2010). Molecular marker survey and expression analyses of the rice submergence-tolerance gene SUB1A. *Theoretical and Applied Genetics, 8*, 1441–1453. https://doi.org/10.1007/s00122-010-1400-z.

Sornaraj, P., Luang, S., Lopato, S., & Hrmova, M. (2015). Basic leucine zipper (bZIP) transcription factors involved in abiotic stresses: A molecular model of a wheat bZIP factor and implications of its structure in function. *Biochimica et Biophysica Acta, 1860*, 46–56. https://doi.org/10.1016/j.bbagen.2015.10.014.

Thatcher, S. R., Danilevskaya, O. N., Meng, X., Beatty, M., Zastrow-Hayes, G., Harris, C., Van Allen, B., Habben, J., & Li, B. (2016). Genome-wide analysis of alternative splicing during development and drought stress in maize. *Plant Physiology, 170*, 586–599. https://doi.org/10.1104/pp.15.01267.

To, T. K., & Kim, J. M. (2014). Epigenetic regulation of gene responsiveness in Arabidopsis. *Frontiers in Plant Science, 4*, 548. https://doi.org/10.3389/fpls.2013.00548.

Tsuji, H., Saika, H., Tsutsumi, N., Hira, A., & Nakazono, M. (2006). Dynamic and reversible changes in histone H3-Lys4 methylation and H3 acetylation occurring at submergence-inducible genes in rice. *Plant and Cell Physiology, 47*, 995–1003. https://doi.org/10.1093/pcp/pcj072.

Valliyodan, B., Van Toai, T. T., Alves, J. D., & Goulart, P. D. F. P. (2014). Expression of root-related transcription factors associated with flooding tolerance of soybean (*Glycine max*). *International Journal of Molecular Sciences, 15*, 17622–17643. https://doi.org/10.3390/ijms151017622.

Verdin, E., & Ott, M. (2015). 50 years of protein acetylation: From gene regulation to epigenetics, metabolism and beyond. *Nature Reviews Molecular Cell Biology, 16*, 258–264. https://doi.org/10.1038/nrm3931.

Voesenek, L. A., & Bailey-serres, J. (2015). Flood adaptive traits and processes: An overview. *New Phytologist, 206*, 57–73. https://doi.org/10.1111/nph.13209.

Wang, H., Wang, H., Shao, H., & Tang, X. (2016). Recent advances in utilizing transcription factors to improve plant abiotic stress tolerance by transgenic technology. *Frontiers in Plant Science, 7*, 67. https://doi.org/10.3389/fpls.2016.00067.

Wei, A., He, C., Li, B., Li, N., & Zhang, J. (2011). The pyramid of transgenes TsVP and BetA effectively enhances the drought tolerance of maize plants. *Plant Biotechnology Journal, 9*, 216–229. https://doi.org/10.1111/j.1467-7652.2010.00548.x.

Xu, K., Xu, X., Fukao, T., Canlas, P., Maghirang-rodriguez, R., Heuer, S., Ismail, A. M., Bailey-Serres, J., Ronald, P. C., & Mackill, D. J. (2006). Sub1A is an ethylene-response-factor-like gene that confers submergence tolerance to rice. *Nature, 442*, 705–708. https://doi.org/10.1038/nature04920.

Yukiyoshi, K., & Karahara, I. (2014). Role of ethylene signalling in the formation of constitutive aerenchyma in primary roots of rice. *AoB Plants, 6*, plu043. https://doi.org/10.1093/aobpla/plu043.

Zhang, Z., Wei, L., Zou, X., Tao, Y., Liu, Z., & Zheng, Y. (2008). Submergence-responsive MicroRNAs are potentially involved in the regulation of morphological and metabolic adaptations in maize root cells. *Annals of Botany, 102*, 509–519. https://doi.org/10.1093/aob/mcn129.

Zhang, B. (2015). MicroRNA: A new target for improving plant tolerance to abiotic stress. *Journal of Experimental Botany, 66*(7), 1749–1761. https://doi.org/10.1093/jxb/erv013.

Zhang, B., & Wang, Q. (2015). MicroRNA-based biotechnology for plant improvement. *Journal of Cellular Physiology, 230*, 1–15. https://doi.org/10.1002/jcp.24685.

Zheng, X., Chen, L., Li, M., Lou, Q., Xia, H., Wang, P., Li, T., Liu, H., & Luo, L. (2013). Transgenerational variations in DNA methylation induced by drought stress in two rice varieties with distinguished difference to drought resistance. *PLoS One, 8*(11), e80253. https://doi.org/10.1371/journal.pone.0080253.

Chapter 4
Adaption to Climate Change: Climate Adaptive Breeding of Maize, Wheat and Rice

Dave Watson

Abstract The advent of climate change, especially the greater frequency of temperature extremes and both erratic and extreme precipitation, will increasingly challenge the ability of maize, wheat and rice agri-food systems to meet growing global demand for food and feed. The challenge to agri-food systems is threefold. First, increasing temperature (especially extreme temperature events) and reduced or erratic rainfall limits the ability of these crops to produce a harvestable product, especially in rain-fed and underdeveloped agri-food systems. Second, industrial agri-food systems are extremely energy intensive and, both directly and indirectly, produce a considerable amount of greenhouse gases, which further exacerbates climate change. Third, changes in temperature and rainfall will require adjustment to cropping geographies and integration of more drought and heat tolerant crops, especially in the predicted climate change hotspots. The good news is that there is significant genetic variation for heat and drought/submergence tolerance in the global maize, wheat and rice gene banks. Either through the application of conventional breeding or the use of new breeding techniques, this genetic diversity offers a way to ameliorate most of the immediate climate change challenges. However, in order to ensure a continuous pipeline of climate-resilient staple crops, it is essential to maintain adequate funding for blue sky and upstream crop breeding, and capacity building of small to medium-sized (SME) seed companies, especially in sub-Saharan Africa, South Asia and Central America.

4.1 Introduction

Feeding the global population will still predominantly rely on the production of the three staple cereal crops, maize, wheat and rice (Shiferaw et al. 2011; Ignaciuk and Mason-D'Croz 2014). However, the advent of climate change, especially the greater frequency of temperature extremes and erratic and extreme precipitation, will

---

D. Watson (✉)
CGIAR Research Program on Maize, International Maize and Wheat Improvement Center (CIMMYT), Texcoco, México

© Springer Nature Switzerland AG 2019   67
A. Sarkar et al. (eds.), *Sustainable Solutions for Food Security*,
https://doi.org/10.1007/978-3-319-77878-5_4

increasingly challenge the ability of maize, wheat and rice agri-food systems to meet the growing global demand. Indeed, the expected greater frequency of temperature extremes (IPCC 2007) will expose these crops to temperatures so far not experienced; especially in strategic agricultural production areas in developing countries (Deryng et al. 2014). For example, Africa is warming faster than the global average (Collier et al. 2008) and by the end of this century, growing season temperatures are predicted to exceed the most extreme seasonal temperatures recorded in the past century (Battisti and Naylor 2009). This chapter examines breeding approaches taken in maize, wheat and rice that aim to mitigate the effects of climate change. It is broken down into five sections: (1) Climate change adaptation through breeding for increased tolerance to drought. (2) Climate change adaptation through breeding for increased tolerance to heat. (3) Climate change adaptation through breeding for increased tolerance to flooding and salinity. (4) Breeding for increased nutrient use efficiency. (5) Breeding for increased carbon dioxide use efficiency.

4.2 Climate Change Mitigation Through Breeding for Increased Tolerance to Drought

4.2.1 Maize

Although maize originated in the tropics, it is particularly sensitive to water stress, especially around silking/female flower emergence (Boyer and Westgate 2004; Lobell et al. 2011a, Lobell et al. 2014; Frey et al. 2015). Water stress during this critical period results in barren ears or reduced kernel numbers per ear (Bolaños and Edmeades 1996). Yield loss due to water stress during this critical period ranges from 30% to 60% (Li 2002; Shaw 1977) (see Fig. 4.1). In many regions, agriculture is predominantly rain-fed and smallholder in nature (AGRA 2014). Historically, in rain-fed areas, water stress is one of the most important yield limiting factors. For example, in SSA, approximately 25% of the maize crop is exposed to frequent

Fig. 4.1 Drought susceptible maize variety devastated by drought in Mutoko district, Zimbabwe. Photo: Peter Lowe/CIMMYT

drought, which results in yield loss of up to 50% (Tesfaye et al. 2016). In china, water stress can result in annual yield reductions of between 20% and 50% (Hu et al. 2004). The development of maize varieties with increased tolerance to drought stress is deemed essential for adaptation to climate change (Easterling et al. 2007; Fedoroff et al. 2010; Hellin et al. 2012; Edmeades et al. 2006; Xoconostle-Cázares et al. 2010).

The good news is that there has been significant investment over the past 50 years in breeding for drought tolerance in maize. This has been undertaken for two distinct reasons. First, due both to climate change and the unpredictability of rainfall in many geographies, drought became recognised as the principal abiotic stress in maize production systems. Second, due to its yield and versatility, over the past 50 years, maize has slowly made inroads into traditional sorghum and even millet producing areas, which struggle with even lower annual rainfall amounts and unpredictability of rainfall.

Internationally, as far as the corporate private sector is concerned, breeding has traditionally focused on the development of high yielding maize varieties targeted towards high potential environments, namely, irrigated and medium to high potential rain-fed production systems where water scarcity has not been an issue. Public sector maize breeding dedicated to developing high yielding germplasm with drought tolerance was initiated by the International Maize and Wheat Improvement Center (CIMMYT) in Mexico in the early 1970s (Ashraf 2010). During this time, CIMMYT screened maize germplasm across a range of rainfall and temperature gradient environments (Edmeades et al. 1997). By 1985, CIMMYT had established a maize breeding program for Eastern and Southern Africa based in Zimbabwe, and, by the late 1990s, had established a breeding network across 11 countries (Bänziger and Diallo 2004). In 1997, CIMMYT initiated a product-orientated maize breeding programme in Southern Africa aimed at increasing maize yields in drought prone and low input areas (Bänziger and Diallo 2001). Breeding for drought resistance has been achieved through a mix of approaches, including conventional, molecular and transgenic approaches (Bänziger et al. 2000). It is also recognised that, due to increasing climate change, geographic information systems (GIS) and crop modelling will play a greater role in targeting breeding programs based on accurate identification of future stresses and hotspots of vulnerability. The prediction of future climates will allow the identification of current spatial analogues of these climates (Burke et al. 2009). These analogue sites can then be used as phenotyping screening sites that represent future environments (Cairns et al. 2013).

Fortuitously, the global gene bank for maize, managed by CIMMYT, provided an excellent source of genetic variation for drought (Cairns et al. 2013). Whilst breeding for drought tolerance has proven difficult, this genetic variation, combined with the application of new genomics tools, has improved both the accuracy and speed of drought tolerant maize breeding and provides hope for meeting the challenge of climate change (Cairns et al. 2012). Some of the key new genomics tools include (1) genomic Selection (GS), (2) genome-wide association analysis (GWAA), (3) genome-wide selection (GWS), (4) recurrent selection (Bolaños and Edmeades 1993a, 1993b; Edmeades et al. 1999; Singh et al. 2016a), (5) marker-assisted breeding (MAB), (6) marker-assisted recurrent selection (MARS) (Gong et al. 2015; Pray et al. 2011), (7) doubled haploid (DH) (Prasanna et al. 2012) and (8) high-throughput precision phenotyping (Cooper et al. 2014).

In the case of drought, transgenic approaches have also been utilised to enhance genetic variation (Pray et al. 2011; Cairns et al. 2013). One such example is the Water Efficient Maize for Africa (WEMA) Project, which is a public–private collaboration between CIMMYT, Monsanto, and the Bill and Melinda Gates Foundation. This project has developed genetically modified maize hybrids with enhanced drought tolerance and Bt-based insect pest resistance, which will be shared, on a royalty-free basis, with seed companies in Kenya, Uganda, Tanzania and Mozambique (Fischer et al. 2014). More recently, work is being initiated in two new areas of research. First, CIMMYT is exploring the role of epigenetics in conditioning the progeny maize varieties to drought. Second, CIMMYT, in collaboration with DuPont Pioneer, is exploring the capacity of gene editing, mainly CRISPR-Cas9, to address key constraints to maize production such drought and heat (Huang et al. 2012; Gong et al. 2014; Yin et al. 2014; Hu et al. 2015a, 2015b).

Substantial gains have been achieved in maize yields (Fischer and Edmeades 2010) (see Fig. 4.2). Indeed, under drought conditions, CIMMYT maize hybrids yields are 40% higher when compared to commercially available hybrids checks (Cairns et al. 2013); Tesfaye et al. 2016). The Drought Tolerant Maize for Africa (DTMA) project has been the most prominent and successful investments in this area. Funded by the Bill & Melinda gates foundation, the Howard G. Buffett Foundation, the United States Agency for International Development (USAID) and the UK Department for International Development (DIFID), the DTMA Project, was managed by CIMMYT and the International Institute of Tropical Agriculture (IITA). The project worked in 13 countries across SSS, namely, Angola, Benin, Ethiopia, Ghana, Kenya, Malawi, Mali, Mozambique, Nigeria, Tanzania, Uganda, Ethiopia, Zambia and Zimbabwe (Tesfaye et al. 2016). Between 2007 and 2013, the DTMA Project released 160 drought tolerant maize varieties (Fisher et al. 2015), through more than 91 small to medium-sized African seed companies. An estimated 2 million hectares are planted with drought tolerant maize varieties every year, ben-

Fig. 4.2 Drought tolerant maize variety harvested in Zimbabwe. Photo: Peter Lowe/CIMMYT

efiting millions of smallholder households (Tesfaye et al. 2016). These varieties possess specific drought tolerant traits such as shorter anthesis-silking interval (ASI), reduced bareness, reduced evapotranspiration, and functional stay green during grain filling (Bruce et al. 2002; Edmeades 2008). They are also resistant to the major pests and diseases and meet local cooking and milling requirements (Cooper et al. 2013).

Primarily in response to increasing drought intensity in the USA and Western Europe, all TNC seed companies are investing heavily in the application of new breeding techniques (NBTs) (including CRISPR/Cas9 and the more recent CRISPR-Cpf1) to develop high yielding and drought resistant maize, wheat and rice varieties (McMichael 2016). NBTs can bring new products to market in just 5–10 years (McMichael 2016).

4.2.2 Wheat

Climate change and shrinking water resources have made drought the most important abiotic stress affecting the productivity of wheat worldwide (Hu and Xiong 2014). Indeed, in India alone, it is suggested that, by 2030, wheat yields will decline by over 2 million metric tons under irrigated conditions due to the impact of climate change and associated water scarcity (Tehripour et al. 2015). Internationally, public sector wheat breeding dedicated to developing high yielding germplasm with drought tolerance was initiated by the CIMMYT, Mexico, in 1983. The seemingly late engagement of CIMMYT was because wheat is already a pretty drought tolerant crop; with 21% of CIMMYT derived germplasm growing successfully in areas with an annual rainfall of <300 mm and the rest grown in areas with an annual rainfall from 300 to 500 mm. As with maize, drought induces early senescence. As such, stay green (where the flag/major leaf remains green and active under drought conditions) is used as one of the principal characteristics associated with drought tolerance (Singh et al. 2016a). In recent years, attention has increasingly focused on the use of pre-breeding approaches to increase yield and tackle key abiotic stresses such as drought and heat (Cossani and Reynolds 2012; Diab et al. 2012; Zhang et al. 2015). In CIMMYT, over 70,000 accessions from the international wheat gene bank have been screened (Lopes et al. 2014; Pinto et al. 2010). CIMMYT's investment in state-of-the-art remote sensing and precision phenotyping infrastructure in Mexico and Morocco (Araus and Cairns 2014), supported by the international wheat improvement network (IWIN) (Reynolds et al. 2015), enables testing of lines across a range of target environments (Braun et al. 2010). More recently, public and private sector wheat breeding institutions that face the challenge of developing high yielding and stress tolerant wheat varieties in the face of climate change have come together through the global Heat & Drought Wheat Improvement Consortium (HeDWIC).

4.2.3 Rice

Given its close association with water, rice is not intuitively associated with drought. Indeed, in high input irrigated systems, drought has traditionally not been an issue. However, due to a combination of global warming and a shortage of irrigation water, many irrigated paddy lands now lack irrigation water at critical crop growth stages, especially during flowering (Ali et al. 2017). More than 13 million hectares of rice production in subtropical regions of China, Pakistan, India and Southeast Asia are increasingly becoming exposed to physical water scarcity (Ali et al. 2017). However, whilst 75% of rice production may derive from irrigated systems, in which yields have tripled over the past 30 years, approximately 50% of rice production is rain-fed (Pray et al. 2011). Both public and private research programs have traditionally neglected drought-prone, rain-fed areas, and it is in these marginal environments where droughts and floods are frequent and many of Asia's poor live (Pray et al. 2011). To date, due to the heterogeneity of target breeding environments, conventional breeding efforts for drought stress have met with limited success (Ali et al. 2017; Pray et al. 2011). Indeed, according to Pray et al. (2011), emerging public sector-derived drought tolerant rice varieties have only yielded 5–10% better than the commercial varieties currently grown. However, since 1985, the International Rice Research Institute (IRRI), funded by the Rockefeller Foundation's International Program on Rice Biotechnology, began investing in biotechnology approaches to rice breeding. This investment quickly led to delineation of the rice genome, development of biotechnology tools and a much better understanding of the molecular basis of rice resistance to biotic and abiotic stresses and the opportunities for genetic engineering of rice (Pray et al. 2011). Significant variability exists with regard to water productivity in the rice gene pool. For example, tropical *japonica* has 25–30% higher water productivity than for *indica* rice (Peng et al. 1998). This presents an opportunity to breed new rice varieties for aerobic rice culture, which would perform even better under the Alternate Wet and Dry (AWD) system (Yadvinder-Singh et al. 2014). Recently, through the application of molecular breeding, increased water productivity in rice has been achieved through the development of early maturing yet high-yielding varieties/hybrids, which reduce the time during which the rice crop transpires water (Tuong 1999). Economists predict high rates of return on investments in drought tolerant rice research and adoption of these varieties (Kostandini et al. 2009). Kostandini et al. (2009) argue that transgenic drought tolerant rice research would generate sizeable profits for technology producers (both private and public), consumers and the adoption of such cultivars would generate major benefits to farmers (Ramasamy et al. 2007; Pray et al. 2011). Ultimately, against the backdrop of climate change, increasing physical water scarcity, and the growing need to increase yield per unit of land and water, the International Rice Research Institute under the RICE CGIAR Research Programme has now prioritised research on drought, submergence, salinity, extreme heat, and the emergence and spread of new pests and diseases (RICE 2016).

4.3 Climate Change Mitigation Through Breeding for Increased Tolerance to Heat

Climate change models clearly demonstrate increased rates of crop failure due to increasing extremes of both heat and water stress. Whilst the need for investment in infrastructure and other socio-economic measures is acknowledged, significant mitigation of the effects of climate change, can be achieved through enhanced drought and heat tolerance in maize, wheat and rice (Challinor et al. 2010; Hossain and Teixeira da Silva 2013; Cairns et al. 2013; Masuka et al. 2017).

4.3.1 Maize

In the case of maize, global warming will progressively see this crop moving further northwards and may extend the season in areas already sown with maize due to earlier sowing (Fischer et al. 2014). Whilst maize may benefit from global warming in cooler parts of Africa, where temperatures regularly exceed 30 °C, yields will tend to suffer, especially from higher day time temperatures (Lobell et al. 2011). Above optimal temperature reduces the length of time maize can intercept radiation and shortens the grain-filling period (Badu-Apraku et al. 1983). In Southern Africa, yield losses of maize under drought stress doubled when temperatures were above 30 °C (Lobell et al. 2011). This is because the average optimum temperatures for temperate, highland tropical and lowland tropical maize lie between 20 °C and 30 °C, 17 °C and 20 °C, and 30 °C and 34 °C, respectively (Badu-Apraku et al. 1983; Brown 1977; Dale 1983; Shaw 1983). In SSA, maximum temperatures currently exceed optimal temperature conditions for lowland tropical maize (34 °C) in Burkina Faso, Chad, Eritrea, Gambia, Mali, Mauritania, Niger, Nigeria, Senegal and Sudan (Cairns et al. 2013). Roughly 65% of present maize-growing areas in Africa would experience yield losses for 1 °C of warming under optimal rain-fed management, with 100% of areas harmed by warming under drought conditions (Lobell et al. 2011). Indeed, by 2050, air temperatures are expected to increase throughout maize mega-environments within sub-Saharan Africa by an average of 2.1 °C (Cairns et al. 2013). In contrast to drought stress, the male reproductive tissue is more sensitive to heat stress than the female reproductive tissue (Dupuis and Dumas 1990) as pollen production is located above the crop canopy where temperatures are highest (Cairns et al. 2012). Late emergence of female silks further reduces the chance of fertilisation (Cicchino et al. 2011).

However, whilst repeatedly called for (Cairns et al. 2013; Deryng et al. 2014; Lobell et al. 2011), explicit breeding for heat tolerance in maize has been a relative recent phenomenon. Work began with initiation of the "Heat Tolerant Maize for Asia" (HTMA) Project—a collaboration between CIMMYT, Purdue University, DuPont Pioneer Hi-Bred, seed companies and South Asian public sector maize programs. Indeed, given the expected convergence of drought and increased temperatures, CIMMYT's pipeline germplasm is selected for both tolerances to drought and extreme temperatures (Cairns et al. 2013). It has become clear that farmers will increasingly require maize varieties with both drought and heat tolerance; especially in climate change hot spots such Southern Africa (Cairns et al. 2013). Whilst there does appear to be significant genetic variability for tolerance to heat stress in sub-tropical and tropical maize germplasm, which could be exploited through conventional breeding (Cairns et al. 2013), research is still required to better understand the interaction of high temperature on maize growth and development (Cairns et al. 2013). In many respects, breeding for heat has been able to piggy-back on the earlier work on drought. Whilst the two are not linked genetically, which causes problems from a breeding perspective, fortuitously, much of the drought tolerant maize developed by CIMMYT and IITA is relatively tolerant to higher temperatures (Cairns et al. 2013). Indeed, breeding for heat tolerance utilises the same genomic and pre-breeding tools that have benefited the breeding work on drought. Ultimately, according to Cairns et al. (2012), if investments can be maintained, the development of drought and heat tolerant maize varieties could match or even exceed the rainfall and temperature related challenges imposed by climate change. According to Kostandini et al. (2013), the rapid development, dissemination, and adoption of drought and heat tolerant maize could generate between US$362 million and US$590 million over a 7-year period, through both yield gains and an increase in yield stability.

4.3.2 Wheat

In the case of wheat, increased temperature is a double edged sword. It is predicted that wheat yields in many parts of the northern hemisphere, such as Canada, Russia, Kazakhstan and Northern USA, will increase in response to global warming between 10% and 15% (OECD-FAO 2009). This yield increase is principally linked to the replacement of lower yielding spring wheats by winter wheats that would, in theory, spread to more northern latitudes due to milder winters (Fischer et al. 2014). At even higher latitudes, increasing temperatures would allow early sowing of spring wheat and the northward expansion of spring wheat into new areas (Fischer et al. 2014). However, whilst some countries in the north may benefit from global warming, some countries, such as France, have already reported yield reductions due to increasing temperatures (Hossain and Teixeira da Silva 2013). Ultimately, potential gains in the northern hemisphere will be overshadowed by estimated losses of between 20% and 30% by 2050 in developing countries; where approximately two-thirds of all wheat is produced (Easterling et al. 2007; Lobell et al. 2008; Rosegrant and Agcaoili 2010). In South Asia's Indo-Gangetic Plain, where nearly 15% of the global wheat production originates, global warming is likely to reduce the wheat production area by over 50%; threatening the food security of 200 million people (CGIAR 2009). In a similar way to maize, temperatures above 34 °C leads to

4 Adaption to Climate Change: Climate Adaptive Breeding of Maize, Wheat and Rice

accelerated senescence, poor grain filling, and, sometimes, complete sterility (Fischer et al. 2014; Singh et al. 2016a, 2016b). Due to the fact that wheat is often grown in high temperature environments, CIMMYT and major national partners have been breeding for heat tolerance since the late 1980s (Lillemo et al. 2005; Gourdji et al. 2013).

4.3.3 Rice

Rice is also extremely susceptible to heat stress throughout most of its growth stages (Jagadish et al. 2010). Even brief exposure to temperatures above ~35 °C around flowering can significantly reduce rice fertility. According to Jagadish et al. (2010), rice fertility in a range of genotypes was reduced between 18% and 71% after just 6 h exposure to a temperature of 38 °C.

4.4 Climate Change Mitigation Through Breeding for Increased Tolerance to Flooding and Salinity

Unlike traditional varieties of maize and wheat, rice is capable of growing well in waterlogged and submerged soils due to specialised root anatomy (Setter et al. 1997; Jackson and Ram 2003). Rice is the only cereal capable of germination under submerged conditions but requires aerobic conditions to develop into healthy plants (Yamauchi et al. 1993; Ismail et al. 2009; Angaji et al. 2010; Biswas and Yamauchi 1997). When mature, depending on the type of flooding, rice can survive up to a couple of months in waterlogged conditions. However, as a consequence of global warming, sea level rise and the frequency and intensity of flooding in many rice growing areas has increased during the last decade (Hossain and Teixeira da Silva 2013; Coumou and Rahmstorf 2012). In several countries in South Asia and South East Asia, this has caused considerable losses in grain production (Singh et al. 2009, 2011; Mackill et al. 2012; Ismail et al. 2013; Colmer et al. 2014). Luckily, rice varieties and landraces possess significant genetic diversity in relation to submergence and waterlogging and opportunities exist to develop high-yielding varieties with varying degrees of tolerance to submergence and waterlogging (Ismail and Mackill 2013; Kirk et al. 2014). In many lowland rain-fed areas, salinity and submergence often occur during the early stages of the rice crop lifecycle, whilst drought and salinity arrive in the later stages of the rice crop lifecycle (Asaduzzaman et al. 2010). When compounded by severe droughts in intensive rain-fed lowlands and water shortages in irrigated fields, rice breeders are increasingly faced with the need to develop varieties with multiple abiotic stress tolerance (Ali et al. 2017). Marker-assisted backcross breeding has been successfully utilised in rice for the transfer of single major effect QTLs. Examples include the identification and deployment of the flooding tolerance gene SUB1 in South and South East Asia, the salinity tolerance gene Saltol, and a number of drought QTLs (Singh et al. 2016b; Ismail et al. 2013).

Maize, on the other hand, is highly susceptible to anaerobic soil conditions during germination, early growth stages (Mano et al. 2002; Zaidi et al. 2004) and reproduction (Zaidi et al. 2007). Maize breeding isn't traditionally associated with submergence and waterlogging. However, the recent expansion of rice–maize systems (Waddington et al. 2006) in parts of Eastern India and South East Asia, reaching approximately 3.5 million hectares (Timsina et al. 2010), coupled with higher levels of rainfall associated with climate change, intermittent water-logging is quickly becoming one of the principal constraints for maize production in tropical Asia. Over 18% of the total maize production area in South and Southeast Asia is affected by temporary floods and waterlogging in early stages of growth, causing annual production losses of 25–30% (Zaidi et al. 2015). Surprisingly, significant genetic variation exists with regard to maize's tolerance of waterlogging (Mano et al. 2002; Rathore et al. 1996, 1997; Zaidi et al. 2002, 2015). This variation includes the ability of some maize, including wild relatives (Anjos e Silva et al. 2007), to access oxygen during flooding through adventitious roots at the soil surface. Other wild relatives form root aerenchyma to allow passage of oxygen from the surface down into the roots and even possess tolerance to the toxic effects of Fe^{2+} H_2S, which increase in concentration under waterlogged conditions (Mano and Omori 2007). Given that the capacity to tolerate waterlogging is likely to be governed by multiple genes, it is likely that marker-assisted selection (MAS) will need to be utilised to make meaningful progress in this area (Zhou 2010). As wheat is far less exposed to submergence and waterlogging, to date, little attention has been given to this aspect of wheat breeding (Hossain and Teixeira da Silva 2013).

4.5 Climate Change Mitigation Through Breeding for Increased Nutrient Use Efficiency

The capacity to feed the almost seven billion inhabitants of our planet would not have been possible without the synthesis of ammonia via the Haber–Bosch process (Smil 1999). However, a by-product of the Haber–Bosch process is the release of huge quantities of CO_2 (approximately 275 million tons per year) due to the combustion of fossil fuels such as natural gas and coal. Approximately 120 teragrams (Tg) of nitrogen are chemically fixed through the Haber–Bosch process every year; of which almost 80% of is destined to be used as agricultural fertilizer (Galloway et al. 2004). Nitrogen is the principal fertiliser produced and is one of the most limiting factors to crop growth (Oldroyd and Dixon 2014). Given the urgent need to produce more food, fuel and fibre, the quantity of nitrogen-based fertiliser produced will continue to increase (Olivares et al. 2013). However, both the production (Dusenbury et al. 2008) and use (Burney et al. 2010; Goglio et al. 2014) of synthetic fertilizers and pesticides in agricultural production increases emissions of carbon and other greenhouse gases. Annual greenhouse gas emissions from agriculture each year are equivalent to as much as 5.8 billion tons of carbon dioxide (CO_2), or 11% of the total emissions from human activity (Subbaraoa et al. 2017). Indeed, according to (Gan et al. 2011), synthetic nitrogen fertilizers contributed to an average 65% of total agricultural emissions. This total comprises 27% from volatilization of NH_3 and NOx and leaching of nitrate into water courses and 38% from the production, transportation, storage, and delivery of nitrogen fertilizers to the farm gate (Liu et al. 2016).

However, the uptake and utilisation of nitrogen fertiliser in maize, wheat and rice production systems is typically poor. Due to nitrate leaching, denitrification of the nitrate ion to nitrous oxide and di-nitrogen gas (Norton and Stark 2011), only about a third of applied nitrogen is actually utilised by crops (Raun and Johnson 1999). Nitrogen fertilizers are a major source of agricultural emissions. About 70% of N applied to crops in fertilizers is either washed away or flushed into the air as nitrous oxide (N_2O), a greenhouse gas 300 times more potent than CO_2 (Subbaraoa et al. 2017). This amounts to around 100 million tonnes lost annually (Galloway et al. 2008). Nutrient excesses are especially large in China, northern India, the USA, and Western Europe, leading to widespread nutrient pollution (Rogers and Oldroyd 2014). Conversely, ammonium N (NH_4+) is less mobile in soil and is not directly subject to denitrification (O'Sullivan et al. 2016). Maintaining fertiliser N in the form of NH_4+, through management of nitrification in soil, is a long-held strategy to improve N use efficiency in agricultural systems (O'Sullivan et al. 2016). Indeed, improving nitrogen use efficiency has the capacity both to make significant energy, and therefore greenhouse gas emission, savings in the production of nitrogen fertiliser and to lower the cost of crop production. Indeed, Glendining et al. (2009) suggests that, by 2050, 2% of global energy will need to be used to produce nitrogen fertilisers. Biological nitrification inhibition (BNI) is one such approach to increasing nitrogen use efficiency. BNI is the ability to decreases nitrification rates by inhibiting microbial ammonia oxidation (O'Sullivan et al. 2016). BNI has been found in traditional wheat varieties (Subbarao et al. 2007b) and rice (Subbarao et al. 2007a; Tanaka et al. 2010; Zakir et al. 2008) and, if tested across a wide genetic spectrum, may also be present in maize. If transferred into high yielding varieties, the production of nitrification inhibitors by crop plants would increase the utilisation of applied nitrogen fertiliser and reduce the need for nitrification inhibitors to be added to synthetic fertilisers. According to Subbaraoa et al. (2017), in the case of wheat, if the work on BNI is successful, farmers would be able to reduce their expenditure on nitrogen fertilisers and reduce nitrogen and NO_2 emissions by up to 30%.

Other recent efforts to improve the nutrient use efficiency of crops include CIMMYT's Improved Maize for African Soils (IMAS) Project. Using convention-breeding approaches, IMAS aims to develop maize varieties that possess a 20% yield advantage over commercial checks. Conversely, maize varieties developed using biotechnological and transgenic approaches are expected to yield 30–50% more when compared to current varieties grown on nitrogen depleted soils or in low fertiliser input regimes.

Whilst still a bit of a holy grail, the race is on to develop maize, wheat and rice varieties able to biologically fix their own nitrogen. This could basically be achieved one of two ways. First, maize, wheat and rice could be engineered to fix atmospheric nitrogen in the same way that legumes, such as peas and beans, fix nitrogen. Second, plant/microbe associations could be better exploited where they naturally exist or developed where they currently do not exist for maize, wheat and rice.

Currently, legumes are the only plant family that possess the capability to fix atmospheric nitrogen. They are able to accomplish this through a symbiotic association with nitrogen-fixing bacteria, which live in nodules attached to the plant roots. This symbiotic relationship is able to fix upwards of 120 kg ha^{-1} (Salvagiotti et al. 2008). Genetically engineering nitrogen fixation in cereals has been a dream since the 1970s. Due to major advances in our understanding of nitrogen fixation, scientists have crept a little closer to realising this dream (Olivares et al. 2013; Rogers and Oldroyd 2014). The formation of root tumors (paranodules), containing nitrogen-fixing organisms, such as azospirilla (Zeman et al. 1992), rhizobia (Christiansen-Weniger 1996), cyanobacteria (Gantar and Elhai 1999) or enterobacteria (Iniguez et al. 2004) in cereal crops, has already been reported. However, reports suggest that there is still much to be done in order to create nitrogen-fixing cereals (Olivares et al. 2013). Indeed, according to Oldroyd and Dixon (2014), this may take another 20 years to achieve. Ideally, if the rates of nitrogen fixation currently seen in legumes could be achieved, cereals could potentially fix the equivalent of up to 50 kg ha^{-1} of nitrogen (Rogers and Oldroyd 2014). However, there is a concern that the ability to fix nitrogen may come at a price, namely, nitrogen-fixing bacteria would extract photosynthate (food) from the plant and reduce plant productivity (Rogers and Oldroyd 2014). This would be an important consideration in the high input/high output maize, wheat and rice production systems. Conversely, perhaps a more feasible aspiration would be nitrogen-fixing cereals targeted towards low input/low output production systems in the developing world, especially SSA (Rogers and Oldroyd 2014). Whilst extremely challenging scientifically, another approach, which is currently being explored, is to genetically engineer the introduction of the nitrogenase enzyme (responsible for fixing atmospheric nitrogen) directly into organelles within cereal plant cells (Dixon et al. 1997; Temme et al. 2012).

Alternatively, it may be possible to better utilise free-living nitrogen fixing bacteria as bio-fertilisers (Olivares et al. 2013). In nature, crop associations with bacterial diazotrophic endophytes, which enter and live inside the roots and lower shoots, exist. Rice is one such example. Whilst it is still unclear as to the exact contribution made to nitrogen balance of cereal crops, fixation of atmospheric nitrogen has already been demonstrated (Yanni and Dazzo 2010).

4.6 Climate Change Mitigation Through Breeding for Increased Carbon Dioxide Use Efficiency

It is likely that CO_2 will reach somewhere between 450 ppm (Olivares et al. 2013) and 480 ppm by 2050 (Fischer et al. 2014). Whilst this increase adds to the challenges associated with global warming, it could also have a positive side effects of increasing the growth rates of wheat and rice (Zaman-Hussain et al. 2013), whilst reducing water use (Cairns et al. 2013). In turn, in an attempt to produce more wheat and rice output from the same area of land and increase profitability, without the need for additional fertiliser or water, scientists have increasingly focused their attention on increasing the photosynthetic efficiency of crops. Several research teams have focused, explicitly, on introducing a CO_2-concentrating mechanism into wheat and rice. Unlike wheat and rice, which possess a C3 metabolism, maize is able to make extremely efficient use of CO_2 through a C4 metabolism. The C4 metabolism is able to concentrate CO_2 in the leaf, through its unique leaf anatomy, suppress photorespiration (Dai et al. 1993) and enhance photosynthetic efficiency (Sage 2004). Using a transgenic approach, researchers aim to introduce the C4 photosynthetic pathway into wheat and rice, increasing yields by 25−50% (Caemmerer et al. 2012). In the case of rice, C4 photosynthesis research focuses on the development of transgenic lines containing multiple transporter genes. These lines are then crossed with other lines containing the C4 metabolism in order to develop transgenic lines with increased C4 photosynthesis. Genes responsible for C4 leaf anatomy will also be transferred into rice (RICE 2016). However, whilst scientifically feasible, the conversion of wheat and rice plants from C3 to C4 remains an aspirational goal. Even if this transformation can be achieved, there could be a number of downsides, including the following: (1) C4 plants are more sensitive than C3 plants to drought and recover from drought more slowly Driever and Kromdijk (2013). (2) Performance of C4 wheat and rice varieties, as compared to traditional C3 varieties, may be reduced at low temperatures due to photoinhibition—restricting the geographies in which C4 wheat and rice would express a yield advantage (Driever and Kromdijk 2013).

4.7 Discussion

Whilst significant opportunities exist to ameliorate the effects of climate change through adaptive plant breeding, there are three principal challenges. First, in order to ensure a continuous pipeline of climate-resilient staple crops, it is essential to maintain adequate funding for both blue sky (development of C4 wheat and rice varieties, biological nitrogen fixation in cereals, etc.) and upstream crop breeding. Indeed, given the time required to develop conventionally bred drought tolerant varieties (10–15 years), there is an urgent need to increase investments in breeding for drought tolerance (Burke et al. 2009; Thornton et al. 2009). Currently, funding of public sector crop breeding research is heavily skewed towards the commercialisation of final products (varieties), in other words, the delivery end of the breeding pipeline, with only limited resources being invested in upstream breeding that will deliver future climate-resilient varieties. In the case of international public maize breeding, CIMMYT estimates that an additional USD$25,000,000 per annum is required for upstream maize breeding in order to ensure that the maize breeding pipeline is able to keep producing even better climate-resilient maize lines and varieties.

Currently, the efforts of international crop breeding institutions such as CIMMYT and the International Institute of Tropical Agriculture (IITA) result in the release and commercialisation of 70 or more climate-resilient maize varieties per annum through small to medium-sized (SME) seed companies in SSA, South Asia and Latin America. Most of these SME companies do not have the human or financial capital to single-handedly develop new climate-resilient varieties and are almost totally dependent on support from CIMMYT and IITA. It is a similar situation for international public sector wheat and rice breeding programmes. In the private sector case, only the global seed corporations have the capacity to develop climate-resilient varieties of maize, wheat and rice. However, new climate-resilient varieties are predominantly designed for the high potential industrial production systems of the USA, Western Europe, Australia, Brazil, Argentina, etc., and not for the marginal rain-fed systems of developing countries, especially in regions such as SSA and South Asia.

Second, accessibility to new climate-resilient varieties remains a challenge in many developing countries, especially SSA and South Asia. For example, compared to other regions in the world, maize seed systems in SSA emerged more slowly (Tripp and Rohrbach 2001) and the supply of improved germplasm still faces bottlenecks (Fisher et al. 2015). Currently, maize seed sales in West and Eastern Africa constitute only 33% of the total seed requirement and 38% of the total seed requirements in Southern Africa (Abdoulaye et al. 2009). Hybrid turnover is very slow in ESA. On average, varieties are changed every 14 years. This contrasts starkly with the turnover rate for maize varieties in the US Corn Belt where, on average, varieties are changed every 4 years (Masuka et al. 2017).

Continued investments in seed systems is required to ensure that new climate-resilient crop varieties can reach the farmers that desperately need them (Cavatassi et al. 2011; Westengen and Brysting 2014; Fisher et al. 2015). Unfortunately, the small to medium-sized seed companies often lack breeding capacity and rely on national and international research institutes (such as CIMMYT and IITA) for new stress-tolerant and high-yielding maize hybrids. In addition to developing the technical breeding skills of the numerous small to medium-sized private seed companies who serve small-holder farmers in more marginal production environments, investments are also required to develop other skills, such as marketing and business management. This type of capacity building is essential to ensure both the financial viability of these seed companies and their ability to deliver quality seed, at the right price and right time, to smallholder farmers in marginal production environments. Currently, international crop breeding centres, such as CIMMYT, IITA and IRRI work tirelessly to build both the human capital and infrastructure of SME seed companies (Langyintuo et al. 2010; Cairns et al. 2013).

Whilst the expensive IPR-protected crop varieties of global corporate seed companies may fit into the high input agri-food systems of both developed and developing countries, it is unlikely that these expensive and increasingly genetically modified hybrids will easily fit into riskier and marginal crop production environments. The public international crop breeding institutes target these riskier and more marginal environments; providing royalty free pre-commercial lines and varieties of maize, wheat and rice to SME seed companies (often along with capacity strengthening). Indeed, in conjunction with global seed corporations, public international crop breeding institutes are developing new royalty free GM-varieties (see IMAS example) for use in developing countries.

Third, whilst climate-resilient germplasm will be relatively easy to incorporate into high-input/high-out agri-food systems, the integration of climate-resilient germplasm into low-input/low-output smallholder-based agri-food systems will be considerably more challenging. Indeed, in order for smallholder farmers, which make up a large percentage of farmers in SSA, South Asia and Central America, to adopt climate-resilient varieties, significant and sustained support will be required. Support from both donors and developing country governments is required in order to secure profitable access to local, national and regional maize, wheat and rice value-chains.

References

Abdoulaye, T., Sanogo, D., Langyintuo, A., Bamire, S. A., & Olanrewaju, A. (2009). *Assessing the constraints affecting production and deployment of maize seed in DTMA countries of West Africa.* Ibadan: IITA.

AGRA. (2014). *Africa agriculture status report: Climate change and smallholder agriculture in sub-Saharan Africa (No. 2).* Nairobi: Alliance for a Green Revolution in Africa, AGRA.

Ali, J., Xu, J.-L., Gao, Y.-M., Ma, X.-F., Meng, L.-J., Wang, Y., et al. (2017). Harnessing the hidden genetic diversity for improving multiple abiotic stress tolerance in rice (Oryza sativa L.). *PLoS One, 12*(3), e0172515. https://doi.org/10.1371/journal.pone.0172515.

Angaji, S., Septiningsih, E. M., Mackill, D. J., & Ismail, A. M. (2010). QTLs associated with tolerance of anaerobic conditions during germination in rice (Oryzasativa L.). *Euphytica, 172*, 159–168. https://doi.org/10.1007/s10681-009-0014-5.

Anjos e Silva, S. D., Maria, J., Claudia, F. L., Antonio, C. O., & Jose, F. (2007). Inheritance of tolerance to flooded soils in maize. *Crop Breeding and Applied Biotechnology., 7*, 165–172.

Araus, J. L., & Cairns, J. E. (2014). Field high-throughput phenotyping: The new crop breeding frontier. *Trends in Plant Science, 19*, 52–61.

Asaduzzaman M., Ringler C., Thurlow J., & Alam S. (2010). *Investing in crop agriculture in Bangladesh for higher growth and productivity, and adaptation to climate change.* In Bangladesh Food Security Investment Forum, May 26–27, 2010, Dhaka. Retrieved from www.bids.org.bd/ifpri/investing6.pdf.

Ashraf, M. (2010). Inducing drought tolerance in plants: Recent advances. *Biotechnology Advances, 28*, 169–183. https://doi.org/10.1016/j.biotechadv.2009.11.005.

Badu-Apraku, B., Hunter, R. B., & Tollenaar, M. (1983). Effect of temperature during grain filling on whole plant and grain yield in maize (Zea mays L.). *Canadian Journal of Plant Science, 63*, 357–363.

Bänziger, M., Mugo, S., & Edmeades, G. O. (2000). Breeding for drought tolerance in tropical maize: Conventional approaches and challenges to molecular approaches. In J. M. Ribaut & D. Poland (Eds.), *Molecular approaches for the genetic improvement of cereals for stable production in water-limited environments* (pp. 69–72). Mexico D.F.: CIMMYT.

Bänziger, M., & Diallo, A. O. (2001). *Progress in developing drought and stress tolerant maize cultivars in eastern and southern Africa.* In Seventh Eastern and Southern Africa Regional Maize Conference, February 11–15, 2001, pp. 189–194.

Bänziger, M., & Diallo, A. (2004). Progress in developing drought and N stress tolerant maize cultivars for eastern and southern Africa. In D. K. Friesen & A. F. E. Palmer (Eds.), *Integrated approaches to higher maize productivity in the new millennium. Proceedings of the 7th Eastern and Southern Africa Regional Maize Conference, Nariobi, Kenya, 5–11 February, 2002* (pp. 189–194). Nairobi: CIMMYT and KARI.

Battisti, D. S., & Naylor, R. L. (2009). Historical warnings of future food insecurity with unprecedented seasonal heat. *Science, 323*, 240–244.

Biswas, J. K., & Yamauchi, M. (1997). Mechanism of seedling establishment of direct-seeded rice (Oryza sativa L.) under lowland conditions. *Botanical Bulletin Academia Sinica, 38*, 29–32.

Bolaños, J., & Edmeades, G. O. (1993a). Eight cycles of selection for drought tolerance in lowland tropical maize. I. Responses in grain yield, biomass, and radiation utilization. *Field Crops Research, 31*, 233–252.

Bolaños, J., & Edmeades, G. O. (1993b). Eight cycles of selection for drought tolerance in lowland tropical maize. II. Responses in reproductive behaviour. *Field Crops Research, 31*, 253–268.

Bolaños, J., & Edmeades, G. O. (1996). The importance of the anthesis-silking interval inbreeding for drought tolerance in tropical maize. *Field Crops Research, 48*, 65–80.

Boyer, J., & Westgate, M. (2004). Grain yields with limited water. *Journal of Experimental Botany, 55*, 2385–2394. https://doi.org/10.1093/jxb/erh219.

Braun, H., Atlin, G., & Payne, T. (2010). Multi-location testing as a tool to identify plant response to global climate change. In *Climate change and crop production* (pp. 115–138). Wallingford: CABI.

Brown, D. M. (1977). Response of maize to environmental temperatures: A review. In *Agrometeorology of the maize (corn) crop* (Vol. 481, pp. 15–26). Geneva: World Meteorological Organization.

Bruce, W. B., Edmeades, G. O., & Barker, T. C. (2002). Molecular and physiological approaches to maize improvement for drought tolerance. *Journal of Experimental Botany, 53*(366), 13–25.

Burke, M. B., Lobell, D. B., & Guarino, L. (2009). Shifts in African crop climates by 2050, and the implications for crop improvements and genetic resources conservation. *Global Environmental Change, 19*, 317–325.

Burney, J. A., Davis, S. J., & Lobell, D. B. (2010). Greenhouse gas mitigation by agricultural intensification. *Proceedings of the National Academy of Sciences of the United States of America, 107*, 12052–12057. https://doi.org/10.1073/pnas.0914216107.

Cairns, J. E., Sonder, K., Zaidi, P. H., Verhulst, N., Mahuku, G., Babu, R., Nair, S. K., Das, B., Govaerts, B., Vinayan, M. T., Rashid, Z., Noor, J. J., Devi, P., San Vicente, F., & Prasanna, B. M. (2012). Maize production in a changing climate: Impacts, adaptation, and mitigation strategies. *Advances in Agronomy, 114*, 1–57.

Cairns, J. E., Hellin, J., Sonder, K., Araus, J. L., MacRobert, J. F., Thierfelder, C., & Prasanna, B. M. (2013). Adapting maize production to climate change in sub-Saharan Africa. *Food Security, 5*, 345–360. https://doi.org/10.1007/s12571-013-0256-x.

Caemmerer, S. V., Quick, P. W., & Furbank, R. T. (2012). The development of C4 rice: Current progress and future challenges. *Science, 336*, 1671–1672.

Cavatassi, R., Lipper, L., & Narloch, U. (2011). Modern variety adoption and risk management in drought prone areas: Insights from the sorghum farmers of eastern Ethiopia. *Agricultural Economics, 42*, 279–292.

CGIAR (Consultative Group on International Agricultural Research). (2009). *CGIAR & climate change. Global climate change: Can agriculture cope?* Mapping the Menace of Global Climate Change (CGIAR at COP15—Dec 2009). Retrieved November 4, 2012, from http://cgiar.biomirror.cn/pdf/cc_mappingthemenace.pdf.

Challinor, A. J., Simelton, E. S., Fraser, E. D. G., Hemming, D., & Collins, M. (2010). Increased crop failure due to climate change: Assessing adaptation options using models and socio-economic data for wheat in China. *Environmental Research Letters, 5*(3), 034012 (8 pp). https://doi.org/10.1088/1748-9326/5/3/034012.

Christiansen-Weniger, C. (1996). Endophytic establishment of Azorhizobium caulinodans through auxin-induced root tumors on rice (Oryza sativa L.). *Biology and Fertility of Soils, 21*, 293–302.

Cicchino, M., Rattalino Edreria, J. I., Uribelarrea, M., & Otegui, M. E. (2011). Heat stress in field-grown maize: Response of physiological determinants of grain yield. *Crop Science, 50*, 1438–1448.

Collier, P., Conway, G., & Venables, T. (2008). Climate change and Africa. *Oxford Review of Economic Policy, 24*, 337–353.

Colmer, T. D., Armstrong, W., Greenway, H., Ismail, A. M., Kirk, G. J. D., & Atwell, B. J. (2014). Physiological mechanisms of flooding tolerance of rice: Transient complete submergence and prolonged standing water. In U. Lüttge, W. Beyschlag, & J. Cushman (Eds.), *Progress in botany* (Vol. 75). Berlin; Heidelberg: Springer. https://doi.org/10.1007/978-3-642-38797-5_9.

Cooper, M., Messina, C. D., Podlich, D., Radu Totir, L., Baumgarten, A., & Hausmann, N. J. (2014). Predicting the future of plant breeding: Complementing empirical evaluation with genetic prediction. *Crop & Pasture Science, 65*, 311–336. https://doi.org/10.1071/CP14007.

Cooper, P. J. M., Cappiello, S., Vermeulen, S. J., Campbell, B. M., Zougmoré, R., & Kinyangi, J. (2013). *Large-scale implementation of adaptation and mitigation actions in agriculture. CCAFS Working paper, No 50*. Copenhagen: CCAFS.

Coumou, D., & Rahmstorf, S. (2012). A decade of weather extremes. *Nature Climate Change, 2*, 491–496.

Cossani, C. M., & Reynolds, M. P. (2012). Physiological traits for improving heat tolerance in wheat. *Plant Physiology, 160*, 1710–1718.

Dai, Z., Ku, M. S. B., & Edwards, G. E. (1993). C4 photosynthesis: The CO2 concentrating mechanism and photorespiration. *Plant Physiology, 103*(1), 83–90.

Dale, R. F. (1983). Temperature perturbations in the mid-western and southeastern United States important for corn production. In C. D. Raper & P. J. Kramer (Eds.), *Crop reactions to water and temperature stresses in humid and temperate climates* (pp. 21–32). Boulder, CO: Westview Press.

Deryng, D., Conway, D., Ramankutty, N., Price, J., & Warren, R. (2014). Global crop yield response to extreme heat stress under multiple climate change futures. *Environmental Research Letters, 9*(3), 034011 (13 pp).

Diab, A., Amin, A., Badr, S., Teixeira da Silva, J. A., Van, P. T., Abdelgawad, B., Adawy, S., & Sammour, R. (2012). Identification and functional validation of expressed sequence tags (ESTs) preferentially expressed in response to drought stress in durum wheat. *International Journal of Plant Breeding, 6*, 14–20.

Dixon, R., Cheng, Q., Shen, G., Day, A., & Dowson-Day, M. (1997). Nif gene transfer and expression in chloroplasts: Prospects and problems. *Plant and Soil, 194*, 193–203.

Driever, S. M., & Kromdijk, J. (2013). Will C3 crops enhanced with the C4 CO2-concentrating mechanism live up to their full potential (yield)? *Journal of Experimental Botany, 64*(13), 3925–3935. https://doi.org/10.1093/jxb/ert103.

Dupuis, I., & Dumas, C. (1990). Influence of temperature stress on in vitro fertilization and heat shock protein synthesis in maize (Zea mays L.) reproductive tissues. *Plant Physiology, 94*, 665–670.

Dusenbury, M. P., Engel, R. E., Miller, P. R., Lemke, R. L., & Wallander, R. (2008). Nitrous oxide emissions from a northern great plains soil as influenced by nitrogen management and cropping systems. *Journal of Environmental Quality, 37*, 542–550. https://doi.org/10.2134/jeq2006.0395.

Easterling, W., Aggarwal, P., Batima, P., et al. (2007). Food fibre and forest products. In M. L. Parry, O. F. Canziani, J. P. Palutikof, P. J. van Linden, & C. E. Hansen (Eds.), *Impacts, adaptation and vulnerability' contribution of working group II to the fourth assessment report of the intergovernmental panel on climate change* (pp. 273–313). Cambridge: Cambridge University Press.

Edmeades, G. O., Bänziger, M., Cortes, M., & Ortega, A. (1997). From stress-tolerant populations to hybrids: The role of source germplasm. In G. O. Edmeades, M. Bänziger, H. R. Mickelson, & C. B. Peña-Valdivia (Eds.), *Developing drought and low-N tolerant maize* (pp. 263–273). Mexico D.F.: CIMMYT.

Edmeades, G. O., Bolanos, J., Lafitte, H. R., Rajaram, S., Pfeiffer, W., & Fischer, R. A. (1999). Traditional approaches to breeding for drought resistance in cereals. In F. W. G. Baker (Ed.), *Drought resistance in cereals* (pp. 27–52). Wallingford: ICSU and CABI.

Edmeades, G., Bänziger, M., Campos, H., & Schussler, J. (2006). *Improving tolerance to abiotic stresses in staple crops: A random or planned process?* (Vol. 2008, pp. 293–309). London: Blackwell Publishing.

Edmeades, G. (2008). Drought tolerance in maize: An emerging reality. In J. Clive (Ed.), *Global status of commercialized Biotech/GM Crops. ISAAA Brief No. 39*. Ithaca, NY: ISAAA.

Fedoroff, N. V., Battisti, D. S., Beachy, R. N., et al. (2010). Radically rethinking agriculture for the 21st century. *Science, 327*, 833–834.

Fischer, R. A., & Edmeades, G. O. (2010). Breeding and cereal yield progress. *Crop Science, 50*, S85–S98.

Fischer, R. A., Byerlee, D., & Edmeades, G. O. (2014). *Crop yields and global food security: Will yield increase continue to feed the world?* Canberra, ACT: Australian Centre for International Agricultural Research. Retrieved from http://aciar.gov.au/publication/mn158.

Fisher, M., Abate, T., Lunduka, R. W., Asnake, W., Alemayehu, Y., & Madulu, R. B. (2015). Drought tolerant maize for farmer adaptation to drought in sub-Saharan Africa: Determinants of adoption in eastern and southern Africa. *Climatic Change, 133*, 283–299. https://doi.org/10.1007/s10584-015-1459-2.

Frey, F. P., Urbany, C., Hüttel, B., Reinhardt, R., & Stich, B. (2015). Genome-wide expression profiling and phenotypic evaluation of European maize in-breds at seedling stage in response to heat stress. *BMC Genomics, 16*, 123. https://doi.org/10.1186/s12864-015-1282-1.

Galloway, J. N., Dentener, F. J., Capone, D. G., Boyer, E. W., Howarth, R. W., Seitzinger, S. P., Asner, G. P., Cleveland, C. C., Green, P. A., & Holland, E. A. (2004). Nitrogen cycles: Past, present, and future. *Biogeochemistry, 70*, 153–226.

Galloway, J. N., Townsend, A. R., Willem, J., Erisman, J. W., Bekunda, M., Cai, Z., Freney, J. R., Luiz, A., Martinelli, L. A., Seitzinger, S. P., & Sutton, M. A. (2008). Transformation of the nitrogen cycle: Recent trends, questions, and potential solutions. *Science, 320*, 889–892.

Gan, Y., Liang, C., Wang, X., & McConkey, B. (2011). Lowering carbon footprint of durum wheat by diversifying cropping systems. *Field Crops Research, 122*, 199–206. https://doi.org/10.1016/j.fcr.2011.03.020.

Gantar, M., & Elhai, J. (1999). Colonization of wheat para-nodules by the N2-fixing cyanobacterium Nostoc sp. strain 2S9B. *The New Phytologist, 141*, 373–379.

Glendining, M. J., Dailey, A. G., Williams, A. G., van Evert, F. K., Goulding, K. W. T., & Whitmore, A. P. (2009). Is it possible to increase the sustainability of arable and ruminant agriculture by reducing inputs? *Agricultural Systems, 99*, 117–125.

Goglio, P., Grant, B. B., Smith, W. N., Desjardins, R. L., Worth, D. E., Zentner, R., & Malhi, S. S. (2014). Impact of management strategies on the global warming potential at the cropping system level. *Science of the Total Environment, 490*, 921–933. https://doi.org/10.1016/j.scitotenv.2014.05.070.

Gong, F. P., Yang, L., Tai, F. J., Hu, X. L., & Wang, W. (2014). "Omics" of maize stress response for sustainable food production: Opportunities and challenges. *OMICS, 18*, 711–729. https://doi.org/10.1089/omi.2014.0125.

Gong, F. P., Wu, X., Zhang, H., Chen, Y., & Wang, W. (2015). Making better maize plants for sustainable grain production in a changing climate. *Frontiers in Plant Science, 6*, 835. https://doi.org/10.3389/fpls.2015.00835.

Gourdji, S. M., Mathews, K. L., Reynolds, M., Crossa, J., & Lobell, D. B. (2013). An assessment of wheat yield sensitivity and breeding gains in hot environments. *Proceedings of the Royal Society B, 280*, 20122190.

Hellin, J., Shiferaw, B., Cairns, J. E., Reynolds, M., Ortiz-Monasterio, I., Bänziger, M., et al. (2012). Climate change and food security in the developing world: Potential of maize and wheat research to expand options for adaptation and mitigation. *Journal of Development and Agricultural Economics, 4*, 311–321.

Hossain, A., & Teixeira da Silva, J. A. (2013). Wheat production in Bangladesh: Its future in the light of global warming. *AoB Plants, 5*, pls042. https://doi.org/10.1093/aobpla/pls042.

Hu, H., & Xiong, L. (2014). Genetic engineering and breeding of drought resistant crops. *Annual Review of Plant Biology, 65*, 715–741. https://doi.org/10.1146/annurev-arplant-050213-040000. PMID: 24313844.

4 Adaption to Climate Change: Climate Adaptive Breeding of Maize, Wheat and Rice   85

Hu, R. F., Meng, E. C., Zhang, S. H., & Sciences, C. A. O. (2004). Prioritization for maize research and development in China. *Scientia Agricultura Sinica, 37*, 781–787.

Hu, X. L., Wu, L. J., Zhao, F. Y., Zhang, D. Y., Wang, W., & Zhu, G. (2015a). Phosphoproteomic analysis of the response of maize leaves to drought, heat and their combination stress. *Frontiers in Plant Science, 6*, 298. https://doi.org/10.3389/fpls.2015.00298.

Hu, X. L., Yang, Y. F., Gong, F. P., Zhang, D. Y., Wang, W., & Wu, L. (2015b). Proteins HSP26 improves chloroplast performance under heat stress by interacting with specific chloroplast proteins in maize (Zea mays). *Journal of Proteomics, 115*, 81–92. https://doi.org/10.1016/j.jprot.2014.12.009.

Huang, H., Mølle, I. M., & Song, S. Q. (2012). Proteomics of desiccation tolerance during development and germination of maize embryos. *Journal of Proteomics, 75*, 1247–1262. https://doi.org/10.1016/j.jprot.2011.10.036.

Ignaciuk, A., & Mason-D'Croz, D. (2014). Modelling adaptation to climate change in agriculture. In *OECD Food, Agriculture and Fisheries Papers, No. 70*. Paris: OECD Publishing.

Iniguez, A. L., Domg, Y., & Triplett, E. W. (2004). Nitrogen fixation in wheat provided by Klebsiella pneumoniae 342. *Molecular Plant-Microbe Interactions, 17*, 1078–1085.

IPCC. (2007). *Fourth assessment report: Synthesis*. Retrieved November 17, 2007, from http://www.ipcc.ch/pdf/assessment-report/ar4/syr/ar4_syr.pdf.

Ismail, A. M., Ella, E. S., Vergara, G. V., & Mackill, D. J. (2009). Mechanisms associated with tolerance to flooding during germination and early seedling growth in rice (Oryzasativa). *Annals of Botany, 103*, 197–209. https://doi.org/10.1093/aob/mcn211.

Ismail, A. M., Singh, U. S., Singh, S., Dar, M. H., & Mackill, D. J. (2013). The contribution of submergence-tolerant (Sub1) rice varieties to food security in flood-prone rain-fed lowland areas in Asia. *Field Crops Research, 152*, 83–93.

Ismail, A. M., & Mackill, D. J. (2013). Response to flooding: Submergence tolerance in rice. In M. Jackson, B. Ford-Lloyd, & M. Parry (Eds.), *Plant genetic responses and climate change, Chapter 15*. London: CABI International.

Jackson, M. B., & Ram, P. C. (2003). Physiological and molecular basis of susceptibility and tolerance of rice plants to complete submergence. *Annals of Botany, 91*, 227–241. https://doi.org/10.1093/aob/mcf242.

Jagadish, S. V. K., Muthurajan, R., Oane, R., Wheeler, T. R., Heuer, S., Bennett, J., et al. (2010). Physiological and proteonomic approaches to address heat tolerance during anthesis in rice (Oryza sativa L.). *Journal of Experimental Botany, 61*, 143–156.

Kirk, G. J. D., Greenway, B. J., Atwell, B. J., Ismail, A. M., & Colmer, T. D. (2014). Adaptation of rice to flooded soils. In U. Lüttge, W. Beyschlag, & J. Cushman (Eds.), *Progress in botany* (Vol. 75). Berlin; Heidelberg: Springer. https://doi.org/10.1007/978-3-642-38797-5_8.

Kostandini, G., Bradford, F. M., Omamo, S. W., & Wood, S. (2009). Ex-ante analysis of the benefits of transgenic drought tolerance research on cereal crops in low-income countries. *Agricultural Economics, 40*, 477–492.

Kostandini, G., La Rovere, R., & Abdoulaye, T. (2013). Potential impacts of increasing average yields and reducing maize yield variability in Africa. *Food Policy, 43*, 213–226. https://doi.org/10.1016/j.foodpol.2013.09.007.

Langyintuo, A. S., Mwangi, W., Diallo, A., MacRobert, J., Dixon, J., & Bänziger, M. (2010). Challenges of the maize seed industry in eastern and southern Africa: A compelling case for private–public intervention to promote growth. *Food Policy, 35*, 323–331.

Li, X. H. (2002). Genetic diversity of drought tolerance at flowering time in elite maize germplasm. *Acta Agronomica Sinica, 28*, 595–600.

Lillemo, M., van Ginkel, M., Trethowan, R. M., Hernandez, E., & Crossa, J. (2005). Differential adaptation of CIMMYT bread wheat to global high temperature environments. *Crop Science, 45*, 2443–2453.

Liu, C., Cutforth, H., Chai, Q., & Gan, Y. (2016). Farming tactics to reduce the carbon footprint of crop cultivation in semiarid areas. A review. *Agronomy for Sustainable Development, 36*, 69. https://doi.org/10.1007/s13593-016-0404-8.

Lobell, D. B., Burke, M. B., Tebaldi, C., Mastrandrea, M. D., Falcon, W. P., & Naylor, R. L. (2008). Prioritizing climate change adaptation and needs for food security in 2030. *Science, 319*, 607–610.

Lobell, D. B., Bänziger, M., Magorokosho, C., & Vivek, B. (2011). Nonlinear heat effects on African maize as evidenced by historical yield trials. *Nature Climate Change, 1*, 42–45. https://doi.org/10.1038/nclimate1043.

Lobell, D. B., Roberts, M. J., Schlenker, W., Braun, N., & Little, B. B. (2014). Greater sensitivity to drought accompanies maize yield increase in the US Midwest. *Science, 344*, 516–519. https://doi.org/10.1126/science.1251423.

Lopes, M. S., Dreisigacker, S., Peña, R. J., Sukumaran, S., & Reynolds, M. P. (2014). Genetic characterization of the Wheat Association Mapping Initiative (WAMI) panel for dissection of complex traits in spring wheat. *TAG: Theorectial and Applied Genetics, 128*, 453–464.

Mackill, D. J., Ismail, A. M., Singh, U. S., Labios, R. V., & Paris, T. R. (2012). Development and rapid adoption of submergence-tolerant (Sub1)ricev arieties. *Advances in Agronomy, 115*, 303–356. https://doi.org/10.1016/B978-0-12-394276-0.00006-8.

Mano, Y., Muraki, M., Komatsu, T., Fujimori, M., Akiyama, F., & Takamizo, T. (2002). Varietal difference in pre-germination flooding tolerance and waterlogging tolerance at the seedling stage in maize inbred lines. *Crop Science, 71*(3), 361–367.

Mano, Y., & Omori, F. (2007). (2007). Breeding for flooding tolerant maize using "teosinte" as a germplasm resource. *Plant Roots, 1*, 17–21.

Masuka, B., Atlin, G. N., Olsen, M., Magorokosho, C., Labuschagne, M., Crossa, J., Bänziger, M., Pixley, K. V., Vivek, B. S., von Biljon, A., Macrobert, J., Alvarado, G., Prasanna, B. M., Makumbi, D., Tarekegne, A., Das, B., Zaman-Allah, M., & Cairns, J. E. (2017). Gains in maize genetic improvement in eastern and Southern Africa: I. CIMMYT hybrid breeding pipeline. *Crop Science, 57*, 1–12. https://doi.org/10.2135/cropsci2016.05.0343.

McMichael, P. (2016). Commentary: Food regime for thought. *The Journal of Peasant Studies, 43*(3), 648–670. https://doi.org/10.1080/03066150.2016.1143816.

Norton, J. M., & Stark, J. M. (2011). Regulation and measurement of nitrification in terrestrial systems. In M. G. Klotz (Ed.), *Methods in enzymology: Research on nitrification and related processes* (Vol. 486, Part A, pp. 343–368). San Diego, CA: Elsevier Academic Press.

OECD-FAO. (2009). *Agricultural outlook 2009–2018*. Retrieved November 4, 2012, from www.agrioutlook.org.

Oldroyd, G. E. D., & Dixon, R. (2014). Biotechnological solutions to the nitrogen problem. *Current Opinion in Biotechnology, 26*, 19–24.

Olivares, J., Bedmar, E. J., & Sanjuán, J. (2013). Biological nitrogen fixation in the context of global change. *Molecular Plant-Microbe Interactions, 26*(5), 486–494. https://doi.org/10.1094/MPMI-12-12-0293-CR.

O'Sullivan, C. A., Fillery, I. R. P., Roper, M. M., & Richards, R. A. (2016). Identification of several wheat landraces with biological nitrification inhibition capacity. *Plant and Soil, 404*, 61–74. https://doi.org/10.1007/s11104-016-2822-4.

Peng, S., Laza, R. C., Khush, G. S., Sanico, A. L., Visperas, R. M., & Garcia, F. E. (1998). Transpiration efficiencies of indica and improved tropical japonica rice grown under irrigated conditions. *Euphytica, 103*, 103–108.

Pinto, R. S., Reynolds, M. P., Mathews, K. L., McIntyre, C. L., Olivares-Villegas, J. J., & Chapman, S. C. (2010). Heat and drought adaptive QTL in a wheat population designed to minimize confounding agronomic effects. *Theoretical and Applied Genetics, 121*, 1001–1021.

Prasanna, B. M., Chaikam, V., & Mahuku, G. (2012). *Doubled haploid technology in maize breeding: Theory and practice*. Mexico, D.F.: CIMMYT.

Pray, C., Nagarajan, L., Li, L., Huang, J., Hu, R., Selvaraj, K. N., Napasintuwong, O., & Babu, R. C. (2011). Potential impact of biotechnology on adaption of agriculture to climate change: The case of drought tolerant rice breeding in Asia. *Sustainability, 3*, 1723–1741. https://doi.org/10.3390/su3101723.

Ramasamy, C., Selvaraj, K. N., Norton, G. W., & Vijayragahavan, V. K. (2007). *Economic and environmental benefits and costs of transgenic crops: An ex-ante assessment*. Coimbatore: Tamil Nadu Agricultural University Press.

Rathore, T. R., Warsi, M. Z. K., Lothrop, J. E., & Singh, N. N. (1996). *Production of maize under excess soil moisture (water-logging) conditions.* In 1st Asian Regional Maize Workshop, PAU (Punjab Agricultural University), Ludhiana, February 10–12, 1996, pp. 56–63.

Rathore, T. R., Warsi, M. Z. K., Zaidi, P. H., & Singh, N. N. (1997). Waterlogging problem for maize production in Asian region. *TAMNET News Letter., 4*, 13–14.

Raun, W. R., & Johnson, G. V. (1999). Improving nitrogen use efficiency for cereal production. *Agronomy Journal, 91*, 357–363.

Reynolds, M. P., Tattaris, M., Cossani, C. M., Ellis, M., Yamaguchi-Shinozaki, K., & Saint, P. C. (2015). Exploring genetic resources to increase adaptation of wheat to climate change. In Y. Ogihara, S. Takumi, & H. Handa (Eds.), *Advances in wheat genetics: From genome to field.* Tokyo: Springer.

RICE. (2016). *RICE CRP proposal.* Las Banjos: IRRI.

Rogers, C., & Oldroyd, G. E. D. (2014). Synthetic biology approaches to engineering the nitrogen symbiosis in cereals. *Journal of Experimental Botany, 65*(8), 1939–1946. https://doi.org/10.1093/jxb/eru098.

Rosegrant, M. W., & Agcaoili, M. (2010). *Global food demand, supply, and price prospects to 2010.* Washington, DC: International Food Policy Research Institute.

Sage, R. F. (2004). The evolution of C4 photosynthesis. *The New Phytologist, 161*(2), 341–370.

Salvagiotti, F., Cassman, K. G., Specht, J. E., Walters, D. T., Weiss, A., & Dobermann, A. (2008). Nitrogen uptake, fixation and response to fertilizer N in soybeans: A review. *Field Crops Research, 108*, 1–13.

Setter, T. L., Ellis, M., Laureles, C. V., Ella, E. S., Senadhira, D., & Mishra, S. B. (1997). Physiology and genetics of submergence tolerance in rice. *Annals of Botany, 79*, 67–77. https://doi.org/10.1093/oxfordjournals.aob.a010308.

Shaw, R. H. (1977). *Water use and requirements of maize – A review. Agrometeorology of the maize (corn) crop* (Vol. 480, pp. 119–134). Geneva: Secretariat of the World Meteorological Organization.

Shaw, R. H. (1983). Estimates of yield reductions in corn by water and temperature stress. In C. D. Raper & P. J. Kramer (Eds.), *Crop reactions to water and temperature stresses in humid and temperate climates* (pp. 49–65). Boulder, CO: Westview Press.

Shiferaw, B., Prasanna, B. M., Hellin, J., & Bänziger, M. (2011). Crops that feed the world 6. Past successes and future challenges to the role played by maize in global food security. *Food Security, 3*, 307–327.

Singh, P. R., Jain, G. P., Singh, N., Pandey, P. K., & Sharma, M. K. (2016a). Effect of recurrent selection on drought tolerance and related morpho-physiological traits in bread wheat. *PLoS One, 11*(6), e0156869. https://doi.org/10.1371/journal.pone.0156869.

Singh, R., Singh, Y., Xalaxo, S., Verulkar, S., Yadav, N., Singh, S., Singh, N., Prasad, K. S. N., Kondayya, K., Rao, P. V. R., Rani, M. G., Anuradha, T., Suraynarayana, Y., Sharma, P. C., Krishnamurthy, S. L., Sharma, S. K., Dwivedi, J. L., Singh, A. K., Singh, P. K., Nilanjay, Singh, N. K., Kumar, R., Chetia, S. K., Ahmad, T., Rai, M., Perraju, P., Pande, A., Singh, D. N., Mandal, N. P., Reddy, J. N., Singh, O. N., Katara, J. L., Marandi, B., Swain, P., Sarkar, R. K., Singh, D. P., Mohapatra, T., Padmawathi, G., Ram, T., Kathiresan, R. M., Paramsivam, K., Nadarajan, S., Thirumeni, S., Nagarajan, M., Singh, A. K., Vikram, P., Kumar, A., Septiningsih, E., Singh, U. S., Ismail, A. M., Mackill, D., & Singh, N. K. (2016b). From QTL to variety: Harnessing the benefits of QTLs for drought, flood and salt tolerance in mega rice varieties of India through a multi-institutional network. *Plant Science, 242*, 278–287. https://doi.org/10.1016/j.plantsci.2015.08.008. PMID: 26566845.

Singh, S., Mackill, D. J., & Ismail, A. M. (2009). Responses of Sub1 rice introgression lines to submergence in the field: Yield and grain quality. *Field Crops Research, 113*, 12–23. https://doi.org/10.1016/j.fcr.2009.04.003.

Singh, S., Mackill, D. J., & Ismail, A. M. (2011). Tolerance of longer-term partial stagnant flooding is independent of the SUB1 locus in rice. *Field Crops Research, 121*, 311–323. https://doi.org/10.1016/j.fcr.2010.12.021.

Smil, V. (1999). Nitrogen in crop production: An account of global flows. *Global Biogeochemical Cycles, 13*, 647–662.

Subbarao, G. V., Rondon, M., Ito, O., Ishikawa, T., Rao, I. M., Nakahara, K., Lascano, C., & Berry, W. L. (2007a). Biological nitrification inhibition (BNI) - Is it a widespread phenomenon? *Plant and Soil, 294*, 5–18.

Subbarao, G. V., Tomohiro, B., Masahiro, K., Osamu, I., Samejima, H., Wang, H. Y., Pearse, S. J., Gopalakrishnan, S., Nakahara, K., Hossain, A., Tsujimoto, H., & Berry, W. L. (2007b). Can biological nitrification inhibition (BNI) genes from perennial Leymus racemosus (Triticeae) combat nitrification in wheat farming? *Plant and Soil, 299*, 55–64.

Subbaraoa, G. V., Arangob, J., Masahiroc, K., Hooperd, A. M., Yoshihashia, T., Andoa, Y., Nakaharaa, K., Deshpandee, S., Ortiz-Monasterio, I., Ishitanib, M., Peters, M., Chirindab, N., Wollenbergf, L., & Latag, J. C. (2017). Genetic mitigation strategies to tackle agricultural GHG emissions: The case for biological nitrification inhibition technology. *Plant Science, 262*, 165. https://doi.org/10.1016/j.plantsci.2017.05.004.

Tanaka, J. P., Nardi, P., & Wissuwa, M. (2010). Nitrification inhibition activity, a novel trait in root exudates of rice. *AoB Plants, 2010*, plq014.

Tehripour, F., Hertel, T. W., Gopalakrishnan, B. N., Sahin, S., & Escurra, J. J. (2015). *Agricultural production, irrigation, climate change and water scarcity in India*. In 2015 AAEA & WAEA Joint Annual Meeting, July 26–28, San Francisco, 2015, California. No:205591. Agricultural and Applied Economics Association & Western Agricultural Economics Association.

Temme, K., Zhao, D., & Voigt, C. A. (2012). Refactoring the nitrogen fixation gene cluster from Klebsiella oxytoca. *Proceedings of the National Academy of Sciences U S A, 109*, 7085–7090.

Tesfaye, K., Sonder, K., Cairns, J. E., Magorokoshod, k., Tarekegne, A., Kassie, G. T., Getaneh, F., Abdoulaye, T., Abate, T., & Erenstein, O. (2016). Targeting drought-tolerant maize varieties in Southern Africa: A geospatial crop modeling approach using big data. *International Food and Agribusiness Management Review Special Issue, 19*(A), 2016.

Thornton, P. K., Jones, P. G., Alagarswamy, G., & Andersen, J. (2009). Spatial variation of crop yield response to climate change in East Africa. *Global Environmental Change, 19*, 54–65.

Timsina J, Jat ML, Majumdar K. (2010). Rice-maize systems of South Asia: Current status, future prospects and research priorities for nutrient management Plant Soil. 335(1): 65–82.

Tripp, R., & Rohrbach, D. (2001). Policies for African seed enterprise development. *Food Policy, 26*, 147–161.

Tuong, T. P. (1999). Productive water use in rice production: Opportunities and limitations. *Journal of Crop Production, 2*, 241–264.

Waddington, S. R., Elahi, N. E., & Khatun, F. (2006). *The expansion of rice-maize systems in Bangladesh*. In The Symposium on Emerging Rice-Maize Systems in Asia. ASA-CSSA-SSSA, International Annual Meetings, Indianapolis, IN, USA, November 12–16, 2006.

Westengen, O. T., & Brysting, A. K. (2014). Crop adaptation to climate change in the semi-arid zone in Tanzania: The role of genetic resources and seed systems. *Agriculture and Food Security, 3*, 3.

Xoconostle-Cázares, B., Ramírez-Ortega, F. A., Flores-Elenes, L., & Ruiz-Medrano, R. (2010). Drought tolerance in crop plants. *American Journal of Plant Physiology, 5*(5), 241–256.

Yadvinder-Singh, Kukal, S. S., Jat, M. L., & Sidhu, H. S. (2014). Improving water productivity of wheat-based cropping systems in South Asia for sustained productivity. *Advances in Agronomy, 127*, 157–258. https://doi.org/10.1016/B978-0-12-800131-8.00004-2.

Yamauchi, M., Aguilar, A. M., Vaughan, D. A., & Seshu, D. V. (1993). Rice (Oryzasativa L.) germplasm suitable for direct sowing under flooded soil surface. *Euphytica, 67*, 177–184. https://doi.org/10.1007/BF00040619.

Yanni, Y. G., & Dazzo, F. B. (2010). Enhancement of rice production using endophytic strains of Rhizobium leguminosarum bv. trifolii in extensive field inoculation trials within the Egypt Nile delta. *Plant and Soil, 336*, 129–142.

Yin, H., Chen, C. J., Yang, J., Weston, D. J., & Chen, J. G. (2014). Functional genomics of drought tolerance in bioenergy crops. *Crit Rev Plant Sci, 33*, 205–224. https://doi.org/10.1080/07352689.2014.870417.

Zaidi, P. H., Rafique, S., Singh, N. N., & Srinivasan, G. (2002). *Excess moisture tolerance in maize - Progress and challenges.* In Proc. 8th Asian Regional Maize Workshop, Bangkok, Thailand, August 5–9, 2002. pp. 398–412.

Zaidi, P. H., Rafique, S., Rai, P. K., Singh, N. N., & Srinivasan, G. (2004). Tolerance to excess moisture in maize (Zea mays L.): Susceptible crop stages and identification of tolerant genotypes. *Field Crops Research, 90*(2–3), 189–202.

Zaidi, P. H., Maniselvan, P., Sultana, R., Yadav, M., Singh, R. P., Singh, S. B., et al. (2007). Importance of secondary traits in improvement of maize (Zea mays L.) for enhancing tolerance to excessive soil moisture stress. *Cereal Research Communications, 35*, 1427–1435.

Zaidi, P. H., Rashid, Z., Vinayan, M. T., Almeida, G. D., Phagna, R. K., & Babu, R. (2015). QTL mapping of agronomic waterlogging tolerance using recombinant inbred lines derived from tropical maize (Zea mays L) germplasm. *PLoS One, 10*(4), e0124350. https://doi.org/10.1371/journal.pone.0124350.

Zakir, H., Subbarao, G. V., Pearse, S. J., Gopalakrishnan, S., Ito, O., Ishikawa, T., Kawano, N., Nakahara, K., Yoshihashi, T., Ono, H., & Yoshida, M. (2008). Detection, isolation and characterization of a root-exuded compound, methyl 3-(4-hydroxyphenyl) propionate, responsible for biological nitrification inhibition by sorghum (Sorghum bicolor). *New Phytologist, 180*, 442–451.

Zaman-Hussain, M., Van Loocke, A., Siebers, M. H., Ruiz-Vera, U., Markelz, R. J. C., & Leakey, A. D. B. (2013). Future carbon dioxide concentration decreases canopy evapotranspiration and soil water depletion by field-grown maize. *Global Change Biology, 19*, 1572. https://doi.org/10.1111/gcb.12155.

Zeman, A. M., Tchan, Y. T., Elmerich, C., & Kennedy, L. R. (1992). Nitrogenase activity in wheat seedlings bearing para-nodules induced by 2,4-dichlorophenoxyacetic acid (2,4-D) and inoculated with Azospirillum. *Research in Microbiology, 143*, 847–855.

Zhang, L., Richards, R. A., Condon, A. G., Liu, D. C., & Rebetzke, G. J. (2015). Recurrent selection for wider seedling leaves increases early biomass and leaf area in wheat (Triticum aestivum L.). *Journal of Experimental Botany, 66*, 1215. https://doi.org/10.1093/jxb/eru468.

Zhou, M. Z. (2010). Improvement of plant waterlogging tolerance. In S. Mancuso & S. Shabala (Eds.), *Waterlogging signaling and tolerance in plants* (pp. 267–285). Heidelberg: Springer.

Chapter 5
Using Genomics to Adapt Crops to Climate Change

Yuxuan Yuan, Armin Scheben, Jacqueline Batley, and David Edwards

Abstract Rising food demand from a growing global population, combined with a changing climate, endangers global food security. Thus, there is a need to breed new varieties and increase the efficiency and environmental resilience of crops. Past intensification of crop production has primarily been achieved using fertilisers, herbicides and insecticides as well as improved agronomic methods. However, these practices often rely on finite resources and lack sustainability, making them impractical to increase production in the long term. The ongoing revolution in genomics offers an unprecedented potential to aid crops in adapting to changing environments and increase yield, while also facilitating the diversification of crop production with minor and newly established crop species. Identifying the genomic basis of climate-related agronomic traits for introgression into crop germplasm is a major challenge, requiring the integration of sequencing technologies and breeding expertise. Here we review state of the art genomic tools and their application for accelerating crop improvement in the face of climate change.

Keywords Breeding · Climate change · Crop improvement · Food security · Genomics · Genome editing

5.1 Introduction

With the world population growing to 9.7 billion by 2050 (United Nations 2015), developing crop varieties with increased yield stability and yield potential is an important step in ensuring global food security (Godfray et al. 2010; Tester and Langridge 2010). Climate change is predicted to increase temperatures, alter geographical patterns of rainfall and increase the frequency of extreme climatic events, causing unpredictable threats to crops and extensive yield loss (Batley and Edwards

---

Yuxuan Yuan and Armin Scheben contributed equally to this work.

Y. Yuan · A. Scheben · J. Batley · D. Edwards (✉)
School of Biological Sciences and Institute of Agriculture, University of Western Australia, Perth, WA, Australia
e-mail: Dave.Edwards@uwa.edu.au

© Springer Nature Switzerland AG 2019
A. Sarkar et al. (eds.), *Sustainable Solutions for Food Security*,
https://doi.org/10.1007/978-3-319-77878-5_5

2016; Ronald 2014). Although the local effects of climate change are hard to predict, the broad climate-related stressors for plants are cold, heat, drought, submergence, pathogens and pests (Abberton et al. 2015). Making crops adapt to changing conditions will be essential to maintain and increase yields. Although yields of crop staples have been increased by traditional crop breeding and agronomic practices over the last 50 years, most yield gains are incremental. Moreover, crop losses caused by extreme weather have been rising steadily over recent decades (Bailey-Serres et al. 2012; Boyer et al. 2013). With the changing climate and a growing human population, reliance on traditional breeding and use of fertiliser and pesticides must now give way to more sustainable and efficient yield gains through genomics-assisted crop improvement (Edwards 2016; Scheben et al. 2016).

The use of molecular markers, especially the application of single-nucleotide polymorphisms (SNPs), has revolutionised plant genetics and facilitated molecular breeding of crops (Duran et al. 2010; Edwards and Batley 2010). While SNPs are major contributors to crop diversity and phenotypic variation, the discovery of large-scale variations (>1 kb) associated with agronomic traits indicated that these are also important aspects of plant diversity (Saxena et al. 2014). DNA sequencing can detect SNPs but cannot produce reads long enough to span regions representing large variations. To address this problem, optical mapping can be used to produce long physical maps spanning complex regions. A further aspect of structural variation is the presence/absence variation of genes within a species, which is more widespread than previously assumed, suggesting there is a need for pangenomics to fully understand the diversity present within a single crop (Golicz et al. 2016a). These advances in characterising crop diversity provide new targets for genomics-assisted breeding methods, accelerating our ability to engineer climate ready crops.

5.2 Advances in Genomic Sequencing Technologies

5.2.1 Second Generation Sequencing

Plant genomes are often larger and contain more repetitive elements than the genomes of animals. Despite the accuracy and relatively long read length of traditional Sanger sequencing, applying Sanger sequencing to study plant genomes is impractical because of low throughput and high sequencing costs. In 2005, Roche released its 454 pyrosequencing platform (Margulies et al. 2005), which triggered a revolution in DNA sequencing. In the subsequent few years, different platforms, such as those produced by Illumina, ABI, Life Technologies, Pacific Biosciences (PacBio), Oxford Nanopore and Complete Genomics/BGI, were commercially released, changing the landscape of genome sequencing.

Depending on the technique used by different companies, second generation sequencing (SGS) approaches are classified as ligation-based approaches and synthesis-based approaches (Goodwin et al. 2016). In sequencing by ligation approaches, a probe sequence is bound to a fluorophore. After hybridisation to a DNA segment and ligation to an adjacent oligonucleotide, emitted fluorescence is captured to distinguish the identity of nucleotide or nucleotides within the probe. In this process, the relative positions of nucleotides are determined. Example platforms which use this technique are the ABI SOLiD (now no longer in production) and Complete Genomics which was recently purchased by BGI (Goodwin et al. 2016). In contrast, sequencing by synthesis approaches use a polymerase to incorporate a nucleotide into a DNA strand. In this technique, a fluorophore or a change of ionic concentration is recorded to detect the nucleotides. Example platforms adopting this approach are Roche/Life 454 (now no longer in production), Illumina sequencing by synthesis and Ion Torrent (Goodwin et al. 2016).

While the genomic data volume produced by these new technologies has increased dramatically over the last decade, analysis of the short length of SGS reads (<300 bp) is challenging, especially for complex genome assembly. To help overcome this issue, diverse methods and algorithms have been developed, mostly based on overlap-layout-consensus (OLC) and de Bruijn graphs (DBG). The OLC method is based on read overlaps, and the overlaps must be determined by a computationally intensive series of pair-wise alignments (Miller et al. 2010). Commonly used tools adopting this method are the Celera Assembler (Myers et al. 2000), Arachne (Batzoglou et al. 2002) and Newbler (Margulies et al. 2005). In DBG, k-mers are used to represent all possible fixed-length strings, sequence reads are dissected into their constituent k-mers and then all k-mers are used to produce a *de Bruijn* graph (Pevzner and Tang 2001; Pevzner et al. 2001). Example tools based on this method are AbySS (Simpson et al. 2009), Velvet (Zerbino and Birney 2008), SOAPdenovo (Li et al. 2010) and Allpath-LG (Gnerre et al. 2011). Some tools combining both OLC and DBG methods have also been developed, including MaSuRCA (Zimin et al. 2013). Because most assemblers use short paired read data, the genomes constructed are often fragmented and show a low weighted median contig size (N50). To address this problem, mate-pair reads are used to bridge long-range genomic regions, using software such as *DeNovoMAGIC* and HiRise. Compared to traditional approaches, the assemblies generated by *DeNovoMAGIC* are more complete (Yuan et al. 2017).

5.2.2 Long Read Sequencing

Although SGS has changed the landscape of genetics, assembling repetitive regions remains a major challenge. Errors and collapsed regions are frequent in genome assemblies, reducing the accuracy of downstream analyses. To overcome these problems, long read sequencing offers solutions by producing reads which span repeat regions (Yuan et al. 2017). Depending on the technique used, long read sequencing is categorised as single molecule based or a short read synthesised long read sequencing technology (Goodwin et al. 2016).

5.2.2.1 Single-Molecule Based Long Read Sequencing

The two most commonly used single-molecule based long read sequencing technologies were developed by PacBio and Oxford Nanopore. PacBio has launched three platforms based on zero-mode wave guides, nanostructures which allow individual DNA molecules to be isolated for optical analysis, each with continued improvements in throughput, sequencing cost and error handling. PacBio RS, the first generation PacBio's sequencing platform, produced sequencing reads longer than 1 kb but with relatively high error rates (Carneiro et al. 2012). In the PacBio RS II, the length improved to an average of 10 kb, with significant improvement in accuracy (Rhoads and Au 2015). The most recent platform, the Sequel claims to produce longer molecules with much increased throughput. Various algorithms have also been developed to improve sequence accuracy, including PacBio Corrected Reads (Koren et al. 2012), the hierarchical genome assembly process (Chin et al. 2013) and the MinHash Alignment Process (Berlin et al. 2015).

Oxford Nanopore is another company which has commercialised single molecule sequencing, initially through its MinIon platform. This platform is the smallest and most portable on the market and can be powered by a USB port through a laptop. It produces reads of around 5–10 kb with an analysis in real time (Chaney et al. 2016). A more advanced platform, the PromethION is being tested in an early access programme and promises to generate larger quantities of data. For field based DNA sequencing, Oxford Nanopore also designed the SmidgION, which can be connected to a smartphone, though this is not yet under commercial production. As with PacBio sequencing, the error rate of Oxford Nanopore sequencing is high (~15%), and different methods and algorithms have been developed to help reduce this including NanoCorr (Goodwin et al. 2015), NanoPolish (Loman et al. 2015), PoreSeq (Szalay and Golovchenko 2015) and MarginAlign (Jain et al. 2015).

5.2.2.2 Short Read Synthesised Long Read Sequencing

SGS based on Illumina technology has a relatively high accuracy, high throughput and low cost, while in contrast, single-molecule based long read sequencing has lower accuracy, throughput and higher cost per nucleotide. Short read synthesised long read sequencing aims to combine the advantages of long read sequencing with those of SGS using custom library preparation and algorithms (Yuan et al. 2017). Illumina synthetic long read sequencing, using protocols adapted from the Moleculo, produces sequence reads with a length around 3–5 kb (Li et al. 2015). In the Chromium system commercialised by 10× Genomics, a microfluidic technique produces synthetic reads of 30–100 kb long (Eisenstein 2015).

5.2.3 Optical Mapping

While there have been many advances in long read sequencing, these platforms cannot produce reads longer than 100 kb, which complicates the study of large structural variants. To address this problem, optical mapping uses the physical location of restriction enzyme sites to facilitate genome assembly and SV detection. A single molecule optical map can be 200 kb or longer (Chaney et al. 2016), and so can resolve much longer repetitive regions than standard DNA sequencing technologies (Stankova et al. 2016).

Previously, the low throughput and high error rates hampered the uptake of optical mapping. With recent improvements, particularly those contributed by BioNano Genomics and OpGen, the application of optical mapping is increasing (Yuan et al. 2017). Optical mapping has been applied to the study of a wide range of plants, including maize (Zhou et al. 2009), rice (Zhou et al. 2007), wheat (Stankova et al. 2016) and *Brassica* (Yang et al. 2016). The major aims of using optical mapping are to improve the contiguity of DNA sequences, correct false read joins, reverse the orientation of false inversions, identify misassemblies and aid large-scale SV detection.

5.3 Pangenome Analysis

The pangenome refers to the entire gene set for a species and includes genes present in all individuals (core genes) and genes present only in some individuals (variable genes). Pangenomics was first applied to the study of bacterial genomes (Tettelin et al. 2005), and the study of pangenomes in more complex genomes is less common, although increasing. In 2014, pangenome studies were carried out on the crops maize (Hirsch et al. 2014), rice (Schatz et al. 2014), soybean (Li et al. 2014) and *Brassica rapa* (Lin et al. 2014). In 2015, Lu et al. (2015) used a high-resolution genetic mapping to anchor sequences to the maize pangenome. One year later, Golicz et al. (2016b) reported a pangenome analysis of *Brassica oleracea*. Recently, a study of the hexaploid bread wheat pangenome showed the diversity within 18 cultivars of this important crop (Montenegro et al. 2017).

Understanding the presence/absence variation in a crop species is important for the breeding of improved crop varieties which are adapted to changing climates. For example, in rice, the variable *Submergence 1A* (*Sub1A*) gene plays an important role in submergence tolerance (Schatz et al. 2014). The *FLC* gene family controls flowering time, with divergence in the number of *FLC* copies contributing to flowering time variation. In *Brassica oleracea*, four *FLC* genes were identified as core genes, while two other FLCs are variable (Golicz et al. 2016b). Pangenome analysis provides an approach to quickly examine the gene set of an entire crop species to identify candidate genes for applied crop improvement.

5.4 Genome-Wide Genotyping

The rapid decrease in the cost of genome-wide genotyping is enabling large-scale assessment of crop diversity to identify genes underlying climate-related traits. Genotyping by sequencing (GBS) leverages the low cost of sequencing to identify thousands to millions of single nucleotide polymorphisms (SNPs) in plant populations (Davey et al. 2011; Deschamps et al. 2012; He et al. 2014; Heffelfinger et al. 2014; Poland and Rife 2012; Golicz et al. 2015; Scheben et al. 2017). GBS approaches vary in the type and volume of data produced and can be divided into whole-genome resequencing (WGR) approaches such as skimGBS (e.g. Bayer et al. 2015) and reduced representation sequencing (RRS) methods (Andrews et al. 2016). While WGR provides high SNP densities across the entire genome, RRS uses restriction enzymes to extract a subset of the genome, thus reducing sequencing costs at the expense of SNP density.

In contrast, high-density genotyping arrays, or "SNP chips," allow for large-scale genotyping using SNP-specific oligonucleotide probes rather than direct sequencing. SNP arrays for most major crops are available with various numbers of SNPs and these also exhibit a trade-off between SNP density and cost. Although the cost per data point is higher using SNP arrays than GBS, SNP arrays continue to be commonly used because of the ease of use and simplicity of data analysis.

The variants detected by genome-wide genotyping can be used for the traditional analysis of quantitative trait loci (QTL), regions of the genome linked to quantitative phenotypic traits, and the more modern approach of genome-wide association studies (GWAS). GWAS make use of past recombination in diverse association panels to identify genes linked to phenotypic traits at a higher resolution than QTL analysis. SNP genotyping arrays have been widely used for GWAS in crops such as rice (Zhao et al. 2011), maize (Li et al. 2013), wheat (Winfield et al. 2016) and soybean (Hwang et al. 2014), and GBS methods are also becoming more common for this type of study (Arruda et al. 2016; Clarke 2016; Gacek et al. 2017).

The application of genome-wide genotyping is leading to substantial increases in our knowledge of the genetics of yield-related traits. Drought causes greater yield losses than any other type of abiotic stress, with losses likely to increase due to the impact of climate change on rainfall patterns. In a study of heat stress in bread wheat, three significant genomic regions on chromosome 2B, 7B and 7D were found to be associated with heat tolerance (Paliwal et al. 2012). In chickpea, QTL "hotspots" for drought tolerance were recently investigated and four candidate genes identified (*Ca_04561, Ca_04562, Ca_04567* and *Ca_04569*) (Jaganathan et al. 2015). Water use efficiency can contribute to drought tolerance, and in soybean Dhanapal et al. (2015) reported 39 unique SNPs associated with this important trait. Valluru et al. (2017) reported 32 significant SNPs for wheat spike ethylene, which reduces fertility and grain weight in field trials. Reducing yield loss under heat and drought stress by decreasing ethylene effects may allow for breeding of greater tolerance to these stressors. Temperature stress commonly occurs together with drought. In barley, a genomic region on chromosome 2H was associated with grain yield under elevated temperature, and another on 7H with stability of grain yield (Ingvordsen et al. 2015). QTL have also been used to identify 20 C-repeat binding factor genes in barley which are the key regulators of cold tolerance genes (Skinner et al. 2006).

Salinization is expected to impact 50% of arable land by 2050 (Wang et al. 2003), and breeding salt-tolerant crops is important to ensure productivity on increasingly saline soils. In rice, a custom genotyping array including SNPs in 6000 stress-related genes identified 20 SNPs linked to the Na+–K+ ratio, which is important for salt tolerance (Kumar et al. 2015).

Pests cause annual global losses of 26–40% in staple crops and include weeds, insects, fungi, bacteria and viruses (Oerke 2006). Reducing yield losses from pests requires diverse resistance mechanisms to counter rapid adaptation of pests. Identifying resistance genes that can be stacked in elite crop varieties is therefore important. For example, QTL analysis located the canola gene *Rlm4*, which confers resistance to the fungal pathogen blackleg (*Leptosphaeria maculans*) (Tollenaere et al. 2012). A recent investigation in canola also identified four blackleg resistance QTL, which can be used together with the known resistance genes in breeding efforts (Larkan et al. 2016). In cassava, QTL analysis identified the single gene *CMD2* underlying resistance to the pathogen cassava mosaic virus (Rabbi et al. 2014).

Despite rapid advances and decreasing costs of genome-wide genotyping, several limitations of this approach to identifying functional crop diversity remain. While WGR is a powerful approach for accurate SNP calling in recombinant populations with a high-quality reference genome (Golicz et al. 2015), the sequencing of crops with large genomes such as wheat is still constrained by high costs. Reduced representation sequencing, on the other hand, lowers per sample costs and has facilitated access to larger genomes including wheat (Poland et al. 2012b). The limitations of RRS are the lower SNP densities and the often high amounts of missing data due to restriction site polymorphisms and the stochastic sampling process. However, methods to predict missing SNPs in large datasets using imputation are increasingly available to help overcome this challenge (Fu et al. 2016). In light of the decreasing costs of sequencing and the high cost of candidate gene validation, the use of WGR to increase the resolution of mapping studies is likely to become more common in the future (Scheben et al. 2017). As the knowledge of gene functions even in major crops remains inconsistent, genome-wide genotyping will play an important role in characterising the variation that can be harnessed for crop improvement.

5.5 Applying Molecular Markers in Crop Breeding

5.5.1 Marker-Assisted Breeding

Genotypic markers associated with agronomic traits can be introgressed into elite crop genetic backgrounds via marker-assisted breeding (MAB). This approach enables stacking of desirable traits into elite varieties to make them adapt to climate change. Although our understanding of crop genetics is rapidly increasing, these gains in knowledge have rarely been translated into gains in breeding (Xu and Crouch 2008). This is because untangling complex genetic traits such as abiotic stress tolerance is laborious and constrained by marker quantity and quality. Nevertheless, the past two decades provide several important examples of successful MAB. Cooper et al. (2014) reported proprietary genomic approaches to facilitate breeding of Optimum AQUAmax (Dupont Pioneer) drought-tolerant maize hybrids, which were shown to produce higher yields under drought stress with no yield penalty under normal conditions in on-farm trials (Gaffney et al. 2015).

An important step for rice breeding using genomic markers was the introgression of the submergence-tolerance gene (*Sub1A*) into elite commercial rice varieties (Septiningsih et al. 2009). Annual yield losses due to flooding can exceed the US $1 billion, but this has been substantially reduced due to the success of cultivars with the *Sub1A* gene (Ismail et al. 2013). In canola, MAB allowed commercial release of pod-shatter resistant PodGuard (Bayer) cultivars (Lambert et al. 2015). The sequencing of the canola genome in 2014 (Chalhoub et al. 2014) helped identify the target gene influencing pod shattering. Novel cultivars were then generated by Bayer via a reverse genetics approach, using untargeted genome mutagenesis and subsequent stacking of desirable mutant alleles of the known target. The developed pod shatter resistant cultivars reduce yield loss caused by pods shattering before harvest due to extreme weather such as heavy rain. Use of MAB also allowed development of a salt-tolerant durum wheat variety capable of excluding Na+ using the *Nax2* locus, which includes a Na+-selective transporter gene, and increases yields on saline soils by 25% without an apparent yield penalty on nonsaline soil (Munns et al. 2012). Markers linked to resistance to pests including cereal cyst nematode, root lesion nematode and crown rot are also used extensively in the breeding programs of the International Maize and Wheat Improvement Center (CIMMYT) (Trethowan et al. 2005). Bacterial blight resistance genes in rice are well characterised, and genes such as *Xa-21* have been incorporated into novel blight-resistant varieties via MAB (Cao et al. 2003)

5.5.2 Genomic Selection

Genomic selection (GS) is a promising approach that improves on MAB by allowing rapid trait improvement without detailed study of individual trait-related genetic loci. GS uses breeding values calculated using a phenotyped and genotyped training population to estimate the performance of crop genotypes. Breeding populations can thus undergo selection based on these breeding values without plants having to be grown out to evaluate phenotypes (Meuwissen et al. 2001). This can reduce the cost of breeding by limiting the size and number of field experiments required to trial crops, while also potentially capturing all of the quantitative trait loci controlling a complex trait. Although GS was first developed for animal breeding and has been extensively used in livestock such as cattle and pigs, the potential of GS in crop breeding has led to increased interest in applications of this approach in plants.

Heffner et al. (2010) estimate that GS can deliver threefold the annual genetic gains for maize achievable using MAB. This underlines the stark increase in genetic gains facilitated by the increased speed of selection cycles with GS (Lin et al. 2016).

Although GS in plants has been carried out largely using computer simulations, researchers have also carried out several empirical studies, particularly in wheat and maize. Poland et al. (2012a) and Rutkoski et al. (2014) applied GBS to sets of elite wheat breeding lines and developed GS models with moderately high prediction accuracies of 0.28–0.45 and 0.45–0.62 for grain yield and stem rust resistance respectively. In maize, GBS performed as well as the more established SNP arrays and showed potential for harnessing variation for breeding populations and indicated the importance of including environment effects in GS models (Crossa et al. 2013; Gorjanc et al. 2016; Zhang et al. 2015). GS can make an important contribution to future crop improvement by accelerating the selection of complex agronomic traits such abiotic stress tolerance and yield. However, GS still faces the limitation of high variability in prediction accuracies. While the improvement of statistical modelling methods may help overcome this limitation, researchers in the field of crop genetics must also commit to data sharing to increase the amount of data available to train models and ensure their accuracy (Spindel and McCouch 2016).

5.6 Genome Editing

Genome editing allows for precise insertion or deletion of DNA in the genome and has the potential to deliver a major advance in crop improvement (Belhaj et al. 2015; Bortesi and Fischer 2015; Liu and Fan 2014; Pennisi 2013; Scheben and Edwards 2017). While genome editing approaches using engineered nucleases such as transcription activator-like effector nucleases (TALENs) and zinc finger nucleases (ZFNs) have been available for the last decade, the recently developed type II clustered regularly interspaced short palindromic repeat (CRISPR)/CRISPR-associated protein (Cas) system (Jinek et al. 2012) is now the most widely used approach. CRISPR/Cas is generally delivered into protoplasts (cells without a cell wall) using a plasmid vector to insert coding sequences for the Cas9 nuclease and a custom guide-RNA (gRNA). Cas9 forms a complex with the gRNA, which hybridises a $G(N)_{19-22}NGG$ target DNA sequence (Gasiunas et al. 2012; Jinek et al. 2012). This results in the Cas9 nuclease cleaving double stranded DNA three nucleotides upstream of the NGG motif. Error-prone repair mechanisms can then lead to a frameshift mutation causing a knockout. Alternatively, precise nucleotide editing such as gene knock-in can be carried out by providing DNA donor template for error-free DNA repair mechanisms. Unlike TALENs and ZFNs, CRISPR/Cas9 only requires the design of a gRNA, and no complex protein engineering. While the target sequence specificity of CRISPR/Cas9 genome editing and thus the number of off-target editing events is controversial (Fu et al. 2013; Jiang et al. 2013; Pattanayak et al. 2013), studies in plants have generally reported low frequencies of off-target editing (Feng et al. 2014; Jiang et al. 2013; Mikami et al. 2016). Target site specificity can be increased by using high-fidelity variants of the Cas9 enzyme that show higher target site specificities. Paired Cas9 nickases that "nick" DNA, cleaving only a single strand of DNA rather than both strands, also reduce off-target events, as editing only occurs when paired nicks are independently produced in complementary positions on the genome (Gasiunas et al. 2012; Ran et al. 2013; Shen et al. 2014). Delivery of CRISPR/Cas9 in the form of a ribonucleoprotein complex that rapidly degrades after editing can also reduce off-target effects (Svitashev et al. 2016).

The CRISPR/Cas9 system has been used in a laboratory setting to edit the genomes of many major crop species, including maize (Shi et al. 2017; Sun et al. 2016; Svitashev et al. 2016, 2015), orange (Jia and Wang 2014), potato (Wang et al. 2015), rice (Jiang et al. 2013; Li et al. 2016; Mikami et al. 2016; Zhang et al. 2014), tobacco (Gao et al. 2015; Jiang et al. 2013), tomato (Brooks et al. 2014; Cermak et al. 2015; de Toledo Thomazella et al. 2016; Soyk et al. 2017) and wheat (Wang et al. 2014). In rice and tomato, CRISPR/Cas9 can introduce heritable homozygous mutations in the first generation of transformants (Brooks et al. 2014; Zhang et al. 2014), highlighting the speed of this approach. Genome editing in plants has the advantage that unwanted transgenes or off-target mutations can be bred out by backcrossing to parental genotypes. The cycle time for breeding programs could be substantially decreased by using CRISPR/Cas9 for genome editing rather than relying on the recombination of allelic variants present in the breeding germplasm or generated via untargeted mutagenesis. Many results achieved by time-intensive reverse genetics approaches such as the breeding of pod-shatter resistant canola (Lambert et al. 2015) could now be achieved more rapidly using genome editing. The first commercialised crop edited using CRISPR/Cas9, a high amylopectin maize cultivar, is due to be released in the next 5 years (Waltz 2016).

CRISPR/Cas9 has mainly been applied to disrupt crop genes, particularly those associated with susceptibility to pests. For instance, the system was recently used to enhance blast resistance in rice by targeting the *OsERF922* gene (Wang et al. 2016). CRISPR/Cas9 was also used to promote powdery mildew resistance in wheat by causing loss of function in the susceptibility gene *TaMLO* (Wang et al. 2014). In cucumber, broad virus resistance was engineered by disrupting the function of the *eIF4E* gene (Zhang et al. 2016). Virus resistance could also be enhanced in tobacco by introducing the CRISPR/Cas9 system as part of a plant immune response to target viral DNA for cleavage (Baltes et al. 2015; Ji et al. 2015).

More complex traits than pest resistance have also been improved using genome editing. In a study on rice, genes controlling yield components were edited, allowing development of lines with increased grain size and grain number. Replacement of the promoter of the maize gene *ARGOS8* with a more powerful promoter using CRISPR/Cas9 allowed researchers to increase drought resistance in maize, which was confirmed in field trials (Shi et al. 2017). Flowering time could be manipulated in tomato using CRISPR/Cas9 to disrupt the flower-repressing *SP5G* gene to generate early-yielding varieties (Soyk et al. 2017). As the frequency of extreme weather events increases in current growing seasons, farmers may be able to profit from plants with flowering times that allow crucial periods of plant development to occur before phases of extreme weather. While current applications of genome editing are generally limited to model plants and major crops, genome editing also has the potential to accelerate the domestication of new crops from wild species or locally important minor crops with valuable climate-related traits. Functional genomics has provided insights into the mechanisms underlying domestication, particularly yield traits controlling number, size and shattering of seeds. Using CRISPR/Cas9 to edit these key domestication traits in potential new crops would enable more rapid expansion of currently limited crop gene pools to maximise use of germplasm adapted to the changing climate.

5.7 Remaining Challenges for Genomic Technologies in Crop Breeding

Although an unprecedented genomic toolbox for crop improvement is available today, several challenges must be overcome to ensure that genomic technologies reach their full potential in helping crops adapt to climate change. While seed companies are increasingly applying genomic technologies to their breeding programs, the limited return on the substantial investments required to implement these technologies may prevent their wider use. Returns on investment are limited because improved seeds can generally be purchased once by farmers and then saved after each growing season for reuse. Hybrid seeds, on the other hand, cannot be reused in the same way and often offer agronomic advantages over non-hybrid seeds. Increased use of hybrid seeds in crops such as wheat, where the development of hybrid varieties lags behind that found in successful hybrid crops such as maize, may ensure that seed companies are incentivized to invest in genomic technologies. A further hurdle for private sector investment in genomic technology is the unclear regulatory status of crop genome editing. Current biotechnology regulation in the USA and Europe does not treat genome edited crops as genetically modified (GM), theoretically allowing easier market approval of these crops. Nevertheless, in this case, regulatory legislation may lag behind technological innovation, and future policy decisions may affect the potential benefits of crop genome editing. Refinement of legislation regulating biotechnology to clarify the status of genome edited crops will help seed companies make investment decisions and may help overcome public scepticism towards the safety of these crops.

Improved crops are likely to have the greatest impact in developing regions such as sub-Saharan Africa where the gap between potential yields and observed yields is highest (Mueller et al. 2012). These substantial yield gaps could, at least in part, be filled by improved seed adapted to local conditions. However, smallholder farmers in developing countries are slow to adopt new varieties and technologies because they lack access to credit and output and input markets, reducing access to improved seeds. Government support schemes and aid from industrialised countries will play an important role in providing these smallholder farmers with opportunities to grow higher-yielding and more reliable adapted crops. Finally, there are major scientific challenges for breeding climate ready crops using genomic technologies, including gaps in the scientific knowledge concerning the precise impacts of climate change on agriculture and a limited understanding of the complex genetic networks underlying many climate-related crop traits. Predicting the effects of climate change across different agricultural regions will allow crop breeders to focus their research on the most critical traits in those staple crops likely to be most severely impacted. At the same time, further insights into the genetic networks controlling these critical traits, and the management and sharing of this information via public databases, will increase the amount of genetic gains possible in crop improvement. While genomic technologies in crop improvement are not a panacea, if the sociopolitical and scientific hurdles to their widespread use can be overcome they stand to make a major contribution to face the challenge climate change poses for agriculture.

Acknowledgements Armin Scheben was supported by an IPRS awarded by the Australian government. Yuxuan Yuan was supported by a SIRF funded by the China Scholarship Council and the University of Western Australia. David Edwards acknowledges support from the Australian Research Council LP140100537, LP130100061, LP130100925 and LP110100200.

References

Abberton, M., Batley, J., Bentley, A., Bryant, J., Cai, H., Cockram, J., Costa de Oliveira, A., Cseke, L. J., Dempewolf, H., De Pace, C., Edwards, D., Gepts, P., Greenland, A., Hall, A. E., Henry, R., Hori, K., Howe, G. T., Hughes, S., Humphreys, M., Lightfoot, D., Marshall, A., Mayes, S., Nguyen, H. T., Ogbonnaya, F. C., Ortiz, R., Paterson, A. H., Tuberosa, R., Valliyodan, B., Varshney, R. K., & Yano, M. (2015). Global agricultural intensification during climate change: A role for genomics. *Plant Biotechnology Journal, 14*, 1095–1098.

Andrews, K. R., Good, J. M., Miller, M. R., Luikart, G., & Hohenlohe, P. A. (2016). Harnessing the power of RADseq for ecological and evolutionary genomics. *Nature Reviews. Genetics, 17*, 81–92.

Arruda, M. P., Brown, P., Brown-Guedira, G., Krill, A. M., Thurber, C., Merrill, K. R., Foresman, B. J., et al. (2016). Genome-wide association mapping of *Fusarium* head blight resistance in wheat using genotyping-by-sequencing. *Plant Genome, 9*.

Bailey-Serres, J., Lee, S. C., & Brinton, E. (2012). Waterproofing crops: Effective flooding survival strategies. *Plant Physiology, 160*, 1698–1709.

Baltes, N. J., Hummel, A. W., Konecna, E., Cegan, R., Bruns, A. N., Bisaro, D. M., & Voytas, D. F. (2015). Conferring resistance to geminiviruses with the CRISPR-Cas prokaryotic immune system. *Nature Plants, 1*, 15145.

Batley, J., & Edwards, D. (2016). The application of genomics and bioinformatics to accelerate crop improvement in a changing climate. *Current Opinion in Plant Biology, 30*, 78–81.

Batzoglou, S., Jaffe, D. B., Stanley, K., Butler, J., Gnerre, S., Mauceli, E., Berger, B., et al. (2002). ARACHNE: A whole-genome shotgun assembler. *Genome Research, 12*, 177–189.

Bayer, P. E., Ruperao, P., Mason, A. S., Stiller, J., Chan, C.-K. K., Hayashi, S., Long, Y., et al. (2015). High-resolution skim genotyping by sequencing reveals the distribution of crossovers and gene conversions in *Cicer arietinum* and *Brassica napus*. *Theoretical and Applied Genetics, 128*, 1039–1047.

Belhaj, K., Chaparro-Garcia, A., Kamoun, S., Patron, N. J., & Nekrasov, V. (2015). Editing plant genomes with CRISPR/Cas9. *Current Opinion in Biotechnology, 32*, 76–84.

5 Using Genomics to Adapt Crops to Climate Change 103

Berlin, K., Koren, S., Chin, C. S., Drake, J. P., Landolin, J. M., & Phillippy, A. M. (2015). Assembling large genomes with single-molecule sequencing and locality-sensitive hashing. *Nature Biotechnology, 33*, 623–630.

Bortesi, L., & Fischer, R. (2015). The CRISPR/Cas9 system for plant genome editing and beyond. *Biotechnology Advances, 33*, 41–52.

Boyer, J. S., Byrne, P., Cassman, K. G., Cooper, M., Delmer, D., Greene, T., Gruis, F., et al. (2013). The US drought of 2012 in perspective: A call to action. *Global Food Security-Agriculture Policy Economics and Environment, 2*, 139–143.

Brooks, C., Nekrasov, V., Lippman, Z. B., & Van Eck, J. (2014). Efficient gene editing in tomato in the first generation using the clustered regularly interspaced short palindromic repeats/CRISPR-Associated9 System. *Plant Physiology, 166*, 1292–1297.

Cao, L. Y., Zhuang, J. Y., Yuan, S. J., Zhan, X. D., Zheng, K. L., & Cheng, S. H. (2003). Hybrid rice resistant to bacterial leaf blight developed by marker assisted selection. *Rice Science, 11*, 68–70.

Carneiro, M. O., Russ, C., Ross, M. G., Gabriel, S. B., Nusbaum, C., & DePristo, M. A. (2012). Pacific biosciences sequencing technology for genotyping and variation discovery in human data. *BMC Genomics, 13*, 375.

Cermak, T., Baltes, N. J., Cegan, R., Zhang, Y., & Voytas, D. F. (2015). High-frequency, precise modification of the tomato genome. *Genome Biology, 16*, 232.

Chalhoub, B., Denoeud, F., Liu, S., Parkin, I. A. P., Tang, H., Wang, X., Chiquet, J., et al. (2014). Early allopolyploid evolution in the post-Neolithic Brassica napus oilseed genome. *Science, 345*, 950–953.

Chaney, L., Sharp, A. R., Evans, C. R., & Udall, J. A. (2016). Genome mapping in plant comparative genomics. *Trends in Plant Science, 21*, 770–780.

Chin, C. S., Alexander, D. H., Marks, P., Klammer, A. A., Drake, J., Heiner, C., Clum, A., et al. (2013). Nonhybrid, finished microbial genome assemblies from long-read SMRT sequencing data. *Nature Methods, 10*, 563–569.

Clarke, S. (2016). *Is genotyping by sequencing a viable alternative to existing methods for genomic selection and GWAS?* In Plant and Animal Genome XXIV Conference. San Diego, CA.

Cooper, M., Gho, C., Leafgren, R., Tang, T., & Messina, C. (2014). Breeding drought-tolerant maize hybrids for the US corn-belt: Discovery to product. *Journal of Experimental Botany, 65*, 6191–6204.

Crossa, J., Beyene, Y., Kassa, S., Perez, P., Hickey, J. M., Chen, C., de los Campos, G., et al. (2013). Genomic prediction in maize breeding populations with genotyping-by-sequencing. *G3 (Bethesda), 3*, 1903–1926.

de Toledo Thomazella, D. P., Brail, Q., Dahlbeck, D., & Staskawicz, B. J. (2016). CRISPR-Cas9 mediated mutagenesis of a DMR6 ortholog in tomato confers broad-spectrum disease resistance. *bioRxiv*, 064824.

Davey, J. W., Hohenlohe, P. A., Etter, P. D., Boone, J. Q., Catchen, J. M., & Blaxter, M. L. (2011). Genome-wide genetic marker discovery and genotyping using next-generation sequencing. *Nature Reviews. Genetics, 12*, 499–510.

Deschamps, S., Llaca, V., & May, G. D. (2012). Genotyping-by-sequencing in plants. *Biology (Basel), 1*, 460–483.

Dhanapal, A. P., Ray, J. D., Singh, S. K., Hoyos-Villegas, V., Smith, J. R., Purcell, L. C., King, C. A., et al. (2015). Genome-wide association study (GWAS) of carbon isotope ratio (delta C-13) in diverse soybean [*Glycine max* (L.) Merr.] genotypes. *Theoretical and Applied Genetics, 128*, 73–91.

Duran, C., Eales, D., Marshall, D., Imelfort, M., Stiller, J., Berkman, P. J., Clark, T., McKenzie, M., Appleby, N., Batley, J., Basford, K., & Edwards, D. (2010). Future tools for association mapping in crop plants. *Genome, 53*, 1017–1023.

Edwards, D. (2016). The impact of genomics technology on adapting plants to climate change. In D. Edwards & J. Batley (Eds.), *Plant genomics and climate change* (pp. 173–178). New York, NY: Springer.

Edwards, D., & Batley, J. (2010). Plant genome sequencing: Applications for crop improvement. *Plant Biotechnology Journal, 8*, 2–9.

Eisenstein, M. (2015). Startups use short-read data to expand long-read sequencing market. *Nature Biotechnology, 33*, 433–435.

Feng, Z., Mao, Y., Xu, N., Zhang, B., Wei, P., Yang, D.-L., Wang, Z., et al. (2014). Multigeneration analysis reveals the inheritance, specificity, and patterns of CRISPR/Cas-induced gene modifications in Arabidopsis. *Proceedings of the National Academy of Sciences of the United States of America, 111*, 4632–4637.

Fu, Y., Foden, J. A., Khayter, C., Maeder, M. L., Reyon, D., Joung, J. K., & Sander, J. D. (2013). High-frequency off-target mutagenesis induced by CRISPR-Cas nucleases in human cells. *Nature Biotechnology, 31*, 822–826.

Fu, L. X., Cai, C. C., Cui, Y. N., Wu, J., Liang, J. L., Cheng, F., & Wang, X. W. (2016). Pooled mapping: An efficient method of calling variations for population samples with low-depth resequencing data. *Molecular Breeding, 36*, 48–48.

Gacek, K., Bayer, P. E., Bartkowiak-Broda, I., Szala, L., Bocianowski, J., Edwards, D., & Batley, J. (2017). Genome-wide association study of genetic control of seed fatty acid biosynthesis in *Brassica napus*. *Frontiers in Plant Science, 7*, 2062.

Gaffney, J., Schussler, J., Loffler, C., Cai, W. G., Paszkiewicz, S., Messina, C., Groeteke, J., et al. (2015). Industry-scale evaluation of maize hybrids selected for increased yield in drought-stress conditions of the US Corn Belt. *Crop Science, 55*, 1608–1618.

Gao, J. P., Wang, G. H., Ma, S. Y., Xie, X. D., Wu, X. W., Zhang, X. T., Wu, Y. Q., et al. (2015). CRISPR/Cas9-mediated targeted mutagenesis in *Nicotiana tabacum*. *Plant Molecular Biology, 87*, 99–110.

Gasiunas, G., Barrangou, R., Horvath, P., & Siksnys, V. (2012). Cas9-crRNA ribonucleoprotein complex mediates specific DNA cleavage for adaptive immunity in bacteria. *Proceedings of the National Academy of Sciences of the United States of America, 109*, E2579–E2586.

Gnerre, S., Maccallum, I., Przybylski, D., Ribeiro, F. J., Burton, J. N., Walker, B. J., Sharpe, T., et al. (2011). High-quality draft assemblies of mammalian genomes from massively parallel sequence data. *Proceedings of the National Academy of Sciences of the United States of America, 108*, 1513–1518.

Godfray, H. C. J., Beddington, J. R., Crute, I. R., Haddad, L., Lawrence, D., Muir, J. F., Pretty, J., et al. (2010). Food security: The challenge of feeding 9 billion people. *Science, 327*, 812–818.

Golicz, A. A., Bayer, P. E., & Edwards, D. (2015). Skim-based genotyping by sequencing. *Methods in Molecular Biology, 1245*, 257–270.

Golicz, A. A., Batley, J., & Edwards, D. (2016a). Towards plant pangenomics. *Plant Biotechnology Journal, 14*, 1099–1105.

Golicz, A. A., Bayer, P. E., Barker, G. C., Edger, P. P., Kim, H., Martinez, P. A., Chan, C. K., et al. (2016b). The pangenome of an agronomically important crop plant Brassica oleracea. *Nature Communications, 7*, 13390.

Goodwin, S., Gurtowski, J., Ethe-Sayers, S., Deshpande, P., Schatz, M. C., & McCombie, W. R. (2015). Oxford Nanopore sequencing, hybrid error correction, and de novo assembly of a eukaryotic genome. *Genome Research, 25*, 1750–1756.

Goodwin, S., McPherson, J. D., & McCombie, W. R. (2016). Coming of age: Ten years of next-generation sequencing technologies. *Nature Reviews. Genetics, 17*, 333–351.

Gorjanc, G., Jenko, J., Hearne, S. J., & Hickey, J. M. (2016). Initiating maize pre-breeding programs using genomic selection to harness polygenic variation from landrace populations. *BMC Genomics, 17*, 30.

He, J., Zhao, X., Laroche, A., Lu, Z. X., Liu, H., & Li, Z. (2014). Genotyping-by-sequencing (GBS), an ultimate marker-assisted selection (MAS) tool to accelerate plant breeding. *Frontiers in Plant Science, 5*, 484.

Heffelfinger, C., Fragoso, C. A., Moreno, M. A., Overton, J. D., Mottinger, J. P., Zhao, H., Tohme, J., et al. (2014). Flexible and scalable genotyping-by-sequencing strategies for population studies. *BMC Genomics, 15*, 979–979.

Heffner, E. L., Lorenz, A. J., Jannink, J. L., & Sorrells, M. E. (2010). Plant breeding with genomic selection: Gain per unit time and cost. *Crop Science, 50*, 1681–1690.

Hirsch, C. N., Foerster, J. M., Johnson, J. M., Sekhon, R. S., Muttoni, G., Vaillancourt, B., Penagaricano, F., et al. (2014). Insights into the maize pan-genome and pan-transcriptome. *Plant Cell, 26*, 121–135.

Hwang, E. Y., Song, Q., Jia, G., Specht, J. E., Hyten, D. L., Costa, J., & Cregan, P. B. (2014). A genome-wide association study of seed protein and oil content in soybean. *BMC Genomics, 15*, 1.

Ingvordsen, C. H., Backes, G., Lyngkjaer, M. F., Peltonen-Sainio, P., Jahoor, A., Mikkelsen, T. N., & Jorgensen, R. B. (2015). Genome-wide association study of production and stability traits in barley cultivated under future climate scenarios. *Molecular Breeding, 35*, 84.

Ismail, A. M., Singh, U. S., Singh, S., Dar, M. H., & Mackill, D. J. (2013). The contribution of submergence-tolerant (*Sub1*) rice varieties to food security in flood-prone rainfed lowland areas in Asia. *Field Crops Research, 152*, 83–93.

Jaganathan, D., Thudi, M., Kale, S., Azam, S., Roorkiwal, M., Gaur, P. M., Kishor, P. B. K., et al. (2015). Genotyping-by-sequencing based intra-specific genetic map refines a "QTL-hotspot" region for drought tolerance in chickpea. *Molecular Genetics and Genomics, 290*, 559–571.

Jain, M., Fiddes, I. T., Miga, K. H., Olsen, H. E., Paten, B., & Akeson, M. (2015). Improved data analysis for the MinION nanopore sequencer. *Nature Methods, 12*, 351–356.

Ji, X., Zhang, H. W., Zhang, Y., Wang, Y. P., & Gao, C. X. (2015). Establishing a CRISPR-Cas-like immune system conferring DNA virus resistance in plants. *Nat. Plants, 1*, 15144.

Jia, H. G., & Wang, N. (2014). Targeted genome editing of sweet orange using Cas9/sgRNA. *PLoS One, 9*, e93806.

Jiang, W. Z., Zhou, H. B., Bi, H. H., Fromm, M., Yang, B., & Weeks, D. P. (2013). Demonstration of CRISPR/Cas9/sgRNA-mediated targeted gene modification in Arabidopsis, tobacco, sorghum and rice. *Nucleic Acids Research, 41*, e188.

Jinek, M., Chylinski, K., Fonfara, I., Hauer, M., Doudna, J. A., & Charpentier, E. (2012). A programmable dual-RNA-guided DNA endonuclease in adaptive bacterial immunity. *Science, 337*, 816–821.

Koren, S., Schatz, M. C., Walenz, B. P., Martin, J., Howard, J. T., Ganapathy, G., Wang, Z., et al. (2012). Hybrid error correction and de novo assembly of single-molecule sequencing reads. *Nature Biotechnology, 30*, 693–700.

Kumar, V., Singh, A., Mithra, S. V., Krishnamurthy, S. L., Parida, S. K., Jain, S., Tiwari, K. K., et al. (2015). Genome-wide association mapping of salinity tolerance in rice (Oryza sativa). *DNA Research, 22*, 133–145.

Lambert, B., Denolf, P., Engelen, S., Golds, T., Haesendonckx, B., Ruiter, R., Robbens, S., et al. (2015). Omics-directed reverse genetics enables the creation of new productivity traits for the vegetable oil crop canola. *Procedia Environmental Sciences, 29*, 77–78.

Larkan, N. J., Raman, H., Lydiate, D. J., Robinson, S. J., Yu, F., Barbulescu, D. M., Raman, R., et al. (2016). Multi-environment QTL studies suggest a role for cysteine-rich protein kinase genes in quantitative resistance to blackleg disease in Brassica napus. *BMC Plant Biology, 16*, 183.

Li, R., Zhu, H., Ruan, J., Qian, W., Fang, X., Shi, Z., Li, Y., et al. (2010). De novo assembly of human genomes with massively parallel short read sequencing. *Genome Research, 20*, 265–272.

Li, H., Peng, Z., Yang, X., Wang, W., Fu, J., Wang, J., Han, Y., et al. (2013). Genome-wide association study dissects the genetic architecture of oil biosynthesis in maize kernels. *Nature Genetics, 45*, 43–50.

Li, Y. H., Zhou, G., Ma, J., Jiang, W., Jin, L. G., Zhang, Z., Guo, Y., et al. (2014). De novo assembly of soybean wild relatives for pan-genome analysis of diversity and agronomic traits. *Nature Biotechnology, 32*, 1045–1052.

Li, R., Hsieh, C. L., Young, A., Zhang, Z., Ren, X., & Zhao, Z. (2015). Illumina synthetic long read sequencing allows recovery of missing sequences even in the "finished" *C. elegans* Genome. *Scientific Reports, 5*, 10814.

Li, J., Meng, X., Zong, Y., Chen, K., Zhang, H., Liu, J., Li, J., et al. (2016). Gene replacements and insertions in rice by intron targeting using CRISPR-Cas9. *Nat. Plants, 2*, 16139.

Lin, K., Zhang, N., Severing, E. I., Nijveen, H., Cheng, F., Visser, R. G., Wang, X., et al. (2014). Beyond genomic variation--Comparison and functional annotation of three Brassica rapa genomes: A turnip, a rapid cycling and a Chinese cabbage. *BMC Genomics, 15*, 250.

Lin, Z., Cogan, N. O. I., Pembleton, L. W., Spangenberg, G. C., Forster, J. W., Hayes, B. J., & Daetwyler, H. D. (2016). Genetic gain and inbreeding from genomic selection in a simulated commercial breeding program for perennial ryegrass. *Plant Genome, 9*.

Liu, L., & Fan, X. D. (2014). CRISPR-Cas system: A powerful tool for genome engineering. *Plant Molecular Biology, 85*, 209–218.

Loman, N. J., Quick, J., & Simpson, J. T. (2015). A complete bacterial genome assembled de novo using only nanopore sequencing data. *Nature Methods, 12*, 733–735.

Lu, F., Romay, M. C., Glaubitz, J. C., Bradbury, P. J., Elshire, R. J., Wang, T., Li, Y., Li, Y., Semagn, K., Zhang, X., Hernandez, A. G., Mikel, M. A., Soifer, I., Barad, O., & Buckler, E. S. (2015). High-resolution genetic mapping of maize pan-genome sequence anchors. *Nature Communications, 6*, 6914.

Margulies, M., Egholm, M., Altman, W. E., Attiya, S., Bader, J. S., Bemben, L. A., Berka, J., et al. (2005). Genome sequencing in microfabricated high-density picolitre reactors. *Nature, 437*, 376–380.

Meuwissen, T. H. E., Hayes, B. J., & Goddard, M. E. (2001). Prediction of total genetic value using genome-wide dense marker maps. *Genetics, 157*, 1819–1829.

Mikami, M., Toki, S., & Endo, M. (2016). Precision targeted mutagenesis via Cas9 paired nickases in rice. *Plant & Cell Physiology, 57*, 1058–1068.

Miller, J. R., Koren, S., & Sutton, G. (2010). Assembly algorithms for next-generation sequencing data. *Genomics, 95*, 315–327.

Montenegro, J. D., Golicz, A. A., Bayer, P. E., Hurgobin, B., Lee, H., Chan, C. K., Visendi, P., Lai, K., Dolezel, J., Batley, J., & Edwards, D. (2017). The pangenome of hexaploid bread wheat. *The Plant Journal, 90*, 1007. https://doi.org/10.1111/tpj.13515.

Mueller, N. D., Gerber, J. S., Johnston, M., Ray, D. K., Ramankutty, N., & Foley, J. A. (2012). Closing yield gaps through nutrient and water management. *Nature, 490*, 254–257.

Munns, R., James, R. A., Xu, B., Athman, A., Conn, S. J., Jordans, C., Byrt, C. S., et al. (2012). Wheat grain yield on saline soils is improved by an ancestral Na+ transporter gene. *Nature Biotechnology, 30*, 360–364.

Myers, E. W., Sutton, G. G., Delcher, A. L., Dew, I. M., Fasulo, D. P., Flanigan, M. J., Kravitz, S. A., et al. (2000). A whole-genome assembly of Drosophila. *Science, 287*, 2196–2204.

Oerke, E. C. (2006). Crop losses to pests. *The Journal of Agricultural Science, 144*, 31–43.

Paliwal, R., Roder, M. S., Kumar, U., Srivastava, J. P., & Joshi, A. K. (2012). QTL mapping of terminal heat tolerance in hexaploid wheat (*T. aestivum* L.). *Theoretical and Applied Genetics, 125*, 561–575.

Pattanayak, V., Lin, S., Guilinger, J. P., Ma, E. B., Doudna, J. A., & Liu, D. R. (2013). High-throughput profiling of off-target DNA cleavage reveals RNA-programmed Cas9 nuclease specificity. *Nature Biotechnology, 31*, 839–843.

Pennisi, E. (2013). The CRISPR craze. *Science, 341*, 833–836.

Pevzner, P. A., & Tang, H. X. (2001). Fragment assembly with double-barreled data. *Bioinformatics, 17*, Suppl. 1, 225–Suppl. 1, 233.

Pevzner, P. A., Tang, H. X., & Waterman, M. S. (2001). An Eulerian path approach to DNA fragment assembly. *Proceedings of the National Academy of Sciences of the United States of America, 98*, 9748–9753.

Poland, J. A., & Rife, T. W. (2012). Genotyping-by-sequencing for plant breeding and genetics. *Plant Genome, 5*, 92–102.

Poland, J., Endelman, J., Dawson, J., Rutkoski, J., Wu, S. Y., Manes, Y., Dreisigacker, S., et al. (2012a). Genomic selection in wheat breeding using genotyping-by-sequencing. *Plant Genome, 5*, 103–113.

Poland, J. A., Brown, P. J., Sorrells, M. E., & Jannink, J. L. (2012b). Development of high-density genetic maps for barley and wheat using a novel two-enzyme genotyping-by-sequencing approach. *PLoS One, 7*, e32253.

Rabbi, I. Y., Hamblin, M. T., Kumar, P. L., Gedil, M. A., Ikpan, A. S., Jannink, J. L., & Kulakow, P. A. (2014). High-resolution mapping of resistance to cassava mosaic geminiviruses in cassava using genotyping-by-sequencing and its implications for breeding. *Virus Research, 186*, 87–96.

Ran, F. A., Hsu, P. D., Lin, C. Y., Gootenberg, J. S., Konermann, S., Trevino, A. E., Scott, D. A., et al. (2013). Double nicking by RNA-guided CRISPR Cas9 for enhanced genome editing specificity. *Cell, 154*, 1380–1389.

Rhoads, A., & Au, K. F. (2015). PacBio sequencing and its applications. *Genomics, Proteomics & Bioinformatics, 13*, 278–289.

Ronald, P. C. (2014). Lab to farm: Applying research on plant genetics and genomics to crop improvement. *PLoS Biology, 12*, e1001878.

Rutkoski, J. E., Poland, J. A., Singh, R. P., Huerta-Espino, J., Bhavani, S., Barbier, H., Rouse, M. N., et al. (2014). Genomic selection for quantitative adult plant stem rust resistance in wheat. *Plant Genome, 7*, 1–10.

Saxena, R. K., Edwards, D., & Varshney, R. K. (2014). Structural variations in plant genomes. *Briefings in Functional Genomics, 13*, 296–307.

Schatz, M. C., Maron, L. G., Stein, J. C., Hernandez Wences, A., Gurtowski, J., Biggers, E., Lee, H., et al. (2014). Whole genome de novo assemblies of three divergent strains of rice, *Oryza sativa*, document novel gene space of aus and indica. *Genome Biology, 15*, 506.

Scheben, A., & Edwards, D. (2017). Genome editors take on crops. *Science, 355*, 1122–1123.

Scheben, A., Yuan, Y., & Edwards, D. (2016). Advances in genomics for adapting crops to climate change. *Current Plant Biology, 6*, 2–10.

Scheben, A., Batley, J., & Edwards, D. (2017). Genotyping-by-sequencing approaches to characterize crop genomes: Choosing the right tool for the right application. *Plant Biotechnology Journal, 15*, 149–161.

Septiningsih, E. M., Pamplona, A. M., Sanchez, D. L., Neeraja, C. N., Vergara, G. V., Heuer, S., Ismail, A. M., et al. (2009). Development of submergence-tolerant rice cultivars: The *Sub1* locus and beyond. *Annals of Botany, 103*, 151–160.

Shen, B., Zhang, W. S., Zhang, J., Zhou, J. K., Wang, J. Y., Chen, L., Wang, L., et al. (2014). Efficient genome modification by CRISPR-Cas9 nickase with minimal off-target effects. *Nature Methods, 11*, 399–402.

Shi, J., Gao, H., Wang, H., Lafitte, H. R., Archibald, R. L., Yang, M., Hakimi, S. M., et al. (2017). ARGOS8 variants generated by CRISPR-Cas9 improve maize grain yield under field drought stress conditions. *Plant Biotechnology Journal, 15*, 207.

Simpson, J. T., Wong, K., Jackman, S. D., Schein, J. E., Jones, S. J., & Birol, I. (2009). ABySS: A parallel assembler for short read sequence data. *Genome Research, 19*, 1117–1123.

Skinner, J., Szucs, P., von Zitzewitz, J., Marquez-Cedillo, L., Filichkin, T., Stockinger, E. J., Thomashow, M. F., et al. (2006). Mapping of barley homologs to genes that regulate low temperature tolerance in *Arabidopsis. Theoretical and Applied Genetics, 112*, 832–842.

Soyk, S., Muller, N. A., Park, S. J., Schmalenbach, I., Jiang, K., Hayama, R., Zhang, L., et al. (2017). Variation in the flowering gene SELF PRUNING 5G promotes day-neutrality and early yield in tomato. *Nature Genetics, 49*, 162–168.

Spindel, J. E., & McCouch, S. R. (2016). When more is better: How data sharing would accelerate genomic selection of crop plants. *The New Phytologist, 212*, 814–826.

Stankova, H., Hastie, A. R., Chan, S., Vrana, J., Tulpova, Z., Kubalakova, M., Visendi, P., et al. (2016). BioNano genome mapping of individual chromosomes supports physical mapping and sequence assembly in complex plant genomes. *Plant Biotechnology Journal, 14*, 1523–1531.

Sun, Y., Zhang, X., Wu, C., He, Y., Ma, Y., Hou, H., Guo, X., et al. (2016). Engineering herbicide-resistant rice plants through CRISPR/Cas9-mediated homologous recombination of acetolactate synthase. *Molecular Plant, 9*, 628–631.

Svitashev, S., Young, J. K., Schwartz, C., Gao, H. R., Falco, S. C., & Cigan, A. M. (2015). Targeted mutagenesis, precise gene editing, and site-specific gene insertion in maize using Cas9 and guide RNA. *Plant Physiology, 169*, 931–945.

Svitashev, S., Schwartz, C., Lenderts, B., Young, J. K., & Mark Cigan, A. (2016). Genome editing in maize directed by CRISPR-Cas9 ribonucleoprotein complexes. *Nature Communications, 7*, 13274.

Szalay, T., & Golovchenko, J. A. (2015). De novo sequencing and variant calling with nanopores using PoreSeq. *Nature Biotechnology, 33*, 1087–1091.

Tester, M., & Langridge, P. (2010). Breeding technologies to increase crop production in a changing world. *Science, 327*, 818–822.

Tettelin, H., Masignani, V., Cieslewicz, M. J., Donati, C., Medini, D., Ward, N. L., Angiuoli, S. V., et al. (2005). Genome analysis of multiple pathogenic isolates of *Streptococcus agalactiae*: Implications for the microbial "pan-genome". *Proceedings of the National Academy of Sciences of the United States of America, 102*, 13950–13955.

Tollenaere, R., Hayward, A., Dalton-Morgan, J., Campbell, E., Lee, J. R. M., Lorenc, M. T., Manoli, S., et al. (2012). Identification and characterization of candidate *Rlm4* blackleg resistance genes in *Brassica napus* using next-generation sequencing. *Plant Biotechnology Journal, 10*, 709–715.

Trethowan, R. M., Reynolds, M., Sayre, K., & Ortiz-Monasterio, I. (2005). Adapting wheat cultivars to resource conserving farming practices and human nutritional needs. *The Annals of Applied Biology, 146*, 405–413.

United Nations. (2015). *World population prospects: The 2015 revision*. New York, NY: Population Division of the Department of Economic and Social Affairs.

Valluru, R., Reynolds, M. P., Davies, W. J., & Sukumaran, S. (2017). Phenotypic and genome-wide association analysis of spike ethylene in diverse wheat genotypes under heat stress. *The New Phytologist, 214*, 271.

Waltz, E. (2016). CRISPR-edited crops free to enter market, skip regulation. *Nature Biotechnology, 34*, 582.

Wang, W., Vinocur, B., & Altman, A. (2003). Plant responses to drought, salinity and extreme temperatures: Towards genetic engineering for stress tolerance. *Planta, 218*, 1–14.

Wang, Y. P., Cheng, X., Shan, Q. W., Zhang, Y., Liu, J. X., Gao, C. X., & Qiu, J. L. (2014). Simultaneous editing of three homoeoalleles in hexaploid bread wheat confers heritable resistance to powdery mildew. *Nature Biotechnology, 32*, 947–951.

Wang, S. H., Zhang, S. B., Wang, W. X., Xiong, X. Y., Meng, F. R., & Cui, X. (2015). Efficient targeted mutagenesis in potato by the CRISPR/Cas9 system. *Plant Cell Reports, 34*, 1473–1476.

Wang, F. J., Wang, C. L., Liu, P. Q., Lei, C. L., Hao, W., Gao, Y., Liu, Y. G., et al. (2016). Enhanced rice blast resistance by CRISPR/Cas9-targeted mutagenesis of the ERF transcription factor gene *OsERF922*. *PLoS One, 11*, e0154027.

Winfield, M. O., Allen, A. M., Burridge, A. J., Barker, G. L. A., Benbow, H. R., Wilkinson, P. A., Coghill, J., et al. (2016). High-density SNP genotyping array for hexaploid wheat and its secondary and tertiary gene pool. *Plant Biotechnology Journal, 14*, 1195–1206.

Xu, Y. B., & Crouch, J. H. (2008). Marker-assisted selection in plant breeding: From publications to practice. *Crop Science, 48*, 391–407.

Yang, J., Liu, D., Wang, X., Ji, C., Cheng, F., Liu, B., Hu, Z., et al. (2016). The genome sequence of allopolyploid Brassica juncea and analysis of differential homoeolog gene expression influencing selection. *Nature Genetics, 48*, 1225–1232.

Yuan, Y., Bayer, P. E., Batley, J., & Edwards, D. (2017). Improvements in genomic technologies: Application to crop genomics. *Trends in Biotechnology, 35*, 547–558.

Zerbino, D. R., & Birney, E. (2008). Velvet: Algorithms for de novo short read assembly using de Bruijn graphs. *Genome Research, 18*, 821–829.

Zhang, H., Zhang, J., Wei, P., Zhang, B., Gou, F., Feng, Z., Mao, Y., et al. (2014). The CRISPR/Cas9 system produces specific and homozygous targeted gene editing in rice in one generation. *Plant Biotechnology Journal, 12*, 797–807.

Zhang, X., Perez-Rodriguez, P., Semagn, K., Beyene, Y., Babu, R., Lopez-Cruz, M. A., Vicente, F. S., et al. (2015). Genomic prediction in biparental tropical maize populations in water-stressed and well-watered environments using low-density and GBS SNPs. *Heredity, 114*, 291–299.

Zhang, J., Ratanasirintrawoot, S., Chandrasekaran, S., Wu, Z., Ficarro, S. B., Yu, C., Ross, C. A., et al. (2016). LIN28 regulates stem cell metabolism and conversion to primed pluripotency. *Cell Stem Cell, 19*, 66–80.

Zhao, K., Tung, C. W., Eizenga, G. C., Wright, M. H., Ali, M. L., Price, A. H., Norton, G. J., et al. (2011). Genome-wide association mapping reveals a rich genetic architecture of complex traits in *Oryza sativa*. *Nature Communications, 2*, 467.

Zhou, S., Bechner, M. C., Place, M., Churas, C. P., Pape, L., Leong, S. A., Runnheim, R., et al. (2007). Validation of rice genome sequence by optical mapping. *BMC Genomics, 8*, 278.

Zhou, S., Wei, F., Nguyen, J., Bechner, M., Potamousis, K., Goldstein, S., Pape, L., et al. (2009). A single molecule scaffold for the maize genome. *PLoS Genetics, 5*, e1000711.

Zimin, A. V., Marcais, G., Puiu, D., Roberts, M., Salzberg, S. L., & Yorke, J. A. (2013). The MaSuRCA genome assembler. *Bioinformatics, 29*, 2669–2677.

Chapter 6
Climate-Resilient Future Crop: Development of C_4 Rice

Hsiang Chun Lin, Robert A. Coe, W. Paul Quick, and Anindya Bandyopadhyay

Abstract Rice is the most important crop in the world. It is a staple food for more than half of the human population and a primary food source for the world's poorest people. Asia currently accounts for 90% of global rice production but it will need to increase this by 50% within the next 30 years. By this time the region will be home to nearly 90% of the global population increase and will likely be experiencing extreme climatic conditions. Agriculture will be challenged by diminishing water resources, reduced nutrient inputs and an increase in abiotic stresses. Rice yield increases have already stagnated and so a new paradigm is needed to meet these future challenges. Most crop plants, like rice and wheat, have a simple and less efficient photosynthetic mechanism (C_3 photosynthesis) that as a consequence results in considerable loss of water through stomatal pores on their leaves that open widely to let in more carbon dioxide. They also make a large amount of photosynthetic protein to maximise their photosynthetic rate that requires a large investment of nitrogen and hence fertiliser application. However, a few plants have evolved a more efficient C_4 photosynthetic pathway that greatly alleviates these problems. The installation of a C_4 photosynthetic pathway into major crops like rice could potentially increase yields by 50%, double the water-use efficiency and reduce fertiliser use by 40%. This is because plants with a C_4 photosynthetic pathway concentrate CO_2 within the leaf prior to photosynthetic fixation leading to increased photosynthetic efficiency and large reductions in the requirement for scarce resources like water and nitrogen (fertiliser). These modifications would be particularly advantageous in future climate scenarios where water scarcity and global temperature are predicted to increase.

Keywords C_4 photosynthesis · Rubisco · Photorespiration · C_4 evolution · Rice

---

H. C. Lin · R. A. Coe · W. P. Quick · A. Bandyopadhyay (✉)
C_4 Rice Centre, International Rice Research Institute (IRRI), Los Baños, Philippines
e-mail: anindya.bandyopadhyay@syngenta.com

© Springer Nature Switzerland AG 2019
A. Sarkar et al. (eds.), *Sustainable Solutions for Food Security*,
https://doi.org/10.1007/978-3-319-77878-5_6

6.1 Rice Production

Almost all of the 700 million tons of rice grains produced each year are consumed directly by three billion people. Rice production has more than doubled from 252 million tons in 1966 to 600 million tons in 2000, closely matching population growth (Mitchell and Sheehy 2006). However, the global population is set to reach nine billion people by 2050, and this means rice yields will have to be increased further by an estimated 50% in the world (Fig. 6.1). In addition, climate change will lead to an increase in high temperatures and weather extremes, which are likely to negatively impact crop yield. While a second green revolution has been called for it is unlikely that current breeding programs alone will be able to meet these challenges. The traditional breeding targets for increased yield potential in rice, grain number and harvest index have been exhausted; further improvements in this area cannot meet future needs. Instead, biomass and photosynthetic performance are the new breeding targets (J. Sheehy et al. 2007). C_4 photosynthesis provides a biological precedent for increasing radiation use efficiency (RUE). It has been predicted that introducing a C_4 photosynthetic pathway into rice could increase yields by as much as 50% (Hibberd et al. 2008; Mitchell and Sheehy 2006). The associated improvements to nitrogen and water use efficiency will leader to greater resilience to abiotic stresses associated with climate change (von Caemmerer et al. 2012; J. Sheehy et al. 2007). In this chapter we outline the potential contribution of C_4 photosynthesis to

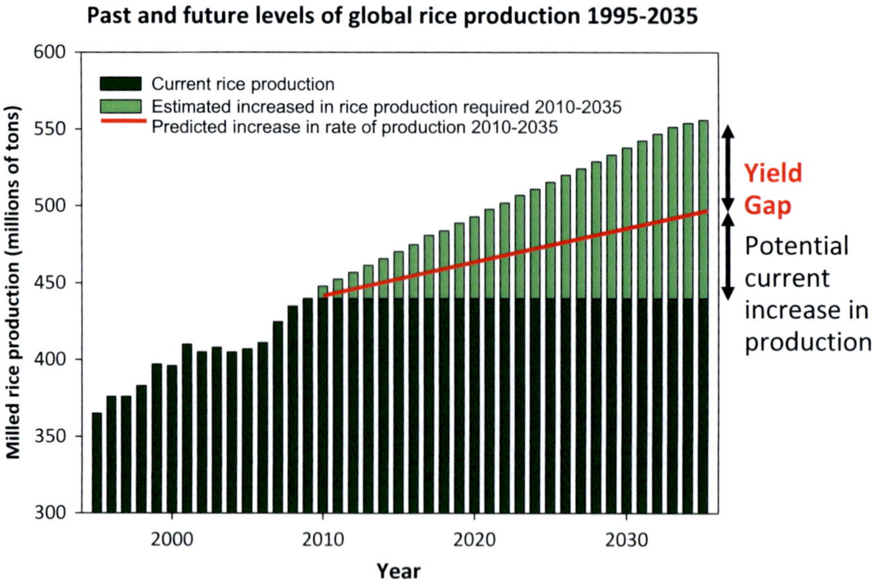

Fig. 6.1 Rice production from 1995 to 2010 and predicted increase in the rate of production required from 2010 to 2035

climate change and the current strategy of The International C₄ Rice Consortium for engineering it into rice.

6.2 The Problem with Photosynthesis

Life on Earth is dependent on photosynthesis to form carbohydrates as the basis of the global food chain. Most plant possess a C_3 photosynthetic pathway, in which carbon dioxide (CO_2) is fixed by an enzyme called ribulose bisphosphate carboxylase oxygenase (Rubisco), this leads to the generation of a three-carbon compound (Fig. 6.2a). A major problem with type of photosynthesis is that Rubisco can also fix oxygen (O_2). When this occurs, a toxic metabolite called phosphoglycolate is formed. This has to be broken down in a series of energy-consuming reactions known as photorespiration, which leads to a net loss of carbon. The ratio of carboxylation (fixation of CO_2) to oxygenation (fixation of O_2) reactions depends on the ratio of available CO_2 and O_2 around Rubisco. When C_3 photosynthesis evolved the

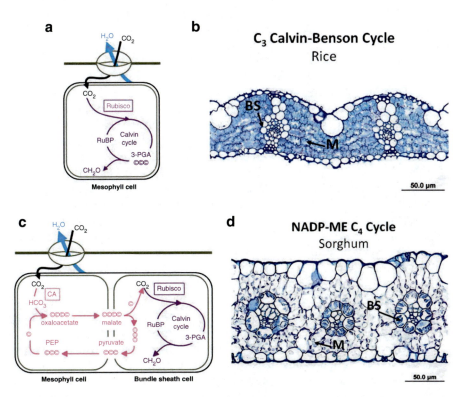

Fig. 6.2 Schematic of a C_3 (**a**) and C_4 (**c**) photosynthetic pathway. Light microscopy images of transverse sections of leaves of rice (**b**) (C_3) and sorghum (**d**) (C_4) (redrawn from von Caemmerer et al. 2012)

concentration of O_2 in the atmosphere was very low and so O_2 was rarely fixed. Today, this oxygenation reaction can lead to loss of up to 25% of carbon fixed by photosynthesis, limiting the yield of many crop species (Leegood 2013). Nature has evolved strategies to overcome the effect of O_2 fixation by Rubisco, the most common is called the C_4 CO_2-concentrating mechanism. In this type of photosynthesis the classical C_3 pathway is moved from the mesophyll cell to a different, highly specialised cell type called the bundled sheath cell (Fig. 6.2). A metabolic pathway is then installed to act like a pump that fills this cell with CO_2, crucially excluding O_2 and thus leading to a reduction in photorespiration. This pathway is called C_4 because CO_2 is first fixed by an oxygen-insensitive enzyme called carboxylase phosphoenolpyruvate carboxylase (PEPCase) into a four-carbon compound (Fig. 6.2c).

6.3 Advantages of the C_4 Photosynthetic Pathway

The C_4 pathway leads to CO_2 concretions inside of the bundle sheath cells of 70 µM compared to 4 µM in mesophyll cells (Jenkins et al. 1989). This leads to a reduction in the oxygenase reaction of Rubisco by more than 80%, although this is highly dependent on the temperature (Kanai and Edwards 2001; Sage and Zhu 2011). As a result, C_4 plants can have up to a twofold increase in photosynthetic rate (Fig. 6.3a) (von Caemmerer et al. 2012), twofold increase in yield (Fig. 6.3b) (Monteith 1978), 1.5 to 3-fold increase in water-use efficiency (WUE) (Fig. 6.3c) (Kocacinar et al. 2008) and a 2.5-fold increase in nitrogen-use efficiency (NUE) (Fig. 6.3d) (Evans and von Caemmerer 2000) compared with ecologically similar C_3 plants. In additional, elevating the concentration of CO_2 within the bundle sheath allows Rubisco to increase its in vivo catalytic activity two- to fivefold in warm climates (Sage and Zhu 2011). As a result C_4 plants also only contain 50–80% of Rubisco of C_3 plants (Ghannoum et al. 2011). This enables them to maintain a higher leaf area production rate at lower leaf nitrogen levels than C_3 species. Water use efficiency is improved because CO_2 is fixed more efficiently and stomata do not have open as widely, leading to lower rates of transpiration. Higher WUE can promote C_4 plants to have longer growing seasons and exhibit more flexible allocation patterns, for example more biomass to shoots in moist environments or to roots in dry environments (Taylor et al. 2010). The benefits of enhanced WUE and NUE in C_4 photosynthesis are the potential to reduce agronomic costs and environmental impacts by lowering water and fertiliser use. In warm climate, C_4 plants also exhibit a 50% greater radiation use efficiency (RUE) than C_3 plants (Figs. 6.3e and 6.4) (Sheehy et al. 2007), because C_4 plants can build up a leaf canopy sooner than C_3 plants of similar growth form. The C_4 pathway is beneficial in environments with high temperature and strong light conditions that promote high rates of photorespiration. Although C_4 species are relatively few compared with the more numerous C_3 plants (~7500 C_4 species to nearly 250,000 C_3 species), they produce nearly a quarter of the primary productivity and dominate the grassland and savannah biomes of warm-temperature to tropical latitudes (Edwards et al. 2010; Sage 1999; Still et al. 2003).

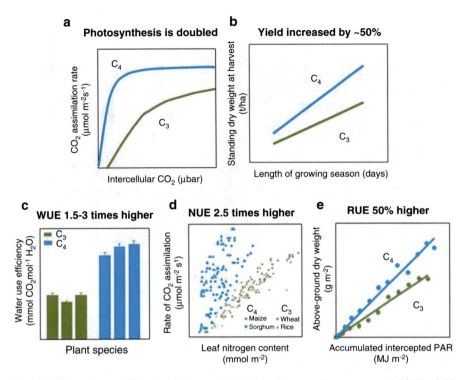

Fig. 6.3 Comparisons of photosynthetic rate (**a**) (redrawn from von Caemmerer et al. 2012), yield (**b**) (redrawn from Monteith 1978), water use efficiency (**c**) (redrawn from Kocacinar et al. 2008), nitrogen use efficiency (**d**) (redrawn from Evans and von Caemmerer 2000) and radiation use efficiency (**e**) (redrawn from Sheehy et al. 2007) between C_4 and C_3 plants

This is because as temperatures increase, the relative solubility of CO_2 reduces, leading to higher rates of photorespiration (Ku and Edwards 1977), and hence C_4 plants have a competitive advantage in these environments.

6.4 Climate-Resilient Crops: The Importance of C_4 Photosynthesis

The atmospheric concentration of CO_2 in 2016 was 400 ppm, the intergovernmental Panel on Climate Change (IPCC) predicts that by 2050 atmospheric CO_2 will be more than 450 ppm. At the same time global temperature will increase by 0.8–2.6 °C. The effect of rising CO_2 on C_3 and C_4 plants has been studied, it has been reported that C_4 photosynthesis would be modestly enhanced (5–25% on average) with a doubling of the atmospheric CO_2 concentration. By comparison, C_3 photosynthesis would be two- to threefold greater (Ainsworth and Long 2005; Ghannoum et al. 2011). Although elevated CO_2 is predicted to favour increased productivity in

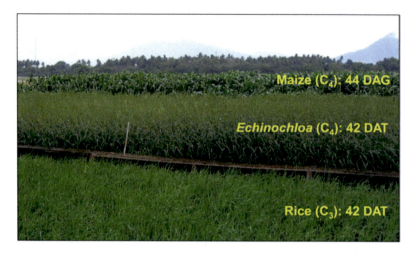

Fig. 6.4 A photograph of the field trial shown maize (C₄), a high yielding rice variety (IR72, C₃), and the major rice weed *Echinochloa glabrescens* (C₄) were grown side by side on well-fertilised soils at the International Rice Research Institute (IRRI) in the Philippines during the dry season of 2006. Photo courtesy of John Sheehy of IRRI

C_3 plants, when combined with increases in temperature and soil water content, the productivity of C_4 species would remain greater than C_3 species (Morgan et al. 2011). For example, rice yield are predicted to decline by 10% for each 1 °C increase in mean minimum temperature during the cropping season (Peng et al. 2004). This provides support for the idea the engineering C_4 photosynthesis into C_3 crops species could help mitigate the effect of climate change and increase crop yields. To achieve this, an understanding of how they C_4 pathway evolved is needed.

6.5 Evolution of the C₄ Pathway

The C_4 photosynthetic pathway has evolved more than 60 times in different plant lineages over the past 30 million years (Sage et al. 2011), it is one of the most remarkable examples of convergent evolution. It is thought that C_4 photosynthesis evolved as a result of seven distinct phases (Fig. 6.5; Sage 2004). During phase 1 and 2, plants are thought to have undergone genetic and anatomical precondition, this may have involved the duplication of genes and a reduction in the distance between mesophyll and bundle sheath cells. Despite these changes plants would still have been classified as C_3. During phase 3 there, the number of organelles within the bundle sheath cells would have increased, leading to plants that could be described as being C_3-like. During phase 4, it is though that there would have been changes to photorespiration with the location of glycine decarboxylase becoming restricted to the bundle sheath cells rather than the mesophyll cell. Concomitant with this would have been an increase in the expression of PEPCase within the

6 Climate-Resilient Future Crop: Development of C₄ Rice

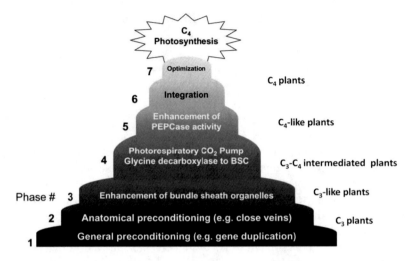

Fig. 6.5 A conceptual model of the main phases of C₄ evolution. Modified from Sage (2004)

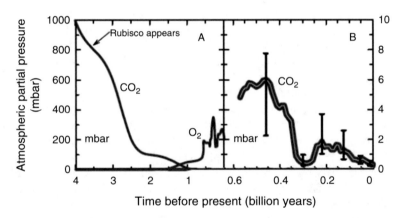

Fig. 6.6 Modelled atmospheric CO_2 partial pressure over the past 4 billion years (**a**) and 0.6 billion years (**b**) during the history of earth. Modelled atmospheric O_2 levels over the past 4 billion years (**a**). Modified from Sage (2004)

mesophyll cells by phase 5 and finally the emergence of C₄-like plants. By phase 6 and 7, plants would have undergone changes in the expression and kinetic properties of photosynthetic enzymes and exhibit the full set of C₄ characteristics.

C₄ photosynthesis is thought to have first evolved in grasses during the Oligocene epoch (24–35 million years ago). However, most C₄ lineages are estimated to have appeared within the last 5 million years (Sage 2004). This was a period when atmospheric CO_2 levels declined to near current CO_2 levels (Fig. 6.6). There were also changes to the global climate with the emergence of dry conditions in the subtropics and temperate zones (Sage 2004). C₄ photosynthesis is thought to have originated in drought regions where the combined effects of heat, high light, drought and/or salinity led to high rates of photorespiration.

6.6 Engineering C_4 Photosynthetic Biochemistry

C_4 photosynthesis represents a complex restructuring of leaf anatomy and physiology as well as numerous biochemical alterations which are likely controlled through dozens of genetic modifications to C_3 photosynthesis (Hibberd and Covshoff 2010). Earlier attempts to generate a single-celled C_4 pathway in rice did not lead to a functional C_4 cycle in leaves of C_3 species (Miyao et al. 2011; Taniguchi et al. 2008, Leegood 2002). Here we outline the background and strategy behind the strategy of the C_4 Rice Consortium, which is to engineer a NADP-dependent malic enzyme (NADP-ME) subtype of C_4 photosynthesis. This has been selected because of all the C_4 subtypes it is the most well-characterised and potentially requires the fewest enzymes and transporters (Kajala et al. 2011; Weber and von Caemmerer 2010). The initial step of this pathway is fixation of CO_2 as bicarbonate (HCO_3^-), this is then incorporated with phosphoenolpyruvate (PEP) by PEPCase into oxaloacetic acid (OAA) within the mesophyll cells. OAA then enters the mesophyll chloroplast where it is reduced to malate by NADP-malate dehydrogenase (MDH) (Fig. 6.7) (Furbank 2011; Hatch et al. 1975). Malate is transferred into, and subsequently decarboxylated by NADP-ME within the bundle sheath cells where the CO_2 concentration around Rubisco can be elevated by up to tenfold atmospheric levels (Hatch 1999). The product of decarboxylation steps, pyruvate, is returned to the mesophyll chloroplast where it is phosphorylated to PEP by pyruvate, orthophosphate dikinase (PPDK).

The first step for engineering C_4 photosynthesis into rice is to transform the classical genes encoding the C_4 biochemical enzymes and metabolite transporters (Fig. 6.7). These must be expressed at high levels in the correct cell types. A subset of genes required for the primary biochemical pathway and the associated metabolite transporters have been cloned and coupled with cell-specific promoters (Kajala et al. 2011). The primary C_4 biochemical enzymes, carbonic anhydrate (CA), PEPCase, MDH and PPDK have been expressed in the mesophyll cells and NADP-ME in the bundle sheath cells. In addition, metabolic transporters are an important component of C_4 engineering, as supporting the increased fluxes of C_4 metabolites across membranes of chloroplast and mesophyll/bundle sheath cells. In particular, the fluxes of metabolites involved in the biochemical inorganic carbon pump of the C_4 cycle, such as malate, pyruvate, oxaloacetate and phospho*enol*pyruvate, must be considerably higher in C_4 plants than in C_3 plants. Candidate C_4 transporters have been identified from differential proteomic and transcriptomic studies (Manandhar-Shrestha et al. 2013; Majeran et al. 2005, 2008, 2010; Bräutigam et al. 2008). Three maize transporters, a putative OAA–malate antiporter (ZmDiT1), a putative dicarboxylate transporter (ZmDiT2g2), a PEP/phosphate translocator (ZmPPT1), and two *Setaria* transporters, a pyruvate transporter bile acid–sodium symporter family protein 2 (BASS2) and a sodium–proton antiporter (NHD1), have been cloned and coupled to the suitable promoters to give cell-specific expression in rice.

One of the initial steps in the evolution of the C_4 pathway is proposed to be the repositioning of photorespiration into the bundle sheath cells of C_3 leaves. Genes

6 Climate-Resilient Future Crop: Development of C₄ Rice

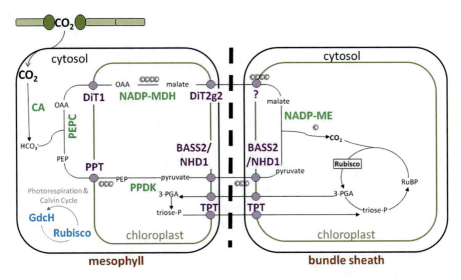

Fig. 6.7 The C₄ cycle in NADP-malic enzyme plants showing the primary biochemical component enzymes and metabolite transporters (Redrawn from Leegood 2013). Carbon dioxide (CO₂) is converted to bicarbonate (HCO₃⁻) by carbonic anhydrase (CA); this is then fixed by phospho*enol*pyruvate carboxylase (PEPCase) catalysing the formation of oxaloacetate (OAA) in the cytosol of mesophyll (M) cells. OAA is imported into the mesophyll chloroplasts by the oxoglutarate/malate transporter (DiT1) transporter, where it is reduced to malate by malate dehydrogenase (MDH). Malate is exported back to the cytosol, possibly by the plastidic dicarboxylate translocator (DiT2g2) and then diffuses into bundle sheath (BS) cells through plasmodesmata along a steep concentration gradient. In the BS cells, malate is transported into the chloroplast by an unknown transporter and oxidatively decarboxylated by NADP-dependent malic enzyme (NADP-ME), yielding CO₂, NADPH and pyruvate. CO₂ is assimilated by Rubisco, yielding two molecules of 3-PGA that can be used in the Calvin cycle in either the BS or M cells. For the latter, 3-PGA and TP are shuttled between the M and BS cells, with TP exported via the triose phosphate translocator (TPT). Pyruvate is transported from the BS cells into the chloroplast of the M cells where it is converted to PEP by pyruvate:phosphate dikinase (PPDK). The system for pyruvate transport, BASS 2/NHD1, a combined BILE ACID:SODIUM SYMPORTER FAMILY PROTEIN 2, pyruvate transporter/NHD1, sodium–proton antiporter. PEP leaves the chloroplast via the PEP/phosphate translocator (PPT) where it can enter a new cycle of the carbon dioxide shuttle

associated with C₃ photosynthesis need to be downregulated in rice mesophyll cells. This is being achieved by using artificial microRNA against the Rubisco small subunit (Rbcs) and H- subunit of glycine decarboxylase (GDCH) genes. This is necessary because in rice with a functional C₄ concentrating mechanism, Rubisco will be sequestered to the bundle sheath cells. The amount of Rubisco can also be reduced by two-thirds as catalytic turnover rate will be higher (von Caemmerer et al. 2012).

To mimic this initial stage, *gdch* knockdown transgenic rice lines have been generated with an artificial microRNA (amiRNAs) designed to preferentially target Os10g37180 (*OsGDCH*) transcripts for degradation in rice mesophyll cells (Lin et al. 2016). GDC H- and P-proteins were undetectable in leaves of *gdch* lines. Plants exhibited a photorespiratory-deficient phenotype with stunted growth, accelerated leaf senescence, reduced chlorophyll and increased glycine accumulation in leaves. In addition, chloroplast coverage of the mesophyll cell area and peripheral cell wall were reduced. The mild, nonlethal phenotype allows these *gdch* lines to be used as a building block in the development of a C_4 rice prototype.

6.7 Engineering C_4 Photosynthetic Anatomy

In most C_4 land plants, C_4 photosynthetic concentration mechanism involves two cell types, mesophyll (M) and bundle sheath (BS) cells, that are arranged in concentric circles around veins—the so-called Kranz anatomy (from the German word for wreath) (Fig. 6.1d). This arrangement generates a consistent interveinal distance of four cells (vein–BS–M–M–BS–vein) and so bundle sheath and mesophyll cells are in close proximity. The Kranz C_4 photosynthesis spatially separates carbon assimilation and reduction to mesophyll and bundle sheath cells, which is accompanied by the spatial distribution of dimorphic chloroplasts to two different cells types. The bundle sheath cells in C_4 plants are also larger than in C_3 plants and have more organelles, chloroplasts, mitochondria, and peroxisomes. Moreover, they are strongly connected with their neighbouring mesophyll cells through plasmodesmata, which facilitates high metabolite flux between these two cell types. In addition, C_4 plants often have suberised bundle sheath cell wall to prevents CO_2 leakage and thus to maintain the CO_2-enriched environment.

The second step for engineering C_4 rice is installing Kranz anatomy and activating bundle sheath organelles. The basic biochemistry of C_4 photosynthesis has been well characterised and the identity of most of the genes regulating this is known. However, the genes controlling for C_4 leaf anatomy or C_4 cell biology are still uncertain. To engineering C_4 leaf anatomy in to rice, we need to first determinate the genes controlling suberisation of bundle sheath cells, increased plasmodesmatal connectivity between mesophyll and bundle sheath cells, and the production of dimorphic chloroplasts and their intracellular positioning. The patterns of leaf vascular systems are highly varied in both C_3 and C_4 species. Typically C_3 leave have 9–10 mesophyll cells between adjacent bundle sheath cells (Fig. 6.1b), whereas in C_4 leaves, this is reduced to 2–3 mesophyll cells (Fig. 6.1d). Measurement of leaf vein density, defined as the total length of veins per unit area, demonstrated that veins are consistently more closely spaced in C_4 species than C_3 species. In order to understand the variation in vein density, two genetic screen approaches were used: a screen for emergence of C_4 characteristics in activation-tagged rice populations (Wan et al. 2009; Hsing et al. 2007; Jeong et al. 2002) and a screen for loss of C_4 characteristics in C_4 sorghum mutant populations (Rizal et al. 2015). The genes controlling vein spacing can been identified from mutant rice plants with reducing numbers of mesophyll cells or mutant sorghum plants with increasing numbers of mesophyll cells between veins. However, where redundancy or complex genetic interactions operate, such activations or knockouts mutant screens have rarely been successful. In the past decade, bioinformatics approaches have become a powerful tool for discovering the gene regulation required for developing Kranz anatomy. For example, a comparative transcriptome study of genes expressed between two types of C_4 plant maize leaf developments, husk leaves (tissues with more C_3-like vein spacing) and foliar leaves (tissues with C_4 vein spacing) (Wang et al. 2013). These data provide us a list of putative genetic regulators for the induction of Kranz anatomy development that can be used for C_4 engineering.

Evolutionary studies suggest that C_3 plants may be in some way have become preconditioned to C_4 photosynthesis (Sage et al. 2011). It is thought that no novel genes are associated with the C_4 pathway. As such we hypothesise that C_4 photosynthesis evolved in C_3 plants as a result of a series of gene activations and deactivations which resulted in more efficient photosynthesis. To test this, the rice and *Setaria viridis* mutant populations are also screened for the gain and loss of C_4 function at low CO_2 concentration (30 ppm), because the carbon-concentrating mechanism of C_4 plants allow them to carry out net carbon assimilation at low CO_2 concentration (Fig. 6.8). S. *viridis* is a new model organism for studying the C_4 pathway in monocots due to its rapid generation time, small stature, high seed production, diploid status, and small genome size (Brutnell et al. 2010). *Agrobacterium*-mediated genetic transformation in *S. viridis* is also developed and reasonable efficiency (Osborn et al. 2016). This system can be used to validate C_4 photosynthetic models an examine enzyme regulation.

Fig. 6.8 CO_2 growth response of rice and *Setaria*. At 17 day after germination rice (*Oryza sativa* IR64) (**a**) and *Setaria viridis* A10 (**b**) grown under ambient conditions within the screenhouse were transferred into nine chambers set at different CO_2 concentration (30–400 ppm) for 14 days

6.8 Conclusions

Engineering C_4 photosynthesis into rice is one of the many possibilities for improving photosynthetic capacity in rice. Future climates are expected to be more variable. A C_4 rice has the potential to deliver a very large improvement in yield for future conditions; it would double water-use efficiency and increase nitrogen use efficiency. Knowledge gained during the C_4 rice engineering will advance our understanding of C_3 and C_4 photosynthesis, and we can also use this knowledge to improve the efficiency of existing C_4 crops for food security and to provide novel biofuel feedstocks.

References

Ainsworth, E. A., & Long, S. P. (2005). What have we learned from 15 years of free-air CO2 enrichment (FACE)? A meta-analytic review of the responses of photosynthesis, canopy properties and plant production to rising CO2. *New Phytologist, 165*(2), 351–372.

Bräutigam, A., Hofmann-Benning, S., & Weber, A. P. M. (2008). Comparative proteomics of chloroplast envelopes from C(3) and C(4) plants reveals specific adaptations of the plastid envelope to C(4) photosynthesis and candidate proteins required for maintaining C(4) metabolite fluxes. *Plant Physiology, 148*(1), 568–579.

Brutnell, T. P., et al. (2010). Setaria viridis: A model for C4 photosynthesis. *The Plant Cell, 22*(8), 2537–2544.

von Caemmerer, S., Quick, W. P., & Furbank, R. T. (2012). The development of C4 rice: Current progress and future challenges. *Science (New York, N.Y.), 336*(6089), 1671–1672.

Edwards, E. J., Smith, S. A., & Crossing Environmental Thresholds. (2010). The origins of C 4 grasslands: Integrating evolutionary and ecosystem science. *Science, 328*(April), 587–590.

Evans, J. R., & von Caemmerer, S. (2000). Would C4 rice produce more biomass than C3 rice? In J. E. Sheehy, P. L. Mitchell, & B. Hardy (Eds.), *Redesigning rice photosynthesis to increase yield* (pp. 53–72). Amsterdam: Elsevier.

Furbank, R. T. (2011). Evolution of the C 4 photosynthetic mechanism: Are there really three C 4 acid decarboxylation types? *Journal of Experimental Botany, 62*(9), 3103–3108.

Ghannoum, O., Evans, J. R., & von Caemmerer, S. (2011). Nitrogen and water use efficiency in C4 plants. In A. S. Raghavendra & R. F. Sage (Eds.), *C4 photosynthesis and related CO2 concentrating mechanisms* (pp. 129–146). Dordrecht: Springer.

Hatch, M. D. (1999). C4 photosynthesis: A historical overview. In R. Sage & R. Monson (Eds.), *C4 plant biology* (pp. 175–196). New York, NY: Academic Press.

Hatch, M. D., Kagawa, T., & Craig, S. (1975). Subdivision of C4-pathway species based on differing C4 acid decarboxylating systems and ultrastructural features. *Australian Journal of Plant Physiology, 2*, 111–128.

Hibberd, J. M., & Covshoff, S. (2010). The regulation of gene expression required for C 4 photosynthesis. *Annual Review of Plant Biology, 61*, 181–207.

Hibberd, J. M., Sheehy, J. E., & Langdale, J. A. (2008). Using C4 photosynthesis to increase the yield of rice—Rationale and feasibility. *Current Opinion in Plant Biology, 11*(2), 228–231.

Hsing, Y. I., et al. (2007). A rice gene activation/knockout mutant resource for high throughput functional genomics. *Plant Molecular Biology, 63*(3), 351–364.

Jenkins, C. L. D., Furbank, R. T., & Hatch, M. D. (1989). Mechanism of C4 photosynthesis - A model describing the inorganic carbon pool in bundle sheath-cells. *Plant Physiology, 91*(4), 1372–1381.

Jeong, D. H., et al. (2002). T-DNA insertional mutagenesis for activation tagging in rice. *Plant Physiology, 130*(4), 1636–1644.

Kajala, K., et al. (2011). Strategies for engineering a two-celled C4 photosynthetic pathway into rice. *Journal of Experimental Botany, 62*(9), 3001–3010.

Kanai, R., & Edwards, G. E. (2001). The biochemistry of C4 photosynthesis. In *C4 plant biology* (pp. 49–87). New York, NY: Academic Press.

Kocacinar, F., McKown, A. D., Sage, T. L., & Sage, R. F. (2008). Photosynthetic pathway influences xylem structure and function in flaveria (Asteraceae). *Plant, Cell and Environment, 31*(10), 1363–1376.

Ku, S.-b., & Edwards, G. E. (1977). Oxygen inhibition of photosynthesis. *Plant Physiology, 59*, 986–990.

Leegood, R. C. (2002). C4 photosynthesis: Principles of CO2 concentration and prospects for its introduction into C3 plants. *Journal of Experimental Botany, 53*(369), 581–590.

Leegood, R. C. (2013). Strategies for engineering C 4 photosynthesis. *Journal of Plant Physiology, 170*(4), 378–388.

Lin, H., et al. (2016). Targeted knockdown of GDCH in rice leads to a photorespiratory-deficient phenotype useful as a building block for C4 rice. *Plant and Cell Physiology, 57*(5), 919–932.

Majeran, W., Cai, Y., Sun, Q., & van Wijk, K. J. (2005). The plant cell functional differentiation of bundle sheath and mesophyll maize chloroplasts determined by comparative proteomics. *Plant Cell, 17*, 3111.

Majeran, W., et al. (2008). Consequences of C4 differentiation for chloroplast membrane proteomes in maize mesophyll and bundle sheath cells. *Molecular & Cellular Proteomics : MCP, 7*(9), 1609–1638.

Majeran, W., et al. (2010). Structural and metabolic transitions of C4 leaf development and differentiation defined by microscopy and quantitative proteomics in maize. *The Plant Cell, 22*(11), 3509–3542.

Manandhar-Shrestha, K., et al. (2013). Comparative proteomics of chloroplasts envelopes from bundle sheath and mesophyll chloroplasts reveals novel membrane proteins with a possible role in c4-related metabolite fluxes and development. *Frontiers in Plant Science, 4*(March), 65.

Mitchell, P. L., & Sheehy, J. E. (2006). Surveying the possible pathways to C 4 rice. In *Charting new pathways to C4 rice* (pp. 399–412). Los Banos: International Rice Research Institute.

Miyao, M., Masumoto, C., Miyazawa, S. I., & Fukayama, H. (2011). Lessons from engineering a single-cell C 4 photosynthetic pathway into rice. *Journal of Experimental Botany, 62*(9), 3021–3029.

Monteith, J. L. (1978). Reassessment of maximum growth rates for C3 and C4 crops. *Experimental Agriculture, 14*, 1–5.

Morgan, J. a., et al. (2011). C4 grasses prosper as carbon dioxide eliminates desiccation in warmed semi-arid grassland. *Nature, 476*(7359), 202–205.

Osborn, H. L., et al. (2016). Effects of reduced carbonic anhydrase activity on co2 assimilation rates in setaria viridis: A transgenic analysis. *Journal of Experimental Botany, 68*(2), erw357.

Peng, S., et al. (2004). Rice yields decline with higher night temperature from global warming. *Proceedings of the National Academy of Sciences of the United States of America, 101*(27), 9971–9975.

Rizal, G., et al. (2015). Two forward genetic screens for vein density mutants in sorghum converge on a cytochrome p450 gene in the brassinosteroid pathway. *Plant Journal, 84*(2), 257–266.

Sage, R. F. (1999). Why C4 photosynthesis? In R. F. Sage & R. K. Monson (Eds.), *C4 plant biology* (pp. 3–16). San Diego, CA: Academic Press.

Sage, R. F. (2004). The evolution of C 4 photosynthesis. *New Phytologist, 161*(2), 341–370.

Sage, R. F., & Zhu, X. G. (2011). Exploiting the engine of C 4 photosynthesis. *Journal of Experimental Botany, 62*(9), 2989–3000.

Sage, R. F., Christin, P. A., & Edwards, E. J. (2011). The C 4 plant lineages of planet earth. *Journal of Experimental Botany, 62*(9), 3155–3169.

Sheehy, J. E., et al. (2007). How the rice crop works and why it needs a new engine. In J. E. Sheehy, P. L. Mitchell, & B. Hardy (Eds.), *Charting new pathways to C4 rice* (pp. 3–26). Los Banos: International Rice Research Institute.

Still, C. J., Berry, J. A., James Collatz, G., & DeFries, R. S. (2003). Global distribution of C 3 and C 4 vegetation: Carbon cycle implications. *Global Biogeochemical Cycles, 17*(1), 6-1–6-14.

Taniguchi, Y., Ohkawa, H., Masumoto, C., Fukuda, T., Tamai, T., Lee, K., Sudoh, S., Tsuchida, H., Sasaki, H., Fukayama, H., & Miyao, M. (2008). Overproduction of C4 photosynthetic enzymes in transgenic rice plants: An approach to introduce the C4-like photosynthetic pathway into rice. *Journal of Experimental Botany, 59*(7), 1799–1809.

Taylor, S. H., et al. (2010). Ecophysiological traits in C3 and C4 grasses: A phylogenetically controlled screening experiment. *New Phytologist, 185*, 780.

Wan, S., et al. (2009). Activation tagging, an efficient tool for functional analysis of the rice genome. *Plant Molecular Biology, 69*(1–2), 69–80.

Wang, P., Kelly, S., Fouracre, J. P., & Langdale, J. A. (2013). Genome-wide transcript analysis of early maize leaf development reveals gene cohorts associated with the differentiation of C4 kranz anatomy. *Plant Journal, 75*(4), 656–670.

Weber, A. P. M., & von Caemmerer, S. (2010). Plastid transport and metabolism of C3 and C4 plants — Comparative analysis and possible biotechnological exploitation. *Current Opinion in Plant Biology, 13*(3), 256–264.

Chapter 7
Crop Diversification Through a Wider Use of Underutilised Crops: A Strategy to Ensure Food and Nutrition Security in the Face of Climate Change

M. A. Mustafa, S. Mayes, and F. Massawe

Abstract Global dependence on only a few crops for food and non-food uses is risky due to the multifaceted challenges that crop production faces. One such challenge is climate change and its effects on food production. Emerging evidence suggests that climate change will cause shifts in crop production areas and yield loss due to more unpredictable and hostile weather patterns. The shrinking list of crops that feed the world, has also been attributed to reported reduced agricultural biodiversity and increased genetic uniformity for yield traits in crop plants. This could lead to crop vulnerability to the dangers of pests and diseases. Part of the solution to these problems lies with crop diversification through a wider use of underutilised and minor crops. Underutilised, minor or neglected crop plants are plant species that are indigenous rather than adapted introductions, which often form a complex part of the culture and diets of the people who grow them. The wider use of underutilised crops would increase agricultural biodiversity (genetic, species and ecosystem) to buffer against crop vulnerability to climate change, pests and diseases and would provide the quality of food and diverse food sources to address both food and nutritional security.

There is evidence to suggest that people are increasingly changing their attitude in favour of crop diversification instead of specialisation on a few major crop species. This chapter provides a background on crop diversification and discusses the potential roles of underutilised crops to address major global concerns such as food

---

M. A. Mustafa · F. Massawe (✉)
The University of Nottingham Malaysia, Semenyih, Selangor, Malaysia

Crops for the Future, Semenyih, Selangor, Malaysia
e-mail: festo.massawe@nottingham.edu.my

S. Mayes
The University of Nottingham Malaysia, Semenyih, Selangor, Malaysia

Crops for the Future, Semenyih, Selangor, Malaysia

Plant and Crop Sciences, Biosciences, University of Nottingham,
Loughborough, Leicestershire, UK

© Springer Nature Switzerland AG 2019
A. Sarkar et al. (eds.), *Sustainable Solutions for Food Security*,
https://doi.org/10.1007/978-3-319-77878-5_7

and nutrition security, agricultural biodiversity, climate change, environmental degradation and future livelihoods.

7.1 Introduction

Agriculture is a largely weather-dependent practice utilising resources such as water, soil and solar radiation and genetic material. Thus, this income-generating process that serves the entire population through the production of food, feed, fuel and fibre, is extremely vulnerable and susceptible to the impacts of climatic change. As populations grow and economies thrive, the demand on agriculture will continue to increase, which must be met with the limited resources available and with minimal negative impacts on the surrounding environment (Berg et al. 2013).

It is estimated that globally 1.5 billion hectares are currently under cultivation, while one third of that (445 million hectares) remains as available for farming yet uncultivated (Arezki et al. 2012), which may not be sufficient to meet the increasing needs of a growing and affluent global population. This, however, does not account for the impact of extreme weather patterns on several key food production areas, which may reduce the scale of available farming regions. The agricultural sphere today is dominated by a handful of crops: rice, wheat, maize and soybean. These crops make up the bulk of our food and non-food uses, and utilise the larger share of the resources available for crop production. Food security has always been fret with difficulty, but our accomplishments to date are largely attributed to our success in growing these key crops.

7.1.1 The Main Crops That Feed the World

The move towards specialisation and intensification in crop production was most apparent at the onset of the Green Revolution, which set a tremendous milestone in ensuring that crop production kept up with the increasing demands of the booming population needs. The focus was on breeding high yielding plant varieties and improving farming techniques through intensive fertiliser inputs and irrigation systems (Massawe et al. 2016). Consequently, performance of the three main cereals—rice, wheat and maize—increased significantly to almost double the pre-Green Revolution yields (Michler and Josephson 2017).

Dedicated research centres focussing on these main crops ensured—and continue to ensure—that food security is achieved and remains a viable goal over the following decades, alas, contributing towards monocultures of genetically uniform crops. Moreover, in the threat of food insecurity, national policies encouraged the increased production of crops that were demonstrated to be high-yielding. Crops that were known to grow well, through available research and technologies, were increasingly adopted under national policies, with national per capita food supplies seeing a massive increase in the share of the three main crops: rice, wheat and maize (Khoury et al. 2014), and further increases anticipated in most developing nations by 2050. The role of rice as a staple is increasing globally, even in regions that depend more on importing rice such as sub-Saharan Africa (Seck et al. 2010). Africa imports almost a third of global rice production, and is struggling to meet the growing demands of its people as it is also a net importer of wheat, maize and soybean (40.3 million tons, 14.1 million tons and 2.1 million tons, respectively, in 2013) (Tadele 2017; Africa Rice Center 2008).

Asia and South America witnessed tremendous advances during the Green Revolution. However, these strategies were not as successful in African countries due to the lack of availability of human and institutional capacity as well as locally relevant seed varieties and technologies (Dawson et al. 2016). In the 1960s, policies supported smallholders' adoption of Green Revolution agricultural techniques through price guarantees and public investment. Meanwhile African smallholders experiencing the "modern Green Revolution" are exposed to volatile pricing due to trade liberalisation (Dawson et al. 2016). Moreover, factors such as high population density, poor infrastructure, climate change and human migration have further aggravated the situation.

7.1.2 Crops to Match Climates of the Future

Some of the main crops experience high volatility in prices, with the price of wheat and soybean increasing by 40% and 70% respectively within a single year (Tadele 2017). This stems from policy issues such as the absence of a central body responsible for stabilising global prices, as well as productivity issues that are increasingly threatened by the effects of extreme weather patterns (Seck et al. 2010). The new challenges of food security are of different nature and scale to those encountered in the 1960s. It is no longer an issue of increasing the cultivation of an identified set of crops, but rather dealing with variable and volatile climates and changes in biological and nonbiological stress factors.

Throughout the years, the main crops have played a significant role in ensuring food security, and will continue to do so, if complemented with a plethora of other crops that have historically played a role in traditional agriculture. Efforts to increase the competitiveness of domestic rice production in sub-Saharan Africa will indeed influence social stability, but must also be complemented with increased production of local crops that match the climates of the future. Planting a variety of crop plants reduces vulnerability to adverse effects of climate change and may also open new market opportunities to farmers (McCord et al. 2015). In addition to ensuring food security, growing a variety of crops provides dietary balance and diversity necessary to address nutritional security in both developed and developing countries.

The importance of diversification is most notable in regions such as sub-Saharan Africa where expanding farming lands to meet the increasing food demands is no longer a viable option (Berg et al. 2013). Such regions—where environmental risks such as drought and desertification are eminent—need to adopt a variety of crops that are adapted to marginal lands and cropping systems that do not exert further negative environmental impacts. This however, must be supported with improved water and land management practices, informed farming practices, pest and disease management and general access to resources, knowledge and technology (McIntyre et al. 2009). Achieving this successfully can address issues related to food security, income generation and poverty alleviation, and sustainable environmental management (FAO 2012; Michler and Josephson 2017).

7.1.3 The Plan: Improved Resilience and Nutritional Security

The United Nations Sustainable Development Agenda 2030 (SDA 2030) calls for action by all countries to adopt 17 goals to end poverty, protect the planet, and ensure prosperity for all (UN 2016). The second Sustainable Development Goal (SDG) highlights several key approaches towards ending hunger; doubling agricultural productivity, targeting food and nutritional security, adopting resilient agricultural systems, maintaining genetic diversity of crops, and limiting extreme volatility in food prices (UN 2016). Adoption of diversified food systems as a strategy can promote sustainable agricultural practices while meeting the increasing food and nutrition demands (FAO 2012).

Moreover, encouraging diversification among farming households enhances employment generation and poverty alleviation by increasing agricultural incomes (Taffesse et al. 2012). Diversified crop production offers farmers stability in dealing with climate change and the increasing unpredictability of weather patterns as well as fluctuating market prices. Considering the economic returns of a varied crop portfolio, farmers gain more protection from price declines for specific produce. This chapter explores crop diversification as a strategy to ensure food and nutritional security in a changing world. We provide an overview of the global challenges that can be addressed through a wider use of underutilised crops.

7.2 Crop Diversification for Improved Resilience and Nutritional Security

Diversification is a spatial or a temporal process of creating a heterogenous farming system through activities within or outside a farm to build resiliency of the ecosystem (Kremen et al. 2012; Lin 2011; Senger et al. 2017). Off-farm diversification occurs at the end of the production chain by developing variable products from the available produce, adding value to products and improving income generation. On the other hand, diversification within farming systems maximises the use of soil, water and biological resources capitalising on the benefits of ecosystem services such as nutrient cycling and pest control (Kremen et al. 2012).

On-farm diversification exists at various structural levels; crop diversification (e.g. polycultures of genetically diverse crops), field diversification (e.g. mixed cropping systems) and even landscape diversification (Kremen et al. 2012; Lin 2011). The various diversification levels address changes in the farming production systems to essentially create more resilient agricultural systems through the incorporation of various production systems. Diversification can also be viewed in various lights based on the functionality of the system, e.g. pest suppression, enhanced productivity (Fig. 7.1). Techniques such as crop rotation or agroforestry contribute towards developing an ecological farming system with numerous benefits such as encouraging natural enemy habitats to increase pest suppression or reducing soil evaporation through shaded systems (Lin et al. 2008).

7.2.1 Structural Levels of Diversification

Crop diversification is the practice of growing more than one crop species within a farming area in the form of rotations or mixed cropping (Makate et al. 2016). It refers to the shift from specialisation and high dependence on a few global crops

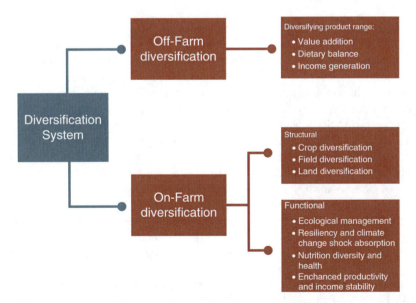

Fig. 7.1 Levels of diversification: on farm diversification can be created at various structural levels, offering multiple benefits that include environmental and ecological management; supporting livelihoods through income stability and minimising risks; supporting food security through encouraging diversity of diet; improving farming output by reducing pest and disease infection and maximising returns (Makate et al. 2016; Pellegrini and Tasciotti 2014; FAO 2012)

grown in monocultures to an expanding list of regional crops cultivated in multiple cropping systems. This concept was initially proposed more than 60 years ago (Heady 1952) to lower production risk and vulnerability. Multiple cropping systems account for almost 15–20% of the food supply; for example, almost 70% of beans grown by farmers in Latin America were intercropped with other crops such as maize and potatoes (Altieri 1999).

Crop diversification through adoption of underutilised crops focusses on the wider cultivation of under-exploited and locally adapted crops to reduce risks associated with crop production, improve food, nutrition and income (Makate et al. 2016). The adoption of locally adapted crops and crop varieties can provide stable yield under the regional environmental conditions and increase the genetic diversity of crops in a farming system (Mayes et al. 2012; Lin 2011). Cultivation of a diverse range of crops not only contributes to local biodiversity but also minimises environmental damage by supporting environmentally sustainable agricultural systems. Various farming communities in the Andes, East Africa, sub-Saharan Africa and South Asia have been cultivating a wide range of domesticated plant species and their wild relatives as a common traditional practice (Thilakarathna and Raizada 2015; Altieri 1999).

7.2.2 Functionality of Diversified Systems

Diversity within an ecosystem serves numerous roles besides the production of food, that include regulation of the hydrological cycle, pest management and nutrient recycling (Altieri 1999). Different species perform different roles in an ecosystem; however, the diversity of species is much higher than the diversity of roles in an ecosystem, and thus varying species may assume duplicate functions. The existence of several species occupying similar roles within a system allows for redundancy to exist in the ecosystem (Lin 2011). As global climates continue to change, it is difficult to predict the impacts that it may have on the functionality of the ecosystem, the response of the multiple species, and the roles that may increase in importance (Vandermeer et al. 1998). Increasing crop diversity allows for the existence of diverse species with varied responses to environmental changes, which builds resiliency within an agricultural system and provides insurance against environmental instabilities and increases the ability to cope with climatic challenges (Yachi and Loreau 1999; McCord et al. 2015). Similarly, the loss of these functions results in systems of reduced quality which are unable of supporting their own nutrient and water recycling (Lin 2011).

Resilience is a natural function within ecosystems, however it is not a frequently applied concept in agricultural systems (Lin 2011). Modern agricultural systems increasingly depend on external input to maintain crop performance and output, which ultimately impacts the environment negatively (Altieri 1999). Communities that are dependent on agriculture rarely possess other alternatives for securing their livelihood or resources for investing in adaptation strategies, ultimately increasing their vulnerability to weather pattern fluctuations (Altieri 1999). Crop diversification reduces the need for economic, labour and chemical interventions, supporting the resilience of agricultural systems to meet the food and non-food demands of a community and protect livelihoods sustainably (Makate et al. 2016).

7.3 Crop Diversification Through Wider Use of Underutilised Crops

The origin of all domestic crops traces back to a series of planned activities of domestication and selective breeding of wild species (Altieri 1999). Progressive series of domestication has led to increased genetic uniformity among the major crops cultivated today. Modern agriculture has contributed towards the decline in diversity of plant species that meet our food and non-food uses (Khoury et al. 2014). For example, in the USA four varieties of potatoes account for 72% of potatoes grown and three varieties of cotton account for 53% of cotton grown globally (Altieri, 1999; National Academy of Science 1972).

Vast populations of diverse and adaptable landraces and wild crop relatives exist around the world harbouring an abundance of genetic resources (Altieri 1999). These are the underutilised crops. Underutilised crops are locally adapted plant species that are strong contributors to the diet and customs of the regions where they are locally cultivated by the farmers (Mayes et al. 2012). The resiliency of most of these crops and their ability to grow on marginal lands and under environmental stresses is a key factor for advocating their wider use (Massawe et al. 2016). Moreover, these crops often have long histories of strengthening food security within their localities, particularly when the big crops fail. Diversification allows growers to efficiently utilise the available agroecological resources, ultimately crafting a localised system that secures food sovereignty (Chaifetz and Jagger 2014; Njeru 2013).

7.3.1 Diversifying Crops for Food and Nutritional Security

Global affluence coupled with globalisation, plays a key role in shaping consumer eating patterns today. Food production does not only have to meet the demands of a growing population, but also of an expanding middle class with preferences towards processed and easily accessible food products (Massawe et al. 2016). Life and food trends are no longer localised and food demands are growing increasingly similar across borders (Khoury et al. 2014; Massawe et al. 2016). Consumer trends are largely influenced by the increase in global trade, urbanisation and proliferation of transnational food corporations, with increasingly shared diets across regions (Hawkesworth et al. 2010). Between 1961 and 2006, the global consumption of rice increased at almost 4.5% annually (Seck et al. 2010). A significant increase in rice consumption was also witnessed in Africa due to urbanisation and changing lifestyles (Africa Rice Center 2008). Moreover, a rampant increase in energy dense diets that largely depend on meat, dairy products and plants oils have had a significant impact on the way we grow food, with an increased reliance on a handful of crops.

Grain yield of the major staple crops has at least doubled in the past five decades (Godfray et al. 2010). This has, to some extent, sustained past population bursts but may not be sufficient to meet the growing demands of a global population of nine billion by 2050 (Godfray et al. 2010). Moreover, prices of grains have been steadily increasing, with approximately 40% increase reported for rice prices over the past years (Africa Rice Center 2008). The increasing cost of rice imports and the economic and political risks that posed, was the motive behind the development of a high-yielding rice variety adapted to African climates; New Rice for Africa or NERICA (Kijima et al. 2011). However, adoption of the technology faced challenges, with an estimated 50% of farmers who had adopted NERICA in Uganda abandoning it within 2 years (Kijima et al. 2011). This high level of dropout was associated with low profitability of the crop due to rainfall variation and poor quality of farmer-produced seeds (Kijima et al. 2011; Sserunkuuma 2008). Kijima et al. (2011) recommended reforms that target suitable areas for farming NERICA as well as extension systems that disseminate appropriate information on farming methods such as seed production.

This calls for alternative measures to enhance food security and sustainable livelihoods which should include promotion and wider use of underutilised and underexploited food sources. Underutilised crops are an essential part of diets of many people in the world but their global importance remains low (Tadele and Assefa 2012). Food security necessitates that the four main aspects of food provision are met; availability, stability, access and utilisation (FAO 2012; Kang et al. 2009). Availability of food does not necessarily reflect that food security is met, and current statistics on global hunger rarely capture the dimensions of stability, access and utilisation of food (Wheeler and von Braun 2013). Enhancing food security through increased production does not address access and utilisation of food at a local level (Altieri et al. 2012). Moreover, food stability is threated by climate change and environmental risks, and this has been frequently witnessed as short-term shocks affecting food security. Crops that are grown on marginal lands generally show resistance to extreme climates and conditions such as drought, waterlogging and degraded soils, thus have the potential to improve food availability and stability.

In addition to increasing the quantity of food, FAO (2012) has urged to improve the diversity of food and nutrient content for nutritional security. FAO/WHO recommend daily intake of 400 g or more of fruits and vegetables to prevent non-communicable diseases (cancer, diabetes, obesity, cardiovascular), as well as other negative health impacts of micronutrient deficiencies such as blindness and birth defects. Achieving nutritional security will require the promotion of diversified diets and improving micronutrient intake (McIntyre et al. 2009) through improved availability of horticultural products and other non-grain crops. Incorporation of vegetables in diets can have a significant impact on micronutrient deficiency. The Agricultural Research Council (ARC) of South Africa promoted the cultivation of vegetables through targeted research on cultivation practices, water-use efficiency and plant nutrition of local vegetables such as amaranthus and okra (Bvenura and Afolayan 2015). This resulted in increased cultivation of the vegetables within certain provinces, and improved diet and health of the targeted communities (Bvenura and Afolayan 2015).

The consumption of staples has increased considerably among the urban poor in developing countries (Pellegrini and Tasciotti 2014). Diets that are prevalent in most urban households throughout Africa and Asia lack the necessary vitamins and mineral, such as iron, as evidenced in the increased prevalence of iron deficiency anaemia in sub-Saharan countries (Frison et al. 2006). Moreover, in today's globalised market, developing countries are largely dependent on food imports to meet the increased demand of staples, and are thus vulnerable to the volatility of global market prices (Pellegrini and Tasciotti 2014).

Meanwhile, rural households may be more dependent on their local production, due to the remoteness of the villages and expensive transportation costs (Pellegrini and Tasciotti 2014). When the local production in such communities encompasses a narrow range of crops, the diets in rural households lack basic micronutrients, resulting in compounding medical conditions (Pellegrini and Tasciotti 2014). Thus, the ultimate issue at hand is access to local production of diversified horticultural products with richer micronutrient content as well as fibre, proteins and lipids (FAO 2012).

Prevalence of malnutrition has been reported in the mountainous regions of Pakistan and Nepal due to local communities abandoning the cultivation of traditional crops such as millet, taro, yam and wild vegetables in favour of a narrow range of crops (Adhikari et al. 2017). Crops such as finger millet have been traditionally cultivated throughout Africa and Asia, offering a rich source of diverse minerals, proteins and vitamins and possessing high nitrogen use efficiency (Onyango 2016). Complementing the current production of major grain crops (wheat, rice, maize, soybean) with a wide range of horticultural and other high value crops can dually improve the quality of diet within the region as well as strengthen livelihoods and improve purchasing power through better economic returns from higher value produce (Makate et al. 2016; Kang et al. 2009). There is a trickle-down impact of crop diversification. Not only do the producers thrive economically, but there are also expanded opportunities for local agro-processing, creating a value chain for the produce with employment and higher value produce (FAO 2012). Moringa is a key example of such multifunctionality, as its fruits, nuts or leaves can be used as food, while other parts of the plant can be used for firewood or to make oil and other cosmetic products (Dou and Kister 2016). Ultimately, physical and economic access to fresh nutritious food as well as economic opportunities will improve when diverse crops are grown regionally.

7.3.2 Diversifying Crops for Climate-Resilient Agriculture

As a weather-dependent activity, agriculture is highly susceptible to the impacts of climate change. The tropical regions are the most vulnerable, particularly in developing nations where a large proportion of the population rely on climate-dependent resources. Southern Africa is expected to experience a rise in temperature of between 1.5 and 3 °C by 2050, as well as more extreme weather patterns such as droughts (Dzama 2016). The drought of 2015 was the worst experienced in 35 years in Southern Africa—Swaziland, Lesotho, Malawi, Namibia and Zimbabwe declared a national disaster (Dzama 2016). Climate change severely affects agriculture, and poverty amplifies the effect by limiting the ability of local population to anticipate the impacts of climate change and adapt accordingly (Berg et al. 2013). Moreover, with the increasing rate of population growth, the ability of these communities to meet their food needs in the face of climate change is challenged further.

Climate change and its consequences call for a rethink of crop production to determine which crops species match prevailing climates. For example, consideration must be given to species that are adapted to marginal lands that in the past have been regarded as unsuitable for crop production. Agriculture has also contributed to environmental degradation and therefore future food must be produced sustainably without further negative environmental impacts. Underutilised crops are grown by subsistence farmers, usually on marginal lands with limited or no agricultural inputs. These are usually local crops grown in their natural environments, expressing adaptive features necessary for climate-resilient agriculture.

Changes in temperature negatively impact crop performance, particularly if the changes are experienced during flowering and fruiting (Lin 2011). Tropical regions are extremely vulnerable and forecasts predict that production of the staple crops will decline, with higher temperatures accelerating phenology, decreasing the crop growth period and resulting in reduction of biomass (Berg et al. 2013). This will directly reduce the yield and quality of crop and decreases in yield are estimated to be proportional to rises in temperature (Berg et al. 2013). Moreover, temperature change may result in soil degradation, as well as increased pest and disease prevalence (Kang et al. 2009). Improving crop production in these areas will be fret with difficulty, and will require improving farming techniques, land and water management and careful consideration of the variety of crops that are cultivated (FAO 2012).

On the other hand, agriculture in temperate zones may benefit from the increase in temperatures and longer growth periods (Berg et al. 2013). Even in tropical regions, some forecasts have predicted improved agriculture due to "carbon dioxide enrichment effect", an effect attributed to increased levels of carbon dioxide encouraging plant growth (Kang et al. 2009). Modelling the impact of climate change (2 and 4 °C increase, 20% decrease and increase of rainfall, and doubling of carbon dioxide) on wheat, maize, rice and soybean showed higher crop yields due to doubling of carbon dioxide (Rosenzweig and Parry 1994). Although the increased carbon dioxide levels could benefit crop performance, the threshold for temperature increase was 4°C, with yield declines reported from that level.

In a study by Hochman et al. (2017) modelling wheat cultivation in Australia, CO_2 enrichment prevented a 4% decline in potential yields of 2015 as compared to 1990, despite 28% decline in rainfall and 1.05 °C increase in temperature during that period. Moreover, the authors did note a decline in potential yields during that period (27% decline), although actual yields remained stable (Hochman et al. 2017). The stability of actual yields was attributed to improved farming practices such as integrated weed management and reduced tillage, which resulted in ability to harvest 55% of potential yield in 2015 as compared to 37% of potential yield in 1990 (Hochman et al. 2017). Nonetheless, the accelerating effects of climate change are estimated to overtake the effects of continuous improvement in farming practises. Given an increase in actual yield to 80% of potential yield by 2041, an estimated decline of 11% in harvested tons of wheat per hectare has been projected by Hochman et al. (2017).

Diversified systems that involve crop rotation or intercropping of diverse crop species are common farming practices in various regions and have demonstrated advantages. In a study conducted in West Bengal by Maitra et al. (2000), intercropping finger millet with pigeon pea and groundnut was found to significantly improve yields. Bambara groundnut has been demonstrated repeatedly to enrich the landscape and offer advantages when intercropped with staples such as maize and taro (Alhassan and Egbe 2014; Jacob et al. 2014; Mabhaudhi and Modi 2013). Agroforestry has positive impacts on multiple levels for communities that adopt this system (Leakey and Asaah 2013). In a project conducted in Cameroon that saw leguminous trees and shrubs intercropped with indigenous fruit and nut trees, the lives of smallholders were transformed and positive impacts on the environment were observed (Leakey and Asaah 2013).

7.3.3 Hydrological Cycle and Crop Performance

Water availability is already under pressure due to population increase and pollution. Adding to that, climate change and the resulting changes in precipitation patterns threatens water availability and the hydrological cycle at large (Kang et al. 2009; Droogers and Aerts 2005). Precipitation patterns are tougher to predict, as compared to changes in temperature (Droogers and Aerts 2005). However, extremes in precipitation patterns—flooding and droughts—are more likely scenarios than changes in the average precipitation (Droogers and Aerts 2005). On a global level, the effects of droughts impact more people than other natural disasters, resulting in the highest economic and environmental costs (Burchfield and Gilligan 2016; Wilhite and Vanyarkho 2000).

Agriculture as a water-dependent activity is largely affected by the incidence of droughts, which is further aggravated by poor soil conservation practices, poor water infrastructure and poverty (Midgley and Methner 2016). Fluctuating precipitation patterns and global temperatures will directly influence soil evaporation and transpiration, changing soil water balance (Kang et al. 2009). Changes in the soil water storage and soil moisture status as well as groundwater levels will require different cropping and irrigation practices (Kang et al. 2009). Using the CERES-wheat model, Eitzinger et al. (2003) demonstrated a direct impact of the changes in soil water balance under different climate scenarios on crop production, negatively affecting sustainable crop production. The drought of 2015/2016 in South Africa was strongly felt by the agriculture sector, leading to 50–100% losses of wheat and 15% of fruit per farm (Midgley and Methner 2016). Meanwhile farms in the same region that adopted land management practices that maintained soil moisture, produced enough yield to avoid losses (Midgley and Methner 2016).

Water scarcity results from an interaction of reduced precipitation patterns, increased soil moisture stress, increased dependence on irrigation systems as well as impact of socio-economic factors (Burchfield and Gilligan 2016). Increased irrigation can increase yield, but may result in higher runoff of heavier soils (Holden and Brereton 2006). In Russia, for example, water availability is expected to increase although occurrences of runoff will also increase, thus equally impacting the food production capacity in the region (Alcamo et al. 2007). As for the likelihood and duration of droughts, it is expected to increase in many regions throughout the world (Burchfield and Gilligan 2016; Touma et al. 2015).

The impacts of climate change are not proportional over time periods, as the projected short term effects of climate change differ from the long-term effects. In arid zones a decrease in precipitation patterns is forecasted during 2020–2049, while an increase is predicted in 2070–2099 (Berg et al. 2013). Planning for short-term climate change by cultivation of drought-tolerant crops will not offer a long-term solution. A combined increase of farming lands by 2.5%, harvesting frequency by 7%, and crop yields by 20% resulted in 28% total crop production over the period of 1985–2005 (Ray et al. 2012). This evidences that expanding our crop production areas and intensifying agricultural activities cannot efficiently meet our growing food needs, and will have a negative impact on food security due to environmental degradation and increased mismanagement of water resources. Slowing the expansion of farming areas is increasingly advocated by researchers and policy makers, exemplified in SDG 15 calling for sustainable management of land and zero deforestation (UN 2016; Byerlee et al. 2014).

In a recent study assessing global patterns of crop yield growth Ray et al. (2012) reported yield stagnation of the big crops in more than a quarter of their production areas, 37% of wheat, 35% of rice, 26% of maize and 23% of soybean. Increasing crop diversification will allow for adaptation to a vast range of scenarios, offering a more pragmatic solution that captures the full story of climate change as we plan for long-term agricultural adaptation. Amaranth, beans and pearl millet are examples of drought-tolerant crops that are adapted to marginal lands, as evidenced by Chivenge et al. (2015) to be more drought tolerant than the big three crops. A diversified portfolio of crops expands the opportunities for farmers who may experience droughts, and improve the resiliency of the agroecosystem towards water scarcity (Burchfield and Gilligan 2016). Thus, developing more crop varieties will allow for better suited agricultural practices, increasing adaptation to changing environmental patterns as well as degraded soils to meet growing food needs (Kang et al. 2009).

7.3.4 Pest and Pathogen Suppression

Stability of farming systems can be clearly reflected in their resiliency towards pest and pathogen infections. Pests and pathogens reduce crop productivity, and increase the economic cost of pest and disease management (Chakraborty et al. 2000). The occurrence of diseases is largely influenced by weather patterns, such as precipitation which directly affects the life cycle and distribution of pathogens, vectors and hosts and the development of disease symptoms (Anderson et al. 2004; Chakraborty et al. 2000). However, it is difficult to predict the impact of climate change on disease as several factors besides climate change can impact the occurrence and severity of diseases, such as the morphology and physiology of the hosts as well as the nature of the pathogen (Scherm and Yang 1995; Chakraborty et al. 2000).

The expansion of warmer zones and the shortening of winters may promote the distribution of virus-transmitting vectors, increasing the chance of disease infection (Anderson et al. 2004). The warmer winters and summers could also encourage occurrence of fungal infections, such as powdery mildew and Cercospora leaf spot (Patterson et al. 1999). Additionally, extremes in weather patterns may tilt the scale to favour the distribution of certain diseases, such as viruses and insects in dry regions and fungal and bacterial infections in wet regions (Anderson et al. 2004). Insects are expected to thrive better under higher temperatures that could favour increased insect population growth and migration (Lin 2011). Moreover, the shorter life-cycle and faster reproduction of pests may help in their adaption to climate change at a faster rate than the existing crops within the habitat (Bale et al. 2002; Lin 2011).

When the emphasis of our farming systems becomes on crop output rather than overall resilience, susceptibility to pest and pathogens manifests (Altieri 1999). The widespread of genetically uniform varieties in monocultures has increased our dependence on external inputs for controlling pests and pathogens as well as the susceptibility to diseases. Even within the same species, disease resistance exists across a spectrum in the different varieties, thus intraspecific crop diversification can offer a barrier for controlling disease transmission (Finckh et al. 2000). As demonstrated by Zhu et al. (2000), intercropping genetically diverse rice in China reduced rice blast by 94% and increased yield by 89%, as compared to monocultures that lacked genetic heterogeneity. Interspecies crop diversification has added benefits, as demonstrated by Midega et al. (2010) in the enhanced control of Striga and increased yield through intercropping millet with greenleaf desmodium. Moreover, mixed cropping systems, particularly through the inclusion of nitrogen fixing crops and/or cover crops, increase soil fertility and improve the functionality of the system at larger (Smith and Read 2008).

Agroforestry has a greater impact as it encourages the abundance of natural enemies leading to an increase in the population of natural enemies such as insectivorous birds that suppress pests (Perfecto et al. 2004). Naturally biodiverse ecosystems possess an inherent self-regulation mechanism and can maintain a diversity of flora and fauna, allowing them to interact with their environment and respond to changes

(Altieri 1999). Such systems function through multiple ways to improve productivity by increasing natural enemies, suppressing weed growth, breaking disease cycles as well as modifying the microenvironment (Makate et al. 2016). Thus, increased diversification contributes towards improving resiliency in agricultural systems, particularly by enhancing the ability for pest and pathogen suppression and overall crop performance (Lin 2011).

7.3.5 Poverty Alleviation and Livelihood Protection

Farmers with small lands are at a disadvantage with regards to investing in adaptation strategies to climate change. They are also increasingly vulnerable to the volatility in the supply and prices of main grain, as experienced in the food crisis of 2008 (Mayes et al. 2012). Crop diversification, particularly among smallholder farmers, serves as an adaptation strategy that reduces the economic risks, increases household incomes and strengthens their purchasing power (Weltin et al. 2017). It improves the capacity to deal with market pressures, as well as better adaptation to the environmental challenges that produce shortages in food production (Saenz and Thompson 2017). The growth of the agricultural sector is a vital tool for escaping poverty in most developing countries, particularly among the rural poor with small or no lands (World Bank 2008). Agricultural growth provides employment as well as improving local access to food at lower prices, developing the rural economy further (Kassie et al. 2011).

Cash crops have an advantage through their higher product value, offering improved household income (Govereh and Jayne 2003). However, specialisation on cash crops poses an even larger threat to household livelihoods when these cash crops substitute food crops (Saenz and Thompson 2017). The benefits of cash crops can be fully realised through their incorporation in diverse farming systems, lowering the production risks and vulnerability to climate changes (Orr 2000). Michler and Josephson (2017) reported a direct link between increasing diversification portfolio and lower probability of remaining in poverty. They examined the households' crop diversity relative to their village practices, including in their analysis a diversity count of 50 different crops such as staple crops, e.g. teff and maize, and cash crops, e.g. sesame, linseed, coffee, chat and enset. In their analysis of the benefits of crop diversification in Ethiopian households, Michler and Josephson (2017) found that the chances of living in poverty decreased by 16.9% for households above poverty line, and by 18.3% for households already in poverty. Their recommendation was for increasing crop diversity, rather than promoting specialisation in cash crops to improve the biodiversity of crops grown.

Several authors reported diversification as a positive farming approach for agricultural growth and enhancing livelihoods in various communities. Van den Berg et al. (2007) drew a positive relationship between diversification and improved income returns for traditional rice-farmers in China, when cultivating vegetable crops. Meanwhile in Vietnam diversified farming systems were also reported to slightly improve incomes for rice farmers (Nguyen 2017). The authors reported improved technical efficiency and output complementarity for the farmers with a diverse crop portfolio (Nguyen 2017). In the Gezira Scheme of Sudan, Guvele (2001) reported that crop diversification reduced variability in income for farmers. Moreover, in Bangladesh rice monocultures were found to lead to lower production of pulses, vegetables, spices, thus negatively impacting the environment and sustainability of the farming system (Rahman 2009). It was reported that the increased cultivation of rice monocultures not only displaced the non-rice crops, but also the traditional varieties of rice, and resulted in more intensive farming and shrinking of net farmed lands (Rahman 2009). Thus, national policies in Bangladesh are now encouraging diversification as a tool towards increasing the profitability of the agricultural sector through better farming practices (Rahman 2009). Taken together, these case studies from various parts of the world highlight the potential of diversification as a contributor towards resilient agricultural development.

7.4 Factors Influencing the Adoption of Crop Diversification

Technological and scientific support for the big crops (e.g. maize, rice, wheat, potato), large-scale adoption of monocultures, as well as economic incentives for select crops have to a great extent stalled the adoption of crop diversification. Global attention is increasingly shifting towards sustainable agricultural systems (Samberg et al. 2016). However, as the prevalence of smallholders vary between Africa, Asia and Latin America, so do the farming practices and dimensions of rural poverty. This should be a key consideration in developing policies and technologies that improve productivity and uplift the rural poor more effectively (Kassie et al. 2011).

7.4.1 Resources and Knowledge Systems

Large-scale cultivation of the big crops generated significant emphasis on the research of these main crops, such as on disease infection and resistance patterns (Anderson et al. 2004). Consequently, neglecting the study of underutilised crops and their role within their communities (Cheng et al. 2017). The domestication and introduction of underutilised crops will be largely influenced by our current knowledge of these crops. Crop improvement programmes are dependent on the availability of knowledge on the variabilities within germplasm collections (Aliyu et al. 2016).

This disparity in our available knowledge can even be seen in the research carried out on the agronomic practises of crop production as well as its response to climate change. Agronomic practices of underutilised crops remain understudied, while rapid developments in the fields of precision agriculture of major crops have been witnessed that allow for optimising irrigation and chemical input (Mayes et al.

2012). Dedicated models have been established for forecasting cropping outcomes under various climate change scenarios, such as CERES-wheat and CERES-rice (Lin 2011). These models are very powerful, with the ability to estimate various climate change scenarios as well different agricultural farming systems, yet very little has been done on studying the response of underutilised crops (Karunaratne et al. 2010). In studies by Karunaratne et al. (2010) and Karunaratne et al. (2011), canopy development and biomass formation of Bambara groundnut was modelled and used to produce simulations of the plant response to abiotic stresses. A crop application model was developed by Karunaratne et al. (2015) on Bambara groundnut and pearl millet to test for productivity and water-use efficiency of different germplasms of these crops under different locations and future climates. It was demonstrated that Bambara groundnut had a wide climatic adaptation, while one variety of pearl millet was found to be more efficient than the traditionally grown Monyaloti variety.

Nonetheless, the availability of general information might not always be a limiting factor in the adoption of crop diversification systems. In a study by Makate et al. (2016), greater access to general information and increase in years of farming experience were found to be inversely related to adoption of diversification systems, with a 29.7% and 5.5% decrease, respectively, in the probability of adopting diversification reported. Alternatively, a positive relationship was reported between land size holding and adoption of crop diversification (Makate et al. 2016). While this study was limited to Zimbabwe, the findings imply that more information and experience does not necessarily translate to greater adoption of diversification practices.

7.4.2 Technology

Agreement within the scientific community that climate change will negatively impact crop production has been a driving force behind the technological advances in agriculture today. Most of these advances have focussed on crop improvement programmes to develop drought-resistant crops through application of biotechnology (Lin 2011). While this approach may be successful in protecting crop performance, it does not account for the larger proportion of farmers who will not be able to adopt these advancements due to their high costs (Lin 2011). Such technologies are developed for widespread use, and show little regard for the local condition and requirements of small-scale farmers (Kremen et al. 2012). A key limitation of these technologies lies in this failure to realise the need for locally adapted techniques and crop varieties. Additionally, as with most scientific advancement, such biotech-led developments cannot be rushed, thus raising concerns on the availability of these technologies in time when they are most needed for climate adaptation.

The advancement of technologies has been specifically targeted towards the big crops, with little relevance to the variety of other crops available. Molecular technologies generated significant data on the main crops, such as gene expression under abiotic stresses (Bonthala et al. 2016). Unfortunately, these technologies such as microarrays have yet to be developed for the research of underutilised crops. However, through the adoption of other technologies such as Next Generation Sequencing, the available knowledge on the main crops can be integrated to generate data on underutilised crops (Mayes et al. 2012).

Technologies can play a pivotal role in reducing rural poverty by improving resilience and closing the potential-actual yield gap. This requires broader investment in agricultural research as well as policy support to allow for more effective research uptake and technology adoption by the communities (Kassie et al. 2011). Moreover, for agricultural research to be beneficial to the rural poor, it must be demand-driven and consider the structural differences between farming communities across the different continents.

7.4.3 Farming Systems

Modern agricultural systems are increasingly dependent on monoculture farming systems that constantly require external input and infrastructure (Altieri 1999). These systems generally view the integration of biodiversity as a competition for resources, hindering productivity (Kremen et al. 2012). Large-scale adoption of such farming systems has contributed to improved biomass production, but at an expensive environmental cost. Such systems reduce biodiversity while degrading natural resources, resulting in soil erosion and increased run-off from chemical fertilisers and pesticides.

The mechanisation of farming practices reduces labour costs and times, but are only effective when single crops of uniform variety are cultivated, as activities such as planting, irrigation and harvesting can be carried out uniformly (Lin 2011). This development has brought about a scale-up in crop production and stability in food prices in the short-term, becoming the norm for modern agricultural practices. As long as national and global interests remain solely on increasing immediate output rather than on the functionality and well-being of the agroecosystem, such practices will continue to take precedence over sustainable agricultural farming systems. Food sovereignty is negatively impacted by national policies that focus on yield, discouraging the cultivation of local crops that are more adapted to the regional climate and less susceptible to global economic impacts (Kijima et al. 2011). Such policies that emphasise intensification and the use of external farming inputs, disadvantage smallholders and show little regard for the local needs of the community (Dawson et al. 2016).

Smallholder farms account for 30% of the global agricultural sphere, playing a key role in sustaining the food needs of vulnerable populations (Samberg et al. 2016). These systems tend to adopt sustainable cropping systems that support social, economic and environmental needs (Samberg et al. 2016; Kremen et al. 2012). It is a carefully planned approach that considers the functional biodiversity and biotic interactions of the system. Thus, are capable of coexisting with threatened and marginal landscape, maximising the benefits of the ecosystem services for a productive and sustainable agricultural system (Kremen et al. 2012).

7.4.4 Investment and Financial Support

National economic incentives have shown a longstanding preference towards monoculture farming systems, with government subsidies made available to select main crops. As an example, the United States offers subsidies for the cultivation of rice, wheat, maize, soybean and cotton (Lin 2011). According to Boody et al. (2009), 89% of the agricultural subsidies in 1995–2002 were allocated to the cultivation of these crops with maize and soybean receiving 56% of the subsidies. Moreover, maximising production of single crops is further encouraged by these subsidies that are allocated based on the acreage of crop farmed.

The US is not alone in the adoption of such policies. African governments have also placed significant effort into enhancing agricultural productivity by developing subsidy programmes for specific crops, particularly maize (Saenz and Thompson 2017). The Zambian government recently acknowledged the impact of this policy on increasing production risk due to increased specialisation on input-dependent crops (Saenz and Thompson 2017). Similarly, Sri Lanka's national emphasis on rice has encouraged its farmers to focus on the cultivation of this water-intensive crop in a region that has suffered numerous drought spells with the advent of climate change (Prasanna et al. 2011; Burchfield and Gilligan 2016). With no subsidies available for other less water demanding crops, households will continue to struggle with the cultivation of rice.

The liberalisation of agricultural policies and markets means that farmers are now exposed to global fluctuations in produce prices, regardless of the domestic conditions (Pellegrini and Tasciotti 2014; Dawson et al. 2016). Thus, small farmers producing the global crops struggle to compete with large producers in the global market, with cheap imported grains flooding the local market. Crop diversification addresses this challenge and minimises the external influences of the big players, encouraging small farmers to produce diverse crops for their local market. However, this external influence is encroaching into the lands of smallholders due to the increasing acquisition of farmland by foreign investors.

Foreign governments, with mounting concerns on food security such as China, Qatar and Saudi Arabia, are acquiring thousands of hectares in poorer countries, mostly in Africa, South America and South East Asia (Cotula et al. 2009). Cotula et al. (2009) have reported a total of almost 2.5 million hectares of land acquired in Sudan, Ethiopia, Madagascar, Mozambique and Tanzania between 2004 and 2009. Opportunities and risks are brought about by these changes. Such investments may increase GDP growth and revenue, improve infrastructure, technology and market access, as well as provide economic development in rural areas (Cotula et al. 2009). Moreover, the increased investment in agriculture may encourage growth of the underinvested agricultural sector in the recipient countries as they shift from their dependence on other commodities, such as oil in Sudan (Arezki et al. 2012).

While some of these land acquisitions cultivate a diverse range of crops such as Jatropha for biofuel and non-food agricultural commodities, the vast interest is on food security and on the main crops. As an example, Saudi agricultural firms are massively investing in Egypt, Ethiopia and Sudan to promote food security through cultivation of strategic crops such as rice, wheat, barley and corn (Cotula et al. 2009). However, as prime arable lands are often sought by foreign investors, small farmers lose access to their land and resources, negatively impacting their food security and livelihood, and pushing these communities into marginal, less-productive lands. Small scale farmers generally adopt farming practices that are more responsive to climate variations and build resiliency in the system, and are less likely to do so if pushed into barren lands with minimal access to irrigation (Arezki et al. 2012). Arezki et al. (2012) further note that host countries with weak land governance tend to be more attractive for foreign investors. Although data on land acquisitions is generally scarce or incomplete, it is clear that better governance policies are required to protect small farmers while reaping the benefits of foreign investment, to ensure sustainable farming practices are adopted.

7.5 Global Strategy

Climate variability negatively impacts agroecological processes such as nutrient cycling, hydrological cycle, and occurrences of pests and diseases, threatening the functionality of crops and ecosystems. These changes will increase the level of biotic and abiotic stresses experienced by the crops and ecosystems. Wider adoption of underutilised crops is a means of enhancing food security by increasing small-holder productivity and improving sustainability and resiliency of farming systems. Crop diversification hinges on the expansion of the portfolio of crops grown, rather than exclusion or neglect of the studied or the under-studied crops. Efforts must be made to integrate these diverse crops into national policies and to encourage their integration into the current food trends, as this will incentivise farmers to produce crops with a market demand.

It is essential to recognise the importance of indigenous knowledge and document the knowledge systems that have accumulated to date, and continue to expand, as well as the advances in technology, practices and policies to date. Future developments need to place trust in the existing farming systems as well as the natural biodiversity. The increasing literature and research on underutilised crops is promising, and so is the development of a research centre, Crops For the Future, dedicated to the study of underutilised crops. Crops For the Future, has pioneered a Global Action Plan for Agricultural Diversification (GAPAD), as an initiative that aims to develop new partnerships among various stakeholders to address today's challenges. Adaptations to combat climate change must be viewed as a process rather than a product with policies taking into full consideration the effects of household cropping practices on poverty reduction.

Glossary

Crop diversification Cultivating more than one variety of crops belonging to the same or different species within a region, using multiple cropping, agroforestry and/or crop rotation systems, with diversity evident in form (e.g. genetic, species, structural), function (e.g. pest suppression, increased production) and scale (temporal and spatial) (Lin 2011; Makate et al. 2016).

Agroforestry The incorporation of trees or shrubs within a cropping system as part of crop diversification to maximise the benefits of interactions between the various biological components.

Crop rotation A temporal approach to crop diversification by systematically varying the crops planted on a given plot between seasons, for example cultivating maize in summer and peas in the following season.

Multiple cropping A spatial approach to crop diversification by systematically cultivating two or more crops in a given plot within the same season, for example, cultivating maize and peas simultaneously on the same piece of land.

Intensification Increase in the productivity of land as determined by the value of agricultural output, which can be market-driven (e.g. production of higher value crops) or technologically driven (e.g. better cropping practices) (Byerlee et al. 2014).

Specialisation Focus on a single activity within a farming system, with the activity providing at least two-thirds of the farm income.

Food sovereignty "The right of a nation or region to produce, distribute or consume food with appropriate productive and cultural diversity" (Altieri 2009).

Vulnerability A measure of a community's exposure to stresses (social and/or environmental), sensitivity to the stresses, and ability to adapt (McCord et al. 2015).

References

Adhikari, L., Hussain, A., & Rasul, G. (2017). Tapping the potential of neglected and underutilised food crops for sustainable nutrition security in the mountains of Pakistan and Nepal. *Sustainability, 9*, 291. https://doi.org/10.3390/su9020291.

Africa Rice Center (WARDA). (2008). *Africa rice trends 2007*. Cotonou: Africa Rice Center (WARDA).

Alcamo, J., Dronin, N., Endejan, M., Golubev, G., & Kirilenko, A. (2007). A new assessment of climate change impacts on food production shortfalls and water availability in Russia. *Global Environmental Change, 7*, 429–444.

Alhassan, G. A., & Egbe, M. O. (2014). Bambara groundnut/maize intercropping: Effects of planting densities in Southern guinea savanna of Nigeria. *African Journal of Agricultural Research, 9*(4), 479–486.

Aliyu, S., Massawe, F., & Mayes, S. (2016). Genetic diversity and population structure of Bambara groundnut (Vigna subterranea (L.) Verdc.): Synopsis of the past two decades of analysis and implications for crop improvement programmes. *Genetic Resources and Crop Evolution, 63*(6), 925–943.

7 Crop Diversification Through a Wider Use of Underutilised Crops: A Strategy... 145

Altieri, M. A. (1999). The ecological role of biodiversity in agroecosystems. *Agriculture, Ecosystems and Environment, 74*, 19–31.

Altieri, M. A. (2009). Agroecology, small farms, and food sovereignty. *Monthly Review, 61*(3), 102–113.

Altieri, M. A., Funes-Monzote, F. R., & Petersen, P. (2012). Agroecologically efficient agricultural systems for smallholder farmers: Contributions to food sovereignty. *Agronomy for Sustainable Development, 32*(1), 1–13.

Anderson, P. K., Cunningham, A. A., Patel, N. G., Morales, F. J., Epstein, P. R., & Daszak, P. (2004). Emerging infectious diseases of plants: Pathogen pollution, climate change and agrotechnology drivers. *Trends in Ecology & Evolution, 19*(10), 535–544.

Arezki, R., Deininger, K., & Seld, H. (2012). The global land rush. *Finance and Development, 49*, 46–49.

Bale, J. S., Masters, G. J., Hodkinson, I. D., Awmack, C., Martijn Bezemer, T., Brown, V. K., Butterfield, J., Buse, A., Coulson, J. C., Farrar, J., Good, J. E. G., Harrington, R., Hartley, S., Hefin Jones, T., Lindroth, R. L., Press, M. C., Symrnioudis, I., Watt, A. D., & Whittaker, J. B. (2002). Herbivory in global climate change research: direct effects of rising temperature on insect herbivores. *Global Change Biology, 8*(1), 1–16.

Berg, A., de Noblet-Ducoudré, N., Sultan, B., Lengaigne, M., & Guimberteau, M. (2013). Projections of climate change impacts on potential C4 crop productivity over tropical regions. *Agricultural and Forest Meteorology, 170*, 89–102.

Bonthala, V. S., Mayes, K., Moreton, J., Blythe, M., Wright, V., May, S. T., Massawe, F., Mayes, S., & Twycross, J. (2016). Identification of gene modules associated with low temperatures response in bambara groundnut by network-based analysis. *PLoS One, 11*(2), e0148771.

Boody, G., Vondracek, B., Andow, D. A., Krinke, M., Westra, J., Zimmerman, J., & Welle, P. (2009). Multifunctional agriculture in the United States. *Bioscience, 55*, 27–38.

Burchfield, E. K., & Gilligan, J. (2016). Agricultural adaptation to drought in the Sri Lankan dry zone. *Applied Geography, 77*, 92–100.

Bvenura, C., & Afolayan, A. J. (2015). The role of wild vegetables in household food security in South Africa: A review. *Food Research International, 76*, 1001–1011.

Byerlee, D., Stevenson, J., & Villoria, N. (2014). Does intensification slow crop land expansion or encourage deforestation? *Global Food Security, 3*(2), 92–98.

Chaifetz, A., & Jagger, P. (2014). 40 Years of dialogue on food sovereignty: A review and a look ahead. *Global Food Security, 3*(2), 85–91.

Chakraborty, S., Tiedemann, A. V., & Teng, P. S. (2000). Climate change: Potential impact on plant diseases. *Environmental Pollution, 108*(3), 317–326.

Cheng, A., Mayes, S., Dalle, G., Demissew, S., & Massawe, F. (2017). Diversifying crops for food and nutrition security - a case of teff. *Biological Reviews, 92*(1), 188–198.

Chivenge, P., Mabhaudhi, T., Modi, A. T., & Mafongoya, P. (2015). The potential role of neglected and underutilised crop species as future crops under water scarce conditions in Sub-Saharan Africa. *International Journal of Environmental Research and Public Health, 12*, 5685–5711.

Cotula, L., Vermeulen, S., Leonard, R., & Keeley, J. (2009). *Land grab or development opportunity? Agricultural investment and international land deals in Africa.* London; Rome: IIED; FAO, IFAD.

Dawson, N., Martin, A., & Sikor, T. (2016). Green revolution in Sub-Saharan Africa: Implications of imposed innovation for the well-being of rural smallholders. *World Development, 78*, 204–218.

Dou, H., & Kister, J. (2016). Research and development on Moringa oleifera - Comparison between academic research and patents. *World Patent Information, 47*, 21–33.

Droogers, P., & Aerts, J. (2005). Adaptation strategies to climate change and climate variability: A comparative study between seven contrasting river basins. *Physics and Chemistry of the Earth, Parts A/B/C, 30*(6–7), 339–346.

Dzama, K. (2016). *Is the livestock sector in Southern Africa prepared for climate change. South African Institute of International Affairs Policy Briefing 153.* Johannesburg: South African Institute of International Affairs.

Eitzinger, J., Stastna, M., & Zalud, Z. (2003). A simulation study of the effect of soil water balance and water stress on winter wheat production under different climate change scenarios. *Agricultural Water Management, 61*, 195–217.

FAO. (2012). *Crop diversification for sustainable diets and nutrition.* Rome: Plant Production and Protection Division (AGP).

Finckh, M. R., Gacek, E., Goyeau, H., Lannou, C., Merz, U., Mundt, C., et al. (2000). Cereal variety and species mixtures in practice, with emphasis on disease resistance. Agronomie. *EDP Sciences, 20*(7), 813–837.

Frison, E., Smith, I. F., Cherfas, J., & Eyzaguirre, P. B. (2006). Agricultural biodiversity, nutrition, and health: Making a difference to hunger and nutrition in the developing world. *Food and Nutrition Bulletin, 27*(2), 167–179.

Godfray, H. C. J., Beddington, J. R., Crute, I. R., Haddad, L., Lawrence, D., Muir, J. F., Pretty, J., Robinson, S., Thomas, S. M., & Toulmin, C. (2010). Food security: The challenge of feeding 9 billion people. *Science, 327*(5967), 812–818.

Govereh, J., & Jayne, T. S. (2003). Cash cropping and food crop productivity: synergies or trade-offs? *Agricultural Economics, 28*(1), 39–50.

Guvele, C. A. (2001). Gains from crop diversification in the Sudan Gezira scheme. *Agricultural Systems, 70*, 319–333.

Hawkesworth, S., Dangour, A. D., Johnston, D., Lock, K., Poole, N., Rushton, J., et al. (2010). Feeding the world healthily: The challenge of measuring the effects of agriculture on health. *Philosophical Transactions of the Royal Society, 365*, 3083–3097.

Heady, E. O. (1952). Diversification in resource allocation and minimization of income variability. *Journal of Farm Economics, 34*, 482–496.

Hochman, Z., Gobbett, D. L., & Horan, H. (2017). Climate trends account for stalled wheat yields in Australia since 1990. *Global Change Biology, 23*, 2071–2081.

Holden, N. M., & Brereton, A. J. (2006). Adaptation of water and nitrogen management of spring barley and potato as a response to possible climate change in Ireland. *Agricultural Water Management, 82*, 297–317.

Jacob, K. D., Charlotte, T. D., Henri, K. K., & Arsene, Z. B. I. (2014). Effect of intercropping bambara groundnut (Vigna subterranea (L.) Verdc) and maize (Zea mays L.) on the yield and the yield component in woodland savannahs of Côte d'Ivoire. *International Journal of Agronomy and Agricultural Research, 5*(1), 46–55.

Kang, Y., Khan, S., & Ma, X. (2009). Climate change impacts on crop yield, crop water productivity and food security – A review. *Progress in Natural Science, 19*, 1665–1674.

Karunaratne, A., Azam-Ali, S. N., Al-Shareef, I., Sesay, A., Jorgensen, S. T., & Crout, N. M. J. (2010). Modelling the canopy development of Bambara groundnut. *Agricultural and Forest Meteorology, 7-8*, 1007–1015.

Karunaratne, A., Azam-Ali, S. N., & Crout, N. M. J. (2011). BAMGRO: A simple model to simulate the response of Bambara groundnut to abiotic stress. *Experimental Agriculture, 47*(3), 489–507.

Karunaratne, A. S., Walker, S., & Azam-Ali, S. N. (2015). Assessing the productivity and resource-use efficiency of underutilised crops: Towards an integrative system. *Agricultural Water Management, 147*, 129–134.

Kassie, M., Shiferaw, B., & Muricho, G. (2011). Agricultural technology, crop income, and poverty alleviation in Uganda. *World Development, 39*(10), 1784–1795.

Khoury, C. K., Bjorkman, A. D., Dempewolf, H., Ramirez-Villegas, J., Guarino, L., Jarvis, J., et al. (2014). Increasing homogeneity in global food supplies and the implications for food security. *Proceedings of the National Academy of Sciences, 111*(11), 4001–4006.

Kijima, Y., Otsuka, K., & Sserunkuuma, D. (2011). An Inquiry into Constraints on a Green Revolution in Sub-Saharan Africa: The Case of NERICA Rice in Uganda. *World Development, 39*(1), 77–86.

Kremen, C., Iles, A., & Bacon, C. (2012). Diversified farming systems: An agroecological, systems-based alternative to modern industrial agriculture. *Ecology and Society, 17*(4), 44.

Leakey, R. R. B., & Asaah, E. K. (2013). Underutilised species as the backbone of multifunctional agriculture – The next wave of crop domestication. *Acta Horticulturae, 979*, 293–310.

Lin, B. B., Perfecto, I., & Vandermeer, J. (2008). Synergies between agricultural intensification and climate change could create surprising vulnerabilities for crops. *Bioscience, 58*(9), 847–854.

Lin, B. B. (2011). Resilience in agriculture through crop diversification: Adaptive management for environmental change. *Bioscience, 61*, 183–193.

Mabhaudhi, T., & Modi, A. T. (2013). Intercropping taro and Bambara groundnut. *Sustainable Agriculture Reviews, 13*, 275–290.

Maitra, S., Ghosh, D. C., Sounda, G., Jana, P. K., & Roy, D. K. (2000). Productivity, competition and economics of intercropping legumes in finger millet (Eleusine coracana) at different fertility levels. *Indian Journal of Agricultural Science, 70*(12), 824–828.

Makate, C., Wang, R., Makate, M., & Mango, N. (2016). Crop diversification and livelihoods of smallholder farmers in Zimbabwe: Adaptive management for environmental change. *Springerplus, 5*, 1135. https://doi.org/10.1186/s40064-016-2802-4.

Massawe, F. J., Mayes, S., & Cheng, A. (2016). Crop diversity: An unexploited treasure trove for food security. *Trends in Plant Science, 21*(5), 365–368.

Mayes, S., Massawe, F. J., Alderson, P. G., Roberts, J. A., Azam-Ali, S. N., & Hermann, M. (2012). The potential for underutilized crops to improve security of food production. *Journal of Experimental Botany, 63*(3), 1075–1079.

McCord, P. F., Cox, M., Schmitt-Harsh, M., & Evans, T. (2015). Crop diversification as a smallholder livelihood strategy within semi-arid agricultural systems near Mount Kenya. *Land Use Policy, 42*, 738–750.

McIntyre, B. D., Herren, H. R., Wakhungu, J., & Watson, R. T. (2009). *International assessment of agricultural knowledge, science and technology for development (IAASTD): Synthesis report with executive summary: A synthesis of the global and sub-global IAASTD reports.* Washington, DC: IAASTD.

Michler, J. D., & Josephson, A. L. (2017). To specialize or diversify: agricultural diversity and poverty dynamics in Ethiopia. *World Development, 89*, 214–226.

Midega, C. A. O., Khan, Z. R., Amudavi, D. M., Pittchar, J., & Pickett, J. A. (2010). Integrated management of *Striga hermonthica* and cereal stemborers in finger millet (*Eleusine coracana* (L.) Gaertn.) through intercropping with *Desmodium intortum*. *International Journal of Pest Management, 56*(2), 145–151.

Midgley, S., & Methner, N. (2016). *Climate adaptation readiness for agriculture: Drought lessons from the Western Cape, South Africa. African Institute of International Affairs Policy Briefing 154.* Johannesburg: South African Institute of International Affairs.

National Academy of Sciences. (1972). *Genetic vulnerability of major crops.* Washington, DC: NAS.

Nguyen, H. Q. (2017). Analyzing the economies of crop diversification in rural Vietnam using an input distance function. *Agricultural Systems, 153*, 148–156.

Njeru, E. M. (2013). Crop diversification: A potential strategy to mitigate food insecurity by smallholders in sub-Saharan Africa. *Journal of Agriculture, Food Systems, and Community Development, 3*, 63–69.

Onyango, A. O. (2016). Finger millet: Food security crop in the arid and semi-arid lands (ASALs) of Kenya. *World Environment, 6*(2), 62–70.

Orr, A. (2000). Green Gold'?: Burley tobacco, smallholder agriculture, and poverty alleviation in Malawi. *World Development, 28*, 347–363.

Patterson, D. T., Westbrook, J. K., Joyce†, R. J. V., Lingren, P. D., & Rogasik, J. (1999) *Climatic Change, 43*(4), 711–727.

Pellegrini, L., & Tasciotti, L. (2014). Crop diversification, dietary diversity and agricultural income: Empirical evidence from eight developing countries. *Canadian Journal of Development Studies, 35*, 211–277.

Perfecto, I., Vandermeer, J. H., Bautista, G. L., Nuñez, G. I., Greenberg, R., Bichier, P., & Langridge, S. (2004). Greater predation in shaded coffee farms: The role of resident Neotropical birds. *Ecology, 85*, 2677–2681.

Prasanna, R. P. I. R., Bulakulama, S. W. G. K., & Kuruppuge, R. H. (2011). Factors affecting farmers' higher grain from paddy marketing: A case study on paddy farmers in North central province, Sri Lanka. *International Journal of Agricultural Management and Development, 2*, 57–69.

Rahman, S. (2009). Whether crop diversification is a desired strategy for agricultural growth in Bangladesh? *Food Policy, 34*, 340–349.

Ray, K. D., Ramankutty, N., Mueller, N. D., West, P. C., & Foley, J. A. (2012). Recent patterns of crop yield growth and stagnation. *Nature Communications, 3*, 1293. https://doi.org/10.1038/ncomms2296.

Rosenzweig, C., & Parry, M. L. (1994). Potential impact of climate change on world food supply. *Nature, 367*, 133–138.

Saenz, M., & Thompson, E. (2017). Gender and policy roles in farm household diversification in Zambia. *World Development, 89*, 152–169.

Samberg, L. H., Gerber, J. S., Ramankutty, N., Herrero, M., & West, P. C. (2016). Subnational distribution of average farm size and smallholder contributions to global food production. *Environmental Research Letters, 11*(12), 124010. https://doi.org/10.1088/1748-9326/11/12/124010.

Scherm, H., & Yang, X. B. (1995). Interannual variations in wheat rust development in China and the United States in relation to the El Nino/Southern Oscillation. *Phytopathology, 85*, 970–976.

Seck, P. A., Tollens, E., Wopereis, M. C. S., Diagne, A., & Bamba, I. (2010). Rising trends and variability of rice prices: Threats and opportunities for sub-Saharan Africa. *Food Policy, 35*, 403–411.

Senger, I., Borges, J. A. R., & Machado, J. A. D. (2017). Using the theory of planned behavior to understand the intention of small farmers in diversifying their agricultural production. *Journal of Rural Studies, 49*, 32–40.

Smith, S. E., & Read, D. J. (2008). *Mycorrhizal symbiosis* (3rd ed.). Amsterdam: Academic Press, Elsevier.

Sserunkuuma, D. (2008). *Assessment of NERICA training impact. A study report prepared for the Japan international cooperation agency (JICA).* Tokyo: JICA.

Tadele, Z. (2017). Raising crop productivity in Africa through intensification. *Agronomy, 7*(1), 22.

Tadele, Z., & Assefa, K. (2012). Increasing food production in Africa by boosting the productivity of understudied crops. *Agronomy, 2*(4), 240–283.

Taffesse, A. S., Dorosh, P., & Asrat, S. (2012). *Crop production in Ethiopia: Regional patterns and trends. Ethiopia strategy support program (ESSP II).* Washington, DC: International Food Policy Research Institute.

Thilakarathna, M., & Raizada, M. (2015). A review of nutrient management studies involving finger millet in the semi-arid tropics of. *Asia and Africa. Agronomy, 5*(3), 262–290.

Touma, D., Ashfaq, M., Nayak, M., Kao, S., & Diffenbaugh, N. (2015). A multi-model and multi-index evaluation of drought characteristics in the 21st century. *Journal of Hydrology, 526*, 196–207.

UN. (2016). *Sustainable development knowledge platform.* Retrieved from https://sustainabledevelopment.un.org/?page=view&nr=164&type=230&menu=2059.

Van den Berg, M. M., Hengsdijk, H., Wolf, J., Ittersum, M. K. V., Guanghuo, W., & Roetter, R. P. (2007). The impact of increasing farm size and mechanization on rural income and rice production in Zhejiang province, China. *Agricultural Systems, 94*, 841–850.

Vandermeer, J., van Noordwijk, M., Anderson, J., Ong, C., & Perfecto, I. (1998). Global change and multi-species agroecosystems: Concepts and issues. *Agriculture, Ecosystems and Environment, 67*, 1–22.

Weltin, M., Zasada, I., Franke, C., Piorr, A., Raggi, M., & Viaggi, D. (2017). Analysing behavioural differences of farm households: An example of income diversification strategies based on European farm survey data. *Land Use Policy, 62*, 172–184.

Wheeler, T., & von Braun, J. (2013). Climate change impacts on global food security. *Science, 341*, 508–513.

Wilhite, D. A., & Vanyarkho, O. (2000). Drought: Pervasive impacts of a creeping phenomenon. In D. A. Wilhite (Ed.), *Drought: A global assessment* (pp. 245–255).

World Bank. (2008). *World development report 2008: Agriculture for development*. Washington, DC: World Bank.

Yachi, S., & Loreau, M. (1999). Biodiversity and ecosystem productivity in a fluctuating environment: The insurance hypothesis. *Proceedings of the National Academy of Sciences, 96*, 1463–1468.

Zhu, Y., Chen, H., Fan, J., Wang, Y., Li, Y., Chen, J., et al. (2000). Genetic diversity and disease control in rice. *Nature, 406*, 718–722.

Chapter 8
Bambara Groundnut is a Climate-Resilient Crop: How Could a Drought-Tolerant and Nutritious Legume Improve Community Resilience in the Face of Climate Change?

Aryo Feldman, Wai Kuan Ho, Festo Massawe, and Sean Mayes

8.1 Introduction

Bambara groundnut (*Vigna subterranea* (L.) Verdc.; www.bamyield.org) is a crop similar in morphology and growth habit to groundnut (*Arachis hypogaea* L.). It was also historically largely displaced by groundnut upon the latter's introduction to sub-Saharan Africa from Latin America (Sprent et al. 2010). Bambara groundnut nevertheless still holds local importance in West Africa, East Africa, Southern Africa and even Southeast Asia (Fig. 8.1). It is held in high esteem for its nutritional qualities by the consumer and its drought tolerance by

---

Electronic supplementary material: The online version of this chapter (https://doi.org/10.1007/978-3-319-77878-5_8) contains supplementary material, which is available to authorized users.

A. Feldman (✉)
Crops For the Future, Semenyih, Selangor, Malaysia
e-mail: aryo.feldman@cffresearch.org

W. K. Ho
Crops For the Future, University of Nottingham Malaysia Campus,
Semenyih, Selangor, Malaysia
e-mail: waikuan@cffresearch.org

F. Massawe
Faculty of Science, University of Nottingham Malaysia Campus,
Semenyih, Selangor, Malaysia
e-mail: festo.massawe@nottingham.edu.my

S. Mayes
Crop Genetics, South Labs, Plant and Crop Sciences, Biosciences, Sutton Bonington
Campus, University of Nottingham, Leicestershire, UK

Biotechnology, Breeding and Seed Systems, Crops For the Future,
Semenyih, Selangor, Malaysia
e-mail: sean.mayes@cffresearch.org

© Springer Nature Switzerland AG 2019
A. Sarkar et al. (eds.), *Sustainable Solutions for Food Security*,
https://doi.org/10.1007/978-3-319-77878-5_8

Fig. 8.1 World map of Bambara groundnut production. The map highlights countries that produce Bambara groundnut (in blue). The crop has greatest local importance in sub-Saharan Africa with the greatest concentration of cultivation being in West Africa where the crop originates. Its more recent introduction into other regions, and Southeast Asia in particular, highlights the crop's adaptability and adoption into different farming systems, consumption patterns and markets. Map created at AMCHARTS.COM. List of countries of production provided by Kewscience, PROTA and BamNetwork

the farmer (Tables 8.1, 8.2, and 8.3; Fig. 8.2). It could therefore be promoted in areas that are currently drought-prone as well as in areas where climate change projections show an increased frequency and intensity in droughts as well as unpredictable rainfall patterns.

There is growing support for Bambara groundnut and underutilised crops in general, partly from the consumer community that demands greater diversity in their diets, but also from the scientific community interested in the 'treasure trove' of beneficial crop traits (Massawe et al. 2016). The dedicated international research centre for underutilised crops, Crops For the Future (CFF; www.cffresearch.org), operates as a focal point or hub for networks of global partners studying and developing these underutilised (also called minor, alternative, orphan and neglected) crops for food and non-food uses. Research on Bambara groundnut by CFF's networks and partner organisations will test whether the status of a minor crop, such as Bambara groundnut, can be elevated by translating the toolkits and methodologies used for improving major crops via fast-tracked, cost-effective, international coordination.

8 Bambara Groundnut

Table 8.1 Comparison of Bambara groundnut yield figures to important legume crops

Crop	Author	Min	Max	Mean	Location	Remarks
Bambara groundnut	Berchie et al. (2010)	1300	2400	1661	Ghana	Mean of five landraces
	Kouassi and Zorobi (2010)	1310	5100	2670	Nigeria	Mean of two genotypes
	Ouedraogo et al. (2008)	110	1440	830	Burkina Faso	Mean of 310 landraces
	Touré et al. (2012)	79	459	278	Ivory Coast	
	Karikari and Tabona (2004)	308	1477	756	Botswana	Mean of 12 landraces for two seasons
	Pungulani et al. (2012)	485	1322	824	Malawi	Mean of eight landraces across three sites
	Makanda et al. (2009)	1100	2300	1590	Zimbabwe	Mean of 20 landraces
	BAMFOOD (2002)	476	1892	1116	Swaziland	Mean of eight landraces across two sites
	BAMFOOD (2002)	269	776	508	Namibia	Mean of eight landraces across two sites
	Average	604	1907	1137		
Bambara groundnut	FAOSTAT	720	731	725	World	Same value as for 'Africa'
Chickpea		586	649	596	World	
Cowpea		276	360	354	World	
Groundnut		883	849	863	World	
Soybean		1156	1128	1138	World	
Chickpea		497	594	547	Africa	
Cowpea		328	255	330	Africa	
Groundnut		831	848	896	Africa	
Soybean		376	395	394	Africa	

Values are for dry seed yield at kg ha^{-1}. FAOSTAT figures are global and continental means

8.2 Resilience Traits

In terms of traits sought after by farmers, Bambara groundnut is most impressive in terms of drought tolerance, encompassing all three composite traits of drought escape, avoidance and tolerance (Mabhaudhi et al. 2013; Chai et al. 2016; Table 8.2). Studies have confirmed drought avoidance and escape as the primary mechanisms through which Bambara groundnut tolerates drought (Mabhaudhi et al. 2013) and that the crop showed low water use (Mabhaudhi et al. 2013). Drought avoidance was achieved through greater stomatal regulation of gas exchange and reduction in canopy size (Mabhaudhi et al. 2013; Chai et al. 2016).

Table 8.2 Comparison of Bambara groundnut drought tolerance figures to important legume crops

Crop	Author	Water treatment	Yield	Remarks
Bambara groundnut	Mabhaudhi and Modi (2013)	Irrigated	2.79	Mean of three landraces under field conditions with a for two seasons; yield = tons ha^{-1}
		Rain-fed	1.51	
		Difference (%)	45.81	
	Mabhaudhi et al. (2013)	60%	1.1	Mean of two landraces under field conditions with a rain-shelter for one season; yield = tons ha^{-1}
		100%	2.59	
		Difference (%)	57.53	
	Mwale et al. (2007a, b)	Irrigated	297.67	Mean of three landraces under greenhouse conditions for one season; yield = gm^{-2}
		Mild drought (no irrigation from 42 DAS)	164.67	
		Difference (%)	44.68	
	Average	Difference (%)	49.34	
Field pea	Daryanto et al. (2015)	46% Water reduction	21	Yield reduction (%)
Groundnut		72% Water reduction	38	
Pigeon pea		62% Water reduction	24	
Lentil		74% Water reduction	22	
Soybean		61% Water reduction	27	
Faba bean		68% Water reduction	32	
Chickpea		50% Water reduction	41	
Cowpea		63% Water reduction	45	
Green gram		58% Water reduction	46	
Common bean		64% Water reduction	61	

Drought tolerance is calculated as proportional differences between yield figures at optimal conditions and at drought treatments

Drought escape was associated with the crop's phenological plasticity (Mabhaudhi et al. 2013). Drought tolerance was associated with Bambara groundnut's ability to maintain relatively high internal leaf water status and photosynthetic rates under drought stress (Chai et al. 2016). The presence of all three drought-tolerance mechanisms makes Bambara groundnut a model drought-tolerant crop capable of producing reasonable yields under mild, intermittent and even severe drought stress.

This makes it an important crop for use in areas that are currently prone to aridity as well as an important future crop in areas where climate change projec-

8 Bambara Groundnut

Table 8.3 Comparison of Bambara groundnut nutrition figures to important legume crops

Crop	Author	Calories (kcal per 100 g)	Carbohydrate (%)	Protein (%)	Fat (%)
Bambara groundnut	Amarteifio and Moholo (1998)		63.5	18.3	6.6
Marama bean			24.1	34.1	33.5
Mung bean			59.8	26.4	1.1
Tepary bean			63.8	24.7	0.9
Bambara groundnut	Akande et al. (2009)		59.1	20.1	7.2
	Bamshaiye et al. (2011)		54.4	19.9	6.9
	Mazahib et al. (2013)		57	21	6.6
Bambara groundnut	Hillocks et al. (2012)	390	61.9	21.8	6.6
Broad bean		341	58.3	26.1	5.7
Chickpea		364	60.6	19.3	6
Cowpea		343	59.6	23.8	2.1
Kidney bean		333	60	23.6	0.8
Soybean		416	30.2	36.5	19.9
Bambara groundnut	Average		61.9	21.8	6.6
Chickpea	USDA	378	63	20.5	6
Cowpea		336	60	23.5	1.2
Groundnut		567	16.1	25.8	49.2
Soybean		446	30.16	36.5	20

Nutritional values are presented in terms of macronutrients and calories

tions show an increased frequency and intensity in dry spells as well as erratic rainfall patterns. A three-year project funded by the International Treaty for Plant Genetic Resources of Food and Agriculture—Benefit Sharing Fund is currently underway (started in July 2016), which brings together a network of four national research institutions in West Africa and Southeast Asia (project: W3B-PR-26/MALAYSIA/2016/AGDT). The project looks at germplasm collection, breeding and multi-locational field evaluation of Bambara groundnut, mainly for the purpose of improving and understanding its drought-tolerance mechanisms, both at the physiological and genetic levels. As an adjunct, the network has been expanded to include partners outside of the core grant-funded project. One such partner is Acacias 4 All, an NGO based in Tunisia, who specifically are combating desertification as the Sahara Desert is encroaching further North with climatic changes.

Although Bambara groundnut is most commonly found in arid and semi-arid zones in the tropical latitudes, predominantly in the African continent, it has nevertheless been shown to be somewhat adaptive to the humid tropics despite its sensitivity to anoxic/hypoxic stress. This is evident in its modest production in Southeast Asia, as demonstrated. The three Nigerian lines in the study performed very reasonably, producing yields at the global average level (~1.2 tons ha^{-1}; aggregate mean

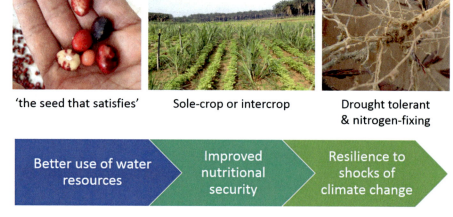

Fig. 8.2 Beneficial traits of Bambara groundnut and potential impact of enhancing its production. Bambara groundnut is described as "the seed that satisfies" due to its balance of crude macronutrients (carbohydrate, protein and fat) that matches human recommended daily intake, as well as micronutrients content of particular vitamins, minerals, and amino and fatty acids (see Subheading Other Beneficial Characteristics). The crop is an ideal companion crop as it makes little demands on the soil in terms of nutrients and water. As it is a legume, it fixes its own nitrogen, which furthers its status as a companion crop. It also displays most if not all drought-tolerant traits (see Subheading Resilience Traits). These three broad characteristics of the underutilised legume impacts the farm by making better use of available water, contributes to nutritional security for the consumer and also aids rural communities in mitigating and adapting to the detrimental effects of climate change

reported, Ouedraogo et al. (2008), Makanda et al. (2009), Berchie et al. (2010), Touré et al. (2012) and Pungulani et al. (2012); Table 8.1) and beyond (up to 2.3 tons ha^{-1}) on generally poor and acidic soils (pH lower than 4). This is in part due to the plant's ability as a legume to fix atmospheric nitrogen into a usable form even under harsh conditions in resource-poor farming systems Nyemba and Dakora (2010).

This ability to grow under extreme soil and climatic conditions cannot be undervalued. Resilience traits that infer the ability to withstand abiotic stresses like drought, salinity and heat, as well as biotic problems, will only be more prized with progressively marked climate changes and losses of optimal agricultural land due to increased soil degradation and urbanisation. The key is to diversify agriculture with resilient non-staple crops, which also serves as a means of risk spreading (Massawe et al. 2016). Bambara groundnut is one such candidate resilient crop that has been validated by suitability models for regions—particularly within sub-Saharan Africa—that are vulnerable to environmental challenges, especially aridity (Karunaratne et al. 2011). BAMnut—a model simulating Bambara groundnut cultivation performance across the globe—displayed the underutilised crop's high potential outside its current geographical focus, namely sub-Saharan Africa, to all continents (besides Antarctica) and predicted that the Mediterranean agro-ecological zone would return the greatest biomass yields (Azam-Ali et al. 2001).

8.3 Other Beneficial Characteristics

Colloquially, Bambara groundnut is considered a complete food in that its balance of macronutrients approximately matches the requirements of the human diet. Mazahib et al. (2013) reported 21% protein, 6.6% fat and 57% carbohydrate in the seed, which was representative of the figures of other researchers (e.g. Akande et al. 2009; Bamshaiye et al. 2011; Table 8.3). Other interesting nutritional features of Bambara groundnut include its relatively high levels of essential amino acids, such as methionine, leucine, tyrosine and arginine (Akande et al. 2009; Mazahib et al. 2013). This makes it a high quality protein source, if not a particularly high crude protein source when compared to some other legumes like soybean (*Glycine max*; 36.5%) and peanut (*Arachis hypogaea*; 25.8%) (USDA, National Nutrient Database for Standard Reference Release 28). Mubaiwa et al. (2017) also cited the crop's high level of fatty acids (e.g. palmitic and linoleic acids), vitamins and minerals (e.g. vitamin A, thiamine, zinc and iron). Indeed, despite the low crude oil content compared again to other legumes like soybean (19.9%) and peanut (49.2%) (USDA, National Nutrient Database for Standard Reference Release 28), Bambara groundnut has high quality and mainly unsaturated fats, including essential fats: linoleic acid, palmitoleic, oleic and caprylic acids, and eicosanoic acid.

In the post-Green Revolution era, there is a growing concern with nutritional security rather than food security; malnutrition versus under-nutrition (indeed, over-nutrition has become an important national and international agenda item); and micronutrients versus calories (Burlingame and Dernini 2012). An obvious way to promote nutritionally complete and balanced diets is through diversifying typical eating habits, which have disproportionately high amounts of the big three cereals (maize (*Zea mays*), rice (*Oryza sativa*) and wheat (*Triticum aestivum*)), by including food derived from nutritionally dense minor crops, particularly legumes that tend to have higher protein contents (Massawe et al. 2016).

Bambara groundnut seeds can be eaten raw, boiled, fried, stewed or roasted, according to local preferences (Fig. 8.3). As well as its versatility as a food, research has pointed out its potential as a vegetable milk (Brough et al. 1993), including as an infant milk formula (Ijarotimi and Keshinro 2012); livestock feed (Akande et al. 2009; Mahala and Mohammed 2010), including as a fish feed ingredient; and as a composite flour ingredient for fortifying food products like bread (Alozie et al. 2009).

Fig. 8.3 Examples of food products that can be made from Bambara groundnut. Images are not shown to relative scale. All food products were made at Crops For the Future. The plate at the top shows popular and low-cost worldwide snacks containing Bambara groundnut flour as the main ingredient, namely (clockwise from top-left): murukku from Tamil Nadu, India; keropok from Malaysia and Indonesia; tortilla chips from Mexico; and another murukku snack. The bottom three images show more high-end food products containing Bambara groundnut (from left to right): biscotti, cereal bars and vegetable milk

8.4 Key Constraints

Dry, hardened legume seeds are notorious for the long time it takes to cook them and Bambara groundnut is at the extreme end of these foods. They generally need 3–4 h of boiling, which although similar to soybean (3.6 h), is significantly longer than common bean (*Phaseolus vulgaris*; 1.5 h), cowpea (*Vigna unguiculata*; 2.4 h) and mung bean (*Vigna radiata*; 0.5 h) (Mubaiwa et al. 2017). Bambara groundnut's long cooking time (also called "poor cookability" or the "hard-to-cook" phenomenon) is regularly cited as a key constraint for greater uptake of Bambara groundnut (e.g. the EU-ACP Science & Technology Programme project, FED/2013/330-241, entitled "Strengthening Capacities and Informing Policies for Developing Value Chains of Neglected and Underutilised Crops in Africa"). Not only is there a greater temporal and human effort resource involved in cooking the food but there is also a greater fuel requirement (which is most often wood in rural communities, where it is most commonly consumed) and so economic cost. This also has important health and environmental implications.

There are several seed traits that may contribute to the legume's long cooking time, chiefly its hardness and poor water absorbability. There is however variation in cookability within Bambara groundnut seed types. For example, the trait (broadly, cookability, and specifically, water absorbability) is negatively correlated with seed coat thickness (Kaptso et al. 2008), size (Mubaiwa et al. 2017) and darkness of coat colour (Ojimelukwe 1998). The latter is due to the higher tannin content in darker seed coats, which reduces the seed's protein availability and so also its cookability (Ojimelukwe 1998).

Among its biological limitations, Bambara groundnut landraces are generally short-day plants whereby reproductive development is often hindered by day lengths longer than 12 h (Linnemann et al. 1995; Brink 1997). While the onset of flowering is rarely affected by long day lengths, a non-permissive day length >13 h can completely prevent pod-set and/or reduce seed yield in some landraces. This has restricted it from being cultivated outside of the Tropics, and also limited the transfer of germplasm from equatorial Africa to Southern African countries where day lengths are greater 13 h during the growing season. An EU Framework 6 Programme project (BAMLINK: INCO-CT-2005-015459) demonstrated the potential to introduce Bambara groundnut at more extreme latitudes. The project team identified (at the Royal Danish Veterinary and Agricultural University) and developed (at the University of Nottingham, UK) material with far less requirement for short days during pod-filling and reasonable seed set in controlled glasshouses with 16 h day length. The FAO Agroecological Mapping Project (Azam-Ali et al. 2001) predicted that, if the short-day requirement could be overcome, the Mediterranean basin would provide the ideal environments for the production of Bambara groundnut given its climatic conditions, chiefly the high cumulative sunshine hours. This was especially evident for regions of Southern Europe and North Africa where the crop's drought tolerance could provide advantages over more familiar legumes, as these regions have been predicted to be one of the most severely affected for reduced rainfall under most climate change scenarios (IPCC 2014). In addition, this would help to directly address agricultural sustainability in regions of Africa, in particular, where rainfall is already declining and is predicted to decrease even further under climate change.

A key factor in global production levels of underutilised crops being much lower is that they have received far less breeding investment. Due to their inferior scale and often limited monetary value, they are for the most part likely to have been selected in low input resilient farming systems that tend to be complex as opposed to high input monocultures, which is how most major crop production occurs. One result of this continual selection over millennia under low input and in the same location in complex systems are crops that are often more resilient to localised biotic and abiotic stresses, and that make suitable companion crops for introduced and major crops. While farmer-based selection has partially bred for yield these crops have not been specifically cross-bred and intensively selected for calorific yield and they often have retained other nutritional advantages (e.g. high levels of particular vitamins and minerals) that could make them an important component of a balanced diet. This may also have been emphasised in recent historical times with the introduction of non-indigenous major crops as part of agriculture, with the minor crops appreciated for their cultural, historical and nutritional contribution to food production in low input systems.

Nevertheless, once these candidate orphan crops are adopted, there remains a significant amount of catching up for their breeders to achieve. In many cases they may only be partly domesticated, having been selected for only some of the Domestication Syndrome Factors that are found in major crops. Underutilised crops are often adapted to their local environments due to continuous growth in the same location and so collecting landraces from different environments may represent valuable material for increasing the range of production zones (Mayes et al. 2011). The question posed is therefore whether it is better to make trait specific improvements in major crops, such as in biofortification, or to improve minor crops so that they will make a greater contribution to the global diet. This is not only a question of cost but is socio-politically charged, incorporating such issues of market aversion to novel fresh produce, consumer concerns over genetic modification technology, government support for certain crops over others, in the form of subsidies or otherwise, and a question of ownership of the original material selected by farmers. In practice, the challenges posed by rising world population and climate change means that both approaches need to be pursued and for low input agricultural systems, a more complex integration of both will be needed.

8.5   Other Constraints

It is common for underutilised species to be relegated in agricultural and consumer communities as a 'poor man's crop', and Bambara groundnut is a prime example of this (Adzwala et al. 2015). This is not least the case as it is a legume and so a more affordable source of protein than meat. This stigma likely contributes to Bambara groundnut being a 'woman's crop' with men often focusing on cereal production in many low-income countries, which tend to fetch a higher market price and are able to be exported particularly in the case of sub-Saharan Africa.

This low social perception translates to less developed value chains when compared to major (export) crops. Although Bambara groundnut can sell for higher prices than other legumes, including groundnut, in local markets (often owing to lower supplies), there are almost no value-added products. There is therefore opportunity for the creation of niche markets for it as a processed food and indeed there are currently moves in this direction. This has come from little governmental support but rather as a result of localised 'affection' for its inherent advantages, such as its aforementioned food and farm qualities, including from urban-migrants who grew up with the crop (Adzwala et al. 2016b).

Bambara groundnut also has potential as an export crop. It is a close relative of the more well-known cowpea (black-eyed pea) and the African diaspora is likely to be an important market. One initiative, Believe in Bambara (https://www.believ-einBambara.com; formerly known as 'Ag for Africa'), is sourcing Bambara groundnut seeds from Ghana to be processed into vegetable milk and sold in U.S.A. (initially in New York City). It is a prime example of how social media can reach out to a more prosperous market that is willing to pay a premium to social enterprises that exhibit values of sustainability and environmental protection to support resource-poor communities and marginalised people. The greater purchasing power of the American market could therefore increase the welfare of Ghanaian women farmers, which has been shown to be possible by cultivating Bambara groundnut in Adzwala et al. (2016a). In addition, in Malaysia, where very limited amounts of the crop are grown, there is interest from a range of stakeholders, from the farmer's association in the Northern state of Kedah, through to multinational companies, such as Nestlé and Mamee Double Decker. This complements breeding companies, such as Green World Genetics, who are evaluating lines of Bambara groundnut for potential release.

8.6 Genetic Improvement

The improvement of underutilised crops is part of the larger aim to diversify agriculture, importantly for resilience and diets. The modern molecular tools that were originally developed for major crops are currently accelerating genetic improvement of these plants. Genomic breeding through marker-assisted selection, driven by the advancement in Next Generation Sequencing and more recently Next Generation Mapping, has given and will increasingly give, in the case of minor crops, a far greater understanding to the underlying DNA information of important phenotypic traits, such as those that cope with biotic and abiotic stresses (Abberton et al. 2016).

The African Orphan Crops Consortium (www.africanorphancrops.org) led by Dr. Howard-Yana Shapiro (Mars Incorporated) has given itself the ambitious task of sequencing the whole genomes of 101 underutilised crop species and 100 lines of each species, including Bambara groundnut. The publication of these reference genomes and variants will be a major boon for plant breeders to develop further tools to more efficiently and accurately improve the genetics of under-researched and under-funded, but locally important nutritious food crops.

The development of molecular markers, for example, can guide selections of plant material according to particular characteristics where phenotypic selection would be too time and/or space consuming, complicated by pleiotropy and/or have poor inheritability. There have been several papers published on how molecular markers in Bambara groundnut can *inter alia* inform an understanding of genetic diversity (Somta et al. 2011; Molosiwa et al. 2015), measure heterozygosity for seed purity (Ho et al. 2016), link germplasm to particular parameters like geographical origin (Olukolu et al. 2012), perform quality control and allow an understanding of the breeding system (Mayes et al. 2015a, b) as well as explore the genetic control of important traits, such as leaf morphology and drought tolerance (Chai et al. 2016, 2017; Ahmad et al. 2016) and allow for comparison between crops, with and without complete genome sequences, to allow translation of positional information into less studied crops (Ho et al. 2017). Moreover, developing a detailed understanding of the problems faced by one underutilised crop, such as Bambara groundnut, could allow generic approaches to be developed in others. With the tremendous reduction in cost of genotyping per sample, high-throughput genotyping has been developed for a number of legume species (Varshney 2016): for instance, a 60k cowpea (*Vigna unguiculata* [L.] Walp) iSelect Consortium Array to screen 51,128 SNPs (Muñoz-Amatriaín et al. 2017) and a flexible throughput Competitive Allele Specific PCR (KASPar) assay for 2005 SNPs in chickpea (*Cicer arietinum*) (Hiremath et al. 2012).

Underutilised crop improvement can be fast-tracked by an integrated, coordinated breeding approach by concurrently tackling several inherent biological limitations at once, while at the same time being informed by *inter alia* socio-economists, molecular geneticists, agronomists and nutritionists. Azam-Ali et al. (2001) took a retrospective look at how research into underutilised species could be accelerated with "the use of a methodological framework…that maximise[s] knowledge, minimise[s] duplication of effort, identif[ies] priority areas for further research and dissemination, and derive[s] general principles for application across underutilised crops in general".

For example, Bambara groundnut's bottleneck of requiring short day lengths for reproductive success is being addressed by a network of international centres. Genetic material developed by the University of Nottingham, UK, that is expected to be suitable for pod yield under longer day lengths will be tested under the Mediterranean agroecological conditions of Tunisia (Acacias for All), Italy (Mediterranean Agronomic Institute of Bari), South Africa (University of KwaZulu-Natal), controlled glasshouses in UK (University of Nottingham) and controlled environment chambers in Malaysia (CFF). This will provide a preliminary demonstration of crop performance to potential growers at these more extreme latitudes. Field trials of developed lines that are less sensitive to day length (and therefore able to produce yield in Southern Europe, and Northern and Southern Africa) will be evaluated for growth, productivity and water use efficiency, particularly in marginal environments. The field trials will also allow an assessment of how the crop would be incorporated into existing mechanised farming systems and whether the crop can be introduced as a sole crop, an intercrop or in rotation with other crops. In addition, food products will be developed to suit local markets at each partner site to test full end-user acceptability. More recently, molecular mapping populations have been constructed at the University of Nottingham and are being used for marker-assisted selection of more breeding lines for photoperiod insensitivity (Kendabie et al. 2015).

The integrated, coordinated approach relies on a wide selection of experts from different academic disciplines. The initial step to carrying it out is probably the most difficult: having all these experts communicating and agreeing to work together for a common research goal, whether it is within a single institution (e.g. CFF; cffresearch.org) or on a far-reaching platform (e.g. BamNetwork; bambaragroundnut.org). Filling the gaps along the Research Value Chain, which encompasses molecular genetics through to policy, needs multidisciplinary/interdisciplinary/transdisciplinary organisation. This is not simply a matter of tearing down structural barriers that have been ingrained in academia for centuries, but more importantly needs to address social and interpersonal complexes, such as by building inclusive teams (Mumuni et al. 2016).

8.7 Conclusion: Planning Underutilised Crop Breeding Programmes

The improvement of under-researched crop species will not only need a methodological framework for coordinated multi-disciplinary research in order to make gains on the ground, where underutilised crops are behind the breeding progress of major crops. It will also need to translate cutting edge tools, techniques and knowledge that have been developed for well-researched species. As the foundational traits that will make the largest improvements are part of the Domestication Syndrome (e.g. reduced seed dormancy and lack of seed dispersal mechanisms), the Domestication Syndrome genes should be prioritised. Minor crops are likely to be only partially domesticated, or indeed not domesticated at all (Mayes et al. 2012), and so overcoming these initial genetic barriers to their widespread production is crucial. More recent Green Revolution traits, such as reducing susceptibility to lodging (as in the case of sorghum (*Sorghum bicolor*)) and responsiveness to crop nutrients will then need to be targeted. This is not to mention the basic breeding targets of high quality and yielding seeds. For example, in the case of Bambara groundnut, time for harvest can be variable between plants even in the same crop of the same farmer landrace material. Finally, ideotype breeding that incorporates a suite of targeted traits, such as drought tolerance and cookability, can make the additional gains in yield and adoption.

The integrated breeding approach for new ideotypes will be greatly assisted firstly by correlated molecular markers. Also, having a core collection of common germplasm that are also distinct for important characteristics would be a valuable resource for future breeding efforts. In a joint effort between the University of Nottingham (UK), CFF (Malaysia) and the International Institute for Tropical Agriculture (Nigeria), and tapping into their wider network, 420 accessions derived from local farmer, gene banks and academic breeder seeds are being characterised for their genotype and phenotype. This 'passport' information for each accession will be freely available and easily accessible to assist researchers, breeders and producers in designing their breeding programmes. By using modern innovative approaches, such as multiple advanced generation inter-cross (MAGIC) populations supported by molecular markers, superior crop types could be bred at a much more efficient rate rather than by consecutive stepwise crosses.

Genomics-assisted breeding, coupled with high-throughput sequencing (which is becoming increasingly cheaper) and phenotyping (which is beginning to become more accessible), should be more prevalent in minor crops. Genomic work in such crops is becoming prioritised in the face of climate change as they display inherent resistance to biotic and abiotic stresses. Being able to accurately narrow down the physiological and molecular genetics of such crucial traits will impact not only underutilised crop breeding, but crop breeding in general, for climate change mitigation and adaptation, and so income, food and nutritional security. Finally, the increased uptake of underutilised crops in terms of production, consumption and research is predicted to diversify the currently globally homogenous diet, which is perceived to be an urgent international priority (Mayes et al. 2012).

References

Abberton, M., Batley, J., Bentley, A., Bryant, J., Cai, H., Cockram, J., Costa de Oliveira, A., Cseke, L. J., Dempewolf, H., De Pace, C., Edwards, D., Gepts, P., Greenland, A., Hall, A. E., Henry, R., Hori, K., Howe, G. T., Hughes, S., Humphreys, M., Lightfoot, D., Marshall, A., Mayes, S., Nguyen, H. T., Ogbonnaya, F. C., Ortiz, R., Paterson, A. H., Tuberosa, R., Valliyodan, B., Varshney, R. K., & Yano, M. (2016). Global agricultural intensification during climate change: A role for genomics. *Plant Biotechnology Journal, 14*, 1095–1098.

Adzwala, W., Donkoh, S. A., Nyarko, G., O'Reilly, P. J., Olayide, O. E., & Awai, P. E. (2015). Technical efficiency of Bambara groundnut production in Northern Ghana. *University of Development Studies International Journal of Development, 2*(2), 37–49.

Adzwala, W., Donkoh, S. A., Nyarko, G., O'Reilly, P. J., Olayide, O. E., Mayes, S., Feldman, A., & Azman, H. R. (2016a). Adoption of Bambara groundnut production and its effects on farmers' welfare in Northern Ghana. *African Journal of Agricultural Research, 11*(7), 583–594.

Adzwala, W., Donkoh, S. A., Nyarko, G., O'Reilly, P. J., & Mayes, S. (2016b). Use patterns and perceptions about the attributes of Bambara groundnut (*Vigna subterranea* (L.) *Verdc.*) in Northern Ghana. *Ghana Journal of Science, Technology and Development, 4*(2), 56–71.

Ahmad, N. S., Redjeki, E. S., Ho, W. K., Aliyu, S., Mayes, K., Massawe, F., Kilian, A., & Mayes, S. (2016). Construction of a genetic linkage map and QTL analysis in Bambara groundnut. *Genome, 59*(7), 459–472.

Akande, K. E., Abubakar, M. M., Adegbola, T. A., Bogoro, S. E., Doma, U. D., & Fabiyi, E. F. (2009). Nutrient compostion and uses of Bambara groundnut (*Vignia subterranea* (L.) Verdcourt). *Continental Journal of Food Science and Technology, 3*, 8–13.

Alozie, Y. E., Iyam, M. A., Lawal, O., Udofia, U., & Ani, I. F. (2009). Utilization of Bambara groundnut flour blends in bread production. *Journal of Food Technology, 7*(4), 111–114.

Amarteifio, J. O., & Moholo, D. (1998). The chemical composition of four legumes consumed in Botswana. *Journal of Food Composition and Analysis, 11*, 329–332.

AMCHARTS.COM. (n.d.). Retrieved July 18, 2017, from https://www.amcharts.com/visited_countries

Azam-Ali, S. N., Sesay, A., Karikari, S. K., Massawe, F. J., Aguilar-Manjarrez, J., Bannayan, M., & Hampson, K. J. (2001). Assessing the potential of an underutilized crop – A case study using Bambara groundnut. *Experimental Agriculture, 37*, 433–472.

BamNetwork. (n.d.). Retrieved July 18, 2017, from https://www.bamnetwork.org

Bamshaiye, O. M., Adegbola, J. A., & Bamishaiye, E. I. (2011). Bambara groundnut: An Under-Utilized Nut in Africa. *Advances in agricultural biotechnology, 1*, 60–72.

Berchie, J. N., Adu-Dapaah, H. K., Dankyi, A. A., Plahar, W. A., Nelson-Quartey, F., Haleegoah, J., Asafu-Agyei, J. N., & Addo, J. K. (2010). Practices and constraints in Bambara groundnuts production, marketing and consumption in the Brong Ahafo and Upper-East regions of Ghana. *Journal of Agronomy, 9*(3), 111–118.

Brink, M. (1997). Rates of progress towards flowering and podding in Bambara groundnut (*Vigna subterranea*) as a function of temperature and photoperiod. *Annals of Botany, 80*(4), 505–513.

Brough, S. H., Azam-Ali, S. N., & Taylor, A. J. (1993). The potential of Bambara groundnut (*Vigna subterranea*) in vegetable milk production and basic protein functionality systems. *Food Chemistry, 47*, 277–283.

Burlingame, B., & Dernini, S. (Eds.). (2012). *Sustainable diets and biodiversity - Directions and solutions for policy, research and action.* Proceedings of the International Scientific Symposium – Biodiversity and Sustainable Diets United Against Hunger, E-ISBN 978-92-5-107288-2.

Chai, H. H., Massawe, F., & Mayes, S. (2016). Effects of mild drought stress on the morpho-physiological characteristics of a Bambara groundnut segregating population. *Euphytica, 208*(2), 225–236.

Chai, H. H., Ho, W. K., Graham, N., May, S., Massawe, F., & Mayes, S. (2017). A cross-species gene expression marker-based genetic map and QTL analysis in Bambara groundnut. *Genes, 8*, e84. https://doi.org/10.3390/genes8020084.

Daryanto, S., Wang, L., & Jacinthe, P.-A. (2015). Global synthesis of drought effects on food legume production. *PLoS One, 10*(6), e0127401.

FAOSTAT. (n.d.). Food and Agriculture Organization of the United Nations Statistics Division. Retrieved July 18, 2017, from http://www.fao.org/faostat

Hillocks, R. J., Bennett, C., & Mponda, O. M. (2012). Bambara nut: A review of utilisation, market potential and crop improvement. *African Crop Science Journal, 20*(1), 1–16.

Hiremath, P. J., Kumar, A., Penmetsa, R. V., Farmer, A., Schlueter, J. A., Chamarthi, S. K., Whaley, A. M., Carrasquilla-Garcia, N., Gaur, P. M., Upadhyaya, H. D., Polavarapu, B. K. K., Shah, T. M., Cook, D. R., & Varshney, R. K. (2012). Large-scale development of cost-effective SNP marker assays for diversity assessment and genetic mapping in chickpea and comparative mapping in legumes. *Plant Biotechnology Journal, 10*(6), 716–732.

Ho, W. K., Muchugi, A., Muthemba, S., Kariba, R., Mavenkeni, B. O., Hendre, P., Song, B., Van Deynze, A., Massawe, F. J., & Mayes, S. (2016). Use of microsatellite markers for the assessment of Bambara groundnut breeding system and varietal purity before genome sequencing. *Genome, 59*(6), 427–431.

Ho, W. K., Chai, W. K., Kendabie, P., Ahmad, N. S., Jani, J., Massawe, K. A., & Mayes, S. (2017). Integrating genetic maps in Bambara groundnut [*Vigna subterranea* (L) Verdc.] and their syntenic relationships among closely related legumes. *BMC Genomics, 18*, 192. https://doi.org/10.1186/s12864-016-3393-8.

Ijarotimi, O. S., & Keshinro, O. O. (2012). Formulation and nutritional quality of infant formula produced from germinated popcorn, Bambara groundnut and African locust bean flour. *Journal of Microbiology, Biotechnology and Food Sciences, 1*(6), 1358–1388.

Intergovernmental Panel on Climate Change. (2014). *Fifth Assessment Report (AR5) - WGII climate change 2014: Impacts, adaptation, and vulnerability.* Retrieved December 31, 2016, from https://ipcc.ch/report/ar5/wg2

Kaptso, K. G., Njintang, Y. N., Komnek, A. E., Hounhouigan, J., Scher, J., & Mbofung, C. M. F. (2008). Physical properties and rehydration kinetics of two varieties of cowpea (*Vigna unguiculata*) and Bambara groundnuts (*Voandzeia subterranea*) seeds. *Journal of Food Engineering, 86*, 91–99.

Karikari, S. K., & Tabona, T. T. (2004). Constitutive traits and selective indices of Bambara groundnut (*Vigna subterranea* (L) Verdc) landraces for drought tolerance under Botswana conditions. *Physics and Chemistry of the Earth, 29*, 1029–1034.

Karunaratne, A. S., Azam-Ali, S. N., Izzi, G., & Steduto, P. (2011). Calibration and validation of FAO-Aquacrop Model for irrigated and water deficient Bambara groundnut. *Experimental Agriculture, 47*(3), 509–527.

Kendabie, P., Massawe, F., & Mayes, S. (2015). Developing genetic mapping resources from landrace-derived genotypes that differ for photoperiod sensitivity in Bambara groundnut (*Vigna subterranea* L.). *Aspects of Applied Biology, 124*, 1–8.

Kew Science. (n.d.). *Plants of the world online.* Retrieved July 18, 2017, from http://powo.science.kew.org/taxon/urn:lsid:ipni.org:names:525534-1

Kouassi, N. J., & Zorobi, I. A. (2010). Effect of sowing density and seedbed type on yield and yield components in Bambara groundnut (*Vigna subterranea*) in woodland savannas of Cote D'Ivoire. *Experimental Agriculture, 46*(1), 99–110.

Linnemann, A. R., Westphal, E., & Wessel, M. (1995). Photoperiod regulation of development and growth in Bambara groundnut (*Vigna subterranea*). *Field Crops Research, 40*(1), 39–47.

Mabhaudhi, T., & Modi, A. T. (2013). Growth, phenological and yield responses of a Bambara groundnut (*Vigna subterranea* (L.) Verdc.) landrace to imposed water stress under field conditions. *South African Journal of Plant and Soil, 30*(2), 69–79.

Mabhaudhi, T., Modi, A. T., & Beletse, Y. G. (2013). Growth, phenological and yield responses of a Bambara groundnut (*Vigna subterranea* L. Verdc) landrace to imposed water stress: II. Rain shelter conditions. *Water SA, 39*(2), 191–198.

Mahala, A. G., & Mohammed, A. A. A. (2010). Nutritive evaluation of Bambara groundnut (*Vigna subterranean*) pods. *Seeds and Hull as Animal Feeds, 6*(5), 383–386.

Makanda, I., Tongoona, P., Madamba, R., Icishahayo, D., & Derera, J. (2009). Evaluation of Bambara groundnut varieties for off-season production in Zimbabwe. *African Crop Science Journal, 16*(3), 175–183.

Massawe, F. J., Mayes, S., & Cheng, A. (2016). Crop diversity: An unexploited treasure trove for food security. *Trends in Plant Science, 21*(5), 365–368.

Mayes, S., Massawe, F. J., Alderson, P. G., Roberts, J. A., Azam-Ali, S. N., & Hermann, M. (2012). The potential for underutilized crops to improve security of food production. *Journal of Experimental Botany, 63*(3), 1075–1079.

Mayes, S., Ho, W. K., Kendabie, P., Chai, H. H., Aliyu, S., Feldman, A., Halimi, R. A., Massawe, F., & Azam-Ali, S. (2015a). Applying molecular genetics to underutilised species – Problems and opportunities. *Malaysian Applied Biology, 44*(4), 1–9.

Mayes, S., Kendabie, P., Ho, W. K., & Massawe, F. (2015b). Increasing the contribution that underutilised crops could make to food security – Bambara groundnut as an example. *Aspects of Applied Biology, 124*, 1–8.

Mazahib, A. M., Nuha, M. O., Salawa, I. S., & Babiker, E. E. (2013). Some nutritional attributes of Bambara groundnut as influenced by domestic processing. *International Food Research Journal, 20*(3), 1165–1171.

Molosiwa, O. O., Aliyu, S., Stadler, F., Mayes, K., Massawe, F., Kilian, A., & Mayes, S. (2015). SSR marker development, genetic diversity and population structure analysis of Bambara groundnut [*Vigna subterranea* (L.) Verdc.] landraces. *Genetic Resources and Crop Evolution, 62*, 1225. https://doi.org/10.1007/s10722-015-0226-6.

Mubaiwa, J., Fogliano, V., Chidewe, C., & Linnemann, A. R. (2017). Hard-to-cook phenomenon in Bambara groundnut (*Vigna subterranea* (L.) Verdc.) processing: Options to improve its role in providing food security. *Food Reviews International, 33*, 167. https://doi.org/10.1080/87559129.2016.1149864.

Mumuni, E., Kaliannan, M., & O'Reilly, P. (2016). Approaches for scientific collaboration and interactions in complex research projects under disciplinary influence. *The Journal of Developing Areas, 50*(5), 383–391.

Muñoz-Amatriaín, M., Mirebrahim, H., Xu, P., Wanamaker, S. I., Luo, M., Alhakami, H., Alpert, M., Atokple, I., Batieno, B. J., Boukar, O., Bozdag, S., Cisse, N., Drabo, I., Ehlers, J. D., Farmer, A., Fatokun, C., Gu, Y. Q., Guo, Y.-N., Huynh, B.-L., Jackson, S. A., Kusi, F.,

8 Bambara Groundnut

Lawley, C. T., Lucas, M. R., Ma, Y., Timko, M. P., Wu, J., You, F., Barkley, N. A., Roberts, P. A., Lonardi, S., & Close, T. J. (2017). Genome resources for climate-resilient cowpea, an essential crop for food security. *The Plant Journal, 89*, 1042–1054. https://doi.org/10.1111/tpj.13404.

Mwale, S. S., Azam-Ali, S. N., & Massawe, F. J. (2007a). Growth and development of Bambara groundnut (*Vigna subterranea*) in response to soil moisture: 1. Dry matter and yield. *European Journal of Agronomy, 26*(4), 345–353.

Mwale, S. S., Azam-Ali, S. N., & Massawe, F. J. (2007b). Growth and development of Bambara groundnut (*Vigna subterranea*) in response to soil moisture: 2. Resource capture and conversion. *European Journal of Agronomy, 26*(4), 354–362.

Nyemba, R. C., & Dakora, F. D. (2010). Evaluating N_2 fixation by food grain legumes in farmers' fields in three agro-ecological zones of Zambia, using ^{15}N natural abundance. *Biology and Fertility of Soils, 46*(5), 461–470.

Ojimelukwe, P. C. (1998). Cooking characteristics of four cultivars of Bambara groundnuts seeds and starch isolate. *Journal of Food Biochemistry, 23*, 109–117.

Olukolu, B. A., Mayes, S., Stadler, F., Ng, N. Q., Fawole, I., Dominique, D., Azam-Ali, S. N., Abbott, A. G., & Kole, C. (2012). Genetic diversity in Bambara groundnut (*Vigna subterranea* (L.) Verdc.) as revealed by phenotypic descriptors and DArT marker analysis. *Genetic Resources and Crop Evolution, 59*, 347–358.

Ouedraogo, M., Ouedraogo, J. T., Tignere, J. B., Balma, D., Dabire, C. B., & Konate, G. (2008). Characterization and evaluation of accessions of Bambara groundnut (*Vigna subterranea* (L.) Verdcourt) from Burkina Faso. *Sciences & Nature, 5*(2), 191–197.

PROTA. (n.d.). *Prota 1: Cereals and pulses/Céréales et légumes secs*. Retrieved July 18, 2017, from http://database.prota.org/PROTAhtml/Vigna%20subterranea_En.htm

Pungulani, L., Kadyampakeni, D., Nsapato, L., & Kachapila, M. (2012). Selection of high yielding and farmers' preferred genotypes of Bambara Nut (*Vigna subterranea* (L.) Verdc) in Malawi. *American Journal of Plant Sciences, 3*, 1802–1808.

Somta, P., Chankaew, S., Rungnoi, O., & Srinives, P. (2011). Genetic diversity of the Bambara groundnut (*Vigna subterranea* (L.) Verdc.) as assessed by SSR markers. *Genome, 54*, 898–910.

Sprent, J. I., Odee, D. W., & Dakora, F. D. (2010). African legumes: A vital but under-utilized resource. *Journal of Experimental Botany, 61*(5), 1257–1265.

Touré, Y., Koné, M., Kouakou Tanoh, H., & Koné, D. (2012). Agromorphological and phenological variability of 10 Bambara groundnut [*Vigna subterranea* (L.) Verdc. (Fabaceae)] landraces cultivated in the Ivory Coast. *Tropicultura, 30*(4), 216–221.

United States Department of Agriculture. (n.d.). *National nutrient database for standard reference release 28*. Retrieved December 27, 2016, from https://ndb.nal.usda.gov/ndb/search/list

Varshney, R. K. (2016). Exciting journey of 10 years from genomes to fields and markets: Some success stories of genomics-assisted breeding in chickpea, pigeonpea and groundnut. *Plant Science, 242*, 98–107.

Part III
Natural Resource and Landscape Management

Chapter 9
The Challenge of Feeding the World While Preserving Natural Resources: Findings of a Global Bioeconomic Model

Timothy S. Thomas and Shahnila Dunston

Abstract In this chapter, we use a modeling approach to answer the question concerning how population growth, income growth, and climate change might affect the ability of the world's farmers to produce enough food for the planet through the year 2050. We also ask how much the cost will be in terms of land taken from forests and other natural vegetation to be used as cropland. In the analysis, we take into account productivity gains due to technological advances (such as more productive seeds) and changes in input intensity in response to changes in commodity prices. We discover that climate change creates a drag on productivity increase, resulting in the need to convert more land to agricultural uses. Yet the increase in cropland due strictly to climate change in 2050 will be only 0.8–1.5% of current cropland. More than 80% of the land converted to cropland by 2050 will be due to increased demands from a larger, wealthier population rather than from a changing climate. The greatest percentage increases in cropland will occur in Africa and Latin America, while the largest percentage change in cropland from climate change alone will be in Latin America.

While climate change will lead to an increase in malnutrition and result in lower food security, projected growth in income and overall agricultural production will lead to a decrease in malnutrition that is of greater magnitude than the increase from climate change.

The model presented in this chapter extends to 2050. If greenhouse gas emissions continue at a relatively high rate, the impact of climate change on food and agriculture should increase greatly in the latter half of the century. Therefore, while the outcomes presented in this chapter suggest modestly negative effects by 2050, the threat that climate change presents to global food security should not be discounted for the longer term.

T. S. Thomas (✉) · S. Dunston
International Food Policy Research Institute (IFPRI), Washington, DC, USA
e-mail: tim.thomas@cgiar.org

© Springer Nature Switzerland AG 2019
A. Sarkar et al. (eds.), *Sustainable Solutions for Food Security*,
https://doi.org/10.1007/978-3-319-77878-5_9

9.1 Introduction

Many around the globe currently face malnutrition and food insecurity, while much of the land that is suitable for agriculture is already under cultivation, leaving only modest amounts of land that could be converted to agriculture. Yet the world's population will continue to grow, perhaps reaching 9.1 billion people by 2050. Furthermore, household income is projected to keep growing, which will generally lead to higher demand for livestock products (meat, milk, and eggs), all of which will require more land not only for the livestock, but for growing the crops to feed the livestock. Then, on top of all this, we are faced with the challenge of a changing climate, which is projected to reduce agricultural productivity.

Despite some of the challenges for improving global food security, we have seen agricultural productivity rise for decades, and there is good reason to believe that productivity will continue to rise. The question, however, is will agricultural land expansion and productivity increases be able to keep up with growth in demand for food? Secondly, how much agricultural expansion is likely to take place, and what type of impact will that have on natural resources—in particular, on forests, shrub lands, and grasslands.

To answer these questions, we will rely on the International Model for Policy Analysis of Agricultural Commodities and Trade (IMPACT, Robinson et al. 2015). IMPACT is a partial equilibrium model of the agricultural sector. The model has 159 regions (most countries with some aggregated regions) and 62 commodities. It also has an integrated water model that provides information on how much water is available for irrigation. This is done at the water basin level, and thus production in IMPACT is determined at geographic level called a food production unit (FPU) which is the intersection of a country/regional boundary and the water basin. There are 320 FPUs in IMPACT. Further, the model uses inputs from crop models which provide yield changes due to climate change (Fig. 9.1).

The strength of the IMPACT model lies in the fact that it uses a structural approach that links physiological/biological studies, crop models, and water models. Information from several Global Circulation Models (GCMs) are used as inputs into crop models (in this case, DSSAT) which then provides inputs for the IMPACT economic model. The model also has assumptions on how yields will grow over time. Expert opinion from all CGIAR centers along with historical information has been used to estimate these growth rates. After solving for supply and demand of commodities, IMPACT also provides scenario results on food security parameters through integrated post-processing modules.

The IMPACT model uses production and demand data from the FAO database that are run through a cross entropy program to assure consistency. The area data obtained from FAO and used in IMPACT deal with only harvested area and not physical area, thus not having a way to account for multiple crops on the same farmland.

In this chapter, we develop a method that combines production statistics from FAO that accounts for harvested area of crops together with satellite data on land

9 The Challenge of Feeding the World While Preserving Natural Resources: Findings...

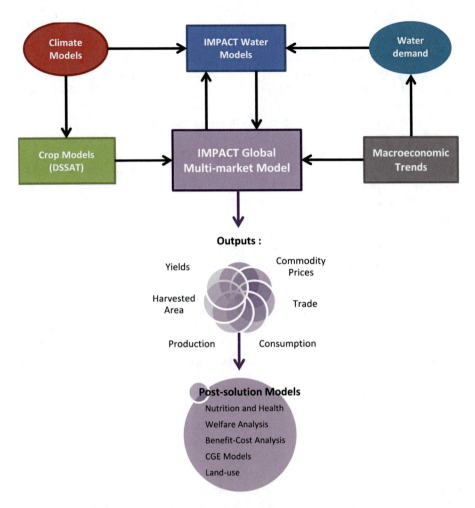

Fig. 9.1 IMPACT system of models. *Source: Robinson et al. (2015)*

cover to tell us the relationship between the two. We use the results of the analysis on historical data to projected future harvested area from IMPACT into future land cover area.

Previous analyses on how satellite data on land cover relates to actual crop coverage and other land uses (such as forests, grassland, and shrub lands) have found that there are various degrees of accuracy depending on the land use type and location and also based on the type of satellite data used (Congalton et al. 2014). Herold et al. (2008) compared four 1 km datasets and found that the IGBP DISCover, UMD, MODIS LCP, and GLC2000 data were comparable across tropical rain forests, drylands, or ice sheets, but varied widely when land use was much more heterogenous such as mixed tree types and shrubland. When examining China, the same four datasets show that the estimates for cropland were similar, but again shrubland and grassland areas showed high variation (Ran et al. 2010).

Another study done by Wu et al. (2008) in China compared cropland from the four datasets to the Chinese 1996 National Land Survey Dataset (NLSD-1996) to compare the allocation of crops. They found that with increasing land scape heterogeneity, results were less accurate. In order to categorize land as "cropland," they used the methodology from Xiao et al. (2002) and assigned 40% of cropland/natural vegetation mosaic to cropland and 60% of pure cropland to cropland. They concluded that the MODIS data was most accurate when compared to the National Land Survey data. It is unclear if this data accounts for multiple cropping but it is known that the survey data is littered with underestimation of cropping areas and inflation of reported average yields (Lin and Ho 2003).

Bai et al. (2014) also found MODIS to be the most consistent among five satellite datasets (GLCC, UMD, GLC2000, GLOBCOVER, an MODIS) as compared to the Geodata Land Cover Dataset for year 2005 (GLDC-2005). Further studies done by Fritz et al. (2009) and Hannerz et al. focused on Africa and found large differences between global land cover data sets. Leroux et al. (2014) focus on the MODIS land cover dataset to estimate its accuracy in mapping crops for sub-Saharan Africa. They compare this to the FAOSTAT agricultural database (only using arable land that does not have multiple cropping) and AGRHYMET data base for Burkina Faso. They found that at the national level, tabulated accuracy of MODIS was good, but at a smaller scale, it tended to perform worse.

Turning our focus to models that project future land use taking into consideration socioeconomic factors—similar to what we do in this chapter—two important ones are the Land Simulation to Harmonize and Integrate Freshwater availability and the Terrestrial Environment (LandSHIFT) model (Schaldach et al. 2011) and the Model of Agricultural Production and its Impact on the Environment (MAgPIE) (Lotze-Campen et al. 2008). Inputs for the LandSHIFT model include data such as population trajectories and the production of agricultural commodities, both crops and livestock. The model outputs land use changes in 5-year time steps. It uses suitability analysis and land allocation routines to determine land use changes. The suitability analysis is based on demand in that production is allocated to the most suitable areas to meet demand within a specific country. The model also uses data of water availability and water stress, thus incorporating climate change into the land use allocation decision.

The MAgPIE model also uses demand for agricultural commodities along with technological improvement and production costs as inputs. Further, it uses information about crop yields, land and water availability from the Lund-Potsdam-Jena managed Land (LPJmL) gridded dynamic vegetation model (Bondeau et al. 2007). For a set of regional food and bioenergy demand the model minimizes the total cost of production and estimates land use patterns, yields, and total production at the grid cell level.

9.2 Socioeconomic and Climate Projections in This Chapter

IMPACT relies on data from Global Climate Models (GCMs)—also referred to as General Circulation Models. GCMs project how a pathway of greenhouse gas (GHG) emissions and accumulation might affect temperature and precipitation in the future. The avenues through which climate affects agriculture inside IMPACT are through direct impact on crop productivity and through the availability of water for irrigation. The productivity effect is computed by the Decision Support System for Agrotechnology Transfer (DSSAT) crop modeling software at every half degree pixel, and then is aggregated up to the major river basin level for each country (Jones et al. 2003).

In this chapter, we use two GCMs that were submitted to the Intergovernmental Panel on Climate Change (IPCC) for their Fifth Assessment Report (AR5). We use the IPSL-CM5A-LR model generated by Institut Pierre-Simon Laplace (IPSL; see Dufresne et al. 2013) and the HadGem2-ES model produced by the Met Office Hadley Centre (Collins et al. 2011; Martin et al. 2011).

The climate models require that assumptions be made about GHG emissions, and these are given by the IPCC's Representative Concentration Pathways (RCPs). There are four pathways that were developed. For this chapter we use the highest emissions scenario which is RCP8.5. We also use a baseline that ignores climate change, which allows us to distinguish the demographic and socioeconomic affects from the climate affects. By choosing a scenario with no climate change, and one that assumes high emissions, we are able to create the broadest range of possible climate effects.

In addition to RCPs and GCMs, the IPCC established five different growth pathways for GDP and population which they refer to as shared socio economic pathways (SSPs). We use SSP2, which might be referred to as the "business as usual" pathway. In it, global population reaches 9.1 billion by 2050. This scenario also has GDP per capita growing to about $25,000[1] as compared to only $9800 in 2010 (Fig. 9.2).

SSP2 assumes moderate growth in population and GDP assumes continuation of historic trends with some regional variation. In it we see high growth in population for sub-Saharan Africa (SSA) and South Asia (SAS) compared to other regions. The East Asia Pacific (EAP) region sees growth until 2030 and then a slight decline to 2.6 million by 2050. This is driven by China, where the population will rise until 2025 and then decline by 2050 to levels lower than 2010. In South Asia, higher populations are dominated by the growth expected in India where population is expected to reach 1.7 billion by 2050.

Under SSP2, income growth from 2010 to 2050 is steadily higher in all regions. SAS and SSA have the greatest increases, where the total GDP is over 7 and 8 times

---

[1] Unless otherwise noted, this and all other dollar amounts are for the purchasing power parity (PPP) values and are denominated in 2005 constant US dollars.

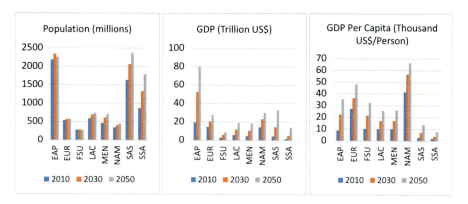

Fig. 9.2 Projections for population, GDP, and GDP per capita by region, 2010 to 2050. *Source: Population data from Samir and Lutz (2017). GDP data from Dellink et al. (2017).* **Note:** *These are for SSP2. SSP = shared socioeconomic pathway. EAP = East Asia and the Pacific; EUR = Europe; FSU = Former Soviet Union; LAC = Latin America and the Caribbean; MEN = Middle East and North Africa; SAS = South Asia; SSA = Sub-Saharan Africa*

as much in 2050 as compared to 2010, respectively. In the more developed regions of Europe and North America, income increases but to a much lesser degree, only doubling by 2050.

Under SSP2, income growth outstrips population growth resulting in higher per capita GDP globally and by region. We see similar patterns of change in the former Soviet Union (FSU), Latin America and the Caribbean (LAC), and the Middle East and North Africa (MEN) in terms of the final value of (GDP per capita ranging from 25,000 to 33,000) and in terms of percentage increases from 2010 to 2050 (160–215%). Again, the increases in SAS and SSA are the largest and increase by 407% and 295%. Although, this is the case, in terms of actual value, the per capita income in these regions remains significantly lower than other regions in 2050 with SAS at 13,000 and SSA at only 7000.

9.3 Changes in Global Food Consumption and Production

The IMPACT model works by solving for international prices that cause global supply to equal global demand. Total demand consists of food demand, feed and processed goods demand, biofuel demand, and intermediate demand from sectors such as biofuels and other industrial uses. Trade links national production and demand to world markets and is a function of domestic production, domestic demand, and stock change. Net Trade for a given country is determined by whether national production exceeds total demand; if the difference is negative, the country imports a commodity; and if it is positive, it exports the commodity.

9 The Challenge of Feeding the World While Preserving Natural Resources: Findings... 177

The increase in population and per capita income under SSP2 are projected to drive up food demand globally. At a regional level, we can see that percent increase is highest in SSA and SAS, following the trends in population growth and income growth. In both these regions food demand more than doubles by 2050 as compared to 2010 (Fig. 9.3).

As a result of the rapid rise in food demand, SAS and SSA are also projected to have increased shares of total global demand, going from 16% to 26% and 9–15%, respectively. Their rising share is seems to be mostly derived from East Asia and the Pacific, where the share falls from 34% to 26% from 2010 to 2050. Other regions remain mostly constant (Fig. 9.3).

The composition of food demanded will also shift due to rising per capita income, allowing previously poorer countries to afford commodities such as meat, milk, and eggs. For example, in South Asia, a very large increase in demand is projected for poultry (Table 9.1). Change in demand for milk, however, is projected to be much lower. Similarly, in SSA, we see that beef and poultry demand is projected to increase rapidly, to approximately 300% by 2050 (Table 9.1).

Some readers may be skeptical about the projected growth of beef consumption in South Asia, since beef is not a common part of the diet of Hindus, which comprise a large percentage of the region's population. But it is important to keep in mind that Table 9.1 reports changes in consumption, and if the region has a low starting level of consumption, the changes in terms of percentage points may seem large, but the actual increase would still be relatively small. Given the non-Hindu population of India, along with the large populations in other countries of the region, including

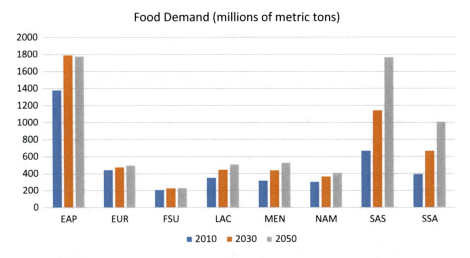

Fig. 9.3 Projected total food demand by region, 2010–2050 (millions of metric tons). *Source: IPCC (2014). **Note**: These are for SSP2, ignoring the effect of climate change. SSP = shared socioeconomic pathway. EAP = East Asia and the Pacific; EUR = Europe; FSU = Former Soviet Union; LAC = Latin America and the Caribbean; MEN = Middle East and North Africa; SAS = South Asia; SSA = Sub-Saharan Africa*

Table 9.1 Projected percentage change in demand for key livestock products by region, 2030 and 2050

	Beef		Milk		Poultry		Eggs	
Region	2030 (%)	2050 (%)	2030 (%)	2050 (%)	2030 (%)	2050 (%)	2030 (%)	2050 (%)
EAP	77	84	103	139	73	98	25	26
EUR	9	18	6	10	17	30	2	4
FSU	10	13	0	−1	37	48	5	4
LAC	26	42	29	48	44	77	25	41
MEN	74	145	32	58	69	131	48	87
NAM	14	26	16	29	27	46	12	24
SAS	88	198	61	90	217	641	107	186
SSA	112	321	62	148	107	294	89	244

Source: IPCC (2014)
Note: These are for SSP2, ignoring the effect of climate change. SSP = shared socioeconomic pathway. EAP = East Asia and the Pacific; EUR = Europe; FSU = Former Soviet Union; LAC = Latin America and the Caribbean; MEN = Middle East and North Africa; SAS = South Asia; SSA = Sub-Saharan Africa

Bangladesh and Pakistan, the relative large percentage growth in consumption of beef is reasonable.

In Fig. 9.4, we see the impact of the increased livestock demand on feed production. Demand for most commodities increases with the biggest coming in the maize sector. Maize feed demand increases from 450 million metric tons to 870 million metric tons globally, and reflects the increased in every region. The increase in feed demand for soybean meal is driven by EAP, Europe (EUR), and North America (NAM), while the increased feed demand related to milk production comes mostly from SAS. The large increases in feed demand applies substantial pressure to expand cultivated area.

9.4 Areas, Yields and Climate Change

We can see the pressure posed by population, income, and increased demand on agricultural area in Fig. 9.5. We see increases in all regions except for Europe, and particularly large increases in sub-Saharan Africa and the East-Asia Pacific region under the hypothetical scenario of no climate change (NoCC), which allows us to focus on the socioeconomic impacts of assumptions in SSP2 without the added burden of climate change.

When climate change is added to the mix, areas increase further. This is true for both climate change scenarios analyzed for this chapter (IPSL and HGEM) although we see that HGEM creates the most pressure in all regions (Fig. 9.6).

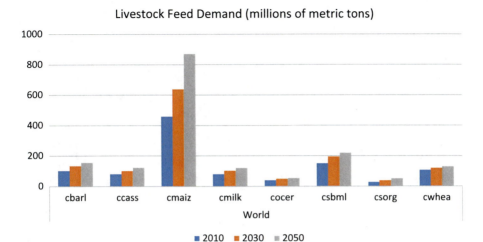

Fig. 9.4 Projected feed demand by region, 2010–2050, millions of metric tons. *Source*: IPCC *(2014)*. **Note**: *These are for SSP2, ignoring the effect of climate change. SSP = shared socioeconomic pathway. cmaiz = maize; cbarl = barley; csbml = soybean meal; cwhea = wheat; ccass = cassava; cmilk = milk; cocer = other cereals; csorg = sorghum*

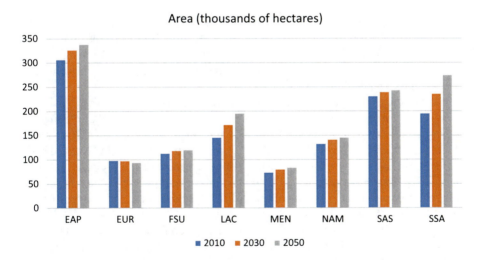

Fig. 9.5 Projected total harvested area by region, 2010–2050, thousands of hectares. *Source*: IPCC *(2014)*. **Note**: *These are for SSP2, ignoring the effect of climate change. SSP = shared socioeconomic pathway. EAP = East Asia and the Pacific; EUR = Europe; FSU = Former Soviet Union; LAC = Latin America and the Caribbean; MEN = Middle East and North Africa; SAS = South Asia; SSA = Sub-Saharan Africa*

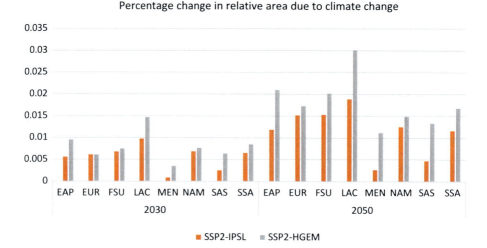

Fig. 9.6 Percent change in harvested area relative to a scenario without climate change, by region, 2010–2050. *Source: IPCC (2014)*. **Note**: *These are for SSP2, and show the relative change of two climate models compared to the case of no climate change. SSP = shared socioeconomic pathway. NoCC = hypothetical future without climate change; IPSL = IPSL-CM5A-LR (Dufresne et al. 2013)—the Institut Pierre Simon Laplace's ESM; HGEM = HADGEM2-ES (Collins et al. 2011; Martin et al. 2011)—the Hadley Centre's Global Environment Model, version 2. EAP = East Asia and the Pacific; EUR = Europe; FSU = Former Soviet Union; LAC = Latin America and the Caribbean; MEN = Middle East and North Africa; SAS = South Asia; SSA = Sub-Saharan Africa*

To be clear, Figs. 9.5 and 9.6 indicate that we expect to see expansion in agricultural area in the decades to come. Figure 9.5 focuses on the effect of increased demand for food and livestock feed from both rising population and changing diets due to higher incomes. Figure 9.6 shows how much additional land will be demanded once climate change is taken into account.

One of the aspects associated with climate change is sea level rise. In principle, countries with much agricultural land at or near sea level would be in danger of losing area devoted to cropland. Countries with large river deltas—like Bangladesh, Vietnam, and Egypt—could lose a large proportion of their land. The IMPACT model has the ability to control for this. However, most of the projected losses in land due to sea level rise and salinization are expected to occur in the latter half of the century, and for the purposes of this chapter, we have not accounted for lost land due to either effect.

If we there were to be a higher area reduction of cropland from sea level rise than we believe there will be by 2050, then we would expect that some of the lost agricultural land would be compensated for by converting forest and other land types into cropland. In such a case, the implications would be that the projected effect of climate change on natural resources presented in this chapter would underestimate what it would be if sea level rise had a more noticeable impact.

While increases in cultivated land will lead to the reduction of forests and other natural landcover, the story could be much worse, if it were not for projected increases in agricultural productivity. Agricultural productivity has risen continuously in the past century and will likely continue into the future. These projections hold true for the major commodities of maize, rice, soybeans, and wheat, as well as for many other commodities.

The projections hold true not just for the world or developed countries, but for regions such as South Asia that have benefited from the Green Revolution, but still have annualized growth rates in maize yields of close to 3% and around 1.5% for both rice and wheat (based on FAOSTAT data through 2014).

Ignoring climate change, the IMPACT model projects that global maize yields will increase from 6.16 metric tons/ha in 2030 to 7.01 metric tons/ha in 2050 (Fig. 9.7), and will rise with similar proportions even with climate change.

Even though we see increases in productivity over time, these increases are dampened by climate change for most crops except for wheat. For wheat, the results show that by 2030 under the IPSL climate scenario, yields will be above the no climate change levels while for the HadGEM climate model, yields will only be slightly lower than under no climate change. By 2050, under both climate scenarios used, yields are higher than the no climate reference case.

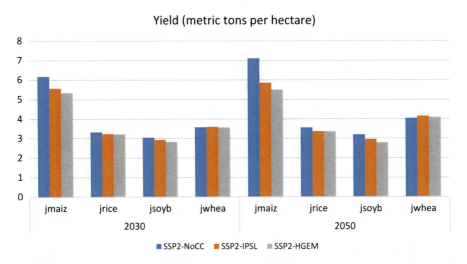

Fig. 9.7 Projected yields for key agricultural commodities, 2030 and 2050, (metric tons per hectare). *Source: IMPACT model (Robinson et al. 2015). Note: These are for SSP2, and show the relative change of 2 climate models compared to the case of no climate change. SSP = shared socioeconomic pathway. jmaiz = maize; jrice = rice; jsoyb = soybeans; jwhea = wheat. NoCC = hypothetical future without climate change; IPSL = IPSL-CM5A-LR (Dufresne et al. 2013)—the Institut Pierre Simon Laplace's ESM; HGEM = HADGEM2-ES (Collins et al. 2011; Martin et al. 2011)—the Hadley Centre's Global Environment Model, version 2*

The objectives of Sustainable Development Goal 2 (SDG2) are to end hunger, achieve food security and improve nutrition, and promote sustainable agriculture (United Nations 2015). IMPACT uses equations derived from Fischer et al. (2005) to estimate the population at risk of hunger and Smith and Haddad (2000) to estimate child malnutrition. The model shows that without climate change, both malnutrition and the population at risk of hunger decreases. The number of malnourished children decreases from 150 million in 2010 to just under 100 million in 2050 while the population at risk of hunger falls from 840 million to just above 410 million in the same time period (Fig. 9.8). Although this is good news, climate change again dampens the decreases we see in both malnutrition and hunger, with the number of people at risk of hunger being 80 million higher in 2050 under the HadGEM climate model.

These equations are only estimates and fail to take into account a number of relevant characteristics of each country, along with important issues that can be influenced by policy and governance, including distributional issues. Yet at the global level, the equations should perform reasonably well. However, if inequality would rise in any given country in the future, then these equations would tend to underestimate the amount of hunger and malnutrition in that country, and if the general trend is for inequality to rise within countries, then the graphs in Fig. 9.8 would underestimate the global numbers.

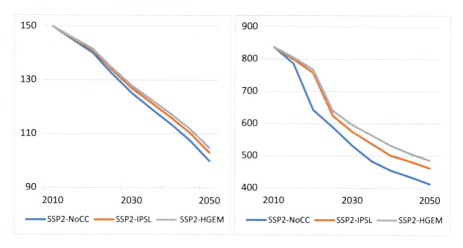

Fig. 9.8 Projected numbers of malnourished children and population at risk of hunger, 2010–2050, under three climate scenarios (millions). ***Source****: IMPACT model (Robinson et al. 2015).* ***Note****: These are for SSP2, and show the relative change of 2 climate models compared to the case of no climate change. SSP = shared socioeconomic pathway. NoCC = hypothetical future without climate change; IPSL = IPSL-CM5A-LR (Dufresne et al. 2013)—the Institut Pierre Simon Laplace's ESM; HGEM = HADGEM2-ES (Collins et al. 2011; Martin et al. 2011)—the Hadley Centre's Global Environment Model, version 2*

9.5 Effect of Food Production on Global Land Use

The IMPACT model is concerned with harvested area, which is different than land area devoted to cultivation. The ratio of harvested area to physical area is the cropping intensity. In many places of the world, two or more crops can be grown each year on the same piece of land. Sometimes this is enabled by the use of irrigation in dry seasons, but often times can be supported without irrigation. It is often facilitated by the use of shorter duration varieties that can be harvested in less time than regular varieties, freeing the land to be planted in an additional crop. This section is devoted to estimating the impact of increased harvested area on the reduction other types of land area such as shrub land, forests, and grasslands especially with changing climates.

We developed a simple methodology to convert cropped area—based on FAOSTAT harvested area statistics—into land cover. For the land cover data, we relied on MODIS MCD12Q1 (Friedl et al. 2010), which were available to use for each year between 2001 and 2013 (the 2010 data is depicted in Fig. 9.9). We used the IGBP definitions (Belward et al. 1999; Scepan 1999), and condensed them into six categories: forest (from the five forest types); shrub land (from the two shrub land types); grassland (from the grassland and two savanna types); cropland (from cropland); mosaic (from mosaic); and other (containing barren and sparse vegetation, wetlands, urban, and snow and ice covered).

We use a multinomial logit on the six condensed land cover types, with FAOSTAT harvested area and a constant term as explanatory variables. We also tried using GDP per capita and year as explanatory variables, but in the end rejected both. The GDP per capita tended to pick up cross-sectional variation in the variable which reflected differences between countries that probably would not reflect the path of countries as they grow in per capita income. The year variable picked up important variation in data across the 13 years in the dataset, but extending the trends for an additional 37 years, out to 2050, potentially projected short-term trends over a longer term that they would be observed. We also computed a fixed effect for each country in a second regression.

We used the country fixed effect together with the parameter for the effect of harvested area times the harvested area computed inside the IMPACT model to project out to 2050. We did this for a scenario that was meant to test productivity and consumption without taking into consideration the effect of climate change, only taking into consideration changes in population, GDP, agricultural technology, and changes in food demands. We also did this for two climate change models, HadGEM (Collins et al. 2011; Martin et al. 2011) and IPSL (Dufresne et al. 2013). These climate models were run under the high emissions assumption (RCP8.5) from the IPCC's AR5 report (IPCC 2014).

Table 9.2 shows the results of applying the multinomial logit regression results to the IMPACT model data for the entire world. Even without climate change, the

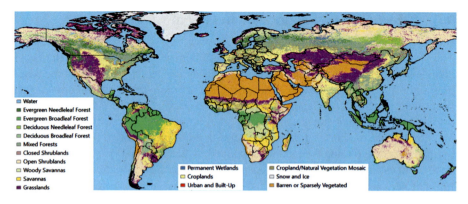

Fig. 9.9 Land cover types, 2010, MODIS. *Source: Friedl et al. (2010)*

Table 9.2 Projected changes in land distribution, 2010–2050

Land type	Thousands of hectares (MODIS 2010)	Without climate change, 2010–2050 Actual change	% Change	2050, % difference from no climate change HadGEM climate model (%)	IPSL climate model (%)
Cropland	12,564,141	971,153	7.6	1.5	0.8
Mosaic	9,040,126	869,529	9.2	0.9	0.7
Forest	29,483,364	−287,793	−1.0	−0.2	−0.1
Shrub land	20,548,493	−694,965	−3.5	−0.5	−0.3
Grass land	38,732,826	−220,021	−0.6	−0.1	0.0
Other	24,354,551	−637,903	−2.6	−0.4	−0.2

Source: Authors

increased demand brought on by a richer and more populated world is projected to lead to an increase of 1.8 billion hectares, increasing pure cropland by almost 8% compared to the 2010 values, and increasing the mosaic (lower density) cropland by just over 9%.

Climate change is projected to lead to even higher conversion of land to cropland, with more land being converted under the HadGEM climate model than under the IPSL model: 1.5% more pure cropland and 0.9% more lower density cropland under HadGEM, compared to 0.8% more and 0.7% more under IPSL. This does not mean that climate change is not an important driver of land use change, but that it only seems to add an additional 10–20% demand for cropland.

It is also important to note that this is only through 2050. Population growth is likely to be at a much lower rate after 2050 than before 2050, while climate change under most scenarios is projected to be higher after 2050 than before 2050.

9 The Challenge of Feeding the World While Preserving Natural Resources: Findings... 185

Therefore, the effect of climate on demand for cropland will likely be of much more importance at the end of the century.

Naturally all this added cropland came from somewhere, and this analysis concludes that will be around 290 million hectares of forest, almost 700 million hectares of shrub land, 220 million hectares of grassland, and 640 million hectares of other land. The reduction in the "other" land category was an unexpected result from the analysis. Upon further investigation, it appears that the largest category inside the "other" category is "barren or sparsely vegetated" areas, and for the 2001–2013 period for which we analyzed MODIS data, this seems to be driven largely by a reduction of barren land in China. Liu et al. (2008) confirms massive afforestation efforts on barren land in China.

It is not China alone, however, where such trends are observed. In South Asia, where much of the land is already cropped, in the 2002–2013 period, the MODIS data suggests that cropland expanded by 2.5%, while "other" land—mostly barren—was reduced by almost 20%. This is not to say that barren land became cropland. Looking carefully at the data, we observe that most of this barren land transitioned to shrub land, and that it was concentrated in Rajasthan in India and Balochistan in Pakistan. The new cropland was less concentrated and came primarily from other land use types.

Between 2010 and 2050, our analysis projects that even without climate change, cropland in South Asia would expand by 9% and cropland mosaic area would expand by 4%. Land would be reduced in barren, forest, shrub land and grass land by between 5% and 10% in each category.

It is worth mentioning that urban areas also are counted in the "other" category. While urbanization is increasing rapidly around the world and is a key factor for labor markets, in terms of land, urban area expansion is fairly small in regard to total land available, and makes only a small contribution, especially considering how large the "barren and sparsely vegetated" areas are.

Even if the projection for barren land conversion noted in our period of analysis might not be reasonably projected to 2050, the 1.8 billion hectares of land converted to agriculture will need to come from one of the other land used types. If most of the "other" land was converted to forest during the 2001–2013 period, then if that conversion had not occurred, it must be true that much of the "other" category should actually have come from the forest category, so change in forest would be much higher than Table 9.2 suggests. As an upper bound estimate, if the 640 million hectares of barren land in the table were instead taken from the forest category, the gross loss of forest would be closer to 925 million hectares, or around 3.3% of total forest land.

As we continue investigating the projections to 2050 by region, perhaps it is best to focus on conversion to cropland rather than determining what it is converted from, recognizing that any amount being converted to cropland must be converted from some other land cover type, which will generally be a type closer to its original vegetation cover. In Table 9.3, we note very high rates of increase for pure cropland in sub-Saharan Africa (SSA) and Latin America and the Caribbean (LAC). In fact, these numbers translate into an increase in SSA of close to 250 million hectares and an increase in LAC of 220 million hectares. While the percentage increase is higher

Table 9.3 Projected changes in cropland and mosaic distribution by region, 2010–2050

	Cropland				Cropland/vegetation mosaic			
			2050, % difference from no climate change				2050, % difference from no climate change	
		2050				2050		
Region	Thousands of hectares (MODIS 2010)	Without climate change (%)	HadGEM climate model (%)	IPSL climate model (%)	Thousands of hectares (MODIS 2010)	Without climate change (%)	HadGEM climate model (%)	IPSL climate model (%)
EAP	2,408,546	5.8	2.0	1.0	1,397,453	11.3	1.9	1.3
EUR	1,805,469	−2.2	0.8	0.7	1,095,182	−0.8	0.4	0.4
FSU	2,016,722	2.0	0.7	0.5	1,240,365	1.3	0.4	0.3
LAC	1,162,592	20.0	2.8	1.7	1,527,446	9.8	1.2	0.7
MEN	593,147	12.5	1.1	−0.2	63,617	6.5	0.8	0.4
NAM	1,698,987	5.7	0.8	0.8	1,139,622	4.6	0.6	0.6
SAS	2,011,202	8.8	2.0	0.7	422,603	3.9	1.0	0.4
SSA	867,477	28.5	1.7	1.4	2,153,837	20.2	0.9	0.7
TOT	12,564,141	7.6	1.5	0.8	9,040,126	9.2	0.9	0.7

Source: Authors
Note: These are for SSP2. SSP = shared socioeconomic pathway. NoCC = hypothetical future without climate change; IPSL = IPSL-CM5A-LR (Dufresne et al. 2013)—the Institut Pierre Simon Laplace's ESM; HGEM = HADGEM2-ES (Collins et al. 2011; Martin et al. 2011)—the Hadley Centre's Global Environment Model, version 2. EAP = East Asia and the Pacific; EUR = Europe; FSU = Former Soviet Union; LAC = Latin America and the Caribbean; MEN = Middle East and North Africa; SAS = South Asia; SSA = Sub-Saharan Africa; TOT = Global.

in the Middle East and North Africa (MENA) than for South Asia, the area of increase is much higher in South Asia, with 180 million hectares being added, against the 80 million for MENA. Europe, on the other hand, is projected to reduce cropland.

The effect of climate on need for additional cropland varies greatly across regions. For example, the HadGEM model suggests that South Asia would increase its pure cropland area an additional 23% because of climate change compared to the conversion without climate change, and for East Asia and Pacific, this would be 31% more. For sub-Saharan Africa, this would only be 6% more.

For low density cropland (mosaic), SSA will increase by 460 million hectares. The next highest regions are East Asia and the Pacific (EAP) and LAC, both of which will increase by around 160 million hectares.

Digging a little deeper into the regional differences as to what kind of land the cropland will be taken from, EAP is projected to have 140 million hectares of forest loss, which is about equal to what the entire rest of the world is projected to lose. When it comes to shrubland, however, the largest portion, 245 million hectares, will be lost by LAC, with 170 million being lost by SSA and 115 million by North

America (NAM). For grassland, almost all of the world's grassland that will be lost are in SSA, 190 million hectares, with the second highest being SAS at 40 million hectares.

9.6 Conclusions

In this chapter, we summarize the projected changes in food demand by region between 2010 and 2050. We note shifts in demand away from East Asia and the Pacific and toward sub-Saharan Africa and South Asia. These largely reflect projected changes in population, but also include projected changes in income which will influence the composition of food consumption, with animal-based protein demand rising, along with all other food categories.

To go along with rising demand, we see how crop productivity should rise, though not sufficiently to compensate for the rise in demand. As a result, we also note the projected rise in harvested area. This would be, in part, from rising cropping intensity, but also in part from land converted for other uses.

Acknowledgments Funding for this work from the CGIAR Research Program on Policies, Institutions, and Markets (PIM) and the Bill and Melinda Gates Foundation is gratefully acknowledged, and the chapter draws on work supported by both organizations as well as the CGIAR Research Program on Climate Change, Agriculture and Food Security (CCAFS). The views expressed here are those of the authors, and may not be attributed to the funding organizations or any other entity.

References

Bai, Y., Feng, M., Jiang, H., Wang, J., Zhu, Y., & Liu, Y. (2014). Assessing consistency of five global land cover data sets in China. *Remote Sensing, 6*(9), 8739–8759.

Belward, A. S., Estes, J. E., & Kline, K. D. (1999). The IGBP-DIS global 1-km land cover data set DISCover: A project overview. *Photogrammetric Engineering and Remote Sensing, 65*, 1013–1020.

Bondeau, A., Smith, P., Zaehle, S., Schaphoff, S., Lucht, W., Cramer, W., Gerten, D., Lotze-Campen, H., Müller, C., Reichstein, M., & Smith, B. (2007). Modelling the role of agriculture for the 20th century global terrestrial carbon balance. *Global Change Biology, 13*, 679–706.

Collins, W., Bellouin, N., Doutriaux-Boucher, M., Gedney, N., Halloran, P., Hinton, T., Hughes, J., et al. (2011). Development and evaluation of an earth–system model—HadGEM2. *Geoscience Model Development, 4*(4), 1051–1075.

Congalton, R. G., Gu, J., Yadav, K., Thenkabail, P., & Ozdogan, M. (2014). Global land cover mapping: A review and uncertainty analysis. *Remote Sensing, 6*(12), 12070–12093.

Dellink, R., Chateau, J., Lanzi, E., & Magne, B. (2017). Long-term economic growth projections in the shared socioeconomic pathways. *Global Environmental Change, 42*, 200–214. https://doi.org/10.1016/j.gloenvcha.2015.06.004.

Dufresne, J.-L., Foujols, M.-A., Denvil, S., Caubel, A., Marti, O., Aumont, O., Balkanski, Y., et al. (2013). Climate change projections using the IPSL–CM5 earth system model: From CMIP3 to CMIP5. *Climate Dynamics, 40*(9/10), 2123–2165.

Fischer, G., Shah, M., Tubiello, F. N., & van Velhuizen, H. (2005). Socio-economic and climate change impacts on agriculture: an integrated assessment. *Philosophical Transactions of the Royal Society B, 360*, 2067–2083. http://rstb.royalsocietypublishing.org/content/360/1463/2067.full.

Friedl, M. A., Sulla-Menashe, D., Tan, B., Schneider, A., Ramankutty, N., Sibley, A., & Huang, X. (2010). MODIS Collection 5 global land cover: Algorithm refinements and characterization of new datasets. *Remote Sensing of Environment, 114*, 168–182.

Fritz, S., McCallum, I., Schill, C., Perger, C., Grillmayer, R., Achard, F., et al. (2009). Geo-Wiki. Org: The use of crowdsourcing to improve global land cover. *Remote Sensing, 1*(3), 345–354.

Herold, M., Mayaux, P., Woodcock, C. E., Baccini, A., & Schmullius, C. (2008). Some challenges in global land cover mapping: An assessment of agreement and accuracy in existing 1 km datasets. *Remote Sensing of Environment, 112*(5), 2538–2556.

IPCC. (2014). Climate change 2014: Synthesis report. In R. K. Pachauri & L. A. Meyer (Eds.), *Contribution of Working Groups I, II and III to the Fifth Assessment Report of the Intergovernmental Panel on Climate Change*. Geneva: Core Writing Team.

Jones, J., Hoogenboom, G., Porter, C., Boote, K., Batchelor, W., Hunt, L., Wilkens, P., et al. (2003). The DSSAT cropping system model. *European Journal of Agronomy, 18*(3–4), 235–265.

Leroux, L., Jolivot, A., Bégué, A., Seen, D. L., & Zoungrana, B. (2014). How reliable is the MODIS land cover product for crop mapping Sub-Saharan agricultural landscapes? *Remote Sensing, 6*(9), 8541–8564.

Lin, G. C., & Ho, S. P. (2003). China's land resources and land-use change: Insights from the 1996 land survey. *Land Use Policy, 20*(2), 87–107.

Liu, J., Li, S., Ouyang, Z., Tam, C., & Chen, X. (2008). Ecological and socioeconomic effects of China's policies for ecosystem services. *PNAS, 105*(28), 9477–9482. https://doi.org/10.1073/pnas.0706436105. Retrieved from www.pnas.org.

Lotze-Campen, H., Müller, C., Bondeau, A., Rost, S., Popp, A., & Lucht, W. (2008). Global food demand, productivity growth, and the scarcity of land and water resources: A spatially explicit mathematical programming approach. *Agricultural Economics, 39*(3), 325–338.

Martin, G., Bellouin, N., Collins, W., Culverwell, I., Halloran, P., Hardiman, S., Hinton, T., et al. (2011). The HadGEM2 family of met office unified model climate configurations. *Geophysical Model Development, 4*, 723–757.

Ran, Y., Li, X., & Lu, L. (2010). Evaluation of four remote sensing based land cover products over China. *International Journal of Remote Sensing, 31*(2), 391–401.

Robinson, S., Mason d'Croz, D., Islam, S., Sulser, T. B., Robertson, R. D., Zhu, T., Gueneau, A., Pitois, G., & Rosegrant, M. W. (2015). *The International Model for Policy Analysis of Agricultural Commodities and Trade (IMPACT): Model description for version 3. IFPRI Discussion paper 1483*. Washington, DC: International Food Policy Research Institute (IFPRI). Retrieved from http://ebrary.ifpri.org/cdm/ref/collection/p15738coll2/id/129825.

Samir, K. C., & Lutz, W. (2017). The human core of the shared socioeconomic pathways: Population scenarios by age, sex and level of education for all countries to 2100. *Global Environmental Change, 42*, 181–192. https://doi.org/10.1016/j.gloenvcha.2014.06.004.

Scepan, J. (1999). Thematic validation of high-resolution global land-cover data sets. *Photogrammetric Engineering and Remote Sensing, 65*, 1051–1060.

Schaldach, R., Alcamo, J., Koch, J., Kölking, C., Lapola, D. M., Schüngel, J., & Pries, J. (2011). An integrated approach to modelling land-use change on continental and global scales. *Environmental Modelling and Software, 26*(8), 1041–1051.

Smith, L., & Haddad, L. (2000). *Explaining child malnutrition in developing countries: A cross-country analysis*. Washington, DC: International Food Policy Research Institute.

United Nations. (2015). *Transforming our world: The 2030 agenda for sustainable development*. Resolution adopted by the General Assembly on 25 September. Retrieved September 10, 2017, from http://www.un.org/ga/search/view_doc.asp?symbol=A/RES/70/1&Lang=E.

Wu, W., Shibasaki, R., Yang, P., Zhou, Q., & Tang, H. (2008). Remotely sensed estimation of cropland in China: A comparison of the maps derived from four global land cover datasets. *Canadian Journal of Remote Sensing, 34*(5), 467–479.

Xiao, X., Boles, S., Frolking, S., Salas, W., Moore Iii, B., Li, C., et al. (2002). Landscape-scale characterization of cropland in China using Vegetation and Landsat TM images. *International Journal of Remote Sensing, 23*(18), 3579–3594.

Chapter 10
Adaptation to Climate Change Through Adaptive Crop Management

Dave Watson

Abstract In order to meet the growing needs of the global food system, whilst at the same time mitigating the effects of climate change and other production limiting factors and reducing negative environmental externalities, maize, wheat and rice agri-food systems will be required to sustainably intensify production. In the irrigated systems of developing countries, significant scope exists for increasing water use efficiency through new soil, water and fertiliser management approaches (such as Conservation Agriculture, Direct Seeding, Alternate Wetting and Drying, Site Specific Nutrient Management and Nutrient Expert) and crop diversification. In addition to the development of weather agro-advisory services and weather index insurance, the options available in rain-fed systems differ markedly from those available in irrigated systems, namely, to optimise every drop of available rainfall or to avoid drought stress situations. Whilst breeding for heat tolerance and diversified cropping systems are likely to be the principal short-term responses to increased mean global temperatures and extreme heat events, changes over the longer term are predicted to be quite dramatic, especially with regard to a productive expansion of temperate crops towards the poles. Whilst significant opportunities exist to ameliorate the effects of climate change, these opportunities generally involve risk. In developed countries, the private sector (seed, fertiliser, pesticide, irrigation, credit and insurance suppliers) is generally at hand to advise farmers how to address the challenges posed by climate change. Conversely, in most developing countries, there are few sources of advice and support for smallholder farmers. This situation leaves many smallholder farmers in developing countries without the advice and support that they desperately need. Ultimately, the most vulnerable farmers and communities, namely, smallholder subsistence and market oriented farmers in marginal environments, are those who face the most extreme climate change-related challenges at the same time as being the least able to adapt. Whilst international agricultural research centres (CGIAR), advanced research and development-focused centres of developed countries, and both local and international NGOs strive to both

---

D. Watson (✉)
CGIAR Research Program on Maize, International Maize and Wheat Improvement Center (CIMMYT), Texcoco, México

© Springer Nature Switzerland AG 2019
A. Sarkar et al. (eds.), *Sustainable Solutions for Food Security*,
https://doi.org/10.1007/978-3-319-77878-5_10

develop and translate evolving crop management approaches; the dissemination of climate smart agricultural practices is extremely slow.

10.1 Introduction

In order to meet the growing needs of the global food system, whilst at the same time mitigating the effects of climate change (Vermeulen et al. 2012) and other production limiting factors (World Bank 2014) and reducing negative environmental externalities, maize, wheat and rice agri-food systems will be required to sustainably intensify production. This chapter focuses on crop management and socio-economic approaches that are able to either boost or stabilise yields in the face of climate change and/or reduce the environmental (especially carbon) footprint of maize, wheat and rice production systems. This chapter is broken down into sub-sections, each focusing on different dimensions of Climate Smart Agriculture (CSA), which has generated significant interest and support (AGRA 2014). Relevant maize, wheat and rice examples are utilised.

According to the CGIAR CRP on Climate Change, Agriculture and Food Security (CCAFS), Climate Smart Agriculture (CSA) is an integrative approach to address the interlinked challenges of food security and climate change that explicitly aims for three objectives:

1. Sustainably increasing agricultural productivity, to support equitable increases in farm incomes, food security and development;
2. Adapting and building resilience of agricultural and food security systems to climate change at multiple levels; and
3. Reducing greenhouse gas emissions from agriculture (including crops, livestock and fisheries).

10.2 Increasing Water Use Efficiency in Maize, Wheat and Rice Agri-Food Systems

10.2.1 Irrigated Systems

Whilst enabling developing countries to feed their growing populations, much of the success of the Green Revolution relied on the substantial increase in ground water pumping for irrigated rice and wheat (Balwinder-Singh et al. 2015). Consequently, over-pumping, especially in South Asia, led rapidly declining and unsustainable water tables (Yadvinder-Singh et al. 2014). Not only have absolute amounts of water available to irrigated agriculture declined but the costs of extracting remaining water has significantly increased. Furthermore, in many areas, the quality of water extracted has deteriorated due to intrusion of saline groundwater (Balwinder-Singh et al. 2015). Increasing water scarcity threatens the sustainability of rice production (Rijsberman 2006). In many Asian countries, per capita availability of water declined by 40–60% between 1955 and 1990, and is expected to decline further by 15–54% over the next 35 years (Lampayan et al. 2015b). By 2025, 15–20 million hectares of rice lands will suffer some degree of water scarcity (Tuong and Bouman 2003). As summarised in Bouman et al. (2007), hot-spots of water scarcity in rice-growing areas have been reported in many countries, including the Philippines, Bangladesh, northern China, the northwest Indo-Gangetic Plains, India's Cauvery delta, and the Chao Phraya delta in Thailand (Lampayan et al. 2015b). The reasons for water scarcity are diverse and location-specific, but include decreasing availability because of increasingly erratic rainfall brought about by climate change, decreasing water quality because of pollution (e.g., chemicals, salts, silts), malfunctioning of irrigation systems, increased competition from other sectors (i.e., urban and industrial users), and excessive ground water extraction (Vörösmarty et al. 2010; Wada et al. 2011). Currently, approximately 70% of the world's freshwater is used for irrigated agriculture, although the use of water for agriculture varies highly from country to country. For example, agricultural water use in several South Asian countries, such as; Afghanistan, Bangladesh, Bhutan, India, Nepal, Maldives, Pakistan and Sri Lanka is almost 95% of extracted water (Babel and Wahid 2008). Here, approximately 40% of the cropland is irrigated, and irrigated agriculture accounts for 60–80% of food production (Yadvinder-Singh et al. 2014). Furthermore, whilst significant competition is expected from alternative water users, agricultural demand is expected to increase further at a rate of 0.7% per year (Doll and Siebert 2002; Rosegrant et al. 2009) to meet future food security requirements. Fischer et al. (2007) estimated an increase in irrigation water (IW) requirements of 50% in developing regions and 16% in developed regions between 2000 and 2080 (Yadvinder-Singh et al. 2014). Rice occupies 30% of the world's irrigated cropland but receives between 34% and 43% (Dawe 2005) of the total world's irrigation water, or 24–30% of the total world's freshwater withdrawals (Bouman et al. 2007). More than 50% of the world's 271 million hectares of irrigated crop area is located in Asia where rice production accounts for 40–46% of the net irrigated crop area (Dawe 2005). Irrigated rice has very low water productivity; consuming 3000–5000 L of water to produce 1 kg of grain. On the other hand, wheat, an intrinsically water use efficient crop, requires 900 L of water to produce 1 kg of grain. Through breeding and good crop management, it may even be possible to increase the efficiency of wheat to produce 1 kg of grain with just 450 L of water. Thus, enhancing the water productivity of rice at field and regional scales is essential (Yadvinder-Singh et al. 2014).

Whilst this backdrop presents a significant challenge, both existing and evolving technologies are at hand to improve water use efficiency in irrigated wheat and rice systems (Vörösmarty et al. 2010; Wada et al. 2011). The most important of these technologies include. (1) Drought-tolerant wheat and rice varieties. (2) Delayed rice transplanting (Balwinder-Singh et al. 2015). (3) Changing to shorter duration rice varieties (Balwinder-Singh et al. 2015). (4) Changing from continuous flooding to alternate wetting and drying (AWD) water management for rice (Balwinder-Singh et al. 2015). (5) Drip and pivot irrigation for wheat. (6) Improved irrigation scheduling (Yadvinder-Singh et al. 2014). (7) Improved surface irrigation infrastructure—drainage canals, etc. (Yadvinder-Singh et al. 2014). (8) Precision laser levelling, which reduces irrigated water usage between 15% and 30% (Jat et al. 2006).

10.2.2 Alternate Wetting and Drying

To address increasing water scarcity, IRRI, in collaboration with national agricultural research and extension system (NARES) partners, developed and promoted the "alternate wetting and drying" (AWD) water management technology (Lampayan et al. 2015b) (see Fig. 10.1). The Alternate Wetting and Drying (AWD) technique was developed to reduce water use, production costs and methane emissions (Sander et al. 2014). Under AWD, the field is intermittently irrigated (Lampayan et al. 2015b). According to Bouman et al. (2007), "Safe-AWD consists of three key elements. (1) Shallow flooding for the first 2 weeks after transplanting (this helps rice plants recover from transplanting shock and suppresses weeds). Starting at about 2–3 weeks after transplanting (3–4 weeks after sowing) the field is left to dry out until the water table reaches a level of about 10–15 cm below the soil surface. Once the threshold is reached, irrigation water should be applied until 3–5 cm of standing water in the field is reached (Nelson et al. 2015). (2) Shallow ponding from heading to the end of flowering, as this is a critical stage where rice plants are very sensitive to water-deficit stress and a time when they have a high growth rate and water requirement. (3) Keeping irrigation water applied whenever the perched water table falls to about 15 cm below the soil surface during all other periods. The threshold of 15 cm will not cause any yield decline since the roots of the rice plants are still able

Fig. 10.1 Alternate wetting and drying. Source: International Rice Reasearch Institute (IRRI)

Alternate Wetting and Drying

to take up water from the perched groundwater and the almost saturated soil above the water table" (Lampayan et al. 2015b). Compared to flooded controls, AWD has reduced irrigation water input between 10% and 40% (Lampayan et al. 2015b); Cabangon et al. 2003; Belder et al. 2004, 2007; Yadvinder-Singh et al. 2009; Bueno et al. 2010; Lampayan et al. 2015a; Nelson et al. 2015). This translates into irrigation water savings of 200–900 mm, whilst maintaining yields at 6400–9200 kg ha^{-1} (Yadvinder-Singh et al. 2014). As expected, significant savings are also made associated with pumping water. This was due to reduced labour costs associated with irrigation and a 20–25% reduction in fuel and oil consumption. Based on a "with and without" AWD comparison, this increased farmer income by 38%, 32%, and 17% in Bangladesh, the Philippines, and southern Vietnam respectively (Lampayan et al. 2015b). Consequently, as long as farmers practice careful water management, AWD results in up to 132% higher profits (Lampayan et al. 2015b).

AWD is now recommended practice in Bangladesh (Palis et al. 2014), China, India, Indonesia, Lao PDR, Myanmar (Lampayan et al. 2014), the Philippines (Rejesus et al. 2013; Lampayan et al. 2014), Thailand, and Vietnam (Rejesus et al. 2013). Due to the growing awareness of AWD as a potential climate change mitigation option, Vietnam and the Philippines now incorporate AWD in their national policy interventions (Nelson et al. 2015).

10.2.3 Crop Diversification

In the irrigated rice–wheat systems in south Asia, significant scope exists for increasing water use efficiency through crop diversification, which are better adapted to local agro-climatic and economic conditions (Liu et al. 2016). Indeed, in many cases, this is already happening. For example, in increasingly water scarce Bangladesh, wheat, maize and sunflower are beginning to replace the traditional double-rice cropping system. According to Weller et al. (2016), this change reduced irrigation water use in the dry season by approximately 70% and decreased CH_4 emissions by 97% without causing economic penalty. Furthermore, the introduction of early maturing rice varieties into two-rice crop systems has allowed farmers to introduce an additional crop such as maize or mustard; increasing income by US$ 600–700 per hectare (RICE 2016). Double- or triple-cropped systems combining stress-tolerant rice and maize varieties with new breeds of fish have been shown to double the production of both rice and fish (RICE 2016). Rice–lentil–mung bean systems are five times more profitable (Yadvinder-Singh et al. 2014).

According to Gangwar and Singh (2011), savings in irrigation water in maize–potato–onion, maize–potato–summer green gram and maize–wheat–summer green gram system as compared to irrigation water used in rice–wheat production was 38.7%, 50.5% and 55.6%, respectively. Indeed, Gangwar and Singh (2011) went further estimating that by replacing 2.6 million hectares under rice–wheat cultivation would result in savings of approximately 1358.5 million m^3 of irrigation water and USD$75 million (Yadvinder-Singh et al. 2014). Even delaying the planting date of rice from mid-May to late June significantly increased the crops water productivity without incurring a loss in yield (Chahal et al. 2007; Jalota et al. 2009). Subsequently, the Government of Punjab in India banned rice transplanting before the 10th of June in order to reduce seasonal evapotranspiration and minimize over exploitation of groundwater.

10.2.4 Direct Seeded Rice (DSR)

Direct seeding of rice (DSR) (see Fig. 10.2) refers to the process of establishing a rice crop from seeds sown in the field rather than by transplanting seedlings from the nursery (Farooq et al. 2011). There are three principal methods of DSR: dry seeding (sowing dry seeds into dry soil), wet seeding (sowing pre-germinated seeds on wet puddled soils), and water standing (seeds sown into standing water). In recent years, interest has grown in expanding the aerobic cultivation of rice to lowland paddy fields in north-west India and the Punjab in Pakistan (Huaqi et al. 2003; Bouman et al. 2005; Waqar et al. 2007; Mahajan et al. 2011). Interest in DSR is based on the assumption that it reduces the need for labour and irrigation and enables early establishment of rice. In turn, this leads to an earlier harvest of the rice crop and, subsequent, more time and higher yields for a second crop (Kumar and Ladha 2011). Kumar and Ladha (2011) reported an average of 21–25% saving on irrigation water with (DSR) compared with traditional paddy rice system. DSR also facilitates conversion of the conventional rice–wheat system to a Conservation

Fig. 10.2 Direct seeded rice. Source: International Rice Reasearch Institute (IRRI)

Direct Seeded Rice

Agriculture (CA) system; described later in the chapter. One of the key benefits of DSR is that the seed drills used for wheat are used to direct sow rice, negating the need for time consuming and costly puddling and manual transplanting of rice and reducing the amount of irrigation water required to establish the rice crop (Miro and Ismail 2013). Conversion of paddy rice–wheat system to a CA-based system has several advantages: (1) It significantly reduces tillage and crop establishment costs. (2) It facilitates early crop establishment. (3) It improves soil structure for wheat. (4) It improves nutrient cycling. (5) It reduces soil evaporation and weeds by surface residue retention. (6) It reduces energy consumption (and greenhouse gas genera-tion) and groundwater pumping. (7) It reduces air pollution from stubble burning (Balwinder-Singh et al. 2011; Gathala et al. 2014). Lastly, investments are required to breed rice with tolerance to waterlogging, which often occurs in DSR systems just after sowing or if the land has not been properly levelled (Kirk et al. 2014). This can lead to poor establishment of the rice crop (Ismail et al. 2013).

10.2.5 Rain-Fed Systems

The options available in rain-fed systems differ markedly from those available in irrigated systems. In addition to developing drought tolerant maize, wheat and rice, this Chapter outlines some of the main climate change mitigation options available that address reduced and erratic rainfall. Crop management approaches designed to mitigate reduced and erratic rainfall fall into two categories: (1) those which attempt to optimise every last drop of available rainfall, and (2) those which attempt to avoid drought stress situations. Water harvesting is one approach to utilising every last drop of available rainfall and is often practiced in resource constrained low-input agricultural production systems. Water harvesting refers to a range of approaches from in situ moisture conservation to the collection and storage of surface runoff water for small-scale irrigation (Critchley and Gowing 2012). In-situ rain water harvesting systems were shown to increase crop yields by 30–50% in Southern Africa (Pachpute et al. 2009). In addition, harvesting of rainwater may help facili-tate crop diversification in order to enhance household food security, dietary status, and economic return (Biazin et al. 2012). For example, in Ethiopia, an increase of just 1% in harvested water increased the output of onion and tomato by 0.12% and 0.23%, respectively (Wakeyo 2012); prompting governments in SSA to invest in water harvesting interventions (Malesu et al. 2012).

In addition to harvesting water, many farmers in rain-fed SSA have already begun to make changes in order to mitigate the effects of climate change; especially reduced or erratic rainfall. One such change has been to adjust planting dates so that they are better aligned with when the rains actual come. This has already been reported across a number of countries in SSA (Fosu-Mensah et al. 2012; Bryan et al. 2013; Sofoluwe et al. 2011; Bele et al. 2014). Many farmers are also beginning to switch to more drought tolerant varieties of their traditional crops. In some cases, farmers are beginning to switch crops altogether. For example, in SSA, whilst maize remains firmly entrenched (Cairns et al. 2013), there are cases where maize has been, and could potentially be, switched for more heat and drought-tolerant crops such as sorghum and millet (Burke et al. 2009). In Kenya, farmers have begun switching from maize to cassava, sweet potatoes, and pigeon peas (Bryan et al. 2013). Whilst not as widely accepted as maize, cassava is especially adapted to climate change in SSA because of its inherent tolerance of drought, heat and nutrient-depleted soils (Jarvis et al. 2012). Indeed, according to Brooks (2014), many Malawian farmers switched to cassava during the 2001/2002 drought and famine. In SSA and SA, evidence is growing that farmers are slowly moving away from monocultures to more diversified cropping (Bryan et al. 2009, 2013; Bele et al. 2014; Westengen and Brysting 2014) (see Fig. 10.3). According to Bryan et al. (2009, 2013), access to irrigation and extension information are the two most important factors involved in switching crop species.

There is also an opportunity to develop weather agro-advisory services, especially through the development and application of phone Apps (AGRA 2014). The rationale for this is that the adoption and success of CSA technologies and practices depend on the effective delivery of agro-advisory services. For example, in Ethiopia and Mali, farmers who received and applied climate agro-advisories increased their income by 10–80% due to a substantial production increase from maize, sorghum, pearl millet, groundnut and cotton. Farmers felt that they were exposed to lower levels of risk and hence were more confident about purchasing and using inputs like improved seed, fertiliser, and pesticides, all of which boost production (AGRA 2014; Hellmuth et al. 2007).

Besides agro-advisory services, the availability of long-term climate data at the national level and the capacity to access and install automatic weather stations helped piloting of weather index insurance as means of managing climate risks in Ethiopia, Kenya and Senegal (AGRA 2014; Tadesse et al. 2015). Whilst still in an

Fig. 10.3 Crop diversification in rain-fed production systems in Karnataka State, India

Crop diversification in Rain-fed production systems in Karnataka State - India

embryonic stage, micro-insurance, particularly index-based insurance, may also prove important in safeguarding fragile agricultural production systems against climate change (Lybbert and Sumner 2012).

10.2.6 Conservation Agriculture

Conservation Agriculture (CA) is being advanced as one of the most likely approaches able to meet increasing demands from the global food system, whilst increasing both resource-use efficiency and increasing profits. CA involves the application of three basic agronomic principles: (1) Minimum or at least reduced soil disturbance; (2) Retention of crop residues, and (3) Crop rotations, especially with legumes (Hobbs 2007; FAO 2011) (see Fig. 10.4). Application of these principles improve soil and water conservation, soil fertility, crop yield, and reduces production costs, soil degradation, runoff and evaporation of precious water and lessens the effect of drought (Verhulst et al. 2010). In the case of reduced and/or erratic rainfall, CA is particularly useful in facilitating increased infiltration of rainfall (Thierfelder and Wall 2009); acting as a reservoir of soil moisture (Ussiri and Lal 2009; Dendooven et al. 2012). Improved soil water status cushions plants against drought stress brought about by intermittent rainfall and drought (Fischer et al. 2002; Thierfelder and Wall 2009, 2010).

Fig. 10.4 Zero-Till wheat sown after rice. Source: Conservation Facebook

Zero-Till wheat sown after rice Source Conservation Agriculture Facebook

CA-based systems are operational on more than 157 million hectares (Kassam et al. 2015). However, much of the success of CA has been on large commercial farms (Derpsch et al. 2016). Smallholder farmers have tended to adopt components of CA; especially Zero/no-tillage (Thierfelder et al. 2017). Indeed, the suitability of CA for smallholders in SSA continues to be contested (Giller et al. 2015, 2011, 2009; Andersson and D'Souza 2014; Baudron et al. 2015). Ultimately, CA is a relatively complex technology, requiring. (1) Significant knowledge. (2) Willingness and ability to maintain biomass on the soil surface (Valbuena et al. 2012). (3) Access to inputs such as machinery, fertiliser and herbicides. (4) Available cash or access to credit to purchase inputs. (5) Access to markets for rotational crops. (6) Capacity to adapt/capacity to innovate. Weed control, especially in the absence of herbicides has also been cited as a principal constraint to adoption (Andersson and D'Souza 2014; Giller et al. 2015). However, challenges aside, CA practices have been widely adopted in rain-fed and irrigated systems in tropical, subtropical, and temperate regions of the world (Pretty et al. 2006; FAO 2008; Hengxin et al. 2008; Hobbs and Govaerts 2010; Jat et al. 2011). Recent ground breaking studies in Malawi and Zambia have highlighted win-win scenarios associated with the adoption of improved maize varieties and sustainable agricultural practices (SAPs). Thierfelder et al. (2016) examine the impact of Conservation agriculture (CA), a combination of SAPs, on maize yields and small farmer incomes over the period 2005–2014. The study determined that the adoption of CA out yielded conventional ridge tilled control plots in Mwansambo and Zidyana Districts between 22% and 31%, respectively, and increased income by 50% and 83%, respectively. This was in part due to the fact that crops were produced with 28–39 less labour days ha^{-1} compared with the conventional practice. Successful extension of CA systems by Total Land Care (TLC), an NGO working in Malawi (and Zambia), using innovation systems approaches, has led to significant out-scaling of this technology to more than 30,000 farmers on more than 14,000 ha in Malawi in the last decade and this is expected to increase. In Zambia, MAIZE (2016), reports that SAPs such as CA are essential in mitigating risks from climate change. For example, it was found that when practicing crop rotation and crop diversification (components of CA), farmers are sowing a diverse range of crops that can perform well under a range of environmental conditions. In addition, due to different sowing dates and maturity periods of these crops, produce is harvested at different times of year, thus reducing the risk of total crop loss if drought strikes. Indeed, the retention of crop residue, another SAP, was found to be a vital factor in "improving the soil and retaining moisture especially in drought prone areas". The results of the Zambia study clearly suggest that "farmers are adopting these SAPs to reduce the effects of droughts" (MAIZE 2016). The study goes on to recommend the need for policy interventions that promote the combined adoption of improved maize varieties and SAPs, such as a maize–legume rotation and residue retention, which can boost yields and farm incomes especially among resource-poor farmers who cannot afford inorganic fertilisers.

10.3 Adapting to a Warmer Climate

Whilst breeding for heat tolerance and diversifying crops are likely to be the principal short-term responses to increased mean global temperatures and extreme heat events, changes over the longer term are predicted to be quite dramatic; especially with regard to a productive expansion of temperate crops towards the poles (Challinor et al. 2010). For example, it is predicted that maize and soybean will replace wheat and oil seed rape (canola) in the southern parts of Canada and that wheat will increasingly move northward and replace barley. In areas where maize and soybean are already grown, increasing temperatures may necessitate their replacement by more heat tolerant crops such as cassava, sorghum, millet, mung bean and cowpea (Fischer et al. 2014).

10.4 Increasing Nutrient Use Efficiency in Maize, Wheat and Rice Production Systems

Increased nutrient use efficiency, especially of synthetic nitrogen fertiliser, is important on a number of fronts. Optimum management of nitrogen fertiliser increases profits and reduces environmental externalities associated with nitrogen loss. In general, due to increasing use of synthetic fertilisers and limited supplies of farmyard manures, fertiliser use in the developed world is increasingly being more precisely applied. As such, given the limited space available, this section will focus on more precise nutrient management approaches currently being applied in developing countries: (1) Site Specific Nutrient Management (SSNM), and (2) Nutrient Expert.

10.4.1 Site-Specific Nutrient Management

The site-specific nutrient management (SSNM) approach was developed by IRRI and partners in the mid-1990s. Pampolino et al. (2007) define site-specific nutrient management (SSNM) as a low-tech, plant need-based approach for optimally applying fertilisers such as nitrogen (N), phosphorus (P) and potassium (K) to rice. The key features are (1) dynamic adjustments in fertiliser management to accommodate field and season-specific conditions, (2) effective use of indigenous nutrients, (3) efficient fertiliser N management through the use of the leaf colour chart, (4) Use of the omission plot technique to determine the requirements for P and K fertiliser, (5) managing fertiliser P and K to both overcome P and K deficiencies and avoid mining of these nutrients from the soil. The rationale for developing SSNM arose from the need to assist rice farmers to increase yields, improve nutrient use efficiency and reduce negative externalities of fertiliser use (Pampolino et al. 2007). Efficient nitrogen management achieved through the application of SSNM avoids a build-up, and loss, of inorganic nitrogen in the soil during times when there are no crops in the ground or when crop demand for nitrogen is low (Pampolino et al. 2007).

Between 1997 and 1999, working with national extension partners, the SSNM was tested and refined in 200 irrigated rice farms in eight strategic irrigated rice production systems across six countries in Asia (Dobermann et al. 2002, 2004). Use of SSNM led to increased nitrogen-use efficiency and profitability (Dawe et al. 2004). Indeed, use of SSNM in the Philippines and Vietnam, where nitrogen use efficiency was historically low, led to both increased partial factor productivity (kg grain kg^{-1} fertiliser N) and a reduction in fertiliser use of 10% and 14% respectively (Pampolino et al. 2007). In turn, the increased yields achieved through SSNM led to net annual benefits of 34 US$ ha^{-1} $year^{-1}$ in Vietnam, 106 US$ ha^{-1} $year^{-1}$ in the Philippines, and 168 US$ ha^{-1} $year^{-1}$ in India (Pampolino et al. 2007). In India, yields increased by 17% and profits per hectare increased by 48% (Pampolino et al. 2007). In addition, through improved nitrogen management, SSNM has demonstrated its ability to increase yields whilst reducing N_2O emissions (Pampolino et al. 2007).

10.4.2 Nutrient Expert

Inefficient use of fertiliser is one of the principal reasons for low maize productivity in South Asia and SSA. In part, this has been due to a lack of appropriate tools and implementation mechanisms. The International Plant Nutrition Institute (IPNI), and the CGIAR CRPs—MAIZE, WHEAT and CCAFS jointly developed Nutrient Expert®, a user-friendly dynamic nutrient management tool, which has become widely recognised as a major climate smart decision support system for yield maximisation whilst optimising nutrient use (see Fig. 10.5). The Nutrient Expert® (NE) tool uses similar principles to the SSNM (Pampolino et al. 2007). Based on key information about the production environment, Nutrient Expert® provides location-based balanced fertiliser recommendations for maize and wheat production systems, which are directly linked to the farmers' financial resources and yield aspirations (Majumdar et al. 2017). Key aspects of the Nutrient Expert® tool include (1) targeted management of fertiliser and organic resources under highly variable soil fertility conditions, (2) fertiliser rates that optimise crop productivity and farmers' income, (3) Timing of N application under highly variable inter- and intra-seasonal rainfall and (4) Spot-application of fertiliser for low fertiliser rates and for P management in P-fixing soils (Majumdar et al. 2017). Nutrient Expert® has been made available on a wide variety of platforms, including versions for the Web and mobile phones and has led to significant improvements in nutrient use efficiency and farmer profitability. For example, in Nepal, Nutrient Expert® was estimated to

Fig. 10.5 Farmers using Nutrient Expert™ in Bihar State, India

Farmers using Nutrient Expert TM in Bihar State - India

have improved yields by about 3 tons ha^{-1} over farmers' fertilisation practices across different farm types, leading to an average additional income of US$ 688 over farmer fertiliser practice (MAIZE 2015). Increasingly, Nutrient Expert® is being used with real time assessment of crop nitrogen status using the Green Seeker hand held tool developed by CIMMYT (Sapkota et al. 2014).

Nutrient Expert® has sparked significant public and private sector interest. In 2013, Nutrient Expert® was recognised by the Bihar Innovation Forum as the Best Innovation for Improving Rural Livelihood in Bihar, India (MAIZE 2013). In 2015, IPNI and Tata Consultancy Services Limited (TCS) signed a memorandum of understanding (MoU) to out-scale the Nutrient Expert® to nearly 100,000 farmers involved in their social initiatives across India. Also in India, DuPont Pioneer has expressed interest to utilise Nutrient Expert® in their Pravakta Programme (serving 30,000 farmers) and the Unnati Abhiyan Programme (serving 1000 farmers). Nutrient Expert® has also caught the attention of the fertiliser company Mosaic, who see the tool as a way of better enabling farmers to benefit from their products (MAIZE 2016). The Integrated Nutrient Management (INM) Division of the Ministry of Agriculture and Farmers Welfare, Government of India have also included Nutrient Expert®, as one of the key tools for disseminating precision nutrient recommendations under national Soil Health Card program. Large-scale training for farm advisers in the use of Nutrient Expert® under a series of public and private sector agreements took place in 2016 (MAIZE AR 2015).

10.5 Discussion

Whilst significant opportunities exist to ameliorate the effects of climate change through adaptive crop management, adjustments to crop management generally involve risk. Risks maybe related to the need to invest farm profits or credit in order to experiment with different crops, adjusting the types, quantities and timing of inputs (seed, fertiliser, water, etc.) or associated with the need to purchase or hire machinery. There may be risk associated with post-harvest processing and markets. In developed countries, the private sector (seed, fertiliser, pesticide, irrigation, credit and insurance suppliers) are generally at hand to advise farmers how to address the challenges posed by climate change. Indeed, much of this advice is linked to the targeted use of inputs through the application of precision agriculture. Increasingly, in the light of climate change, the amelioration of risk is being managed through index insurance. For example, in 2015, John Deere Insurance Company and BASF began collaborating to offer "Risk Advantage" Insurance (ETC 2016). Conversely, in most developing countries, there are few sources of advice and support for smallholder farmers. The private sector is much weaker and less coordinated. Likewise, in most developing countries, public sector extension is generally under resourced and unable to provide the nuanced guidance that smallholder farmers require. Ultimately, the most vulnerable farmers and communities, namely, smallholder subsistence and market oriented farmers in marginal environments, are those who face the most extreme climate change-related challenges at the same time as being the least able to adapt.

Whilst international agricultural research centres (CGIAR), advanced research and development-focused centres of developed countries, and both local and international NGOs strive to both develop and translate evolving crop management approaches, the dissemination of climate smart agricultural practices is extremely slow. One of the highest profile interventions is the Climate Smart Village approach. The concept of Climate Smart Villages was developed and implemented by the CGIAR Consortium Research Programme—Climate Change, Agriculture and Food Security (CCAFS). "Climate-Smart Villages are sites where researchers (other CGIAR centres etc.), farmers' cooperatives, government officials and private sector partners come together to identify the most appropriate climate-smart interventions in agriculture based on its agro-ecological, climate risk profile and socio-economic conditions. These interventions include water smart practices (rainwater harvesting, laser levelling), weather smart activities (ICT-based agromet services, index-based insurance), nitrogen and carbon smart practices (precision fertilisers, catch-cropping), energy smart (zero tillage, residue management) and knowledge smart activities (farmer-farmer learning, seed banks and cooperatives). These interventions work together increase an agriculture community's resilience to climatic stresses" (CCAFS 2014). Successful climate-change mitigation approaches generated in Climate Smart Villages are expected to be scaled out to other similar locations and scaled up to garner support at sub-national and national levels. Whilst many farmers and local NGOs experiment with grass-roots solutions to climate change, farmers and rural communities need technical, institutional and financial support to adapt to climate change. This can come from either strengthened public or private sector extension (both have their own pros and cons).

References

AGRA. (2014). *Africa agriculture status report: Climate change and smallholder agriculture in sub-Saharan Africa (No. 2)*. Nairobi: Alliance for a Green Revolution in Africa, AGRA.

Andersson, J. A., & D'Souza, S. (2014). From adoption claims to understanding farmers and contexts: A literature review of conservation agriculture (CA) adoption among smallholder farmers in southern Africa. *Agriculture, Ecosystems & Environment, 187*, 116–132.

Babel, M. S., & Wahid, S. W. (2008). *Freshwater under threat in South Asia. UNEP Report*. Nairobi: United Nations Environment Programme (UNEP). ISBN 978-92-807-2949-8. 29 pp.

Balwinder-Singh, Gaydon, D. S., Humphreys, E., & Eberbach, P. L. (2011). Evaluating the performance of APSIM for irrigated wheat in Punjab, India. *Field Crops Research, 124*, 1–13.

Balwinder-Singh, Humphreys, E., Gaydonca, D. S., & Yadavb, S. (2015). Options for increasing the productivity of the rice–wheat system of north west India while reducing groundwater depletion. Part 2. Is conservation agriculture the answer? *Field Crops Research, 173*(2015), 81–94.

Baudron, F., Thierfelder, C., Nyagumbo, I., & Gérard, B. (2015). Where to target conservation agriculture for African smallholders? How to overcome challenges associated with its implementation? Experience from eastern and southern Africa. *Environments, 2*(3), 338–357.

Belder, P., Bouman, B. A. M., Cabangon, R. J., Guoan, L., Quilang, E. J. P., Yuanhua, L., Spiertz, J. H. J., & Tuong, T. P. (2004). Effect of water-saving irrigation on rice yield and water use in typical lowland conditions in Asia. *Agricultural Water Management, 65*, 193–210.

Belder, P., Bouman, B. A. M., & Spiertz, J. H. J. (2007). Exploring options for water savings in lowland rice using a modelling approach. *Agricultural Systems, 92*, 91–114.

Bele, M. Y., Sonwa, D. J., & Tiani, A. M. (2014). Local communities' vulnerability to climate change and adaptation strategies in Bukavu in DR Congo. *Journal of Environment & Development, 23*(3), 331–357.

Biazin, B., Sterk, G., Temesgen, M., Abdulkedir, A., & Stroosnijder, L. (2012). Rainwater harvesting and management in rain-fed agricultural systems in sub-Saharan Africa—A review. *Physics and Chemistry of the Earth, 47–48*, 139–151.

Bouman, B. A. M., Peng, S., Castaneda, A. R., & Visperas, R. M. (2005). Yield and water use of irrigated tropical aerobic rice systems. *Agricultural Water Management, 74*, 87–105.

Bouman, B. A. M., Lampayan, R. M., & Tuong, T. P. (2007). *Water management in irrigated rice: Coping with water scarcity* (p. 53). Manila: International Rice Research Institute.

Brooks, S. (2014). Enabling adaptation? Lessons from the new 'Green Revolution' in Malawi and Kenya. *Climatic Change, 122*, 15–26.

Bryan, E., Deressa, T. T., Gbetibouo, G. A., & Ringler, C. (2009). Adaptation to climate change in Ethiopia and South Africa: Options and constraints. *Environmental Science & Policy, 12*, 413–426.

Bryan, E., Ringler, C., Okoba, B., Roncoli, C., Silvestri, S., & Herrero, M. (2013). Adapting agriculture to climate change in Kenya: Household strategies and determinants. *Journal of Environmental Management, 114*, 26–35.

Bueno, C. S., Bucourt, M., Kobayashi, N., In ubushi, K., & Lafarge, T. (2010). Water productivity of contrasting rice genotypes grown under water-saving conditions in the tropics and investigation of morphological traits for adaptation. *Agricultural Water Management, 98*, 241–250.

Burke, M. B., Lobell, D. B., & Guarino, L. (2009). Shifts in African crop climates by 2050, and the implications for crop improvements and genetic resources conservation. *Global Environmental Change, 19*, 317–325.

Cabangon, R. J., Lu, G., Tuong, T. P., Bouman, B. A. M., Feng, Y., & Zichuan, Z. (2003). Irrigation management effects on yield and water productivity of inbred and aerobic rice varieties in Kaefeng. In *Proc. of the First International Yellow River Forum on River Basin Management* (Vol. 2, pp. 65–76). Zhengzhou, Henan: The Yellow River Conservancy Publishing House.

Cairns, J. E., Hellin, J., Sonder, K., Araus, J. L., MacRobert, J. F., Thierfelder, C., & Prasanna, B. M. (2013). Adapting maize production to climate change in sub-Saharan Africa. *Food Security, 5*, 345–360. https://doi.org/10.1007/s12571-013-0256-x.

CCAFS. (2014). *Climate-smart villages: A community approach to sustainable agricultural development*. Retrieved from https://ccafs.cgiar.org/publications/climate-smart-villages-community-approach-sustainable-agricultural-development.

Chahal, G. B. S., Sood, A., Jalota, S. K., Choudhury, B. U., & Sharma, P. K. (2007). Yield, evapotranspiration and water productivity of rice (Oryza sativa L.)–wheat (Triticum aestivum L.) system in Punjab-India as influenced by transplanting date of rice and weather parameters. *Agricultural Water Management, 88*, 14–27.

Challinor, A. J., Simelton, E. S., Fraser, E. D. G., Hemming, D., & Collins, M. (2010). Increased crop failure due to climate change: Assessing adaptation options using models and socio-economic data for wheat in China. *Environmental Research Letters, 5*(2010), 034012 (8pp). https://doi.org/10.1088/1748-9326/5/3/034012.

Critchley, W., & Gowing, J. (Eds.). (2012). *Water harvesting in Sub-Saharan Africa*. London: Routledge.

Dawe, D. (2005). Increasing water productivity in rice-based systems in Asia—Past trends, current problems, and future prospects. *Plant Production Science, 8*, 221–230. https://doi.org/10.1626/pps.8.221.

Dawe, D., Dobermann, A., Witt, C., Abdulrachman, S., Gines, H. C., Nagarajan, R., Satawathananont, S., Son, T. T., Tan, P. S., & Wang, G. H. (2004). Nutrient management in the rice soils of Asia and the potential of site-specific nutrient management. In A. Dobermann, C. Witt, & D. Dawe (Eds.), *Increasing productivity of intensive rice systems through site-specific nutrient management* (pp. 337–358). Enfield, NH; Los Baños: Science Publishers; International Rice Research Institute.

Dendooven, L., Patiño-Zúñiga, L., Verhulst, N., Luna-Guido, M., Marsch, R., & Govaerts, B. (2012). Global warming potential of agricultural systems with contrasting tillage and residue management in the central highlands of Mexico. *Agriculture, Ecosystems and Environment, 152*, 50–58.

Derpsch, R., Lange, D., Birbaumer, G., & Moriya, K. (2016). Why do medium-and large-scale farmers succeed practicing CA and small-scale farmers often do not? – Experiences from Paraguay. *International Journal of Agricultural Sustainability, 14*(3), 269–281.

Dobermann, A., Witt, C., Dawe, D., Abdulrachman, S., Gines, H. C., Nagarajan, R., Satawathananont, S., Son, T. T., Tan, P. S., Wang, G. H., Chien, N. V., Thoa, V. T. K., Phung, C. V., Stalin, P., Muthukrishnan, P., Ravi, V., Babu, M., Chatuporn, S., Sookthongsa, J., Sun, Q., Fu, R., Simbahan, G. C., & Adviento, M. A. A. (2002). Site-specific nutrient management for intensive rice cropping systems in Asia. *Field Crops Research, 74*, 37–66.

Dobermann, A., Abdulrachman, S., Gines, H. C., Nagarajan, R., Satawathananont, S., Son, T. T., Tan, P. S., Wang, G. H., Simbahan, G. C., Adviento, M. A. A., & Witt, C. (2004). Agronomic performance of site-specific nutrient management in intensive rice-cropping systems of Asia. In A. Dobermann, C. Witt, & D. Dawe (Eds.), *Increasing productivity of intensive rice systems through site-specific nutrient management* (pp. 307–336). Enfield, NH; Los Baños: Science Publishers; International Rice Research Institute.

Doll, P., & Siebert, S. (2002). Global modeling of irrigation water requirements. *Water Resources Research, 38*, 1–10. https://doi.org/10.1029/2001WR000355.

ETC. (2016). *Software vs. Hardware vs. Nowhere: Deere & Co. is becoming 'Monsanto in a Box*. Val David, QC: ETC Group.

FAO. (2008). *FAO statistical yearbook*. Rome: The Food and Agriculture Organization. Retrieved from http://faostat.fao.org.

FAO. (2011). *What is conservation agriculture? FAO conservation agriculture*. Rome: The Food and Agriculture Organization. Retrieved from http://www.fao.org/ag/ca/1a.html.

Farooq, M., Siddique, K. H. M., Rehman, H., Aziz, T., Dong-Jin, L., & Wahid, A. (2011). Rice direct seeding: Experiences, challenges and opportunities. *Soil and Tillage Research, 111,* 87–98.

Fischer, R. A., Santiveri, F., & Vidal, I. R. (2002). Crop rotation, tillage and crop residue management for wheat and maize in the sub-humid tropical highlands: I. Wheat and legume performance. *Field Crops Research, 79,* 107–122.

Fischer, E. M., Seneviratne, S., & Schr, C. (2007). Contribution of land-atmosphere coupling to recent European summer heat waves. *Geophysical Research Letters, 34,* 606–707.

Fischer, R. A., Byerlee, D., & Edmeades, G. O. (2014). *Crop yields and global food security: Will yield increase continue to feed the world?* Canberra, ACT: Australian Centre for International Agricultural Research. Retrieved from http://aciar.gov.au/publication/mn158.

Fosu-Mensah, B. Y., Vlek, P. L. G., & MacCarthy, D. S. (2012). Farmers' perception and adaptation to climate change: A case study of Sekyedumase District in Ghana. *Environment, Development and Sustainability, 14,* 495–505.

Gangwar, B., & Singh, A. K. (2011). *Efficient alternative cropping systems* (p. 339). Meerut: Project Directorate for Farming Systems Research, Modipuram.

Gathala, M. K., Kumar, V., Sharma, P. C., Saharawat, Y. S., Jat, H. S., Singh, M., Kumar, A., Jat, M. L., Humphreys, E., Sharma, D. K., Sharma, S., & Ladha, J. K. (2014). Optimizing intensive cereal-based cropping systems addressing current and future drivers of agricultural change in the Northwestern Indo-Gangetic Plains of India. *Agriculture, Ecosystems and Environment, 187,* 33–46.

Giller, K. E., Witter, E., Corbeels, M., & Tittonell, P. (2009). Conservation agriculture and smallholder farming in Africa: The heretic's view. *Field Crops Research, 114,* 23–34.

Giller, K. E., Corbeels, M., Nyamangara, J., Triomphe, B., Affholder, F., Scopel, E., et al. (2011). A research agenda to explore the role of conservation agriculture in African smallholder farming systems. *Field Crops Research, 124*(3), 468–472.

Giller, K. E., Andersson, J. A., Corbeels, M., Kirkegaard, J., Mortensen, D., Erenstein, O., et al. (2015). Beyond conservation agriculture. *Frontiers in Plant Science, 6,* 870.

Hellmuth, M. E., Moorhead, A., Thomson, M. C., & Williams, J. (2007). Climate risk management in Africa: Learning from practice. In M. E. Hellmuth, A. Moorhead, M. C. Thomson, & J. Williams (Eds.), *Climate and society: Climate risk management in Africa: Learning from practice* (Vol. 1). New York, NY: IRI.

Hengxin, L., Hongwen, L., Xuemin, F., & Liyu, X. (2008). The current status of conservation tillage in China. In T. Goddard, M. A. Zoebisch, Y. T. Gan, W. Ellis, A. Watson, & S. Sombatpanit (Eds.), *No-till Farming Systems* (pp. 413–428). Bangkok: WASWC. World Association of Soil and Water Conservation, Special Publication No. 3.

Hobbs, P. R. (2007). Conservation agriculture: What is it and why is it important for future sustainable food production. *Journal of Agricultural Science, 145,* 127–137.

Hobbs, P. R., & Govaerts, B. (2010). How conservation agriculture can contribute to buffering climate change. In M. P. Reynolds (Ed.), *Climate change and crop production* (pp. 117–199). Wallingford: CABI.

Huaqi, W., Bouman, B. A. M., Zhao, D., Changgui, W., & Moya, P. F. (2003). Aerobic rice in northern China: Opportunities and challenges. In B. A. M. Bouman, H. Hengsdijk, B. Hardy, P. S. Bindraban, T. P. Tuong, & J. K. Ladha (Eds.), *Water-wise rice production. Proceedings of a Thematic Workshop on Water-wise Rice Production, 8–11 April 2002, Los Baños, Philippines* (pp. 207–222). Los Baños: International Rice Research Institute.

Ismail, A. M., Singh, U. S., Singh, S., Dar, M. H., & Mackill, D. J. (2013). The contribution of submergence-tolerant (Sub1) rice varieties to food security in flood-prone rain-fed lowland areas in Asia. *Field Crops Research, 152,* 83–93.

Jalota, S. K., Singh, K. B., Chahal, G. B. S., Gupta, R. K., Chakraborty, S., Sood, A., Ray, S. S., & Panigraphy, S. (2009). Integrated effect of transplanting date, cultivar and irrigation on yield, water saving and water productivity of rice (Oryza sativa L.) in Indian Punjab: Field and simulation study. *Agricultural Water Management, 96*, 1096–1104.

Jarvis, A., Ramirez-Villegas, J., Herrera Campo, B. V., & Navarro-Racines, C. (2012). Is cassava the answer to African climate change adaptation? *Tropical Plant Biology, 5*(1), 9–29.

Jat, M. L., Chandna, P., Gupta, R., Sharma, S. K., & Gill, M. A. (2006). *Laser land leveling: A precursor technology for resource conservation. Rice-Wheat Consortium Technical Bullletin Series No. 7.* New Delhi: Rice-Wheat Consortium for the Indo-Gangetic Plains 48 pp.

Jat, M. L., Saharawat, Y. S., & Gupta, R. (2011). Conservation agriculture in cereal systems of South Asia: Nutrient management perspectives. *Karnataka Journal of Agricultural Sciences, 24*, 100–105.

Kassam, A., Friedrich, T., Derpsch, R., & Kienzle, J. (2015). Overview of the worldwide spread of conservation agriculture. Field actions science reports. *The Journal of Field Actions, 8*, 1–10.

Kirk, G. J. D., Greenway, B. J., Atwell, B. J., Ismail, A. M., & Colmer, T. D. (2014). Adaptation of rice to flooded soils. In U. Lüttge, W. Beyschlag, & J. Cushman (Eds.), *Progress in botany* (Vol. 75). Berlin: Springer. https://doi.org/10.1007/978-3-642-38797-5_8.

Kumar, V., & Ladha, J. K. (2011). Direct seeding of rice: Recent developments and future research needs. *Advances in Agronomy, 111*, 297–313.

Lampayan, R. M., Bouman, B. A. M., Flor, R. J., & Palis, F. G. (2014). Developing and disseminating alternate wetting and drying water saving technology in the Philippines. In A. Kumar (Ed.), *Mitigating water-shortage challenges in rice cultivation: Aerobic and alternate wetting and drying rice water-saving technologies.* Manila: IRRI, Asian Development Bank.

Lampayan, R. M., Samoy-Pascual, K. C., Sibayan, E. B., Ella, V. B., Jayag, O. P., Caban-gon, R. J., & Bouman, B. A. M. (2015a). Effects of alternate wetting and drying (AWD) threshold level and plant seedling age on crop performance, water input and water productivity of transplanted rice in Central Luzon. *Paddy and Water Environment, 13*, 215. https://doi.org/10.1007/s10333-014-0423-5.

Lampayan, R. M., Rejesus, R. M., Singleton, G. R., & Bouman, B. A. M. (2015b). Adoption and economics of alternate wetting and drying water management for irrigated lowland rice. *Field Crops Research, 170*(2015), 95–108.

Liu, C., Cutforth, H., Chai, Q., & Gan, Y. (2016). Farming tactics to reduce the carbon footprint of crop cultivation in semiarid areas. A review. *Agronomy for Sustainable Development, 36*, 69. https://doi.org/10.1007/s13593-016-0404-8.

Lybbert, T. J., & Sumner, D. A. (2012). Agricultural technologies for climate change in developing countries: Policy options for innovation and technology diffusion. *Food Policy, 37*, 114–123.

Mahajan, G., Timsina, J., & Singh, K. (2011). Performance and water use efficiency of rice relative to establishment methods in northwestern Indo-Gangetic Plains. *Journal of Crop Improvement, 25*, 597–617.

MAIZE. (2013). *MAIZE CRP annual report.* Mexico: CIMMYT.

MAIZE. (2015). *MAIZE CRP annual report.* Mexico: CIMMYT.

MAIZE. (2016). *MAIZE CRP annual report.* Mexico: CIMMYT.

Majumdar, K., Zingore, S., Garcia, F., Correndo, A., Timsina, J., & Johnston, A. M. (2017). Improving nutrient management for sustainable intensification of maize. In D. J. Watson (Ed.), *Achieving sustainable cultivation of maize.* Cambridge: Burleigh Dodds Publishing.

Malesu, M. M., De Leeuw, J., & Oduor, A. (2012). *Water harvesting experiences from the SearNet (2003–2012).* Retrieved January 15, 2016, from http://whater.eu/pluginfile.php/137/mod_page/content/37/Malesu_WaterharvestingExperiencesfromtheSearnet2003-2012_IRC.

Miro, B., & Ismail, A. M. (2013). Tolerance of anaerobic conditions caused by flooding during germination and early growth in rice (OryzasativaL.). *Frontiers in Plant Science, 4*, 269.

Nelson, A., Wassmann, R., Sander, B. O., & Palao, L. K. (2015). Climate-determined suitability of the water saving technology "alternate wetting and drying" in rice systems: A scalable methodology demonstrated for a province in the Philippines. *PLoS One, 10*(12), e0145268. https://doi.org/10.1371/journal.pone.0145268.

Pachpute, J. S., Tumbo, S. D., Sally, H., & Mul, M. L. (2009). Sustainability of rainwater harvesting systems in rural catchment of Sub-Saharan Africa. *Water Resources Management, 23*(13), 2815–2839.

Palis, F. G., Lampayan, R. M., & Bouman, B. A. M. (2014). Adoption and dissemination of alternate wetting and drying technology for boro rice cultivation in Bangladesh. In A. Kumar et al. (Eds.), *Mitigating water-shortage challenges in rice cultivation: Aerobic and alternate wetting and drying rice water-saving technologies*. Manila: IRRI and Asian Development Bank.

Pampolino, M. F., Manguiat, I. J., Ramanathan, S., Gines, H. C., Tan, P. S., Chi, T. T. N., Rajendran, R., & Buresh, R. J. (2007). Environmental impact and economic benefits of site-specific nutrient management (SSNM) in irrigated rice systems. *Agricultural Systems, 93*(2007), 1–24.

Pretty, J., Noble, A. D., Bossio, D., Dixon, J., Hine, R. E., Penning de Vries, F. W. T., & Morison, J. I. L. (2006). Resource-conserving agriculture increases yields in developing countries. *Environmental Science & Technology, 40*, 1114–1119.

Rejesus, R. M., Martin, A. M., & Gypmantasiri, P. (2013). Meta-impact assessment of the irrigated rice research consortium. In *Special IRRI Report*. Los Baños: International Rice Research Institute.

RICE. (2016). *RICE CRP proposal*. Las Banjos: IRRI.

Rijsberman, F. R. (2006). Water scarcity: Fact or fiction? *Agricultural Water Management, 80*, 5–22.

Rosegrant, M. W., Ringler, C., & Zhu, T. (2009). Water for agriculture: Maintaining food security under growing scarcity. *Annual Review of Environment and Resources, 34*, 205–222. https://doi.org/10.1146/annurev.environ.030308.090351.

Sander, B. O., Samson, M., & Buresh, R. J. (2014). Methane and nitrous oxide emissions from flooded rice fields as affected by water and straw management between rice crops. *Geoderma, 235–236*, 355–362. https://doi.org/10.1016/j.geoderma.2014.07.020.

Sapkota, T. K., Majumdar, K., Jat, M. L., Kumara, A., Bishnoia, D. K., McDonald, A. J., & Pampolino, M. (2014). Precision nutrient management in conservation agriculture based wheat production of Northwest India: Profitability, nutrient use efficiency and environmental footprint. *Field Crops Research, 155*(2014), 233–244.

Sofoluwe, N. A., Tijani, A. A., & Baruwa, O. I. (2011). Farmers' perception and adaptation to climate change in Osun State, Nigeria. *African Journal of Agricultural Research, 6*(20), 4789–4794.

Tadesse, M., Shiferaw, B., & Erenstein, O. (2015). Weather index insurance for managing drought risk in smallholder agriculture: Lessons and policy implications for Sub-Saharan Africa. *Agricultural and Food Economics, 3*, 26.

Thierfelder, C., & Wall, P. C. (2009). Effects of conservation agriculture techniques on infiltration and soil water content in Zambia and Zimbabwe. *Soil and Tillage Research, 105*, 217–227.

Thierfelder, C., & Wall, P. C. (2010). Investigating Conservation Agriculture (CA) Systems in Zambia and Zimbabwe to mitigate future effects of climate change. *Journal of Crop Improvement, 24*, 113–121.

Thierfelder, C., Rusinamhodzi, L., Setimela, P., Walker, F., & Eash, N. S. (2016). Conservation agriculture and drought-tolerant germplasm: Reaping the benefits of climate-smart agriculture technologies in Central Mozambique. *Renewable Agriculture and Food Systems, 156*, 99–109.

Thierfelder, C., Chivenge, P., Mupangwa, W., Rosenstock, T. S., Lamanna, C., & Eyre, J. (2017). How climate-smart is conservation agriculture (CA): Its potential to deliver on adaptation, mitigation and productivity on smallholder farms in southern Africa. *Food Security., 9*, 537. https://doi.org/10.1007/s12571-017-0665-3.

Tuong, T. P., & Bouman, B. A. M. (2003). Rice production in water scarce environments. In J. W. Kijne, R. Barker, & D. Molden (Eds.), *Water productivity in agriculture: Limits and opportunities for improvement* (pp. 53–67). Wallingford: CABI Publishing.

Ussiri, D. A. N., & Lal, R. (2009). Long-term tillage effects on soil carbon storage and carbon dioxide emissions in continuous corn cropping system from an alfisol in Ohio. *Soil and Tillage Research, 104*, 39–47.

Valbuena, D., Erenstein, O., Homann-Kee Tui, S., Abdoulaye, T., Claessens, L., Duncan, A. J., et al. (2012). Conservation agriculture in mixed crop–livestock systems: Scoping crop residue trade-offs in sub-Saharan Africa and South Asia. *Field Crops Research, 132*, 175–184.

Verhulst, N., Govaerts, B., Verachtert, E., Castellanos-Navarrete, A., Mezzalama, M., Wall, P., et al. (2010). Conservation agriculture, improving soil quality for sustainable production systems? In R. Lal & B. A. Stewart (Eds.), *Advances in soil science: Food security and soil quality* (pp. 137–208). Boca Raton, FL: CRC Press.

Vermeulen, S., Aggarwal, P., Ainslie, A., Angelone, C., Campbell, B., Challinor, A., Hansen, J., Ingram, J., Jarvis, A., Kristjanson, P., Lau, C., Nelson, G., Thornton, P., & Wollenberg, E. (2012). Options for support to agriculture and food security under climate change. *Environmental Science & Policy, 15*, 136–144.

Vörösmarty, C. J., McIntyre, P. B., Gessner, M. O., Dudgeon, D., Prusevich, A., Green, P., et al. (2010). Global threats to human water security and river biodiversity. *Nature, 467*, 555–561. https://doi.org/10.1038/nature09549.

Wada, Y., LPH, V. B., & Bierkens, M. F. P. (2011). Modelling global water stress of the recent past: On the relative importance of trends in water demand and climate variability. *Hydrology and Earth System Sciences, 15*, 3785–3808. https://doi.org/10.5194/hess-15-3785-2011.

Wakeyo, M. B. (2012). *Economic analysis of water harvesting technologies in Ethiopia*. Retrieved from http://edepot.wur.nl/240909.

Waqar, A., Jehangir, I. M., Shehzad, A., Mustaq, A. G., Maqsood, A., Riaz, A. M., Muhammad, R. C., Asad, S. Q., & Hugh, T. (2007). *Sustaining crop water productivity in rice–Wheat systems of south asia: A case study from the Punjab, Pakistan*. Colombo: International Water Management Institute, IWMI.

Weller, S., Janz, B., Jörg, L., Kraus, D., Racela, H. S. U., Wassmann, R., Butterbach-Bahl, K., & Kiese, R. (2016). Greenhouse gas emissions and global warming potential of traditional and diversified tropical rice rotation systems. *Global Change Biology, 22*, 432–448. https://doi.org/10.1111/gcb.13099.

Westengen, O. T., & Brysting, A. K. (2014). Crop adaptation to climate change in the semi-arid zone in Tanzania: The role of genetic resources and seed systems. *Agriculture & Food Security, 3*, 3.

World Bank. (2014). *World development report 2014*. Washington, DC: World Bank.

Yadvinder-Singh, Humphreys, E., Kukal, S. S., Singh, B., Kaur, A., Thaman, S., Prashar, A., Yadav, S., Kaur, N., Dhillon, S. S., Smith, D. J., Timsina, J., & Gajri, P. R. (2009). Crop performance in a permanent raised bed rice–wheat cropping system in Punjab, India. *Field Crops Research, 110*, 1–20.

Yadvinder-Singh, Kukal, S. S., Jat, M. L., & Sidhu, H. S. (2014). Improving water productivity of wheat-based cropping systems in South Asia for sustained productivity. *Advances in Agronomy, 127*, 157–120. https://doi.org/10.1016/B978-0-12-800131-8.00004-2.

Chapter 11
Water Management Technology for Adaptation to Climate Change in Rice Production: Evidence of Smart-Valley Approach in West Africa

Aminou Arouna and Aristide K. A. Akpa

Abstract Low productivity is the main characteristic of agriculture in sub-Saharan Africa. Adverse effects of climate change increasingly reduce both productivity and production. Rice plays an important role in the food security of population. However, rice production faces many constraints, including low water control and soil fertility management. In order to improve water control and soil management and increase the productivity of local rice production in the context of climate change, a new technology (smart-valley approach) was introduced in Benin and Togo since 2010. The aim of this study is to assess the adoption, the diffusion and impact of smart-valley approach. Data were collected from 590 rice farming households in Benin and Togo. Results revealed that land tenure, total available area, paddy price and production in the lowland increase the adoption of smart-valley approach. Adoption of smart-valley approach increased from 110 ha in 2012 to 474 ha in 2014. In addition, the adoption enables producers to increase the yield by 0.9 tons ha^{-1}, the net income by USD 267 per hectare under the condition of climate change. The study suggests that large diffusion and training on the technology would help for adaptation to climate change and improving their livelihood of smallholder rice farmers.

Keywords Smart-valley approach · Climate change · Rice · Diffusion · Impact · West Africa

---

A. Arouna (✉)
Africa Rice Center (AfricaRice), Bouaké, Côte d'Ivoire
e-mail: a.arouna@cgiar.org

A. K. A. Akpa
Africa Rice Center (AfricaRice), Cotonou, Bénin

© Springer Nature Switzerland AG 2019
A. Sarkar et al. (eds.), *Sustainable Solutions for Food Security*,
https://doi.org/10.1007/978-3-319-77878-5_11

11.1 Introduction

Climate change is increasingly one of the most serious threats which will have significant impacts on natural resources, ecosystem, and biodiversity. Climate change is likely to trigger food insecurity, human migration, economic, and social depression, environmental and political crisis, thereby affecting development (IPCC 2007; World Bank 2010). The impacts of a changing climate are already being felt, with more droughts, more floods, more strong storms, and more heat waves—taxing individuals, firms, and governments, drawing resources away from development (World Bank 2010). Agriculture is the primary medium through which climate change will impact people, ecosystems, and economies. Agriculture is expected to pay a significant cost of the damage caused by climate change (FAO 2009). Without extensive adaptation the effects of climate change on agriculture is expected to exacerbate Africa's deepening food crisis, narrowing channels of food access and slowing efforts to expand food productivity. In agricultural production, rice is one of the crops whose productivity may be highly affected by climate change without new technologies for adaptation. Indeed, rice consumes water more than any other crop.

Rice is the most important basic crop in developing countries, and the staple food of more than half of the world's population. Worldwide, more than 3.5 billion people depend on rice to satisfy more than 20% of their daily calorie intake (IRRI et al. 2010). In sub-Saharan Africa (SSA), rice consumption is continuously increasing due to population growth and changing of consumption habit. Rice consumption has increased at an average annual growth rate of 4% over the last decade (USDA 2016). Despite different policies to boost the local production, especially after the 2007/2008 global food crisis, local production on in SSA is not sufficient to meet demand of increasing population. The annual gap of about 11 million tons of milled rice is largely fulfilled by importation from Asian countries.

In SSA, there is a big difference between potential and actual yields (2.00 tons ha^{-1} in SSA against 8–10 tons ha^{-1} in Asia and Egypt) (AfricaRice 2011). However, various and suitable ecologies are available for rice production in the region. Potential areas for rice production in inland valley were evaluated at about 250 million hectares offering important opportunities for irrigation. Inland valleys are quite spread in West Africa and provide advantageous biophysical conditions for expansion and sustainable rice intensification (AfricaRice 2010).

However, current rice productivity in inland valleys is often low because of various human and natural constraints such as poor water control, weed invasion, low soil fertility, increasing soil degradation, labor deficiency, limited access to information, technology, resource and credit, and exposure to the risk of water-borne diseases, e.g., malaria and bilharzia. These constraints substantially limit rice productivity especially with increasing effect of climate change.

Improving water control as one adaptation measure to climate change is a precondition to enhance and sustain rice production in the inland valleys. The smart-valley approach[1], soil and water management technology for rice production in inland valleys, is an opportunity for rice production intensification through improved water control and soil fertility management. Rice production based on smart-valley approach can help to overcome soil fertility issues through improvement of the fertilization process and increasing water resources management. Indeed, smart-valley approach entails good agricultural practices such as land leveling, bunding, and puddling in combination with good water management (AfricaRice 2015). The potential of smart-valley for rice cultivation is enormous in SSA. Over on 250 million hectares of inland valleys available, only 20 million hectares developed with smart-valley technology would produce an additional food for more than 300 million people (Wakatsuki et al. 2011). Wakatsuki and Masunaga (2005) reported that the potential for rice production base smart-valley approach in West Africa is huge to stimulate the long-awaited green revolution. This revolution may be achieved through agro-ecological conditions in the central region of West Africa that are quite similar to those major rice-producing regions (northern part) of Thailand. This performance would help African countries to overcome their heavy dependence on rice imports and achieve self-sufficiency in rice (AfricaRice 2010).

Smart-valley approach was introduced in Benin and Togo through the Sawah, Market Access and Rice Technologies for Inland Valleys (SMART-IV) project launched by Africa Rice Center (AfricaRice) and national partners in 2010. The project started within the installation of demonstration sites in both countries. Since 2012, the smart-valley approach run into dissemination phase due to its low cost of its implementation and good appreciation from farmers (AfricaRice 2015). During the course of the SMART-IV project different surveys were carried out to show the agronomic characteristics of smart-valley approach (AfricaRice 2015; Zwart 2014; Rodenburg 2013). However, no recent study showed the adoption of smart-valley approach, its dissemination and impact in both Benin and Togo. Therefore, the objective of this paper is threefold: analyze the adoption and dissemination process of smart-valley approach; analyze the determinants of its adoption; and assess its impact on yield, income, poverty, and food security of adopters in the context of climate change.

11.2 Description of the Smart-Valley Approach

The new smart-valley approach was developed and validated during years 3–5 (2013–2015) of the SMART-IV project. This development approach can be implemented in inland valleys that are fully rain-fed or where water additional water resources are available for irrigation. The approach is entirely participatory throughout five steps from the sensitization to the design and implementation of the system. Instead of defining a blueprint for development of a system, this approach allows adaptation under actual local conditions to meet farmers' demands.

---

[1] The Smart-valley approach was introduced in Ghana and Nigeria under the name of "Sawah systems".

Smart-valley approach is a participatory, low-cost and easy to replicate for water and land management system for rice production. It also offers the advantage of being implementable within one agricultural season. More specifically, the smart-valley approach includes the following three pillars: drainage canals, irrigation infrastructure (where water resources are available), and bunded and leveled rice fields in the inland valleys (AfricaRice 2015). Figure 11.1 shows a field of smart-valley approach. Farmers can build and maintain those low-cost water control structures entirely on their own (Rodenburg 2013). Smart-valleys lead to greater water storage in the fields and less field run-off through bunding and drainage facilities. The climate change risks for farmers, such as drought and flooding are reduced and they are more willing to invest in agricultural inputs such fertilizers and good seeds. The project SMART-IV creates demonstration sites and trains both field technicians and lead-farmers. Additionally, farmers are guided in the adoption of good agricultural practices aiming at improving rice production and quality. The field phase of the project actually began in 2012. During the 2012/2013 growing season, over 60 demonstration plots were installed in Benin and Togo. Also, more than 100 field technicians and hundreds of farmers were trained (Zwart 2014). Initial studies suggest a significant uptake in the technology (Zwart 2014; Rodenburg 2013). However, no rigorous analysis on the adoption, dissemination, and impact of smart-valley approach has been performed.

Fig. 11.1 A field of smart-valley approach in Benin

Fig. 11.2 Map of Benin and Togo

11.3 Methodology

11.3.1 Study Area

The study was conducted in Benin and Togo which are the pilot countries of SMART-IV project. Benin and Togo are French-speaking and neighboring country in West Africa (Fig. 11.2).

Togo and Benin are quite similar in terms of agro-ecologic condition and economic development. In 2015, the Gross Domestic Product (GDP) per capita was estimated at USD 827.20 and 547.97 in Benin and Togo, respectively (Trading Economics 2016). In both countries, agriculture occupies some 80% of the population (FAO 2016) through the sector contributes to only 36.3% and 42% of GDP in Benin and Togo, respectively (CIA World Factbook 2016). Rice is one of the major staple foods in both countries and domestic production does not meet national supply. However, the climate Change is increasingly in the two countries. The trend of rainfall over the last 25 years (1991–2015) shows that great variability of climate parameters in both Benin and Togo. Climate change in Benin is related to both rainfall and temperature. Over the last 25 years, the annual rainfall has decreased in Benin (Fig. 11.3). Indeed, the precipitation index[2] depicts a negative trend in Benin. In contrary, the precipitation

---

[2]The precipitation index is calculated as ($X_i - \overline{X} / \delta$) where X_i is year value, $\overline{X}$ is the series average, and δ is standard deviation.

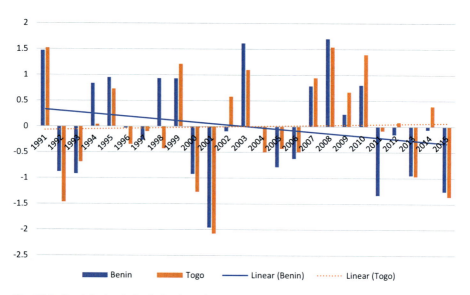

Fig. 11.3 Precipitation index in Benin and Togo

index depicts a slight positive trend in Togo (Fig. 11.3). However, the precipation indexes[3] for temperature shows both minimum and maximum temperatures have increased in the two countries (Figs. 11.4 and 11.5). It results that over the last 25 years, climate change in the two countries were observed through the decrease of rainfall (especially in Benin) and the increasing of both minimum and maximum temperatures have increased in the two countries. This is in line with the estimation of Gnangle et al. (2011) demonstrating an increasing of temperature and decreasing of rainfall in Benin.

Benin and Togo are ideal settings for the study of Smart-valley approach, before its possible rollout to other countries in West Africa. Chabi et al. (2010) identified a total of 181 inland valleys over 1124 ha in the central region of Benin using satellite imagery. About 60% and <5% of those plots, respectively, cover between 0–5 ha and 20–25 ha in size. As such, Benin has an important potential for rice production of which barely 10% is currently utilized. Nation-wide, the amount of lowlands available is estimated to 205,000 ha, to which over 110,000 ha can be added with irrigation. Togo has inland valleys that are potentially suitable for rice production. It is estimated that there is about 185,000 ha of inland valleys in Togo. Water management, technology transfer to rice producers, soil fertility management, as well as an extension are essential to harnessing that potential of inland valleys in both countries.

---

[3] The temperature is calculated as the precipitation index.

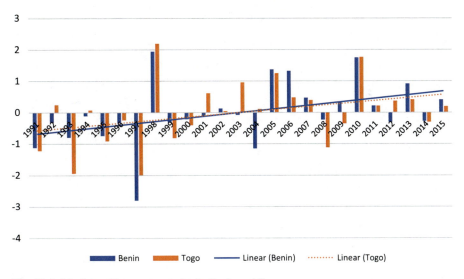

Fig. 11.4 Maximum Temperature index in Benin and Togo

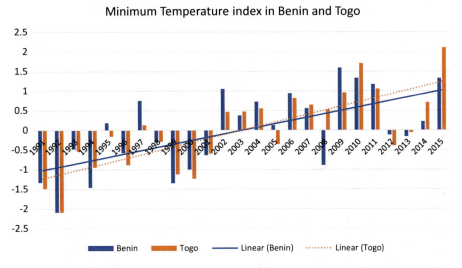

Fig. 11.5 Minmum temperature index in Benin and Togo

11.3.2 Sampling, Data Collection and Processing

For data collection on the diffusion process of smart-valley approach, the snowball technique was used (Fig. 11.6). It consists to get information on adoption status from one actor depending on information provided by previous actor. The principle of this technique is to start data collection from the lead-farmers or other

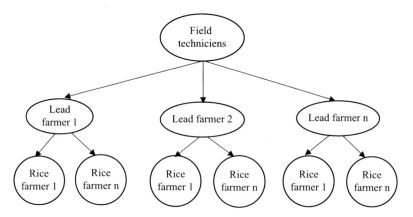

Fig. 11.6 Illustration of snowball approach

farmers who participated to the demonstration to identify adopters of the innovation using the social network. For impact assessment, two-stage stratified random sampling was used. In the first stage, 39 villages were selected randomly from the list of rice producing villages in Benin and 20 villages in Togo. In each village, ten rice-producing households were randomly selected. A total of 590 rice farming households were surveyed.

Socioeconomic and demographic data were collected. Data were collected using a web application developed by AfricaRice (Mlax) on tablet computer. The computer-based data collection avoided many biases associated with paper-based questionnaires, such as mistakes in answer recording, changing of values of variables, recoding test answers for numerical variables, etc. Data were analyzed using STATA 14 software.

11.3.3 Data Analysis

Adoption of smart-valley approach depends on knowledge about its existence. When a farmer has a complete information on the innovation (smart-valley approach), he is able to decide whether to adopt it or not. However, this decision is influenced by its socioeconomic and demographic characteristics. In addition, the decision also depends on the utility that he is expecting to get from the adoption. The dependent variable is the adoption status of the farmer, with 1 for farmers who adopt smart-valley approach, and 0 for non-adopters. In this study, a given farmer is considered as adopter if there is at least one rice plot improved using smart-valley approach among its rice plots. In practice, the Probit or Logit model is often used. Both models give similar results, but this study used the Probit model.

To assess the impact of smart-valley approach, Local Average Treatment Effect (LATE) estimator proposed by Abadie (2003) was used. Indeed, LATE takes into account bias due to both observed and unobserved characteristics. Impact assessment aims to estimate on average the situation of smart-valley adopters if they did not decide to adopt it. A simple method is to determine the difference for the selected outcome. Outcomes selected for this study are yield, income, technical efficiency and food security[4]. But the interpretation of the difference as a causal relationship between adoption status lead to selection bias (Heckman 2010). An unbiased difference will be determined if the groups are similar, and the difference is only the adoption of smart-valley approach. In order to solve this problem of selection bias and generate unbiased bias impact, Imbens and Angrist (1994) proposed the local average treatment effect (LATE) which is the average impact for the subpopulation of potential adopters. The estimation of the LATE requires the use of the instrumental variable method (Imbens and Wooldridge 2009; Heckman and Vytlacil 2005; Abadie 2003). This method assumes the existing of variable z called instrument that directly affects the adoption status, but indirectly the outcomes y_1 and y_0, once the independent variables (x) are controlled.

The instrumental variable (z) used in this study is the knowledge of smart-valley approach. Indeed, the choice of this variable is due to the fact that knowledge affects the decision whether to adopt the smart-valley approach. So only farmer having information on the technology are able to adopt it. However, knowledge alone does not directly influence the outcome. Knowledge therefore meets the definition of the instrumental variable given by Abadie (2003) and Heckman (2010).

11.4 Determinants of Adoption of Smart-Valley Approach

The adoption model of the smart-valley approach was estimated using the Probit model. The maximum likelihood test indicated that the model is globally significant at 1%. Five variables influencing positively or negatively the adoption of smart-valley approach were identified (Table 11.1). These variables are total available area, availability of inland valleys, secure land tenure, selling price of paddy, and membership in farmers' association.

The total available area affects positively and significantly adoption of smart-valley approach. Availability of agricultural area for the household increases the adoption. The price represents an important determining factor for any economic actor to take a decision. It determines the profitability of the activities. Price of paddy rice in the market has a positive effect on the adoption of smart-valley approach. Producers tend to adopt new agricultural technologies in expecting to get advantage from market price and achieve economics scale. High market price encourages rice farmers to invest in smart-valley approach.

---

[4] The analysis of the food and nutritional situation of rice farming households is based on the food consumption score (FCS). The FCS is a composite indicator developed by the World Food Programme (WFP). It is usually calculated to take into account dietary diversity, frequency, and relative nutritional intake of foods and food groups consumed by a given household (WFP 2009).

Safe access to land remains an important aspect for adoption of new agricultural technologies. Results showed positive effect of the land tenure (inheritance and purchase) on adoption of smart-valley approach. Producers with secure access to land will take the risk to invest in soil and water management. This is consistent with the finding of Oladele et al. (2010) who reported that the smart-valley approach increases if the adopter has access to land by inheritance, purchase, or agrees for a long-term lease contract for the land with the land owner. The types of tenure included in the model (inheritance and purchase) increase the probability of adoption by 19% and 89%, respectively.

Smart-valley approach induces higher yield in inland valleys growing environment. This is a suitable ecology for better water control. The results reveal that the availability of inland valleys for the household influenced positively the adoption of smart-valley approach. Against expectations, the variable "membership of farmer group/association" influences negatively the adoption of smart-valley approach. This is in contradiction to the result obtained by Jagwe et al. (2010) and Mathenge et al. (2010) who found that groups and associations represent an important platform for social capital improvement through which small producers might easily have adopt innovations.

Table 11.1 Determinants of the adoption of the smart-valley approach

Independent variables	Coefficients	Standard error	Marginal effect
Age (Year)	−0.001	0.014	0.00
Rice production experience (Year)	−0.010	0.018	−0.002
Gender (Male = 1, Female = 0)	−0292	0.308	−0.067
Production in the inland valleys (1 = yes, 0 = no)	1228[a]	0.300	0.365[a]
Access to land by inheritance (1 = yes, 0 = no)	1141[b]	0.449	0.189[a]
Access to land by purchase (1 = yes, 0 = no)	3268[a]	0.583	0.890[a]
Access to credit (1 = yes, 0 = no)	−0234	0.432	−0.045
Total farm area available (ha)	0044[a]	0.013	0.009[a]
Having receive agricultural training (1 = yes, 0 = no)	−0214	0.258	−0.043
Association membership (1 = yes, 0 = no)	−0435[c]	0.253	−0.098
Paddy market price (FCFA)	0.007[a]	0.003	0.001[a]
Yield (tons ha^{-1})	0.144	0.127	0.030
Constant	3.482	0.995	–
Number of Observations	590		
Log maximum likelihood	−74.44		
Wald chi-square (9)	103.92[a]		
McFadden Pseudo-R^2	0.411		

[a]= significant at 1%
[b]= significant at 5%
[c]= Significant at 10%

11.5 Diffusion of Smart-Valley Approach

Based on characteristics defined by Rogers' diffusion theory (Rogers 2003), smart-valley approach diffusion follows different channels and has different categories of actors (Fig. 11.7). Differences between channels depend on the number of actors. The diffusion was leading by the office of lowlands improvement (Direction du Genie Rural, DGR) in Benin and National Research Institute (ITRA) in Togo. Four main actors were involved in the diffusion of the smart-valley approach: research institute, technicians (research, extension, and NGO), lead-farmers, and rice farmers. Farmer to farmer played important role in the diffusion of the smart-valley approach. This could be the main push factor for the large diffusion of the smart-valley approach. The innovators and early adopters were mainly lead-farmers and are in channels 1, 3, and 4. Field survey confirmed level of importance of each channel used for diffusion (Fig. 11.8). It was found that channel 1 was the most used channel. Indeed, 32% of farmers confirmed that they adopt the smart-valley approach through this channel. It was followed by channels 2 and 4, within a proportion of 23%. The channel 3 was used by 19% while the channel 5 was used by only 3% of farmers. These results showed that lead-farmers play important role in adoption of innovation in Africa. This can be explained by the fact that lead-farmers meet the characteristics of innovators and early adopters as defined by Rogers

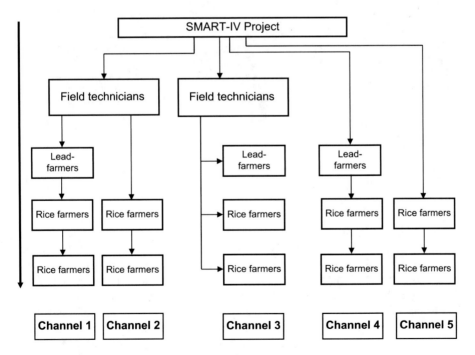

Fig. 11.7 Diffusion channels identified in the field

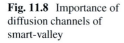
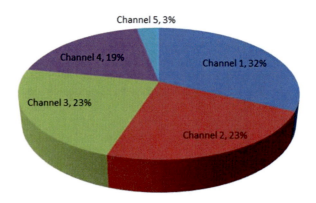

Fig. 11.8 Importance of diffusion channels of smart-valley

Fig. 11.9 Google Earth image of sites of Zoungo (Benin)

(Rogers 2003). Indeed, lead-farmers are willing to take risks, have the highest social status, have financial liquidity, are social and have closest contact to scientific sources and interaction with other innovators.

During the field survey, all fields under smart-valley approach were tracked using the Global Positioning System tool (GPS) in order to get the actual area of the fields, their location and map (Fig. 11.9) in December 2014. The SMART-IV project through the field technicians developed demonstration fields and allow farmer to adopt. For instance, in village of Zoungo in Benin, the demonstration site installed by the researcher of the project was about 1–2 ha on site of Zoungo 1 (left part of the road on Fig. 11.9). But during the survey, the size of the developed area increased to 10 ha on site of Zoungo 1. Due the lack of suitable land on site of Zoungo 1, others farmers developed their field using smart-valley approach on site of Zoungo 2 (right part of the road on Fig. 11.9).

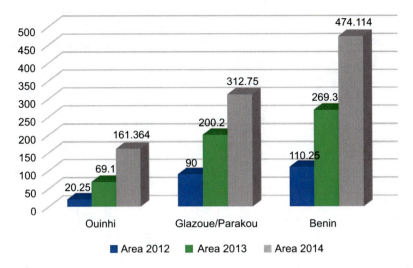

Fig. 11.10 Cumulated area developed in Benin between 2012 and 2014

The demonstration sites were installed in different regions with large rice production potential, and offering suitable conditions for smart-valley approach development. From 2012, the innovation ran into the dissemination phase. Total area developed under smart-valley technology in Benin in 2012 was about 110 ha. This has increased reaching 474 ha in 2014 in Benin (Fig. 11.10). This indicated the quick diffusion of the smart-valley approach, which could be due to its lost cost and other characteristics matching the expectation of farmers.

Results showed that smart-valley approach has been adopted in both Benin and Togo. About 1486 rice farmers (668 male and 818 female) have adopted the smart-valley approach in 2014. Smart-valley approach has been adopted on 233 ha (101 ha in Benin and 132 ha in Togo). This value is obtained by tracking of the rice farms using GPS. However, the area reported by farmers is much greater. For instance, the area reported by farmer in Benin is about 474 ha for 2014 (Fig. 11.10). The difference may be explained by the fact that some fields considered by farmers as smart-valleys were discarded during the tracking. Smart-valley approach has also spread in space. About 139 sites were used to develop smart-valley approach in 2014.

11.6 Impact of Smart-Valley Approach on Rice Yield, Income, and Food Security

The results of impact of smart-Valley technology adoption on rice yield, income, and food security are presented in Table 11.2. The method with interaction was used to generate an impact on yield which is estimated at 0.92 tons ha^{-1}. This result

Table 11.2 Impact of smart-valley technology on rice yield, income, and food security

Parameter LARF	Yield (tons ha^{-1})	Income (USD)	Food security (FCS)
Overall			
LATE	0.92 (44.68)[a]	267.35 (17.57)[a]	10.46 (3.50)[a]
Naive method			
Difference	0.38 (2.82)[a]	239.86 (4.13)[a]	9.04 (3.91)[a]
Adopters	2.05 (17.56)[a]	347.12 (6.45)[a]	71.07 (35.13)[a]
Non-adopters	1.67 (24.96)[a]	107.26 (4.21)[a]	62.02 (55.45)[a]
Impact by gender			
Men			
LATE	0.95 (31.11)[a]	306.04 (11.49)[a]	8.75 (2.31)[b]
Naive method			
Difference	0.36 (2.45)[b]	243.44 (4.15)[a]	10.62 (4.50)[a]
adopters	2.12 (16.79)[a]	381.95 (7.33)[a]	72.06 (34.64)[a]
Non-adopters	1.76 (22.79)[a]	138.50 (5.17)[a]	61.44 (54.85)[a]
Women			
LATE	0.88 (153.78)[a]	257.95 (202.56)[a]	20.57 (40.10)[a]
Naive method			
Difference	0.15 (0.54)	197.68 (1.18)	6.07 (1.23)
adopters	1.58 (6.51)[a]	156.32 (0.98)	70.35 (15.31)[a]
Non-adopters	1.43 (10.77)[a]	−41.35 (−0.82)	64.28 (35.99)[a]

[a]= significant at 1%. Values in parenthesis are z-statistics
[b]= significant at 5%;

implied that adoption of smart-valley has a positive and significant impact on the rice yield of current and potential adopters. In addition, the results show an average a yield of 2 tons ha^{-1} for adopters of smart-valley, and 1.67 tons ha^{-1} for non-adopters. This low yield for adopters was due to the adverse effects of climate change. So without adoption of smart-valley approach as an adaptation measure all farmers would lose on average 920 kg of paddy rice per hectare. Non-adopters have, therefore, a potential to increase their efficiency by adopting the smart-valley approach. It results that smart-valley approach can be used as an adaptation measure to climate change. This adaptation measure will also help farmers to improve their income and livelihood.

The study also broke down the global impact by gender. Results showed heterogeneity of the impact among producers. The impact varies according to gender. The impact is greater for men than for women (0.95 tons ha^{-1} for men and 0.88 tons ha^{-1} for women). High for men can be explained by the labor intensity that requires the smart-valley approach. Indeed, with low level of mechanization in Benin and Togo, men labor will apply better smart-valley approach than women labor.

Estimation showed that the impact of adoption of the smart-valley on the rice income is USD 267 per hectare. Indeed, the smart-valley approach has a positive impact on rice income under climate change. Non-adopters farmers have the potential to increase net operating income through adoption of the smart-valley technology. However, the impact is not homogeneous between men and women. Table 11.2 showed that the impact on men is higher than women. It was USD 258 per hectare and USD 306 per hectare for women and men, respectively. The difference could be explained by the fact that women are getting better price for their paddy compare to men. These results are in line with the estimation of Arouna et al. (2017).

Results of estimates of the impact of smart-valley on food and nutritional security of households are also presented in Table 11.2. Results showed that the adoption of the smart-valley approach allowed adopters to improve their food and nutrition security. Indeed, adoption of the smart-valley approach increases the food consumption score (FSC) by 10.46. This increase is highly significant (at 1% level) and shows that the adoption of the smart-valley technology improves food and nutritional security of households. This can be explained by the fact that rice is a staple in both Benin and Togo. Therefore, by increasing yield, the adoption of smart-valley approach allows rice farmers to have more food for their households in the condition of climate change. However, two potential constraint of the smart-valley need to be considered by future research. Indeed, mosquito breeding might be a problem in irrigated rice in general and with the adoption of the smart-valley approach in particular. Therefore, future researches need to test the new irrigation method such as alternative wet and dry method possibility to reduce mosquito breeding in irrigation rice field. In addition, after continuous use of the same land under smart-valley, the problem of barren or less useful needs to be considered. Although, this problem is less problematic in Benin and Togo due to low rate of salinity, future research need to address this issue especially in the new countries such Sierra Leone and Liberia where AfricaRice aims to extend the smart-valley approach. Indeed, salinity is a major stress for rice in these two countries.

11.7 Conclusion

With increasing negative effect of climate change, smallholder farmers need to adopt new technologies. This study analyzed the adoption of the smart-valley approach for water and soil management as adaptation measure in the context of climate change. The study also assessed the diffusion and impact of the new smart-valley approach. Surveys were conducted in Benin and Togo where the smart-valley approach was introduced. Results showed that five main variables affect the adoption of the smart-valley approach: total area available, tenure of land, price of paddy, membership of farmers' association and availability of inland valley. The smart-valley approach was well disseminated due to its low cost and the participatory approach of dissemination. The total area improved using the smart-valley approach rise from 110 ha in 2012 to 474 ha in 2014. Estimates of the impact on the indicators selected were significantly positive. Indeed, the adoption of the smart-valley technology increases the yield, income, and food security for adopters by 0.92 tons ha^{-1}, USD 267 per hectare and 10.46, respectively. These results showed that the adoption of smart-valley approach can help farmers to mitigate the negative effect of climate change. These results indicate that non-adopters households have potential to improve their livelihood with the adoption of smart-valley approach.

To ensure sustainability and large adoption of the smart-valley approach, training on site selection, construction of irrigation canals and use the tiller will need to be promoted. Training sessions and access to input and output markets will achieve higher impact. Indeed, improved technologies represent a prerequisite for changing agricultural practices in the context of climate change, but with a set of favorable factors such as an effective extension service, secure land tenure, an efficient input distribution system, appropriate training and an economic incentive for producers. However, mosquito breeding might be a problem with the large adoption of the smart-valley approach. Therefore, future researches need to test the possibility of reduction of mosquito breeding in irrigation rice field using innovative irrigation method such as alternative wet and dry method.

References

Abadie, A. (2003). Semi-parametric instrumental variable estimation of treatment response models. *Journal of Econometrics, 113*(2), 231–263.

Africa Rice Center (AfricaRice). (2010). *Launched a project to exploit the potential of African rice lowlands.* Retrieved January, 2017, from http://www.warda.cgiar.org/warda/adrao/newsrel-smartiv-aug10.asp.

Africa Rice Center (AfricaRice). (2011). *Acquired from the rice crisis: Policies for food security in Africa.* Cotonou: AfricaRice. 26 p.

Africa Rice Center (AfricaRice). (2015). *Final technical report of the Sawah, Market Access and Rice Technologies for Inland Valleys (SMART-IV) project.* Cotonou: AfricaRice. 109 pages with DVD.

Arouna, A., Akpa, A. K. A., & Adegbola, P. Y. (2017). Impact de la technologie smart-valley pour l'amenagement des basfonds sur le revenu et le rendement des petits producteurs de riz au Benin. *Cahiers du CBRST, 12,* 47–66.

Chabi, A., Oloukoi, J., Mama, V. J., & Kiepe, P. (2010). Inventory by remote sensing of inland valleys agroecosystems in central Benin. *Cahiers Agricultures, 19,* 446–453.

CIA World Factbook. (2016). *Benin and Togo.* Retrieved July 19, 2017, from https://www.cia.gov/library/publications/theworld-factbook/geos/bn.html.

Food and Agriculture Organization of the United Nations. (2016). *Integrated production and pest management programme in Africa.* Retrieved July 10, 2017, from http://www.fao.org/agriculture/ippm/projects/benin/en/.

Food and Agriculture Organization of the United Nations (FAO). (2009). *Climate change in Africa: The threat to agriculture.* Accra: FAO Regional Office for Africa.

Gnangle, C. P., Glele, R. K., Assogbadjo, A. E., Vodounon, S., Yabi, J., & Sokpon, N. (2011). Tendances climatiques passées, modélisation, perceptions et adaptations locales au Bénin. *Climatologie, 8,* 27–40.

Heckman, J. (2010). Building bridges entre structural and program evaluation approaches to Evaluating Policy. *Journal of Economic Literature, 48*(2), 356–398.

Heckman, J., & Vytlacil, E. (2005). Structural equations, treatment effects, and econometric policy evaluation. *Econometrica, 73,* 669–738.

Imbens, G. W., & Angrist, J. D. (1994). Identification and estimation of local average treatment effects. *Econometrica, 62,* 467–476.

Imbens, G. W., & Wooldridge, J. M. (2009). Recent developments in the econometrics of program evaluation. *Journal of Economic Literature, 47*(1), 5–86.

Intergovernmental Panel on Climate Change (IPCC). (2007). In S. Solomon, D. Qin, M. Manning, Z. Chen, & M. Marquis (Eds.), *Climate change 2007: The physical science basis. Contribution of working Group 1 to the Fourth Assessment Report of the Intergovernmental Panel on Climate Change.* Cambridge: Cambridge University Press.

International Rice Research Institute (IRRI), Africa Rice Center & International Center for Tropical Agriculture (CIAT). (2010). *Global Rice Science Partnership (GRISP), CGIAR Thematic Area 3: Sustainable crop productivity increase for global food security, Proposal for a CGIAR Research Program on Rice-Based Production Systems.* Los Banos: IRRI.

Jagwe, J., Machethe, C., & Ouma, E. (2010). Transaction costs and smallholder farmers' participation in banana markets in the Great Lakes Region of Burundi, Rwanda and the Democratic Republic of Congo. *African Journal of Agricultural and Resource Economics, 6*(1), 1–16.

Mathenge, M., Place, F., Olwande, J., & Mithöfer, D. (2010). *Participation in agricultural markets among the poor and marginalized: Analysis of factors influencing participation and impact on income and poverty in Kenya.* Nairobi: University Egerton.

Oladele, O. I., Bam, R. K., Buri, M. M., & Wakatsuki, T. (2010). Missing prerequisites for green revolution in Africa: Lessons and challenges of sawah rice eco-technology development and dissemination in Nigeria and Ghana. *Journal of Food, Agriculture and Environment, 8*, 1014–1018.

Rodenburg J (2013) Inland valleys: Africa's future food baskets. In: Realizing Africa's rice promise, Marco CSW, Johnson DE, Ahmadi N, Tollens E, Jalloh A CAB International. London, pp 276-293.

Rogers, E. M. (2003). *Diffusion of innovations* (5th ed.). New York, NY: The Free Press, A Division of Simon & Schuster.

Trading Economics. (2016). *Africa GDP per capita.* Retrieved June 25, 2017, from https://trading-economics.com/country-list/gdp-per-capita?continent=africa.

United States Department of Agriculture (USDA). (2016). *Production, supply and distribution online.* Retrieved February 15, 2017, from http://apps.fas.usda.gov/psdonline/psdQuery.aspx.

Wakatsuki, T., & Masunaga, T. (2005). Ecological engineering for sustainable food products and the restoration of degraded watersheds in tropics of low pH soils: Focus on West Africa. *Soil Science & Plant Nutrition, 51*, 629–636.

Wakatsuki, T., Buri, M. M., Obalum, S. E., Bam, R., Oladele, O. I., & Ademiluyi, S. Y. (2011) *Farmers personal irrigated sawah systems to realize the green revolution and Africa's rice potential.* In First International Conference on Rice for Food, Market and Development (Rice-Africa), March 3–5, 2011. Abuja, Nigeria.

World Bank. (2010). *World development report 2010: Development and climate change.* Washington, DC: World Bank.

World Food Programme (WFP). (2009). *Global Analysis of Vulnerability, Food Security and Nutrition (AGVSAN).* 152 p.

Zwart, S. (2014). *Where is my technology going? Mapping of adoption of technologies and assessing diffusion pathways using geospatial technologies.* Retrieved May 12, 2017, from https://smartiv.wordpress.com/.

Chapter 12
An Agroecological Strategy for Adapting to Climate Change: The System of Rice Intensification (SRI)

Norman Uphoff and Amod K. Thakur

Abstract Farmers around the world have to cope with the adverse effects of climate change in their efforts to provide food security for themselves and their families and, to the extent possible, for others. Agricultural production methods developed for less-challenging and more-predictable climatic conditions need to be rethought and revised. The ideas and methods that constitute the system of rice intensification (SRI) developed in Madagascar, and now being extrapolated to other crops beyond rice, are enabling farmers to get more production from their available resources by making reductions in seed, water, and agrochemical inputs. Fortuitously, SRI crops are also more resistant and resilient to the hazards of climate change. When SRI methods are used in irrigated rice production, there is also a reduction in greenhouse gas emissions. The main factors that are contributing to SRI crops' ability to adapt to and mitigate climate-change effects are the enhancement of the growth and functioning of their root systems and at the same time the abundance and diversity of life forms enhanced in the soil by SRI management. Root systems and the soil biota were both ignored by Green Revolution technology. This chapter reviews what is known about SRI as an agroecological approach to enhancing food security under climate-stressed conditions.

Keywords Climate change · Climate resilience · Food security · Improved crop phenotypes · Reduction in greenhouse gas emissions · System of rice intensification

---

N. Uphoff (✉)
SRI International Network and Resources Center (SRI-Rice), International Programs, College of Agriculture and Life Sciences, Cornell University, Ithaca, NY, USA

Government and International Agriculture, Cornell University, Ithaca, NY, USA
e-mail: ntu1@cornell.edu

A. K. Thakur
ICAR–Indian Institute of Water Management, Bhubaneswar, Odisha, India

© Springer Nature Switzerland AG 2019
A. Sarkar et al. (eds.), *Sustainable Solutions for Food Security*,
https://doi.org/10.1007/978-3-319-77878-5_12

12.1 Overview

Irrigated rice is grown in countries around the world in fairly standard ways wherever topographic, soil, and water conditions permit. Paddy fields are kept continuously flooded, and farmers transplant fairly mature seedlings in clumps of three or four (or more) with high plant density, often haphazardly but preferably in rows. Keeping paddy fields flooded suppresses the growth of many but not all weeds; so weed control is a major challenge. This is often done with herbicides where these are available and affordable, but most often by hand. Farmers are encouraged to apply as much chemical fertilizer as they can afford, and also additional agrochemicals as recommended for control of various insect pests and disease. Because weeds are inhibited by submerging paddy fields under water, it is widely believed that if a little water is good, then more is better. Similarly, farmers believe that using more plants and more fertilizer in rice production will increase their output.

These standard practices are, however, encountering rising economic and environmental costs. Growing limitations on water availability are constraining irrigated rice production in many countries, and with greater water scarcity in many agricultural sectors (with less reliability of water supply and growing concerns about degrading soil and water quality), the sustainability of these standard practices is increasingly questionable, especially as climates become increasingly less favorable for crop production. Fortunately, as discussed below, we are learning that using more water, more seeds, and more fertilizer as a strategy to increase rice production may in fact be counterproductive.

Over the past 15 years it has been demonstrated that there are more efficient ways to grow irrigated rice than with the currently favored practices. By making some simple changes in their cultivation methods, farmers can raise the productivity of the land, labor, water, seeds, and capital that they presently invest in growing rice. Moreover, the crop plants that result from these modifications in rice cultivation are better able to adapt to, and resist, the multiple stresses of climate change. As a bonus, these agroecologically informed practices can help mitigate the forces driving climate change by reducing the net emissions of greenhouse gases (GHGs) from paddy fields, thereby lowering the global warming potential (GWP) of irrigated rice production. These effects—productivity increase, adaptation to climate stresses, and mitigation of climate-change dynamics—qualify this methodology now widely known as the System of Rice Intensification (SRI) as climate-smart agriculture (CSA) (FAO 2010; Styger and Uphoff 2016; Thakur and Uphoff 2017).

SRI methods which modify the management of rice plants, soil, water, and nutrients have two significant effects: enhancing (a) the growth of the rice plants' root systems and (b) the abundance and diversity of the soil biota, which can be referred to simply as "the life in the soil." These methods and effects help farmers to grow more productive and more robust rice plants (phenotypes) from given varieties (genotypes), from both improved cultivars and unimproved landraces, by eliciting more complete expression of the plants' genetic potentials. An example of the kind of phenotypical difference that can be elicited with best use of SRI practices is shown in Fig. 12.1.

12 An Agroecological Strategy for Adapting to Climate Change: The System of Rice... 231

Fig. 12.1 Farmer Edward Sohn in Grand Gedee County, Liberia, showing rice plants of same variety grown in adjacent fields, the plant on right grown with SRI methods. Picture courtesy of Robert Bimba, CHAP

SRI plants with larger, better-functioning root systems and with the enhancement of soil systems' biological resources (Uphoff et al. 2006) are able to produce more tillers and larger panicles, to have more grain filling and often heavier grains, with less senescence of roots and leaves, with more photosynthesis throughout the crop cycle, and they thereby achieve greater production of both grain and biomass, as discussed below.

The SRI-grown plants that give higher yields also consume less water and are better able to tolerate the stresses of pests and diseases and to withstand various extremes in weather: drought, storms, flooding, hot spells, and cold snaps. These claims may surprise many readers, but a growing body of scientific literature supports these conclusions.

The number of farmers in over 60 countries who are benefiting from recommended SRI methods is approaching 20 million, so these results are not a matter of hypothesis or speculation. They are being repeatedly confirmed in practice. Moreover, the agroecological insights on which SRI theory and practice are based are being extended and extrapolated to other crops (Adhikari et al. 2017). So the discussion which follows has more relevance broadly than just to rice, one of the world's most important crops for food security.

12.2 Agroecological Principles for More Climate-Resilient Rice Production

Agroecologically grounded management for irrigated rice is based on certain principles that represent sound agronomic practice. Each is valid in its own right, but there are also significant interactions between and among these principles and their associated practices, for example, as seen with the interaction between the age of seedlings when transplanted and the flooded vs. unflooded status of nurseries and main fields (Mishra and Salokhe 2008). Beneficial interactions among SRI principles have been documented in large-scale factorial trials that assessed the effects of the respective practices when used in different combinations (Uphoff and Randriamiharisoa 2002).

12.2.1 Early, Careful, and Quick Establishment of Healthy Plants

When establishing an irrigated rice crop by transplanting, very young seedlings are recommended, just 8–12 days old and certainly less than 15 days old, grown in an unflooded, garden-like nursery. Young transplants retain more genetic potential for vigorous growth than do older seedlings. Transplanting should be quick and careful, to avoid desiccation and trauma to the roots. Minimizing "transplant shock" enables seedlings to resume their growth quickly, and subsequently start tillering and rooting.

To reduce labor time for raising a nursery and then transplanting, some farmers are now establishing their SRI crop through direct seeding in accordance with other SRI principles and adapted practices. While the yield with direct seeding may not be as great as with transplanting, there can be compensating economic savings. The principle to be observed is that farmers should establish healthy root systems in the soil, recognizing that roots are the foundation for plants' survival and success.

12.2.2 Reduced Plant Density

Under SRI, seedlings are transplanted singly with wider spacing and in a square pattern to maximize the growth potential of each respective plant. Lowering the number of plants per m^2 by 80–90% minimizes competition among their roots, and shading within their canopies is reduced. This gives individual rice plants more access to soil nutrients so the plants can achieve better root and shoot growth and higher grain yield. Crowding of plants, which is common in conventional rice production, inhibits these processes due to greater competition for nutrients and more shading of lower leaves.

12.2.3 Improved Soil Conditions

Chemical fertilizers can be used beneficially with SRI; but these provide only nutrients for short-term plant benefit. They do little or nothing to improve the structure and functioning of soil systems, which are the key to greater and more sustainable crop production. Soil fertility depends not only on its available nutrient stocks but also on the multiple services and protections of the soil biota (Uphoff et al. 2013). Life in the soil is enhanced by providing it with sufficient organic matter and by enhancing the soil's aeration. We are beginning to see SRI practices converge with the climate-smart practices of Conservation Agriculture (CA) (e.g., Sharif 2011; Lu et al. 2013). This enhances soil fertility according to the same agroecological principles and dynamics as are operative for SRI.

12.2.4 Reduced and Controlled Water Applications

Continuous flooding of paddy fields creates hypoxic soil conditions that suffocate rice plants' roots. Flooding also impedes the survival and services of beneficial soil organisms such as growth-promoting rhizobacteria and various mycorrhizal fungi. Aerobic soil conditions when combined with the other SRI principles lead to larger root systems and ones that do not senesce (Kar et al. 1974). Keeping soils mostly unflooded provides what can be considered passive soil aeration, supplemented in SRI with a fifth principle.

12.2.5 Active Soil Aeration

When rice paddies are not kept flooded, weed growth will be a greater problem. Controlling weeds with multiple passes of a simple mechanical weeder, rather than manual weeding or with herbicides, has the advantage of aerating the soil's surface while burying or uprooting weeds. It has a green-manure effect while promoting growth of the roots and soil biota. Growing rice plants in a square pattern with wide spacing permits perpendicular weedings that more thoroughly aerate the soil (Fig. 12.2).

Frequent weedings before the canopy closes and even further weeding when it becomes difficult can still add enough to yield so that the weeding becomes a benefit, and not just a cost. Motorized weeders are now being developed to save labor time and expense. Cultivating rice on permanent raised beds, with furrow irrigation, as promoted under Conservation Agriculture, can enhance soil biological activity enough so that aeration is sufficient through such activity rather than by mechanical intervention.

A significant development has been that farmers in various rainfed areas have begun to adapt SRI principles to rice production there, with similar increases in crop productivity and resilience (e.g., Sinha and Talati 2007; Kabir and Uphoff 2007). So

Fig. 12.2 Farmers in Indonesia weeding their SRI field in perpendicular directions to control weeds and aerate the soil around their square-planted rice plants in all sides

the above principles are not limited to irrigated rice cultivation. Nor do they apply just for growing rice because many other crops are now also found to respond positively to some of SRI practices. This derived knowledge for improving other kinds of food production is known as the System of Crop Intensification (SCI) (Abraham et al. 2014; Araya et al. 2013; Adhikari et al. 2017).

12.3 Origin of SRI Ideas and Their Evolution and Spread

The interactive, synergistic effects of these principles for crop management have become better understood through research and farmer experience than they were some 30 years ago when the practices expressing these principles were assembled in Madagascar (Laulanié 1993), or since they began being disseminated beyond Madagascar some 15 years ago (Uphoff 2015). For example, we now know much more about how soil microorganisms can enhance the length, branching, volume, and surface area volume of roots (Yanni et al. 2001) as there is positive feedback between the two main SRI effects denoted above. The scientific foundations of SRI principles and practices are now much better elaborated than they were at first (Stoop 2011; Toriyama and Ando 2011; Thakur et al. 2016).

The spread of the use of SRI principles and practices was initially slow because some of these ideas are so counterintuitive.[1] How can one get more production from just 10–20% as many plants? With no flooding of paddies? With less or no use of chemical fertilizers? These ideas departed radically from the science and practice of the Green Revolution, which was based on developing and introducing new varieties bred to be responsive to chemical fertilizers, pesticides, herbicides, and irrigation water, and then to apply increasing amounts of agrochemicals and water. The idea of producing more output with reduced input makes no sense unless one understands the demonstrable contributions that can be made in plant performance and health by improving the root systems and by promoting the plant–soil–microbial interactions that are fundamental to SRI effects.

By the end of 2013, governments and NGOs in China, India, Indonesia, Cambodia, and Vietnam, where two-thirds of the world's rice is produced, were promoting SRI based on their own researchers' evaluations and on farmers' experience. By now we estimate that 18–20 million farmers in these and other countries are using some or all of SRI's recommended methods on probably about seven million ha. Average yield increase, calculated from 2013 data from these five countries, is about 1.66 tons per ha, with less water requirements, lower costs of production, and more climate resilience (Uphoff 2017).

Exact figures are not possible to report because SRI is not a material technology like a seed variety or agrochemical spray, which is either used or not. SRI is a matter of degree—how many of the recommended practices are used, and how well?[2] Since SRI is open-access, in the public domain with no patents or royalties involved, precise numbers are of little relevance. What is more important are trends, and these have been broadly positive.

The efficacy of these practices has been demonstrated in over 60 countries of Asia, Africa, and Latin America, and an increasing number of international institutions are beginning to support SRI spread.[3] While SRI methods and results have been somewhat controversial in the past, the evidence that more productive and resilient phenotypes can be evoked by alternative management practices has been accumulating, and their merits have been gaining acceptance internationally.[4] There

---

[1] It can also be noted that this innovation goes contrary to various academic, institutional and commercial interests, but we leave consideration of this issue to others.

[2] A matrix for scaling SRI practice as a matter of degree is proposed in Table 2 of Wu and Uphoff (2015).

[3] On spread among and within countries, see SRI-Rice website: http://sri.cals.cornell.edu/countries/; also websites of the World Bank (http://info.worldbank.org/etools/docs/ library/245848/), the International Fund for Agricultural Development (http://www.ifad.org/english/sri/), and IICA, the Inter-American Institute for Agricultural Cooperation (http://www.iica.int/en/press/news/sri-advancing-latin-america-help-address-climate-change). See endorsement of SRI by FAO (2015) and an earlier one by NGOs Africare, Oxfam America, and Worldwide Fund for Nature (Africare/Oxfam America/WWF 2010). In 2002, Prof. Yuan Long-ping, known as "the father of hybrid rice," hosted the first international conference on SRI at his research station in Sanya, China (Yuan 2002).

[4] See listing of over 1000 published articles at http://sri.cals.cornell.edu/research/JournalArticles. html. The most conclusive evaluation has been a meta-analysis of Chinese researchers' published

seem, however, to be some lingering negative effects of a flurry of critiques of SRI published more than a decade ago (Dobermann 2004; Sinclair 2004; Sinclair and Cassman 2004; Sheehy et al. 2004; McDonald et al. 2006), even though the databases and methodologies for the critiques were themselves flawed (Stoop and Kassam 2005; Uphoff et al. 2008), and the "well-established principles" that were asserted as making SRI methods less productive than claimed have been shown to be empirically incorrect (Thakur et al. 2010a). There have been no recent critiques in the scientific literature, so the controversy over SRI appears to be dissipating, although this is a slow process.

12.4 Studies Showing More Productive Phenotypes Are Possible from Available Genotypes

For the past 50 years, since the advent of the Green Revolution, most emphasis in crop improvement programs has been placed on modifying the genetic potentials of plants, together with increasing the use of purchased external inputs. However, SRI experience and research are showing that crops' productivity and hardiness can be increased by making modifications in the growing environments for rice, both above and below ground.

Evidence is accumulating that SRI practices have significant beneficial impacts on microbial populations within the rhizosphere (Zhao et al. 2010; Anas et al. 2011; Lin et al. 2011; Gopalakrishnan et al. 2013; Mishra and Uphoff 2013; Doni et al. 2017). This in turn would affect the dynamics within the soil–plant microbiome, which in recent years has started to receive the attention and research that it deserves (Turner et al. 2013; Mendes et al. 2013; Uphoff et al. 2013; Schlaeppi and Bulgarelli 2015). The microbiome around, on, and in plants influences their growth and health in ways that parallel the impacts which the human microbiome has on our own growth and health (Cho and Blaser 2013).

Evaluations of SRI have shown that most rice genotypes can be made more productive by modifying the conditions under which they grow. Comparison research at the Indian Institute of Water Management of the Indian Council for Agricultural Research (ICAR) has found that rice plants grown with SRI methods have:

- Increased tillering, more grains per panicle, greater grain filling, and heavier grains.
- Larger, deeper, and better distributed root systems, with more xylem exudation.

---

studies comparing SRI methods with what they considered as the best management practices (Wu and Uphoff 2015). The analysis showed SRI methods giving a yield advantage of 10–30%. The wide range reflected differences in how many of the recommended SRI methods were used in each trial, and the extent to which they were used as recommended. When using just a few of the recommended practices, there was a 4% yield *dis*advantage in the trials evaluated, which indicated that synergy among practices is an important element in SRI impact.

- A more open plant architecture aboveground with more erect and larger leaves; the leaf area index (LAI) is higher and there is more light interception: 14% more at panicle initiation—with only one-sixth as many plants per m^2.
- Chlorophyll content in the leaves is greater, and the plants' rate of photosynthesis is correspondingly higher.
- Of particular relevance in a water-constrained world, SRI plants have been found to use water more efficiently, fixing 3.6 μmol of CO_2 per mmol of water transpired, compared to 1.6 μmol of CO_2 per mmol of water with conventional cultivation; twice as much "crop per drop."

These morphological and physiological effects culminate in higher yield from fewer plants (Thakur et al. 2010a, b, 2011, 2013), with the plant phenotypes themselves being better able to withstand climate stresses, as discussed below. Similar results have been reported from research carried out by other ICAR and ICRISAT scientists (Mahender Kumar et al. 2013; Gopalakrishnan et al. 2013).

Scientists at the China National Rice Research Institute conducted similar evaluative research during two seasons using two popular hybrid varieties which showed the same kind of positive phenotypical effects. They compared "standard rice management" commonly employed in China with SRI-based "new rice management" (Lin et al. 2009). Two years of trials showed that with *reductions* in plant population, with *reduced* water application, and with *less use* of chemical fertilizer, the varieties evaluated gave significantly higher yields.

Studies that have examined root growth and functioning under SRI management have showed significant differences in rice plant roots' structure and physiology, as well as in their nitrogen uptake and utilization under SRI management (Barison and Uphoff 2011; Gopalakrishnan et al. 2013; Mishra and Salokhe 2008; Mishra and Uphoff 2013; Thakur et al. 2010a, 2013; Zhao et al. 2009). Modifications of plants' growing environment affect their features and capabilities, with different expression of their genetic potential. The genome of plants, like that of other organisms, is more like a *playbook* for sports than it is a *blueprint* for constructing a predetermined structure. Next steps in ontogenesis are not mechanistically followed but are influenced by nature's equivalent of factors in a game like the present score, how successful any previous moves have been, and how much time there is left to in the game. What occurs is more conditional than predetermined. The interaction between an organism's genetic potentials and its environments thus makes each organism unique, and it will bring out potentials either more or less fully.

12.5 SRI Contributions to Food Security

The various morphophysiological changes in rice plants under SRI crop management culminate in more productive and resilient phenotypes that can be elicited from given genetic potentials (Thakur et al. 2011). Resulting yield numbers can vary, even widely, because SRI results are driven more by biological dynamics than by agrochemical inputs or by a genetic blueprint. The impacts of SRI practices are variable in part because they reflect microbiological activity and influences such as have been documented by Chi et al. (2005, 2010) and Doni et al. (2016). The numbers, species and activity of microbes are all affected by the interacting effects of moisture, temperature, pH, substrate, and other factors. With this qualification noted, the following generalizations can be made about the contributions that SRI management can make to increased food production.

12.5.1 Higher Yields

With SRI management, the same varieties planted on the same soil can usually give 20–50% higher yield, and sometimes increases of 100% or more, the percentage depending in part on what were the previous usual yields. Large percentage increases are easier to achieve from a low base than a high one. An evident basis for such increases is easily seen in the larger root systems, the greater number of tillers, the more numerous and longer panicles, and the reduced and slower senescence of rice plants' roots and leaves.

There has been controversy over some of the highest yields reported for SRI (e.g., Dobermann 2004; Diwakar et al. 2012), but these are of interest primarily for assessing how large the gap may be between what is currently being achieved and possible production. Here we are concerned only with averages, since these have a more decisive effect on food security than do outliers (positive deviation). Whether or not SRI methods outperform "best management practices" (McDonald et al. 2006) is not really relevant for resource-limited households that are struggling to ensure their own food security. Rigorous evaluation such as in a meta-analysis of Chinese experimental data (Wu and Uphoff 2015), shows that there are significant yield improvements achieved with SRI management.

12.5.2 Greater Milling Outturn

More edible food is produced from SRI-grown paddy rice because as a rule there are fewer unfilled grains in SRI panicles, and during milling there is less breakage of grains. The amount of polished rice produced per bag or per bushel of SRI paddy when it is milled, to remove husks and bran layers, is usually about 10–15% more than usual because there is less chaff and fewer broken grains with SRI paddy rice. This represents a bonus of edible food that becomes available over and above the recorded increases of harvested paddy rice. Grain quality has also often been found to be higher, for example, due to reduced chalkiness (Xu et al. 2005).

12.5.3 Lower Costs of Production and Greater Net Income

When farmers can reduce their reliance on purchased inputs of inorganic fertilizer and pesticides, this lowers their production costs and raises their net incomes by more than the actual increase in yield. A large-scale evaluation of SRI in India by the International Water Management Institute (IWMI), covering a sample of >2200 farmers in 13 states, found that even though few of the SRI farmers were using all of the recommended methods, or using them all as recommended, farmers' average cost of production for paddy rice was lower by $28 per ton (Palanisami et al. 2013). Yield increases varied according to how many of the methods were followed, and how well, but generally there was a significant positive impact on household incomes which added to their food security, over and above their greater ability for self-provisioning of their staple food.

There has been controversy over whether SRI methods necessarily increase farmers' labor costs. It is certainly true that SRI is more labor-intensive than capital-intensive, unless a mechanized version of SRI is adopted (Sharif 2011). However, for farmers, the relevant standard for comparison is their current practice. It should be kept in mind that learning any new methods will invariably require some additional time and effort at first. Most rice farmers have found that once they have learned SRI methods, these do not require a greater expenditure of labor, or more money to hire labor. Indeed, some farmers have considered the labor-saving effects of SRI to be an attractive feature (e.g., Li et al. 2005).

The characterization of SRI as "labor-intensive," and possibly too labor-intensive for farmers to adopt, derives from an early evaluation of the economics of SRI that was based on data from Madagascar (Moser and Barrett 2003). In this country, rice is traditionally grown with labor-*extensive* methods, so any improvement in rice production is bound to require some additional labor. In most of Asia, where 90% of the world's rice is produced and where rice production is relatively labor-intensive, SRI has usually proved to be labor-neutral or even labor-saving. Indeed, even in Madagascar SRI becomes labor-saving once the new methods have been mastered (Barrett et al. 2004).

An evaluation of SRI in India found that the only negative effect of its adoption was its lower labor requirements, which reduced the wages paid to agricultural laborers by 50%, something that the researchers classified as a social cost (Gathorne-Hardy et al. 2016). They reported that SRI methods increased average paddy yields by 60%, while net greenhouse gas emissions were reduced by 40% per ha. Because less water was needed, there was a 60% reduction in groundwater extraction and 74% less consumption of fossil fuel. Farmers using the new methods had their net income per hectare increased by about three times (Gathorne-Hardy et al. 2016)

12.6 SRI Adaptations to Climate Stress

Achieving increases in production will not assure food security unless that production is resistant to the less-favorable climatic conditions—shifts in mean temperature and rainfall, and increases in "extreme events," which are becoming more frequent with changes in prevailing climate. While interest in SRI methods was initially spurred by yield increases, both average and maximum, there is evidence, both anecdotal and experimental, that SRI-grown rice phenotypes are better able to withstand most of the stresses which are increasing with climate change (Thakur and Uphoff 2017).

- Lower water requirements, which becomes more important as rainfall patterns change and water becomes more and more a limiting factor in agricultural production.
- Buffering against drought stress, and often flooding tolerance.
- Resistance to storm damage due to wind and rain causing lodging of crops.
- Tolerance of cold temperatures.
- A shorter crop cycle which reduces crops' exposure to biotic and abiotic stresses.
- Resistance to insect pests and diseases.

Along with this, there is increasing evidence that SRI management contributes to net reductions in the emission of greenhouse gases (GHG), so that SRI practices not only support crops' adaptation to climatic stresses but also help to mitigate climate change.

12.6.1 *Increased Water Productivity, Water Saving, and Water Use Efficiency*

Substantial reductions in water consumption are possible when rice paddy fields are no longer kept continuously flooded, as is standard practice wherever water supplies permit. When using less water is combined with other SRI methods, farmers' water saving is rewarded by higher yield which together with lower costs of production gives greater farmer income.

A meta-analysis of SRI water productivity has been done of data from 29 published studies, including 251 comparison trials conducted across eight countries, for which enough data were reported in the literature for detailed quantitative evaluations to be done (Jagannath et al. 2013). The studies compared a selection of SRI practices against whatever the researchers considered to be "best practices." Most of the evaluations of SRI performance were in fact not done with a full set of recommended practices because only 30% of those trials employed >80% of the SRI methods. The advantages seen with SRI practices thus would probably have been greater than calculated if all of its methods had been fully and well utilized in the comparison trials.

The analysis found that SRI water management together with some if not all of the other SRI practices reduced total water requirements for rice production by 22%, from 15.33 million liters per ha to 12.03 million liters, considering both rainfall and irrigation issues, while giving higher yields. The use of irrigation water per ha was on average 35% less with SRI management. Because yields were increased with less water per ha, the total water use efficiency (WUE) was raised by 52% with SRI, producing 0.6 g of paddy rice per liter of water compared with 0.39 g. Irrigation water use efficiency was increased by even more; even with incomplete SRI management, this raised IWUE by 78% (Jagannath et al. 2013). Similar numbers have been reported recently from empirical studies in India by Geetha Lakshmi et al. (2016).

The greater water productivity of SRI methods for irrigated rice production was seen across differences in agroclimatic zones, in soil texture, in soil pH, in wet vs. dry seasons, and with short vs. medium vs. long-duration varieties (Jagannath et al. 2013). Greater water saving and water productivity with SRI has been confirmed by studies in countries as varied as China, Afghanistan, Iraq, Indonesia, Sri Lanka, and Kenya (Zhang et al. 2013; Thomas and Ramzi 2011; Hameed et al. 2011; Sato and Uphoff 2007; Namara et al. 2008; Ndiiri et al. 2013).

12.6.2 Drought Resistance

It should not be surprising that rice plants with deeper roots are better able to tolerate water stress. Roots on SRI plants do not senesce (die back) as roots do when they grow under hypoxic flooded conditions (Kar et al. 1974). This effect of drought resistance is reinforced when there is abundant life in the soil since the soil biota represents a reservoir of water in the soil in addition to their making the soil better-structured and more porous, thus better able to absorb and retain water.

Evidence of this effect on a large scale can be seen from the Sichuan Province of China where according to the Provincial Department of Agriculture data, SRI use went from 1133 ha in 2004 to 383,533 in 2012 (Zhang et al. 2013). The average yield increase with SRI methods was 1.67 tons per ha, amounting to an additional >2.8 million tons of paddy rice produced during this period, with a 25% reduction in the amount of irrigation water consumed. Of relevance here are the data from 2006 and 2010 which were major drought years in Sichuan during the period 2004–2012. In these 2 years, the SRI yield advantage averaged 1.808 tons per ha, which was 10.5% higher than the SRI yield advantage of 1.636 tons per ha in the other seven years with more normal rainfall (Zhang et al. 2013). SRI rice plants under water stress conditions were able to produce relatively better than under normal conditions.[5]

Drought resistance in SRI-grown plants has been documented at field level in an IWMI evaluation in Sri Lanka. The main 2003/2004 season when the study was done had 75 days of severe drought. SRI-grown plants formed panicles on 80% of their tillers, compared with only 70% on the rice plants grown with usual farmer methods (Namara et al. 2008). Even though the latter had ten times more rice plants

---

[5]A similar effect is reported by researchers at the Indian Agricultural Research Institute (IARI) evaluating System of Wheat Intensifications (SWI) methods against IARI recommended practices for wheat over two *rabi* seasons. The SWI yield advantage in 2011–2012, with typical weather, was 30%; in 2012–2013 when there were unusually high temperatures and then flooding, the advantage of SWI was 46% (Dhar et al. 2015).

per m^2, in the SRI fields the number of panicle-bearing tillers per m^2 was 30% higher, and the harvested yield was 38% greater. SRI phenotypes thus were better able under drought conditions to translocate their photosynthate from their leaves and culm to their grains.

12.6.3 Resistance to Storm Damage

When rice plants have larger, deeper, non-senescing root systems, they are better able to withstand the pounding of rain and wind during storms, which are becoming more frequent and more severe with climate change. This phenomenon has not been much studied, although when measurements have been done, this effect has been seen (Chapagain and Yamaji 2010). Figure 12.3 below shows examples from Vietnam, Taiwan, and Indonesia of SRI crops' resistance to lodging.

12.6.4 Resistance to Colder Temperatures

The well-developed root systems of rice plants grown with SRI methods also help them to tolerate cold spells. This was seen from integrated pest management (IPM) trials conducted at the Andhra Pradesh state agricultural university (ANGRAU) in India in 2005–2006. The trials were designed to assess differences in pest resistance, but they happened to generate unplanned-for data on cold resistance when the trial plots were hit by a short but severe cold period when night-time air temperatures dropped below 10 °C for 5 successive days, as seen in Table 12.1. This chill caused the plots with standard rice crop management to produce almost no grain, while the adjacent SRI plots had a yield of >4 tons per ha (Sudhukar and Reddy 2007).

There will, of course, be limits as to how low the temperature for any rice crop can fall without causing damage. But climate change is creating conditions where crops are subject to irregular weather patterns that can bring on extreme temperatures for unexpected periods. SRI rice plants, with their more robust structure and physiology, have more capacity to cope with such conditions. These characteristics should also help SRI plants to maintain themselves under some hotter-than-normal temperatures that would stifle rice plants grown with the conventional cultivation methods.

Fig. 12.3 (continued) root systems, and on the right, rice grown with conventional methods; Adjacent paddy fields in East Java, Indonesia (bottom) which had been affected in 2009 by a tropical storm following a brown planthopper pest attack; the field on the left was planted with a modern variety and managed with fertilizer and agrochemical inputs, while a traditional aromatic variety was being grown in the field on right using organic SRI methods; this field gave a yield of 8 tons/ha (800 kg from 1000 m^2), while the harvest from the field on left was very little

12 An Agroecological Strategy for Adapting to Climate Change: The System of Rice... 243

Fig. 12.3 Paddy fields in the Mekong Delta in Vietnam (top) after they had been hit by a tropical storm; the rice field in the foreground on left is a crop grown with usual methods, and above it on the right is an SRI-managed rice field (Dill et al. 2013). Two paddy fields in Taiwan (middle) after a storm had passed over them; on the left is SRI-grown rice resisting lodging because of stronger

Table 12.1 Yield results as affected by temperature, Andhra Pradesh, India, 2005–2007

Season	Normal methods (tons per ha)		SRI methods (tons per ha)
Rabi (winter) 2005–2006	2.25		3.47
Kharif (monsoon) 2006	0.21[a]		4.16
Period in kharif season	**Mean max. temp. (°C)**	**Mean min. temp. (°C)**	**No. of sunshine hours**
1–15 November	27.7	19.2	4.9
16–30 November	29.6	17.9	7.5
1–15 December	29.1	14.6	8.6
16–31 December	28.1	12.1[b]	8.6

Source: Sudhukar and Reddy (2007)

[a]Low yield in this season was due to cold injury arising from the temperatures shown below

[b]There was sudden drop in minimum temperatures for 5 days, 16–21 December (9.2–9.8 °C)

12.6.5   Shorter Growing Season

With weather becoming more variable and with associated pest and disease problems becoming more severe, farmers can benefit from having a shorter crop cycle and getting harvests out of the field sooner (especially if the harvests themselves are larger). Farmers often report that with SRI management, their rice crops can be harvested in 5–10, sometimes even 15 fewer days after transplanting, while their nursery time is also reduced.

The most systematic data on this effect have come from Nepal where the Morang District Agricultural Development Office gathered data on planting and harvest dates, as well as on yield and other parameters, from 413 farmers who used SRI with eight varieties in the 2005 main season. These varieties' advertised maturity ranged from 120 to 155 days (average 141 days), while the SRI crops were actually harvested in 115–133 days (average 126 days). The average SRI paddy yield was 6.3 tons per ha, while the yield with conventional rice production methods was just half as much, 3.1 tons per ha. SRI rice crops had thus 2 weeks less exposure to biotic and abiotic stresses and gave a more than doubled yield (Uphoff 2011). Similar reductions in the crop cycle under SRI management are reported by Uzzaman et al. (2015) from trials in Bangladesh.

12.6.6   Resistance to Pests and Diseases

The phenotypical advantages of SRI-grown rice crops are seen also in terms of their reduced susceptibility to infestation and damage by pests and diseases. This constraint on agriculture is being heightened by the changes in precipitation and temperature that are occurring with climate change. The first evaluation of SRI practices' inhibition of insect pests and associated diseases was done at Tamil Nadu Agricultural University in 2002 (see Uphoff 2015, Table 2), followed by a number of further studies in India (e.g., Padmavathi et al. (2009); Karthikeyan et al. (2010); Pathak et al. (2012), Visalakshmi et al. (2014)). The research has progressed to where disease resistance with SRI management has been found to be associated with beneficial soil organisms (Doni et al. 2016), but a number of effects are probably involved. Farmers in most countries where SRI has been introduced report that their rice crops are less affected by the common pests and diseases. Possibly this can be explained in part by the theory of trophobiosis proposed by a French scientist (Chaboussou 2004).

These various manifestations of resistance to climate-related stresses—water shortage and drought, wind, and rain damage from storms, temperature extremes, and pest and disease incidence—can be traced back to the effects induced by SRI management practices on the rice plants' belowground and aboveground structures and on associated microbiological activity in the soil and plants.

12.7 Reducing Net Greenhouse Gas Emissions as a Bonus

This book focuses on how farmers, their cropping systems, and our public policies can countervail the adverse impacts of climate change by *adapting* practices and strategies to the multiple stresses that this change introduces into our agriculture and daily lives. Evidence is accumulating that SRI management practices can also help to *mitigate* global warming by abating the net emissions of GHGs from irrigated rice paddies, as summarized in Sect. 4.1 of Uphoff (2015).

Methane (CH_4) emissions are much reduced by halting the continuous flooding of paddy fields, while there is little or sometimes no increase in nitrous oxide (N_2O) emissions when SRI soils are kept in mostly aerobic condition and when little if any inorganic N fertilizer is applied. A multidisciplinary study of SRI in Andhra Pradesh state of India, doing a life cycle analysis of GHG emissions that followed IPCC guidelines and included CO_2 emissions, calculated that SRI practices reduced net GHG emissions by >25% per ha and, given the higher yield, by >60% per kg of rice produced (Gathorne-Hardy et al. 2013). Reductions in global warming potential with SRI practices has also been reported by researchers in Vietnam (Dill et al. 2013, Table 7), Korea (Choi et al. 2014), and India (Jain et al. 2014).

12.8 Constraints and Limits to Scaling Up SRI

Given the many economic and environmental benefits reported above, readers may have two questions in mind. Why has not SRI use spread more widely? And what are the limitations on its adoption? The first question still perplexes us after many years of working with SRI. Various academic, institutional, and commercial interests are probably not very favorably disposed toward SRI ideas and their use because these represent a paradigm shift with some actors and financial disadvantage for others. But this is a subject that has not been researched, and we prefer to deal with evidence to the extent that we can.

What is certain is that SRI methods are both novel and counterintuitive, departing from age-old practices. Farmers are not necessarily more conservative than researchers or policy-makers, but one has to reckon with "resistance to change" among all three groups. Despite claims and norms to the contrary, academics can be very conservative. That SRI has not been more rapidly and more widely taken up is not evidence that the many benefits documented by several hundred researchers and reaped by millions of farmers are not actual and accurate. Our observations here are as factual as can be made.

The question regarding constraints is more answerable. The SRI effect of giving more productive and robust phenotypes from a given genotype has been demonstrated in a wide variety of agroecological systems, from Timbuktu on the arid edge of the Sahara desert (Styger et al. 2011) to Aceh in the humid jungles of Indonesia (Cook and O'Connor 2009), and from sea level in Madagascar to elevations of 2700 m above-sea-level in Nepal. A meta-analysis of published data from eight countries (Jagannath et al. 2013) showed SRI raising crop and water productivity in all climatic systems covered, from tropical to temperate, not equally in all but giving advantages wherever SRI methods were systematically evaluated. The methods were found to be more productive with all different soil types and over the full range of soil pH, from acid to alkaline, as well as in both wet and dry seasons and with short, medium, and long-maturing varieties. Since SRI is based on widely applicable agronomic principles rather than being a fixed set of practices, it has great flexibility and replicability.

The methods need not be used fully or perfectly to confer productivity gains. The "ideal type" of SRI practice has been validated in extensive factorial trials (Uphoff and Randriamiharisoa 2002). However, significant gains are achievable with less than complete use of the methods (Palanisami et al. 2013; Wu and Uphoff 2015). There are, however, some factors that can limit the applicability or productivity of these methods, and these need to be taken into account.

12.8.1 Water Control

This is the probably the most significant constraint for utilizing SRI methods since their success will be affected by how well mostly aerobic soil conditions can be maintained, rather than continuous flooding. While SRI utilizes smaller amounts of water—by one-third to one-half—the smaller supply of water needs to be reliable and sufficient. A certain amount of water-stress can promote deeper root growth in plants, but they cannot survive too much stress as all plants require water. Water control does not need to be perfect for SRI. For best results, getting most water saving and highest water productivity, there should be some control and precision for water applications. But SRI crops have considerable tolerance and robustness if their roots get well established (and are not suffocated) in their first weeks.

12.8.2 Weed Control

Without continuous flooding of rice paddies, weed growth and ensuing competition can become a constraint on production if it is not dealt with. For SRI, it is advised that weed control be accomplished with a simple mechanical hand weeder, which can be motorized. This practice, in contrast with weeding or applying herbicides to control weeds, gives the rice crop the benefit of *active soil aeration* as it churns weeds into the soil to decompose. By promoting the growth of roots and beneficial soil organisms, such weeding becomes an investment rather than just a cost, since yield increments from mechanical weeding can be 0.5–2 tons per ha. Weeders also save labor time and discomfort (Vent et al. 2016).

12.8.3 Access to Biomass

Limited access can present a constraint on using SRI methods most effectively. Sometimes there is not enough available supply of biomass—crop residues, green manures, lopping, weeds, farmyard manure, or any other source of organic material—that can be acquired, decomposed, and applied to improve the soil's structure and functioning. Raising the soil's content of organic matter will give best results, but farmers need to consider their costs in terms of time and money as well as the sheer lack of organic material in some places. Chemical fertilizers can be used instead of or with organic sources of nutrients (compost, manure, mulch) where the latter are in short supply.

12.8.4 Pest and Disease Control

Whenever there is more plant biomass (roots, canopy, grains), there is always some risk that this will also attract and sustain more pests and pathogens. So farmers using SRI methods need to be prepared to deal with these constraints on production, preferably with integrated pest management or other actions that minimize the use of biocides. But chemical crop protection is not necessarily ruled out by SRI. Repeatedly, farmers report that their SRI crops have less need for agrochemical applications because the crops are less susceptible to pest and disease damage.

12.8.5 Labor Supply and Motivation

The most binding constraint for using SRI is to have sufficient, and sufficiently motivated, labor to carry out more precise and careful crop management than with current practice. Contrary to a common stereotype, SRI is not necessarily more labor-intensive, as discussed in Sect. 12.5.3. Whether it requires more or less labor than farmers are investing at present depends mostly on whether they are cultivating rice with minimum inputs of labor, as in Madagascar, or land is energetically cultivated to get the most production possible out of it because land is scarcer than is labor.

Even where current rice cultivation is labor-extensive, it has been seen that over time, as confidence and skill are gained, SRI methods can become labor-saving, reducing requirements for this resource as well as for water, seeds, chemical fertilizer and other inputs. There needs to be a willingness to be experimental and to make adjustments; to capitalize upon existing potentials in the plants and in the soil systems that they grow in; to be very empirical and not be bound to past practices and ways of thinking.

Because labor supply is a constraint in many countries and places, over the past 10 years there has been a lot of innovation and experimentation with mechanizing the most labor-requiring SRI operations. There are now better, more labor-saving means for establishing plants wide apart and regularly spaced, for transplanting seedlings mechanically or for doing direct-seeding, for weeding fields in ways that actively aerate the soil as well as eliminate weeds. While SRI was developed without mechanization, this has become more widespread and it is likely to continue. Examples of mechanization for SRI are already reported from Costa Rica and Pakistan (Montero 2010; Sharif 2011).

12.9 Concluding Remarks

The scope and impact of climate change over the coming decades cannot be known. But it is quite certain that our agriculture, for large-scale and small-scale producers alike, will have to change rather substantially from the technologies and practices developed and benefited from in the preceding century. These made large advances upon the productivity of cropping systems and methods evolved over centuries before that. But the twentieth century modes of production would need revision and improvement in this century even if there were no climate change intruding. Population size, growth and spatial distribution, man–land ratios, technical changes in our energy, transportation and communication infrastructures, capital accumulation relative to labor, all these and other factors affect our options for food production (Hayami and Ruttan 1985), and they continue to change.

By demonstrating that the production of food can be raised by making reductions in inputs—using less seed, fertilizer, and water inputs and reversing reliance on synthetic agrochemicals—there can be a welcome shift in agricultural thinking and practice, to producing more food with a smaller, lighter footprint on the environment. But we must reckon and cope with a situation that is far from static, and indeed dynamic in undesirable ways. In this chapter we have reviewed what has been learned about, and from, the System of Rice Intensification over the past 15 years. Even if widely adopted/adapted, SRI still cannot meet all of the demands that will be placed upon the agricultural sector, and it is anyway not a "silver bullet" that will correct the massive imbalances and reverse the adverse momentum built up by decades of ignoring the needs and vulnerabilities of our global ecosystem and its myriad subsystems.

For life on earth to have a long future, we will have to rely to some extent on its recuperative and regenerative powers, incalculable but probably also underestimated. But this hopeful prospect does not and cannot relieve us of the onus of making ecologically sound, future-oriented decisions about agricultural research and development in the present. SRI and its associated broader family of ecologically based production strategies grouped under the rubric of *agroecology*—conservation agriculture, integrated pest management, agroforestry, aquaculture, and horticulture when integrated into diversified farming systems, holistic rangeland management—as well as the derived cropping systems that come under the heading of System of Crop Intensification—for wheat (Dhar et al. 2015), sugarcane (Gujja et al. 2017), finger millet, mustard, legumes, and other crops—all represent available knowledge that can put agricultural production on track for a more sustainable sector in the decades ahead.

There is much research and experimentation to be done to utilize fully and sustainably the potentials that SRI experience to date has illuminated. The Green Revolution has made its contributions to food security over the past 50 years, but it has been encountering diminishing returns for the past 20 years. SRI departs from the input-dependent strategy of the Green Revolution, mobilizing and utilizing potentials that already exist in crop genomes. The time and cost needed to breed new climate-resilient varieties is less important if modifications in the management of plants, soil, water, and nutrients, not just for rice but also other crops, can directly and immediately improve agronomic and economic crop performance. Most important for the achievement and preservation of long-term food security is that our agriculture become truly "green" and enhance rather than detract from the natural resource base upon which we all depend.

References

Abraham, B., Arayu, H., Berhe, T., Edwards, S., Gujja, B., Khadka, R. B., Koma, Y. S., Sen, D., Sharif, A., Styger, E., Uphoff, N., & Verma, A. (2014). The system of crop intensification (SCI): Reports from the field on improving agricultural production, food security, and resilience to climate change for multiple crops. *Agriculture and Food Security, 3*, 4.

Adhikari, P., Arayu, H., Aruna, G., Balamatti, A., Banerjee, S., Barah, B. C., Baskaran, P., Behera, D., Berhe, T., Boruah, P., Dhar, S., Edwards, S., Fulford, M., Gujja, B., Ibrahima, H., Kabir, H., Kassam, A., Khadka, R. B., Koma, Y. S., Natarajan, U. S., Perez, R., Sen, D., Sharif, A., Singh, G., Styger, E., Thakur, A., Tiwari, A., Uphoff, N., & Verma, A. (2017). System of crop intensification for diversified, sustainable agriculture. *International Journal of Agricultural Sustainability, 16*, 1–28.

Africare/Oxfam America/WWF. (2010). *Farmers leading the way from crisis to resilience: Global farmer perspectives on the system of rice intensification.* http://sri.ciifad.cornell.edu/publications/articles/Global_Farmer_Perspectives_OxfamWWFAfricare.pdf.

Anas, I., Rupela, O. P., Thiyagarajan, T. M., & Uphoff, N. (2011). A review of studies on SRI effects on beneficial organisms in rice soil rhizospheres. *Paddy and Water Environment, 9*, 53–64.

Araya, H., Edwards, S., Asmelash, A., Legasse, H., Zibelo, G. H., Assefa, T., Mohammed, E., & Misgina, S. (2013). SCI: Planting with space. *Farming Matters, 29*, 34–37.

Barison, J., & Uphoff, N. (2011). Rice yield and its relation to root growth and nutrient-use efficiency under SRI and conventional cultivation: An evaluation in Madagascar. *Paddy and Water Environment, 9*, 65–78.

Barrett, C. B., Moser, C., Barison, J., & McHugh, O. V. (2004). Better technology, better plots or better farmers? Identifying changes in productivity and risk among Malagasy rice farmers. *American Journal of Agricultural Economics, 86*, 869–888.

Chaboussou, F. (2004). *Healthy crops: A new agricultural revolution*. Charnley: Jon Anderson.

Chapagain, T., & Yamaji, E. (2010). The effects of irrigation method, age of seedling and spacing on crop performance productivity and water-wise rice production in Japan. *Paddy and Water Environment, 8*, 81–90.

Chi, F., Shen, S. H., Cheng, H. P., Jing, Y. X., Yanni, Y. G., & Dazzo, F. B. (2005). Ascending migration of endophytic rhizobia, from roots to leaves, inside rice plants and assessment of benefits to rice growth physiology. *Applied and Environmental Microbiology, 71*, 7271–7278.

Chi, F., Yang, P. F., Han, F., Jing, Y. X., & Shen, S. H. (2010). Proteomic analysis of rice seedlings infected by *Sinorhizobium meliloti* 1021. *Proteomics, 10*, 1861–1874.

Cho, I., & Blaser, M. J. (2013). The human microbiome: At the interface between health and disease. *Nature Reviews: Genetics, 13*, 260–270.

Choi, J. D., Kim, G. Y., Park, W. J., Shin, M. H., Choi, Y. H., Lee, S., Kim, S. J., & Yun, D. K. (2014). Effect of SRI water management on water quality and greenhouse gas emissions in Korea. *Irrigation and Drainage, 63*, 263–270.

Cook, G., & O'Connor, T. (2009). Rice aplenty in Aceh. *Caritas News, 2009*, 9–11 http://www.caritas-europa.org/module/FileLib/RiceaplentyinAceh.pdf.

Dhar, S., Barah, B. C., Vyas, A. K., & Uphoff, N. (2015). Evaluation of System of Wheat Intensification (SWI) practices as compared to other methods of improved wheat cultivation in the north-western plain zone of India. *Archives in Agronomy and Soil Science, 62*, 994–1006.

Dill, J., Deichert, G., & Le, T. N. T. (Eds.). (2013). *Promoting the system of rice intensification: lessons learned from Trà Vinh Province, Vietnam*. Hanoi: GIZ and International Fund for Agricultural Development.

Diwakar, M. C., Kumar, A., Verma, A., & Uphoff, N. (2012). Report on the world record SRI yields in kharif season 2011 in Nalanda district, Bihar state, India. *Agriculture Today, 15*, 54–56.

Dobermann, A. (2004). A critical assessment of the system of rice intensification (SRI). *Agricultural Systems, 79*, 261–281.

Doni, F., Zain, C. R. C. M., Isahak, A., Fathurrahman, F., Sulaiman, N., Uphoff, N., & Yusoff, W. M. W. (2016). Relationships observed between Trichoderma inoculation and characteristics of rice grown under System of Rice Intensification (SRI) vs. conventional methods of cultivation. *Symbiosis, 72*(1), 45–59. https://doi.org/10.1007/s13199-016-0438-3.

Doni, F., Zain, C. R. C. M., Isahak, A., Fathurrahman, F., Anhar, A., Yusoff, W. M. W., & Uphoff, N. (2017). Synergistic effects of System of Rice Intensification (SRI) management and *Trichoderma asperellum* SL2 increase the resistance of rice plants (*Oryza sativa* L.) against sheath blight (*Rhizoctonia solani*) infection. *Australasian Plant Pathology* Accepted. Journal of Crop Science and Biotechnology, instead of Australasian Plant Pathology

FAO. (2010). *'Climate-Smart' agriculture: Policies, practices and financing for food security, adaptation and mitigation*. Rome: UN Food and Agriculture Organization.

FAO. (2015). *Save and grow: Maize, rice and wheat – A guide to sustainable crop production* (pp. 44–47). Rome: UN Food and Agriculture Organization.

Gathorne-Hardy, A., Narasimha Reddy, D., Venkatanarayana, M., & Harriss-White, B. (2013). A Life Cycle Assessment (LCA) of greenhouse gas emissions from SRI and flooded rice production in SE India. *Taiwan Water Conservancy, 61*, 110–125.

Gathorne-Hardy, A., Narasimha Reddy, D., Venkatanarayana, M., & Harriss-White, B. (2016). System of Rice Intensification provides environmental and economic gains but at the expense of social sustainability: A multidisciplinary analysis in India. *Agricultural Systems, 143*, 159–168.

Geetha Lakshmi, V., Tesfai, M., Lakshmanan, A., Borrell, A., Nagothu, U. S., Arasu, M. S., Senthilraja, K., Mainkandan, N., & Sumathi, S. (2016). System of rice intensification: Climate-smart rice cultivation system to mitigate climate change impacts. In U. S. Nagothu (Ed.), *Climate change and agricultural development: Improving resilience through climate-smart agriculture, agroecology and conservation* (pp. 234–261). New York, NY: Routledge.

Gopalakrishnan, S., Mahender Kumar, R., Humayun, P., Srinivas, V., Ratna Kumari, B., Vijayabharathi, R., Singh, A., Surekha, K., Padmavathi, C., Somashekar, N., Raghuveer Rao, P., Latha, P. C., Subba Rao, L. V., Babu, V. R., Viraktamath, B. C., Vinod Goud, V., Loganandhan, N., Gujja, B., & Rupela, O. (2013). Assessment of different methods of rice (*Oryza sativa* L.) cultivation affecting growth parameters, soil chemical, biological and microbiological properties, water saving, and grain yield in rice-rice system. *Paddy and Water Environment, 12*, 79–87.

Gujja, B., Natarajan, U. S., & Uphoff, N. (2017). Sustainable sugarcane initiative: A new methodology, its overview, and key challenges. In P. Rott (Ed.), *Achieving sustainable cultivation of sugarcane*. Cambridge: Burleigh-Dodds.

Hameed, K. A., Mosa, A. J. K., & Jaber, F. A. (2011). Irrigation water reductions using System of Rice Intensification compared with conventional cultivation methods in Iraq. *Paddy and Water Environment, 9*, 121–127.

Hayami, Y., & Ruttan, V. W. (1985). *Agricultural development: An international perspective* (2nd ed.). Baltimore, MD: Johns Hopkins University Press.

Jagannath, P., Pullabothla, H., & Uphoff, N. (2013). Meta-analysis evaluating water use, water saving, and water productivity in irrigated production of rice with SRI vs. standard management methods. *Taiwan Water Conservancy, 61*, 14–49.

Jain, N., Dubey, R., Dubey, D. S., Singh, J., Khanna, M., Pathak, H., & Bhatia, A. (2014). Mitigation of greenhouse gas emissions with system of rice intensification in the Indo-Gangetic plains. *Paddy and Water Environment, 12*, 355–363.

Kabir, H., & Uphoff, N. (2007). Results of disseminating the System of Rice Intensification (SRI) with farmer field school methods in northern Myanmar. *Experimental Agriculture, 43*, 463–476.

Kar, S., Varade, S. B., Subramanyam, T. K., & Ghildyal, B. P. (1974). Nature and growth pattern of rice root system under submerged and unsaturated conditions. *Il Riso, 23*, 173–179.

Karthikeyan, K., Sosamma, J., & Purushothaman, S. M. (2010). Incidence of insect pests and natural enemies under SRI method of cultivation. *Oryza, 47*, 154–157.

Laulanié, H. (1993). Le système de riziculture intensive malgache. *Tropicultura, 11*, 110–114 Tropicultura 29:183–187.

Li, X. Y., Xu, X. L., & Li, H. (2005). *A socio-economic assessment of the System of Rice Intensification (SRI): A case study of Xinsheng Village, Jianyang County, Sichuan Province. Report for College of Humanities and Development*. Beijing: China Agricultural University http://sri.ciifad.cornell.edu/countries/china/cnciadeng.pdf.

Lin, X. Q., Zhu, D. F., Chen, H. Z., Cheng, S. H., & Uphoff, N. (2009). Effect of plant density and nitrogen fertilizer rates on grain yield and nitrogen uptake of hybrid rice (*Oryza sativa* L.). *Journal of Agrobiotechnology and Sustainable Development, 1*, 44–53.

Lin, X. Q., Zhu, D. F., & Lin, X. J. (2011). Effects of water management and organic fertilization with SRI crop practices on hybrid rice performance and rhizosphere dynamics. *Paddy and Water Environment, 9*, 33–39.

Lu, S. H., Dong, Y. J., Yuan, J., Lee, H., & Padilla, H. (2013). A high-yielding, water-saving innovation combining SRI with plastic cover on no-till raised beds in Sichuan, China. *Taiwan Water Conservancy, 61*, 94–109.

Mahender Kumar, R., Raghuveer Rao, P., Somasekhar, N., Surekha, K., Padmavathi, C. H., Srinivas Prasad, M., Ravindra Babu, V., Subba Rao, L. V., Latha, P. C., Sreedevi, B., Ravichandran, S., Ramprasad, A. S., Muthuraman, P., Gopalakrishnan, S., Vinod Goud, V., & Viraktamath, B. S. (2013). SRI – A method for sustainable intensification of rice production with enhanced water productivity. *Agrotechnology, S11*, 009. https://doi.org/10.4172/2168-9881.S11-009.

McDonald, A., Hobbs, P. R., & Riha, S. J. (2006). Does the System of Rice Intensification outperform conventional best management? A synopsis of the empirical record. *Field Crops Research, 96*, 31–36.

Mendes, R., Garbeva, P., & Raaijmakers, J. M. (2013). The rhizosphere microbiome: Significance of plant beneficial, plant pathogenic and human pathogenic organisms. *FEMS Microbiology Reviews, 37*(5), 634–663.

Mishra, A., & Salokhe, V. M. (2008). Seedling characteristics and the early growth of transplanted rice under different water regimes. *Experimental Agriculture, 44*, 1–19.

Mishra, A., & Uphoff, N. (2013). Morphological and physiological responses of rice roots and shoots to varying water regimes and microbial densities. *Archives of Agronomy and Soil Science, 59*, 705–731.

Montero, O. (2010). *Using the system of rice intensification at El Pedregal in Costa Rica.* Retrieved June 6, 2017, from http://sri.ciifad.cornell.edu/countries/costarica/ElPedregalEng.html.

Moser, C. M., & Barrett, C. B. (2003). The disappointing adoption dynamics of a yield-increasing low external-input technology: The case of SRI in Madagascar. *Agricultural Systems, 76*, 1085–1100.

Namara, R. E., Bossio, D., Weligamage, P., & Herath, I. (2008). The practice and effects of System of Rice Intensification (SRI) in Sri Lanka. *Quarterly Journal of International Agriculture, 47*, 5–23.

Ndiiri, J. A., Mati, B. M., & Uphoff, N. (2013). Water productivity under the System of Rice Intensification from experimental plots and farm surveys in Mwea, Kenya. *Taiwan Water Conservancy, 61*, 63–75.

Padmavathi, C., Mahender, K. R., Subha, R. L. V., Surekha, K., Srinivas, P. M., Ravindra, B. V., & Pasala, I. C. (2009). Influence of SRI method of rice cultivation on insect pest incidence and arthropod diversity. *Oryza, 46*, 227–230.

Palanisami, K., Karunakaran, K. R., Amarasinghe, U., & Ranganathan, C. R. (2013). Doing different things or doing it differently? Rice intensification practices in 13 states of India. *Economic and Political Weekly, 48*, 51–58.

Pathak, M., Shakyar, R. C., Dinesh, S., & Shyam, S. (2012). Prevalence of insect pests, natural enemies and diseases in SRI (System of Rice Intensification) of rice cultivation in North East Region. *Annals of Plant Protection Sciences, 20*, 375–379.

Sato, S., & Uphoff, N. (2007). *A review of on-farm evaluations of system of rice intensification in Eastern Indonesia. CAB review of agriculture, veterinary science, nutrition and natural resources.* Wallingford: CABI.

Schlaeppi, K., & Bulgarelli, D. (2015). The plant microbiome at work. *Molecular Plant Microbe Interactions, 28*, 212–217.

Sharif, A. (2011). Technical adaptations for mechanized SRI production to achieve water saving and increased profitability in Punjab, Pakistan. *Paddy and Water Environment, 9*, 111–119.

Sheehy, J. E., Peng, S. B., Dobermann, A., Mitchell, P. L., Ferrer, A., Yang, J. C., Zou, Y. B., Zhong, X. H., & Huang, J. L. (2004). Fantastic yields in the system of rice intensification: Fact or fallacy? *Field Crops Research, 88*, 1–8.

Sinclair, T. R. (2004). *Agronomic UFOs waste valuable scientific resources. Rice Today* (Vol. 43). Los Baños: International Rice Research Institute.

Sinclair, T. R., & Cassman, K. G. (2004). Agronomic UFOs. *Field Crops Research, 88*, 9–10.

Sinha, S. K., & Talati, J. (2007). The impact of system of rice intensification (SRI) on paddy productivity: Results of a study in Purulia District, West Bengal, India. *Agricultural Water Management, 87*, 55–60.

Stoop, W. A. (2011). The scientific case for the system of rice intensification and its relevance for sustainable crop intensification. *International Journal of Agricultural Sustainability, 9*, 443–455.

Stoop, W. A., & Kassam, A. (2005). The SRI controversy: A response. *Field Crops Research, 91*, 357–360.

Styger, E., & Uphoff, N. (2016). *The System of Rice Intensification (SRI): Revisiting agronomy for a changing climate. Practice brief. Global alliance for climate-smart agriculture.* Rome: FAO.

Styger, E., Aboubacrine, G., Attaher, M. A., & Uphoff, N. (2011). The system of rice intensification as a sustainable agricultural innovation: Introducing, adapting and scaling up SRI practices in the Timbuktu region of Mali. *International Journal of Agricultural Sustainability, 9*, 67–75.

Sudhukar, T. R., & Reddy, P. N. (2007). *Influence of system of rice intensification (SRI) on the incidence of insect pests* (pp. 16–17). Agartala: 2nd National SRI Symposium October 3–5, Slides. Retrieved September 17, 2017, from http://www.slideshare.net/SRI. CORNELL/0729-influence-of-sri-on-the-incidence-of-insect-pests.

Thakur, A. K., & Uphoff, N. (2017). How the System of Rice Intensification can contribute to climate-smart agriculture. *Agronomy Journal, 109*, 1163–1182.

Thakur, A. K., Uphoff, N., & Antony, E. (2010a). An assessment of physiological effects of system of rice intensification (SRI) practices compared to recommended rice cultivation practices in India. *Experimental Agriculture, 46*, 77–98.

Thakur, A. K., Rath, S., Roychowdhury, S., & Uphoff, N. (2010b). Comparative performance of rice with system of rice intensification (SRI) and conventional management using different plant spacings. *Journal of Agronomy and Crop Science, 196*, 146–159.

Thakur, A. K., Rath, S., & Kumar, A. (2011). Performance evaluation of rice varieties under System of Rice Intensification (SRI) compared with conventional transplanting system. *Archives of Agronomy and Soil Science, 57*, 223–238.

Thakur, A. K., Rath, S., & Mandal, K. G. (2013). Differential responses of system of rice intensification (SRI) and conventional flooded-rice management methods to applications of nitrogen fertilizer. *Plant and Soil, 370*, 59–71.

Thakur, A. K., Uphoff, N., & Stoop, W. A. (2016). Scientific underpinnings of the System of Rice Intensification (SRI): What is known so far? *Advances in Agronomy, 135*, 147–179.

Thomas, V., & Ramzi, A. M. (2011). SRI contributions to rice production dealing with water management constraints in northeastern Afghanistan. *Paddy and Water Environment, 9*, 101–109.

Toriyama, K., & Ando, H. (2011). Towards an understanding of the high productivity of rice with system of rice intensification (SRI) management from the perspectives of soil and plant physiological processes. *Soil Science and Plant Nutrition, 57*, 636–649.

Turner, T. R., James, E. K., & Poole, P. S. (2013). The plant microbiome. *Genome Biology, 14*, 6 http://genomebiology.com/2013/14/6/209.

Uphoff, N. (2011). Agroecological approaches to 'climate-proofing' agriculture while raising productivity in the 21st century. In T. Sauer, J. Norman, & M. Sivakumar (Eds.), *Sustaining soil productivity in response to global climate change* (pp. 87–102). Hoboken, NJ: Wiley-Blackwell Table 7.2.

Uphoff, N. (2015). *The System of Rice Intensification (SRI): Responses to frequently-asked questions.* Ithaca, NY: SRI-Rice. http://sri.cals.cornell.edu/aboutsri/SRI_FAQs_Uphoff_2016.pdf

Uphoff, N. (2017). Developments in the System of Rice Intensification (SRI). In T. Sasaki (Ed.), *Achieving sustainable cultivation of rice* (Vol. 2). Cambridge: Burleigh-Dodds.

Uphoff, N., & Randriamiharisoa, R. (2002). Reducing water use in irrigated rice production with the Madagascar System of Rice Intensification. In B. A. M. Bouman, H. Hengsdijk, B. Hardy, P. S. Bindraban, T. P. Tuong, & J. K. Ladha (Eds.), *Water-wise rice production: Proceedings of the international workshop, 8–11 April 2002* (pp. 71–88). Los Baños: International Rice Research Institute.

Uphoff, N., Ball, A., Fernandes, E. C. M., Herren, H., Husson, O., Laing, M., Palm, C., Pretty, J., Sanchez, P., Sanginga, N., & Thies, J. (Eds.). (2006). *Biological approaches to sustainable soil systems*. Boca Raton, FL: CRC Press.

Uphoff, N., Kassam, A., & Stoop, W. (2008). A critical assessment of a desk study comparing crop production systems: The example of 'the system of rice intensification' versus 'best management practices. *Field Crops Research, 108*, 109–114.

Uphoff, N., Chi, F., Dazzo, F. B., & Rodriguez, R. J. (2013). Soil fertility as a contingent rather than inherent characteristic: Considering the contributions of crop-symbiotic soil microbiota. In R. Lal & B. Stewart (Eds.), *Principles of sustainable soil management in agroecosystems* (pp. 141–166). Boca Raton, FL: CRC Press.

Uzzaman, T., Sikder, R. K., Asif, M. I., Mehraj, H., & Jamal Uddin, A. F. M. (2015). Growth and yield trial of sixteen rice varieties under System of Rice Intensification. *Scientia Agriculturae, 11*, 81–89.

Vent, O., Sabarmatee, & Uphoff, N. (2016). The System of Rice Intensification and its impacts on women: Reducing pain, discomfort and labour in rice farming while enhancing household food security. In A. Fletcher & W. Kubik (Eds.), *Women in agriculture worldwide* (pp. 55–75). London: Routledge.

Visalakshmi, V., Rao, P. R. M., & Satyanarayana, N. H. (2014). Impact of paddy cultivation systems on insect pest incidence. *Journal of Crop and Weed, 10*, 139–142.

Wu, W., & Uphoff, N. (2015). A review of System of Rice Intensification in China. *Plant and Soil, 393*, 361–381.

Xu, F. Y., Ma, J., Wang, H. Z., Liu, H. Y., Huang, Q. L., & Ma, W. B. (2005). Rice quality under the cultivation of SRI. *Acta Agronomica Sinica, 31*, 577–582 in Chinese.

Yanni, Y. G., Rizk, R. Y., Abd El-Fattah, F. K., Squartin, A., Corich, V., Giacomini, A., de Bruijn, F., Rademaker, J., Maya-Flores, J., Ostrom, P., Vega-Hernandez, M., Hollings-worth, R. I., Martinez-Molina, E., Mateos, P., Velazquez, E., Wopereis, J., Triplett, E., Umali-Garcia, M., Anarna, J. A., Rolfe, B. G., Ladha, J. K., Hill, J., Mujoo, R., Ng, P. K., & Dazzo, F. B. (2001). The beneficial plant growth-promoting association of *Rhizobacterium leguminosarum* bv. *trifolii* in rice roots. *Australian Journal of Plant Physiology, 28*, 845–870.

Yuan, L. P. (2002). A scientist's perspective on experience with SRI in China for raising the yields of super hybrid rice. In N. Uphoff, E. Fernandes, L. P. Yuan, J. M. Peng, S. Rafaralahy, & J. Rabendrasana (Eds.), *Assessments of the System of Rice Intensification (SRI)* (pp. 23–25). Ithaca, NY: Cornell International Institute for Food, Agriculture and Development http://ciifad.cornell.edu/sri/proc1/sri_06.pdf.

Zhang, J. G., Chi, Z. Z., Li, X. Y., & Jiang, X. L. (2013). Agricultural water savings possible through SRI for water management in Sichuan, China. *Taiwan Water Conservancy, 61*, 50–72.

Zhao, L. M., Wu, L. H., Li, Y. S., Lu, X. H., Zhu, D. F., & Uphoff, N. (2009). Influence of the System of Rice Intensification on rice yield and on nitrogen and water use efficiency with different N application rates. *Experimental Agriculture, 45*, 275–286.

Zhao, L. M., Wu, L. H., Li, Y. S., Animesh, S., Zhu, D. F., & Uphoff, N. (2010). Comparisons of yield, water use efficiency, and soil microbial biomass as affected by the System of Rice Intensification. *Communications in Soil Science and Plant Analysis, 41*, 1–12.

Chapter 13
Efficient Desalinated Water Pricing in Wetlands

Oscar Alfranca

Abstract Pricing desalinated water in wetlands can be inefficient whenever positive and negative externalities are not integrated in final prices. Externalities values are usually related to non-market magnitudes, and great difficulties exist for their precise calculation. In order to provide efficient prices, a methodology is proposed in which automatic prices could be directly estimated with the use of the Travel Cost methodology. In this chapter a dynamic model reflecting eventual differences between optimal social prices for environmental uses of desalinated water and private water prices is proposed. This model is based on a short-run dynamic model, in which socially efficient prices are calculated. In order to apply this methodology to real desalinated water problems in wetlands it is suggested that the calculation of the travel cost, along with the value transfer method price for desalinated water should have to be centralized in one office of the park receiving all information from visitors. Direct links of this office with the technical services of the park should have to be established.

Keywords Travel cost · Value transfer method · Externality · Environmental valuation

13.1 Introduction

Wetlands are natural resources which constitute an essential source of benefits for towns and rural areas, and a rising interest exists in the restoration of wetland areas and in constructed wetlands with the use of reclaimed water. Desalinated water suitably treated for subsequent use provides an opportunity to achieve this result and restore wetlands. Desalinated water can be used not only to restore natural wetland and constructed wetlands but also to create wetlands in peri-urban parks. This chapter proposes a method for desalinated water pricing in these particular cases.

Economic benefits of water reuse are well known when water is used for irrigation in agriculture, in urban or peri-urban gardening, in urban applications (such as

O. Alfranca (✉)
Departament d'Enginyeria Agroalimentària I Biotecnologia, Campus del Baix Llobregat
Universitat Politècnica de Catalunya, Barcelona, Spain
e-mail: oscar.alfranca@upc.edu

© Springer Nature Switzerland AG 2019
A. Sarkar et al. (eds.), *Sustainable Solutions for Food Security*,
https://doi.org/10.1007/978-3-319-77878-5_13

street cleaning), or in crop fields that can result in a net increase of environmental services and therefore a net benefit to society. However, when wastewater is used in the context of environmental restoration projects, the economic value is not so evident, and usually wastewater prices which are commonly applied, are not efficient prices, but just imprecise and ambiguous approximations to a private cost in a production process. The main reason to use these approximations is commonly the enormous difficulties in the calculation of efficient prices for wastewater environmental uses. In this situation, no clear economic value for water exists, and methods to calculate the economic valuation of water should be applied.

Recreational activities are a very relevant output of wetlands. It is essential to point out that the outcomes of any valuation process depend on what the various stakeholders value, whose values count, who benefits, and the way in which social and ecological systems interlinkages are accounted for. Values and the process of valuation reflect socially and culturally constructed realities linked to worldviews, mind-sets and belief systems shaped by social interactions, as well as political and power relations operating within a realm of local, regional, and global interdependencies (Wilk and Cliggett 2006; Hornborg et al. 2007). Thus, the choice of environmental valuation methods also involves selecting the sociocultural context which emerges from the understanding of what values are (or should be), and how they should be elicited. Valuation methods suggest models of nature and their interactions and define whether values are revealed, discovered, or constructed (Vatn and Bromley 1994). From this perspective, valuation methods act as value-articulating institutions by defining a set of rules concerning valuation processes (Jacobs 1997).

An essential point in the application of environmental valuation methods is the degree of resources and stakeholders involvement which is required. Hence, it is always necessary to point out that values are always estimated, and no certainty exists about the accuracy of the calculated prices to approximate real prices.

There are three main categories of valuation methods: those based on markets, those based on revealed preferences, and those based on stated preferences. Methods based on markets estimate the value using market prices to value goods and services, or estimating the value through the avoided cost of prevented environmental damage, the costs of substitutes or mitigation and restoration options as indicators of value. Revealed preferences methods are the travel cost method (to estimate the value of a protected area through the amount of time and money people spend to visit it) and the hedonic pricing method (using changes in property prices due to changes in the surrounding environment as an indicator of landscape value). Stated preferences methods, such as the contingent valuation approach, are based on asking people about the willingness to pay for environmental protection (such as an improvement in water quality in wetlands) or to accept a compensation for a reduction in the environmental quality. A classical criticism to the contingent valuation method is that given that the choice of answers are hypothetical and does not require actual payments, respondents may say that they are willing to pay more than they would, if they were faced with real choices.

An efficient method for desalinated water pricing should have to consider the availability of information (and in detail enough, taking also into consideration the costs associated with data collection), so that the relevant calculations needed for taking into account the principle of recovery of the costs of water services could be applied. Hence, necessary estimates of the volume, prices and costs associated with water services, and also estimates of relevant investment including forecasts of such investments and judgements about the most cost-effective combination of measures in respect of water uses could be properly formulated. These principles are recognized in the European Water Framework Directive.

Three fundamental economic concepts to reflect the nonexistence of efficient markets are needed in desalinated water pricing for environmental uses in wetlands: externality, public good, and ecosystemic service. An externality exists when firms or individuals impose costs or benefits on other agents in society and when these costs or benefits are not compensated, or cannot be appropriately priced or allocated using market mechanisms. In some cases, externalities can be priced or be assigned in the market; nevertheless, consumption and uses of these goods are non-rival and non-excludable. A good is non-rival or indivisible when a unit of the good can be consumed by one individual without detracting in the slightest from the consumption opportunities still available to others from the same unit (Cornes and Sandler 1986). A public good is both non-rival (the consumption of a unit does not reduce the units available for others) and non-excludable (it is not possible to include some while excluding others from this good). In some cases, wetlands behave like public goods, and generating environmental and social externalities. Constructed wetland parks in peri-urban areas could be an example. The existence of externalities is one of the main issues related to environmental uses of wetlands and created wetlands, because many of these wetland services also hold the characteristics of a public good. Public goods are those that once they are provided (in the case of wetlands, by nature), all get to consume it equally. Public goods present the two needed properties: non-rivalness and non-exclusion. Wetlands are non-rival because its use or consumption by one agent does not diminish the amount available to other agents. The condition of non-exclusion implies that it is prohibitively expensive to exclude someone from using or consuming the wetland.

The Millennium Ecosystem Assessment defines ecosystem services as the benefits people derive from ecosystems. These ecosystems provision services or goods like food, wood, and other raw materials; plants, animals, fungi, and microorganisms provide essential regulating services such as pollination of crops, prevention of soil erosion, and water purification, and also a wide vast collection of cultural services (usually related to recreation) (Millenium Ecosystem Assessment 2005). Ecosystem services are the benefits of nature to households, communities, and economies (Boyd and Banzhaf 2007). So ecosystem services of wetlands should be those beneficial outcomes for the natural environment that result from ecosystem functions of wetlands. Some examples of wetland ecosystem functions are provision of wildlife habitat and scenic views, carbon cycling, agricultural production, or trapping nutrients. Valuing ecosystem services from wetlands implies the recognition of related institutions, as manifestations of values, institutions influence preferences, choices, and actions of various stakeholders linked to wetlands. Valuation techniques provide a tool for self-reflection, alerting us to the consequences of our choices and behavior on various dimensions of natural and human capital (Zavestoski 2004). It is therefore an important institution in itself to engendering changes in the way societies respond to the crises of continued wetland loss and degradation.

In order to price desalinated water in wetlands efficiently all these roles and values of wetlands providing key ecosystem services should have to be fully appreciated and integrated into the different scales of water pricing decision making. The incomplete understanding of the diverse decision levels (local, regional, and international) could provide imperfect protection to these wetland ecosystem services, whose values are not usually reflected in water markets (such as water purification, flood and storm protection or nutrient cycling).

Regarding the multiple services offered by wetlands, conducting a primary valuation of the total economic value of wetlands is considered an expensive and time-consuming effort, without which is not commonly receiving a similar reward. When human and monetary resources are a main restriction or when valuation data are limited or simply nonexistent, the value transfer technique could be a useful tool. The value transfer method commonly refers to the replacement of monetary environmental values estimated at one location (study site) to another site (policy site) through the application of any economic valuation technique (Brouwer 2000; Brouwer et al. 1999; Troy and Wilson 2006; Johnston and Rosenberger 2010). This method is the academic debate over the validity of the method continues (Brander et al. 2006; Russi et al. 2013), and it is usually recognized that a primary valuation research could be generally a better strategy for gathering information about the value of ecosystem goods and services. However, when this strategy is not practical, value transfer represents a meaningful alternative and an acceptable basis to develop environmental management and policy strategies.

Valuing all services generated by wetland restoration would yield the value of total ecosystem value changes. However, due to data and model limitations, only partial estimations are usually obtained, and therefore, that not all attributes from wetlands are actually integrated in the valuation estimation. This mistake could induce to wetlands undervaluation, particularly in productive uses of wetlands generating marketable goods and services (such as agriculture or tourism). That is, mistakes in the calculation of the value of wetland services should lead to their omission in public decision making processes related to the conservation of wetlands. Without information on the economic value of wetland services that can be compared directly against the monetary value of alternative public investments, the importance of wetland natural capital has tended to be ignored (Brander et al. 2013). Wetlands in agricultural landscapes provide valuable ecosystem services that contribute to human wellbeing, including provisioning (e.g., food, fuelwood, water), regulating (e.g., flood control, water quality, water supply), habitat (e.g., biodiversity), and cultural services, like recreation, aesthetic or the mere existence of the wetland (Russi et al. 2013).

The main objective of this chapter is to characterize a clear and accessible method for desalinated water pricing which could jointly consider both private and social costs and benefits. This price should be efficient and therefore, not only private costs, but also all related externalities should have to be reflected in prices. The proposal consists mainly of a mechanism based in a combination of the travel cost method and the value transfer method, for desalinated water pricing in wetlands, which could become a tool for the automatic correction of water market deviations from equilibrium, and keep desalinated prices close to an environmental optimal.

There are two main purposes in this chapter. First, price correction from deviations could be automatic, and second, price adjustments from market deviations could be made from a multidisciplinary perspective. The analysis of this function and the main assumptions that are needed to acquire a tool for desalinated water pricing in wetlands can be found in Sect. 13.3. Then, a discussion of this function and the main critiques of the method are proposed. Finally, the chapter concludes.

13.2 Methods

In the first part of this section, a description of the main environmental valuation methods applied to wetland valuation is introduced. In the second part, a method for desalinated water pricing in wetlands is proposed.

13.2.1 Environmental Valuation Methods for Wetlands

In this section, an exposition and some applications of environmental valuation techniques to the valuation of wetlands is introduced. A general review of environmental valuation methods can be found in Freeman (1993), and a comprehensive revision of the literature on wetlands valuing can be also found in Woodward and Wui (2001), Boyer and Polasky (2004), Brander et al. (2006), Turner et al. (2008), Ghermandi et al. (2010), and Drechsel et al. (2015).

The travel cost approach uses information about the number of trips to particular sites and the cost of these travels to infer how much individuals are willing to pay for the access to the site. Travel cost studies to wetlands are applied mainly to estimate recreation values, in which people travel to wetlands to hunt, fish, hike, or simply watch wildlife, among other activities. Whenever externalities exist, no market could reflect efficiently the value of these kinds of goods. Therefore, empirical valuation techniques need to be used. Given that in this chapter a recreational use of wetlands is taken into account, the application of the travel cost method for wetlands is proposed as an empirical approach.

In the papers by Seguí et al. (2009) and Alfranca et al. (2011), externalities, costs, and benefits are estimated using the travel cost method. All costs and benefits (both social and private) are considered in the economic and technological analysis of wastewater pricing in the constructed wetland. The general concern of the wastewater pricing method proposed in Seguí et al. (2009) is that it allows for an improvement in the design of investment decisions. Traditionally, an economic-financial analysis of wastewater reclamation and reuse systems focuses exclusively on the study of costs and private benefits. The methodology that is proposed in this paper incorporates not only the private costs and benefits, but also the related spillovers. In this research, the use value of the reclaimed water in the Natural Wildlife Park of Aiguamolls de l'Emporda, in Northeastern Spain, is estimated. The travel cost technique is applied to calculate the external value of the Park. According to the final results, the efficient price of the reclaimed water in the Park should have to be between 0.75 and 1.20 euros/m^3.

Desalinated water prices for a created wetland located in a peri-urban park in Catalonia (Spain), are calculated in Alfranca et al. (2011). The wetland, which covers an area of 1 ha, was constructed in 2003 and receives a secondary treated wastewater flow of between 100 and 250 m^3/day. Wetland externalities were evaluated using the travel cost method. The value of the wetland is expressed in terms of the price of the water that flows through the system, which is estimated to range from 0.71 to 0.75 h/m^3. The value of positive externalities (1.25 euros/m^3) was greater than private costs (from 0.50 to 0.54 h/m^3). These results constitute empirical evidence that created wetlands in peri-urban parks can be considered to be a source of positive externalities when used in environmental restoration projects focusing on the reuse of treated wastewater. This study also illustrates the small influence of the hydraulic infrastructure depreciation costs on the private costs of constructed wetlands (less than 10%), and the low investment costs of constructed wetlands in comparison with operation and maintenance costs (less than 10% of total private costs).

Regarding the use of desalinated water for recreation uses in wetlands, some papers have considered the joint use of travel cost and contingent valuation. A review of design challenges and applications can be found in (Wendong et al. 2014). In some cases, the Travel Cost Method (TCM) generates higher estimates of value than the Contingent Valuation Method (CVM), even though the latter is commonly associated with potential problems of hypothetical and strategic bias. Both methods are proposed in Rolfe and Dyack (2011) to estimate the recreational values associated with the Coorong on the Murray River in south-eastern Australia, and values per adult visitor per recreation day are estimated to be $149 with the TCM, and $116 with the CVM. In the paper, no methodological and framing issues to explain these value differences are tested. In summary, while no single methodological or framing issue could be identified that would reconcile the difference between TCM and CVM prices, a combination of factors that could induce the systematic adjustments in consumer surplus values might appear. The evidence in this study suggests that the most important of these factors are the consideration of substitute sites, strategic responses and the treatment of uncertain responses within the CVM. In Alora and Nandagari (2015), a positive difference between the evaluations of the economic value of Pilikula Lake, using TC than CV are obtained. This result is mainly attributed to the condition of the country in which these methods are applied. So, according to the authors, positive differences for the TC method are mainly due to the fact that India can be considered to be a developing country when compared to high income developed countries. Therefore, the essential point to justify disparities stems from the observed behavior of respondents in actual markets, whereas CVM is based on expressed or stated preferences (and all social aspects related to this estimation). A main result from this paper is that getting information from respondents through revealed preferences and stated preferences are difficult tasks that could be much attributed to economic and social variables, and mainly to the visitor's reserve to expose the willingness to pay values, given which are the local political and socioeconomic factors.

The hedonic prices (HP) method applies observed market prices for multi-attribute goods with many characteristics that contribute to its value to uncover the value of particular characteristics for which there is no readily available signal of value. A main hypothesis in these models is that houses are not homogenous, and could differ in respect to a variety of characteristics. Some of them are rather conventional when pricing properties (such as number of bathrooms, size of the kitchen, or the existence of a garden), but other characteristics could be location-specific (such as the existence of wetlands or the biodiversity of a Park located not far from the neighborhood).

The HP method can also be used for the environmental valuation of wetlands. The HP method has been widely used for the valuation of properties. A fundamental hypothesis in this method is that individuals are informed on all alternatives and are free to choose a house anywhere in the market. In the model, house prices can be influenced by different location-specific characteristics. This method is based on the hypothesis that houses are not homogenous goods, and hence that house prices could differ according to a variety of characteristics, such as the number of bedrooms, proximity to parks, or location-specific environmental variables such as the existence of an urban wetland park.

The hedonic approach can be used to value wetlands given that house prices contain a capitalized value related to wetland proximity. Hence, buyers need to pay for the wetland value in the form of higher house prices (Alora and Nandagari 2015). Some papers that have applied the hedonic method to estimate the value of urban wetlands in the form of higher house prices are Loomis and Feldman (2003), Boyer and Polasky (2004), and Lupi Jr. et al. (1991). Results in some cases allow concluding that the growing number of new houses has determined artificial wetlands to add extra environmental appeal to these properties, and that significant relationships exist between sales prices and proximity to wetlands. More specifically, in Perth (Western Australia), results indicate that a significant relationship exists between property prices and distances to wetlands (Doss and Taff 1996).

Another way to value services provided by wetlands is to estimate the replacement cost of delivering these services, if wetlands no longer existed. The most famous example of valuing ecosystem services involves pricing the cost of replacing the water filtration services provided by undeveloped watersheds with a drinking water filtration plant. The high replacement cost ($6–$8 billion in 1996) led public officials to protect the watersheds in the Catskills as a cheaper means of providing clean drinking water for New York City (Mahan et al. 2000).

On the other hand, in some cases, there is no observable behavior (such as the purchase of a house or a travel to a recreational site), direct production links to commodities, or replacement alternatives that could be used to generate estimates of value. For such cases, the only approaches capable of generating estimates of value should have to be stated preference methods. The most commonly used stated preference method is contingent valuation, in which respondents are asked whether they would be willing to pay a specified amount for some environmental amenity. By modifying the size of the payment amount across different respondents, one can trace out the demand curve for the environmental amenity and estimate the mean willingness to pay of people in the sample for that amenity.

Contingent valuation techniques have been widely applied in the estimation of an extensive collection of environmental services, and a thorough revision of wetland valuation literature can be found in Tapsuwan et al. (2009). Conjoint analysis also uses surveys to gather information but, unlike contingent valuation, it is not exclusively focused on generating estimates of willingness to pay. Conjoint analysis asks people what tradeoffs they are willing to make between different sets of choice attributes. In the context of wetlands, a survey may ask whether a wetland that provided better habitat for fish was preferable to a wetland that offered more bird habitat and better flood control. If one of the choice attributes is cost, then willingness-to-pay estimates can also be generated.

Critics of contingent valuation also point out that because the choice is hypothetical and does not require a real payment, respondents may say they are willing to pay more than they would if they were faced with an actual choice. A classical discussion on this point can be found in Diamond and Hausman (1994), Chichilnisky and Heal (1998), and Carson and Haanemann (2005).

The paper by Sharma et al. (2015) presents a combination of direct market price and unit adjusted value transfer methods to estimate the economic value of the major direct and indirect uses of the wetland ecosystem services provided by the Koshi Tappu Wildlife Reserve, in Nepal. The value transfer method constitutes the application ecosystem services values from an original study site, to a similar yet different study site. The value transfer is increasingly recognized as a useful method for assessing ecosystem services when it is not feasible due to budget and time constraints (too expensive or time-consuming) to produce primary economic valuation studies. The method is based on the hypothesis that the economic value of ecosystem goods or services at a study site can be determined with sufficient accuracy by analyzing existing valuation studies in some other comparable sites. There are three main approaches for making this transference: transferring the unit value, transferring the benefit function, and employing meta-analysis (Wilson et al. 2006).

Although a significant amount of literature has addressed the ecological issues of restoration outcomes at the scale of individual wetlands (Navrud and Ready 2007; Troy and Wilson 2006), little research can be found on the aggregate, dynamic behavior of wetlands, via permitted actions, at the landscape level. Therefore, problems appear when wetlands are destroyed and their compensatory mitigation payments are initiated, given that changes and modifications in the landscape should imply a temporary net loss of wetlands. The main issue at stake is whether it is possible or not to prevent this from occurring, or to compensate for it applying preventive methods and policies at different scales (either local or international).

13.2.2 Empirical Model: A Proposal

In this section a dynamic model reflecting eventual differences between optimal social prices for environmental uses of desalinated water prices is proposed. This model is based on a very simple short-run dynamics model, with socially efficient prices, which are calculated using an empirical valuation methodology, for recreational uses of wastewater in wetlands (such as the travel method cost could be), and with the help of the value transfer method.

In order to apply this model, the calculation of the travel cost price for desalinated water for wetlands should have be centralized in one office of the park in which this information is introduced, once visitors arrive to the Park:

1. Transport costs. This variable should be estimated considering the transport vehicle to reach the park. So if it is a car, then the volume of fuel consumed during the journey should be a good indicator. In some cases in which no special vehicle is needed, the opportunity cost of time could be accepted.
2. Recreational costs. This variable is calculated from the need to have some food during the journey or at the recreational site, or the need to spend several nights in accommodation either during the journey or at the recreational site, given the local nature of the park.
3. Opportunity costs. This variable should represent the opportunity cost of leisure time, and it is based on the hypothesis that a person who devotes time for leisure activities is paying for it, because no wages are received during this time. Salaries are usually expected as a proxy variable that reflects the opportunity costs of taking part in this recreational activity.

Once travel costs are calculated, the evolution in the market could be calculated with the differences that could exist from the evolution of the market. This difference could be calculated from different time periods. The higher the frequency, the more efficient the price policy (the ideal should be that changes in market and externalities were automatically incorporated into prices).

These changes could be characterized with a very simple equation such as:

$$\Delta P_t = P_t - P_{t-1} \tag{13.1}$$

Where P_t is the optimal wastewater price for wetland environmental uses, and P_{t-1} is the same price in the previous period. P_t represents the optimal price whenever market is efficient and all externalities are considered in price formation. Given that prices are calculated both including private market conditions and a subsidy estimated applying the Travel Cost method, it could be possible that in some cases these prices could be different. ΔP_t is a measure of these differences between the two prices, and will be equal to zero if markets behave properly and externalities are compensated.

So once an efficient price is calculated using the travel cost method, an automatic mechanism is applied in this equation: This very simple model presents a dynamic form:

$$P_t = b_0 + b_1 P_{t-1} + b_2 P_{t-2} + P_{t-1} + u_t \tag{13.2}$$

Nevertheless, since P_t and P_{t-1} will rarely be in equilibrium, a more general relationship between prices is modelled in which this hypothesis about the existence of a short-run disequilibrium between prices is accepted. This relationship takes the form:

$$P_t = b_0 + b_1 P_{t-1} + b_2 P_{t-2} + \mu P_{t-1} + u_t \tag{13.3}$$

in which $0 < \mu < 1$. Adding and subtracting $b_1 P_{t-1}$ from the right side of (13.2), then yields:

$$P_t - P_{t-1} = b_0 + b_1 P_{t-1} - b_1 P_{t-2} + b_1 P_{t-2} + b_2 P_{t-2} - (1 - \mu) P_{t-1} + u_t \tag{13.4}$$

or

$$\Delta P_t = b_0 + b_1 \Delta P_{t-1} + (b_1 + b_2) P_{t-2} - \lambda P_{t-2} + u_t \tag{13.5}$$

Where $\lambda = 1 - \mu$. Equation (13.3) can be reparametrized as follows:

$$\Delta P_t = \mu + b_0 \Delta P_{t-1} - \lambda (P_{t-1} - \beta_1 P_{t-2}) + u_t \tag{13.6}$$

Now a new parameter is defined: $\beta_t = (b_1 + b_2)/\lambda$, and Eq. (13.4) can also be reparametrized as

$$\Delta P_t = b_1 \Delta P_{t-1} - \lambda (P_{t-1} - \beta_0 - \beta_1 P_{t-2}) \tag{13.7}$$

where $\beta_0 = b_0/\lambda$, is a second new parameter. This term is formed by an exogenous constant b_0 and the λ term. The value of λ indicates the percentage of this water market disequilibrium which is corrected in the current period (it could be 1 week if wastewater prices were fixed weekly). If the sign of the coefficient is positive this implies that the disequilibrium should increase every week, in this percentage.

The constant β_0 reflecting all variables that even they have an influence on prices, they act exogenously (such as cultural, social, or climatic variables). The λ term directly provides a measure of the short-run effects of changes in the price of the socially efficient price of wastewater on the price of present wastewater prices. The value of λ indicates the intensity and the speed of the correction for a market disequilibrium and the intensity and the speed of the correction for a market disequilibrium. The value of λ indicates the percentage of this water market disequilibrium which is corrected in the current period (it could be 1 week if wastewater prices were fixed weekly). If the sign of the coefficient is positive this implies that the disequilibrium should increase every week, in this percentage. The value of λ indicates the intensity and the speed of the correction in a market disequilibrium.

In Eq. (13.5), β_1 appears in the equilibrium relationship being the long run elasticity of P_t with respect of P_{t-1}. Coefficient b_1 appears in the short-run disequilibrium relationship, and reflects the immediate response of P_t to changes in P_{t-1}. That is, it can be considered to be short-run elasticity of wastewater price.

13 Efficient Desalinated Water Pricing in Wetlands

Table 13.1 Desalinated water prices

Units	Private price[a]	Urban wetland, social price[b]	Natural wetland park, social price[c]
Euros/m^3	0.57	0.75	1.25

[a]Aparicio et al. (2015)
[b]Alfranca et al. (2011)
[c]Seguí et al. (2009)

The term $(P_{t-1} - \beta_0 - \beta_1 P_{t-2})$ can be understood to be a device for the correction of socially efficient prices which are deviated from equilibrium. Given that usually water trading in water markets is rather "lengthy, cumbersome, and lacking in transparency" (BenDor et al. 2009), and since λ lies between zero and unity, only part of this disequilibrium should be made up for in the present period. This is an especially big concern during droughts, when the speed in adjustment is critical.

13.3 Private Market Versus Social Environmentally Efficient Desalinated Water Prices: An Example in the South East of Spain

In this section, a proposal is made in which, on the basis of the travel cost method, the benefit transfer methodology is used to quantify the ecosystem services which could be obtained from the use of desalinated water in wetlands. The estimation is made by transferring available information from studies already completed in another location and/or context. More specifically, in this study, a combination of private water prices and desalinated water social prices is made. Private prices for desalinated water are directly obtained from observations in an irrigated area of South-Eastern Spain (an orchard territory and vegetable garden area). In this example, social price for desalinated water (i.e., price plus externality linked to the desalination process) is taken from prices in a similar Mediterranean area which are estimated in Alfranca et al. (2011) (urban externality) and in Seguí et al. (2009) (natural park externality). These estimated environmental prices and real market prices are used to calculate the socially efficient price for desalinated water in constructed wetlands, in Murcia (Table 13.1). So the application of this method should allow to keep efficient prices for environmental uses of desalinated water. Estimation of these environmentally efficient prices could be improved using directly an estimation method, and not just an approximation to a similar value.

13.4 Discussion and Conclusion

The selection of the right environmental valuation methodology to estimate an efficient desalinated water price is both a problematic and challenging issue in desalinated water pricing policies. The main conflict relies in the need to consider private decision variables, along with institutional variables (such as social and environmental conditions). Final decisions of desalinated water companies considering both vectors together should be different than a linear sum of private and social variables, given that interactions might exist between both kinds of variables. This interaction could make the decision process nonlinear and hardly predictable.

The proposed method for pricing could be applied without problems under perfect competition conditions, but obliges to be especially careful when relevant external variables switching frequently exist that could affect not only environmental conditions but also the behavior of markets. In order to apply this methodology to real desalinated water problems in wetlands it is suggested that the calculation of the travel cost price for desalinated water should have to be centralized in one office of the park receiving all information from visitors. Direct links of this office with the technical services of the park should have to be established.

Desalinated water market structure is a fundamental factor which could have a very relevant influence on water pricing, unless some public intervention is made. A conflict exists in desalinated water production between the need for big efficient infrastructures, and inefficiency in pricing related to natural monopoly circumstances. Desalinated water prices could be hardly calculated under strong oligopolistic conditions, and the water market for wetlands could present inefficiencies, both from private and social perspectives, and hence the significance of this method for desalinated water pricing.

Changes in the quality of the desalinated water composition could be extremely difficult to reflect when pricing because of the eventual changes in the environmental values of desalination water components, unless the estimation procedures could be easily updated.

The proposed procedure for automatically pricing of environmental uses in wetlands could be useful under some general expected conditions. Nevertheless, problems could appear (both private and social) when some unpredictable modifications exist, both in water markets and in environmental conditions of wetlands. If this was the case, then some public pricing policy intervention could be needed.

The main interest of the proposed methodology for the desalination systems management is that all costs and benefits (both social and private) are considered in the economic and technological analysis. That is, it takes into consideration not only private impacts from the use of desalination water in wetlands but also spillovers, which have been proved to be relevant for efficient water pricing.

References

Alfranca, O., Garcia, J., & Varela, H. (2011). Economic valuation of a created wetland fed with treated wastewater located in a peri-urban park in Catalonia, Spain. *Water Science and Technology, 63*(5), 891–898.

Alora, J., & Nandagari, L. (2015). Evaluation of economic value of Pilikuka Lake using travel cost and contingent valuation methods. *Aquatic Procedia, 4*, 1315–1321.

13 Efficient Desalinated Water Pricing in Wetlands
267

Aparicio, J., Alfranca, O., Jimenez-Martinez, J., Garcia-Arostegui, J.L., Candela, L., & Lopez, J.L. (2015). Groundwater salinity process, mitigation measures and economic assessment: An example from an intensive agricultural area. 42th IAH Congress, 13–18 September 2015, Rome, Italy. Abstract Book, 290.

BenDor, T., Shoeltes, J., & Doyle, M. W. (2009). Landscape characteristics of a stream and wetland mitigation banking program. *Ecological Applications, 19*(8), 2078–2092.

Boyd, J., & Banzhaf, S. (2007). What are ecosystem services? The need for standardized environmental accounting units. *Ecological Economics, 63*(2–3), 616–626.

Boyer, T., & Polasky, S. (2004). Valuing urban wetlands: A review of non-market valuation studies. *Wetlands, 24*(4), 744–755.

Brander, L., Brower, R., & Wagtendonk, A. (2013). Economic valuation of regulating services provided by wetlands in agricultural landscapes: A meta-analysis. *Ecological Engineering, 56*, 89–96.

Brander, L. M., Florax, R. J. G. M., & Vermaat, J. E. (2006). The empirics of wetland valuation: A comprehensive summary and a metaanalysis of the literature. *Environmental and Resource Economics, 33*(2), 223–250.

Brouwer, R. (2000). Environmental value transfer: State of the art and future prospects. *Ecological Economics, 32*, 137–152.

Brouwer, R., Powe, N. A., Turner, R. A., Bateman, I. J., & Langford, I. H. (1999). Public attitudes to contingent valuation and public consultation. *Environmental Values, 8*(3), 325–347.

Carson, T., & Haanemann, W. (2005). Contingent valuation. In K.-G. Mäler & J. R. Vincent (Eds.), *Handbook of environmental economics* (pp. 821–936). London: North Holland.

Chichilnisky, G., & Heal, G. (1998). Economic returns from the biosphere. *Science, 391*, 629–630.

Cornes, R., & Sandler, T. (1986). *The theory of externalities, public goods, and club goods* (2nd ed.). Cambridge: Cambridge University Press.

Diamond, P. A., & Hausman, J. A. (1994). Contingent valuation: Is some number better than no number? *The Journal of Economic Perspectives, 8*(4), 45–64.

Doss, C. R., & Taff, S. J. (1996). The influence of wetland type and wetland proximity on residential property values. *Journal of Agricultural and Resource Economics, 21*, 120–129.

Drechsel, P., Qadir, M., & Wichelns, D. (Eds.). (2015). *Wastewater*. London: Springer.

Freeman, M. (1993). *The measurement of environmental and resource values*. Washington, DC: Resources for the Future.

Ghermandi, A., van den Bergh, J. C., Brander, L. M., de Groot, H. L. F., & Nunes, P. A. L. D. (2010). The values of natural and human-made wetlands: A meta-analysis. *Water Resources Research, 46*, 1–12.

Hornborg, A., McNeill, J., & Martinez-Alier, J. (2007). *Rethinking environmental history: World-system history and global environmental change*. Lanham, MD: Altamira Press.

Jacobs, M. (1997). Environmental valuation, deliberative democracy and public decision-making. In J. Foster (Ed.), *Valuing nature: Economics, ethics and environment* (pp. 211–231). London: Routledge.

Johnston, R., & Rosenberger, R. (2010). Methods, trends and controversies in contemporary benefit transfer. *Journal of Economic Surveys, 24*(3), 479–510.

Loomis, P., & Feldman, M. (2003). Estimating the benefits of maintaining adequate lake levels to homeowners using the hedonic property method. *Water Resources Research, 39*, 21–26.

Lupi, F. Jr., Graham-Thomasi, T., & Taff, S. (1991). A hedonic approach to urban wetland valuation. Staff paper 91–98. Department of Agricultural and Applied Economics, University of Minnesota Minneapolis, MN.

Mahan, B. L., Polasky, S., & Adams, R. M. (2000). Valuing urban wetlands: A property price approach. *Land Economics, 76*, 100–113.

Millennium Ecosystem Assessment. (2005). Synthesis reports. In *Millennium ecosystem assessment*. Washington, DC: Island Press.

Navrud, S., & Ready, R. (2007). *Environmental value transfer: Issues and methods* (p. 290). Dordrecht: Springer.

Rolfe, J., & Dyack, B. (2011). Valuing recreation in the Coorong, Australia, with travel cost and contingent behaviour models. *The Economic Record, 87*, 282–293.

Russi, D., ten Brink, P., Farmer, A., Badura, T., Coates, D., Förster, J., Kumar, R., & Davidson, N. (2013). *The economics of ecosystems and biodiversity for water and wetlands. Final consultation draft.* London: TEEB.

Seguí, L., Alfranca, O., & García, J. (2009). Techno-economical evaluation of water reuse for wetland restoration: A case study in a natural park in Catalonia, Northeastern Spain. *Desalination, 246,* 179–189.

Sharma, B., Rasul, G., & Chetri, N. (2015). The economic value of wetland ecosystem services: Evidence from the Koshi Tappu Wildlife Reserve, Nepal. *Ecosystem Services, 2015*(12), 84–93.

Tapsuwan, S., Ingram, G., Burton, M., & Brennan, D. (2009). Capitalized amenity value of urban wetlands: A hedonic property price approach to urban wetlands in Perth, Western Australia. *Australian Journal of Agricultural and Resource Economics, 53,* 527–545.

Troy, A., & Wilson, M. A. (2006). Mapping ecosystem services: Practical challenges and opportunities in linking GIS and value transfer. *Ecological Economics, 60,* 435–449.

Turner, R., Georgiou, S., & Fisher, B. (2008). *Valuing ecosystem services. The case of multifunctional wetlands.* London: Earthscan.

Vatn, A., & Bromley, D. (1994). Choices without prices without apologies. *Journal of Environmental Economics and Management, 26,* 129–148.

Wendong, T., Bays, J. S., Meyer, D., Smardon, R., & Levy, Z. F. (2014). Constructed wetlands for treatment of combined sewer overflow in the US: A review of design challenges and application status. *Water, 2014*(6), 3362–3385.

Wilson, A., Hoehn, J. P., & Hoehn, J. P. (2006). Valuing environmental goods and services using benefit transfer: The state-of-the art and science. *Ecological Economics, 60,* 325–342.

Wilk, R., & Cliggett, L. (2006). *Economies and cultures: Foundations of economic anthropology* (2nd ed.). Boulder: Westview Press.

Woodward, R. T., & Wui, Y. S. (2001). The economic value of wetland services: A meta-analysis. *Ecological Economics, 37,* 257–270.

Zavestoski, S. (2004). Constructing and maintaining ecological identities: The strategies of deep ecologists. In S. Clayton et al. (Eds.), *Identity and the natural environment: The psychological significance of nature* (pp. 297–316). Cambridge, MA: The M.I.T. Press.

Chapter 14
Drip Irrigation Technology: Analysis of Adoption and Diffusion Processes

Francisco Alcon, Nuria Navarro, María Dolores de-Miguel, and Andrea L. Balbo

Abstract Increasing concerns about water scarcity have promoted the adoption and diffusion of irrigation technologies, such as drip irrigation, which allow farmers to use water in a more efficient way, while saving water resources. While some dry regions have embraced drip irrigation, this technology remains scarcely deployed on a global scale. In this chapter we provide an overview of the processes underlying the adoption and diffusion of innovations, with a focus on the specific context of the adoption and diffusion of drip irrigation technology within the agricultural community of Cartagena, in Southeast Spain. Our final aim is to inform policy makers charged with the designing of initiatives aimed at saving water and at increasing climate change resilience in agricultural contexts. Our main insights suggest that effective policies focused on irrigation technology uptake should consider social, economic, technological and environmental factors affecting adoption and diffusion dynamics, and specifically those factors that define perceptions of water scarcity, such as water prices and availability of water.

Keywords Review · Analytic framework · Irrigation water · Water-saving technology · Spain

---

F. Alcon (✉) · M. D. de-Miguel
Departamento de Economía de la Empresa, Universidad Politécnica de Cartagena,
Murcia, Spain
e-mail: francisco.alcon@upct.es; md.miguel@upct.es

N. Navarro
Centro Integrado de Formación y Experiencias Agrarias (CIFEA),
Molina de Segura, Murcia, Spain
e-mail: nuria.navarro@murciaeduca.es

A. L. Balbo
Research Group Climate Change and Security (CLISEC), Center for Earth System Research
and Sustainability (CEN), University of Hamburg, Hamburg, Germany
e-mail: balbo@cantab.net

© Springer Nature Switzerland AG 2019
A. Sarkar et al. (eds.), *Sustainable Solutions for Food Security*,
https://doi.org/10.1007/978-3-319-77878-5_14

14.1 Introduction

Water is an ever-scarcer key resource, necessary to sustain ecosystems as well as economic and social activities. Worldwide, agriculture accounts for 70% of all water withdrawal. Irrigated agriculture thus plays a crucial role in the global food production system (WWAP 2012) and agricultural water management has become an international priority. Water availability for agriculture has reached critical levels in arid and semi-arid regions, where water shortages are predicted to worsen due to climate change and its derived effects, such as increased frequency and intensity of droughts, lower precipitation, as well as foreseeable increases in water demand for irrigation (IPCC 2007).

In water-scarce areas, where limited water resources must be allocated to various productive uses while preserving the environment and ecosystems, the sustainable use of water resources is perhaps the major policy challenge of our times (Falkenmark 2000). To address this challenge, policy initiatives have been promoted from the two perspectives of supply and demand (Alcon et al. 2014). Supply-focused initiatives aim at increasing water resources availability. On the other hand, demand-focused initiatives foster water-saving management practices through the adoption of irrigation technologies of improved but variable water use efficiency, such as furrows (50–60% efficiency), sprinklers (70–80%) and drip irrigation (90%) (Dasberg and Or 1999).

Relative to traditional irrigation systems, drip irrigation, defined as the application of water through point or line sources at small operating pressures, has the potential to conserve water, improve crop quality, and increase crop production by using controlled irrigation doses and frequencies (Dasberg and Or 1999). Drip irrigation enhances water use efficiency by reducing water losses caused by deep percolation, soil evaporation and runoff. Through drip irrigation weed growth can be reduced, salinity problems mediated and the use of fertilisers optimised. Drip irrigation is generally more energy-efficient than sprinklers, and is adaptable to difficult soils and terrains (Skaggs 2001). At the same time, drip irrigation has a number of limitations, the main constraints for its deployment being the high initial installation costs[1] and the intensive maintenance requirements (emitter clogging, etc.). In addition, drip irrigation limits plant root development, favours salt accumulation near plants and reduce soil capacity to absorb CO_2 (Puy et al. 2016).

Overall, in spite of these limitations, drip irrigation technologies drastically improve the effectiveness of water use in agriculture while maintaining production levels. Thus, drip irrigation can play a key role for the improved use of scarce hydric resources worldwide (Cason and Uhlaner 1991).

---

[1] Investment and operational costs are variable depending on the technology selected. As an example, in Spain, investment and maintenance cost was in 5000€ ha^{-1} and 1232€ ha^{-1}, respectively in 2006 for fruits (Alcon et al. 2013), while in Burkina Faso the investment cost reached 7132$ ha^{-1} and the operational was 1544$ ha^{-1} for herbaceous crops (IFC 2014)

Nevertheless, drip irrigation has been scarcely deployed to date. For instance, the OECD countries with the highest share of irrigated land area (Spain, the Czech Republic, Greece and Italy) have drip irrigation adoption rates lower than 40% (OECD 2010). Some countries outside the OECD show higher drip irrigation adoption rates. In Israel for example, over 50% of irrigated land is now under drip irrigation (OECD 2011). In Jordan and Cyprus adoption rates add up to 60% and 95% respectively. In general, arid and drought-prone regions have shown the widest adoption of such water-saving technologies as drip irrigation (Alcon et al. 2011).

After providing an overview of the processes underlying the adoption and diffusion of innovations, this chapter focuses on the specific context of adoption and diffusion of drip irrigation technology within the agricultural community of Cartagena, in Southeast Spain. By identifying the social, economic, technological and environmental factors that affect irrigation technology adoption and diffusion we wish to support the improvement of policy initiatives that foster water saving and increase climate change resilience in agriculture.

14.2 Adoption and Diffusion of Innovation: Concepts and Approaches

Besides bringing competitive advantages to companies (Dieperink et al. 2004), innovations have an impact on social and cultural factors related to economic development (Freeman 1995). Originated is response to demand or as a need, a given innovation reaches the market in the shape of a technology, a technique or an organisational method or process. Over time, the diffusion of such an innovation follows existing communication channels among the members of a social system (Rogers 2003). The adoption of an innovation refers to a single individual decision on the acceptance of an innovation, whereas diffusion refers to the process of acceptance by a group of individuals through time. Diffusion is thus defined as the process of adopting a technology by the members of a social system or the process through which innovations are disseminated within and beyond a productive system (Feder and Umali 1993).

The innovation–decision–diffusion process comprises a set of phases, including knowledge, persuasion, decision, implementation and confirmation. From first exposure to adoption and confirmation, embracing a given innovation may bring a series of benefits to those implementing it (Pannell et al. 2006).

In the agricultural sector, innovations generally reach the market in the shape of new technologies. While becoming familiar with a newly available technology, farmers go through an adaptation process based on a sequence of decisions leading to its adoption or rejection (Gatignon and Robertson 1991). The amount of time it takes for the farmer to make a decision will depend on several factors, including: uncertainties associated to the proposed innovation, formal and informal knowledge gathered on its efficiency, the authority and credibility of the informants, as well as farmer's characteristics and background. Farmers who apply a new technology expect to benefit from its implementation, thus contributing to a global improvement of their social welfare (Rogers 2003).

While most innovations reaching the market are cost-effective, their spreading has often been slower than expected. In most cases, the slow speed for the adoption and diffusion of a given innovation can be attributed to the lack of information on the expectations that it creates among the potential adopting actors and to poor communication regarding its potential contribution to the achievement of their goals (Pannell et al. 2006). One of the main benefits expected from the implementation of an innovation is a significant decrease of collateral risks associated to the development of the business activity. In agricultural contexts, water scarcity and availability is perhaps the major one of such risks, becoming especially relevant in arid and semi-arid climates, where the adoption of water-saving technologies is one of the goals of national irrigation policies aimed at the sustainable and rational use of water resources.

Different approaches exist for the analysis of the adoption and diffusion of technologies (Feder and Umali 1993; Lindner 1987; Pannell et al. 2006; Rogers 2003), which are summarised in Fig. 14.1.

In a first set of approaches the adoption process is classified from the point of view of innovativeness. Here, studies at the micro level (i.e. individual adopters) focus on the characteristics of the potential adopters and their behaviour when facing innovations. Research studies carried out within this approach have been divided into two types depending on the perspective from which the analysis is undertaken: static, focused on intensity, and dynamic, focused on adjustment (Lindner 1987):

- Cross section studies (static) focus on the reasons why an individual accepts or rejects an innovation. They try to explain those factors and processes that have led the individual to the final decision on whether to adopt or reject an innovation.
- Temporal studies (dynamic) focus on the time elapsing between the development of an innovation and its adoption by an individual. They try to explain why some individuals adopt an innovation more quickly than others.

In a second set of approaches, the diffusion process is studied in broader areas of time and space, estimating the adoption rate of an innovation within the community of potential adopters and its dependence on the characteristics of the technology. As for the micro level, these macro level studies have been divided into two types depending on the perspective approached from both static and dynamic perspectives:

- Cross section studies are aimed at finding the diffusion rate of a technology once the imbalance has been readjusted. They try to explain why some innovations are widely adopted, while others are not.

14 Drip Irrigation Technology: Analysis of Adoption and Diffusion Processes

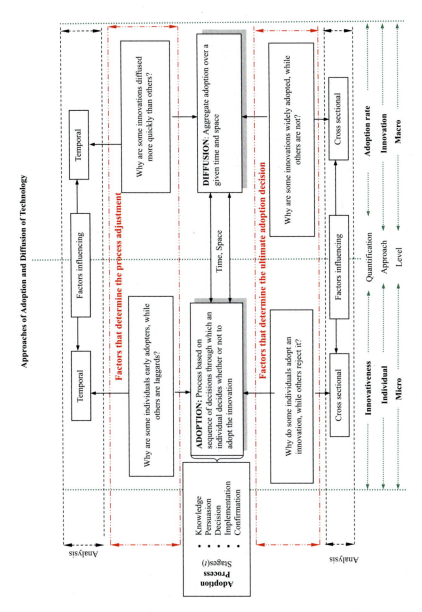

Fig. 14.1 Approaches of adoption and diffusion of technology

- Temporal studies estimate the delay between the first and the last members adopting an innovation within the same group. They try to explain why some innovations diffuse more quickly than others within a given group.

14.3 Adoption and Diffusion of Innovation: Analytical Models

Innovation adoption rates have been analysed mainly using temporal data to describe individuals on the basis of their adoption time. To do so, the density function of adopters can, for example, be divided into five categories using the mean and the standard deviation of the adopters' population, according to the method proposed by Rogers (2003). The resulting adopter categories were: innovators, early adopters, early majority, later majority and laggards. These categories were employed in the irrigation scheme of Cartagena (Murcia, SE Spain) by Alcon et al. (2006) to analyse the diffusion pattern of drip irrigation technology. Although variations of this method have been developed (Bass 1969; Mahajan et al. 1990; Karlheinz and Even 2000), the fundamentals remain the same.

14.3.1 Diffusion Models

Diffusion models aim at describing the number of adopters across time and at predicting the development of the diffusion process. Such models are based on mathematical functions that explain the degree of penetration and the maximum level of adoption of a given technology within a given social system (Van den Bulte 2000). Diffusion models follow the general form:

$$\frac{dN(t)}{dt} = g(t)\big[M - N(t)\big]$$

(14.1)

where $dN(t)/dt$ = diffusion rate in time t, $N(t)$ = cumulative adoption in t, M = total adopters and $g(t)$ = diffusion coefficient that would define the shape of the curve.

Diffusion models can be of internal, external or mix influence. All model equations described below allow plotting the percentage of cumulative adoption $N(t)$ along the time line t, specifying a maximum percentage of adoption. Therefore, different technologies fit better in different contexts.

1. Internal influence diffusion models, also called logistic models, are based on the assumption that the diffusion is produced by information and experience accumulation within the community (autochthonous), reducing the initial uncertainty about the technology. Here, early adopters influence later potential adopters,

similar to the propagation of an epidemic by contagion. The logistic model is defined by Mansfield (1961):

$$N(t) = \frac{M}{1 + e^{-(a+bt)}}$$ (14.2)

2. Where, a is the integration constant and b the adoption rate. In these models, the maximum adoption rate is found in correspondence with the inflection point ($dN/dt = 0$), i.e. when the innovation is adopted by 50% of the population. According to Banks (1994) it can be reached at $t^* = a/b$ and $N(t^*) = M/2$. External influence diffusion models, proposed by Fourt and Woodlock (1960), assume that information reaching potential adopters proceeds from external sources (allochthonous), such as social media or external agents. External influence models assume that adoption rate only depends on the number of potential adopters across time and it is defined by:

$$N(t) = M\left[1 + e^{-(a+bt)}\right]$$ (14.3)

3. Mix influence diffusion models, also known as "Bass models", include both models previously described, taking into account innovators (external influence) as well as imitators (internal influence) in a sigmoidal function as:

$$N(t) = \frac{M\left(1 - e^{-(p+q)t}\right)}{\left[(q/p)e^{-(p+q)t} + 1\right]}$$ (14.4)

where p is the external influence coefficient and q the internal influence coefficient.

At higher values of p and q higher diffusion speed is achieved, the maximum adoption being rate reached at time $t^* = \ln(q/p)/p + q$ where the cumulative adoption is $N(t^*) = M(1/2 - p/2q)$ (Mahajan et al. 1990).

14.3.1.1 Innovativeness: Logit and Probit Models

Innovativeness has been approached in individual models where the technology adoption decision is considered at a single moment in time. Such cross-section studies explain the adoption of a new technology as a function of its expected utility compared with that gained from an existing technology. These models have been applied to a number of studies on drip irrigation technology where the variable to be explained was adoption itself (Dinar and Yaron 1990; Feder and Umali 1993; Shrestha and Gopalakrishnan 1993; Green et al. 1996; Green and Sunding 1997).

The most common approach to explain this categorical variable has been the use of logit and probit models, which provide similar results, though based on different error distributions. Logit models allow the endogenous variable Y_i, bounded between 0 and 1, to be related with a series of explicative variables X_{ki} through a logistic distribution function (Cramer 2003). The goal of logit models is to find the relationship between the endogenous variable and a set of independent variables. This model generates the coefficients to predict a logit transformation of the probability of the factors of interest being present:

$$\log \text{it}(p) = \beta_k X_k + \varepsilon_i \tag{14.5}$$

where p is the probability of adoption and β_k is a vector of regression coefficients. Extensions of this model have been also used to analyse the joint adoption of two complementary decisions. This was explored by Moreno and Sunding (2005) for irrigation technology and land allocation and by Engler et al. (2016) for irrigation technology and scheduling.

14.3.1.2 Duration Models

A different approach, introduced in the 1980s, uses duration models to explain adjustment processes. (In such models, duration analysis explains the time elapsed between the moment when an individual becomes a potential adopter (T) and the moment when the full adoption occurs. In these works, technology adoption is presented as a dynamic process, explained by cross-section and other time-dependent external variables). The adoption is modelled by the hazard function (Lancaster 1990). Assuming that $F(t)$ denotes the cumulative distribution of adoption ($F(t) = Pr(T \leq t)$), then the survival function $S(t)$ is the reverse of the cumulative distribution function of T (i.e. $S(t) = 1 - F(t)$) which defines the probability of not adopting at time t. The hazard function ($h(t)$) specifies the rate at which adoption occurs through time as defined by Jenkin (1995):

$$\begin{aligned} H(t) &= \lim_{dt \to 0} \frac{P(t \leq T < t + dt \mid T \geq t)}{dt} \\ &= \lim_{dt \to 0} \frac{F(t + dt) - F(t)}{dt(1 - F(t))} = \frac{f(t)}{S(t)} \end{aligned} \tag{14.6}$$

This approach focuses on the timing of innovation adoption, considering the impact of variables which intensity changes over time. As such, duration analysis has been deemed suitable to analyse the adoption of agricultural technologies in general (Burton et al. 2003 and D'Emden et al. 2006) and that of drip irrigation in particular (Alcon et al. 2011).

14.4 Factors Explaining the Adoption of Innovations

In this section we focus on the factors that influence the adoption of innovations and on their relation to the adoption process. We pay special attention to the context of irrigation technology.

Based on existing scholarship, we have proposed above a classification of the different factors that have been used to explain technology adoption (Rogers 2003; Foltz 2003; Pannell et al. 2006; Feder and Umali 1993; Mohammadzadeh et al. 2014). A revised classification is proposed hereafter, clustering all analysed factors under five major groups (Fig. 14.2): (1) farmer characteristics, (2) economic factors, (3) farm characteristics, (4) characteristics of the technology, and (5) environmental factors in which the adoption process develops.

Farmer characteristics have been widely analysed in adoption literature proving significant in many instances over the adoption decision. Regarding irrigation technologies for example, it is expected that young farmers adopt earlier than elder ones (Skaggs 2001; Alcon et al. 2011). More experienced farmers, that are usually the elders in the community, tend to delay the adoption (Dinar and Yaron 1990; Shrestha and Gopalakrishnan 1993; Engler et al. 2016). Farmers with higher levels of education tend to be early adopters (Sidibé 2005). Also, personal inclination to cooperate promotes innovation adoption among individuals as it provides access to collective investment (Dinar and Yaron 1990, Sidibé 2005). Personal beliefs about the technology can also affect the innovativeness level (Skaggs 2001; Sidibé 2005). Finally, the probability of adopting irrigation technologies is proportional to land ownership (Moreno and Sunding 2005).

Economic characteristics arise from farmers' goal to maximise profitability of the farm. In this sense, the adoption of irrigation technology implies an investment and a consequent variation of inputs, use and outputs produced. The size of the business (measured as annual turnover), and the access to capital (e.g. as loans) can be considered the two most important economic variables. Generally, richer farmers have been more innovative thanks to their investment capacity and access to credit (Dinar and Yaron 1990; Skaggs 2001; Foltz 2003). Lower labor costs have also shown a positive effect on the adoption decision (Negri and Brooks 1990). However, the most important economic factor influencing the adoption of irrigation technolo-

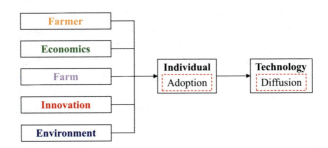

Fig. 14.2 Factors affecting technology adoption

gies has been shown to be water price, increasing as a consequence of water scarcity. Several studies have demonstrated that hydrological scarcity leads to higher shadow prices for water, pushing farmers towards the adoption of water-saving technologies (Dinar and Yaron 1990; Caswell et al. 1990; Negri and Brooks 1990; Green et al. 1996; Green and Sunding 1997; Carey and Zilberman 2002; Foltz 2003; Alcon et al. 2011).

Farm characteristics (physical, technical and locational) may facilitate the adoption of irrigation technologies. Soil characteristics, such as soil slope or texture, as well as microclimatic context have been claimed to influence innovation adoption rate (Green et al. 1996; Green and Sunding 1997; Negri and Brooks 1990). In some cases, crop type preferences have also influenced irrigation adoption decision-making (Moreno and Sunding 2005).

Innovation characteristics defined by Rogers (2003) may be summarised as follows: (1) the *relative advantage* of the innovation over existing alternatives, (2) the *complexity*, understood as the degree of difficulty that is perceived in using the new technology; (3) the *compatibility*, or degree to which an innovation is perceived as being consistent with the cultural values, previous experience, needs and resources of the adopters; (4) the *trialability*, that is the possibility to appreciate with real examples the advantages of the technology; and (5) the *observability*, or the degree to which the results of the innovation can be observed by other potential adopters. For irrigation technologies, relative advantage has been shown to influence adoption decision to the greatest extent (Moreno and Sunding 2005; Engler et al. 2016), together with technology cost (Caswell et al. 1990; Negri and Brooks 1990). Trialability and observability (i.e. the possibility to test) have become increasingly important in the past years in the adoption of several technologies (Pannell et al. 2006; López-Becerra et al. 2016).

Environment factors, other than those related with the farmers, the farm or the technology, may also affect the diffusion of irrigation technologies. These factors refer to the economic, social and political environment within which the farmers are embedded. For example, water allocation and water price policies, defined by political actors, play a key role on the adoption of irrigation technologies. In general, farms with endowments from different water sources, such as rivers or aquifers, have a lower perception of water scarcity and are less likely to adopt irrigation technologies with respect to farmers that depend on a single water source (Moreno and Sunding 2005). In one instance, Alcon et al. (2011) showed that farmers that use surface water, complemented with groundwater resources, tend to adopt early to reach some certainty in water supply. Finally, the contact with change agents (opinion leaders within the agricultural community), the exposure to mass media and access to interpersonal communication channels would also promote early adoption (Rogers 2003). Frequent contact with relevant information sources thus seems to contribute to reducing the degree of uncertainty when deciding on the adoption of new irrigation technologies, favouring early adoption (Engler et al. 2016).

14.5 Drip Irrigation Technology Adoption and Diffusion in Spain

Here we present the example of adoption and diffusion of drip irrigation technology in a water-scarce area of Spain, previously studied by Alcon et al. (2006, 2011), to illustrate the definitions, concepts and approaches used in the analysis of adoption and diffusion processes. The study was developed at *Campo de Cartagena*, a major irrigation community found in one of the most water-stressed areas of Europe. First introduced as early as 1975, drip irrigation technology is presently used by more than 95% of farmers within the Cartagena community (Fig. 14.3).

Diffusion analysis to model the drip irrigation adoption pattern was explored using the logistic model (see above Diffusion model 1, Alcon et al. 2006). Inter-firm diffusion of technology was studied according to individual farmer adoption time. The results obtained highlighted that "word of mouth" and farmer's visual perceptions fuelled the adoption process. In fact, access to knowledge acquired from previous adopters reduces perceived uncertainty around a new technology, motivating later adopters.

Thus, Fig. 14.4 and Table 14.1 present the key points defining the density (n) and cumulative (N) diffusion curves. Diffusion rates reached their maximum growth ratio between 1987 and 1988 (12 and 13 years after first introduction), about two-thirds down the overall diffusion period of 18 years. The percentage of cumulative adoption (N) at density function inflection points refers to the year in which the maximum diffusion speed is reached, i.e. 1982, and the year when the establishment period starts, i.e. 1993.

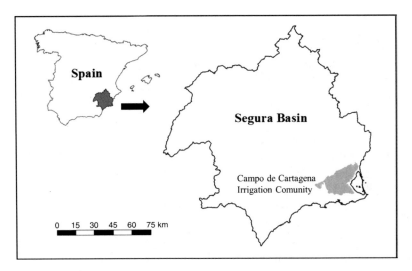

Fig. 14.3 Location of the Campo de Cartagena Irrigation Community in south-eastern Spain

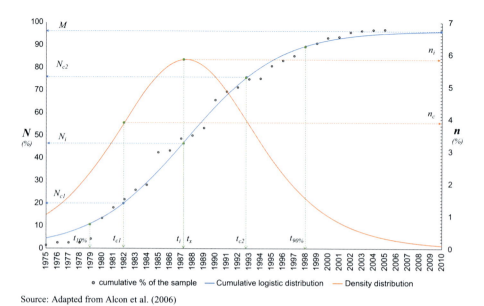

Source: Adapted from Alcon et al. (2006)

Fig. 14.4 Inter-firm diffusion of drip irrigation technology in the Cartagena Irrigation Community
Source: Adapted from Alcon et al. (2006)

Table 14.1 Features of the logistic distribution of drip technology adoption

Description	Symbol	Value
Integration constant	a	3.02
Diffusion coefficient	b	0.24
Total adopters (%)	M	96.45
Value of t at cumulative inflection point (year)	t_i	12.44
Value of N at cumulative inflection point (%)	N_i	48.23
Maximum value of n (slope) at cumulative inflection point	n_i	5.86
Value of t at density inflection points	t_{c1}	7.02
	t_{c2}	17.86
Value of N at density inflection points (%)	N_{c1}	20.38
	N_{c2}	76.07
Value of n (slope) at cumulative inflection point	n_c	3.91
Mean value of density distribution	t_x	12.44
Standard deviation of density distribution	s	7.46
Diffusion time	$t_{10-90\%}$	18.08

Source: Adapted from Alcon et al. (2006)

A Duration model (see above) that measures proportional hazard was used to analyse the decision-making process followed by farmers in adopting drip irrigation technology (Alcon et al. 2011). Table 14.2 reports the description of the variables used and the maximum likelihood estimation of the variables determining the adop-

14 Drip Irrigation Technology: Analysis of Adoption and Diffusion Processes

Table 14.2 Duration model of drip irrigation adoption

	Coeff.	Std. err.	$P > |z|$	Variable definitions
$\alpha 1$	2.45	0.46	0.00	
$\alpha 2$	0.83	0.39	0.03	
Age	−0.03	0.01	0.01	Age of farmer (years)
Study	0.26	0.11	0.02	Study levels: no studies = 0 Left education at 14/16 = 1 Left education at 18 = 2 3 year at university study = 3 5 year at university study = 4
Coop	0.46	0.19	0.01	If a member of a cooperative = 1, = 0 otherwise
Labour	0.04	0.10	0.65	Number of household members working in the farm (persons full time/year)
On-farm income	0.41	0.26	0.12	if income from agriculture is the main source of household income = 1, = 0 otherwise
Water price$_t$	0.11	0.02	0.00	Real price of water (€/100 m^3, constant 2005€)
Credit	0.07	0.03	0.05	If farmer has had credit availability (personal valuation 0–10)
Size	−0.00	0.00	0.31	The size of the farm (ha)
Fruits	0.37	0.18	0.04	If farmer crop fruits = 1, = 0 otherwise.
Trial	0.36	0.21	0.08	If farmer has tested technology in part of his farm prior to adoption or rejection = 1, = 0 otherwise
Info	0.91	0.25	0.00	If main information source is specialized personnel in agriculture (technology suppliers or other inputs, agricultural extension services, cooperative technicians or research centres) = 1, = 0 other farmers
Availaw$_t$	0.00	0.00	0.00	Water availability at time t (m^3/ha)
Groundwater	0.38	0.22	0.09	if farmer has groundwater allocation = 1, = 0 otherwise
Year_85$_t$	0.72	0.18	0.00	Dummy variable to measure the effects of expectations of becoming a European Union member in 1986, = 1 in 1985, = 0 otherwise
$p - 1$	0.97	0.28	0.00	
Ln (λp)	−7.03	0.84	0.00	
U	−0.61	0.33	0.07	
V	0.53	0.19	0.00	
Log likelihood	−827.58			
Number of obs.	326			
LR Gamma test	9.84			
Prob. > =chibar^2	0.00			
Pseudo R^2	0.18			

Source: Alcon et al. (2011)

t time-dependent variables

tion of drip irrigation technology. Positive coefficients imply that the variable has a positive impact on the probability of adoption and vice versa. The change in probability for a unit change in the exogenous variable is given by the exponential of the coefficient. For example, the hazard ratio for the variable denoting groundwater allocation (GROUNDWATER) is $e^{0.38} = 1.46$, which indicates that farmers with groundwater allocation have a 46% greater probability of adopting drip irrigation than those that use only surface water.

When results are interpreted on the basis of the five major groups described above (farmer, economic, farm, innovation and environment) a number of results from previous studies published within the adoption literature are confirmed. Specifically, age, study, membership of farmer groups, specific information sources, trialability of the technology and credit availability, all contribute to increased innovation uptake.

1. Farmer's characteristics influence the decision to adopt a new technology. Younger and least experienced farmers show more interest in adopting the irrigation technology. More educated farmers are also more likely to adopt.
2. Regarding economic aspects, availability of credit, through association and greater business size, also favours adoption.
3. Farm characteristics did not show any significant difference in adoption decision.
4. As for technological aspects, the trialability has been shown to play an important role on the adoption of drip irrigation technology in Cartagena.
5. Regarding environmental factors, direct knowledge of the information sources and of their reliability also increased the likelihood of adoption. Also, awareness of water availability and knowledge of the existence of an alternative water supply (ensuring benchmark water availability per year) increased adoption probabilities.

In Cartagena, water availability favoured the adoption of drip irrigation technology. Individual preferences and perceptions of agriculture and technology were also crucial in the adoption process. Alcon et al. (2011) confirmed the key role of benchmark water availability and water price in determining the speed of adoption of drip irrigation. However, and somewhat counter intuitively, more water allotment in a given year has been shown to increase adoption speed, possibly highlighting the importance for farmers to be reassured on the economic profitability of investing in the drip irrigation technology. Overall, time and time-dependent variables were shown to have an important influence on the adoption process.

14.6 Conclusions

In this chapter we have reviewed the different methodological and modelling approaches used to investigate adoption and diffusion processes within the framework of irrigation related technologies. We have illustrated the state of the art using the example of drip irrigation uptake at *Campo de Cartagena*, a large community of irrigators in the region of Murcia, Southeast Spain, which has one of the driest climates in Europe. Our proposed interpretation aims at highlighting how the analysis of adoption and diffusion dynamics of drip irrigation technology can provide information to policy-makers that design initiatives aimed to save water and increase climate change resilience in agriculture. In fact, a good understanding of the social, economic, technological and environmental factors affecting adoption and diffusion dynamics is paramount to actually implement technologies that improve water productivity and conservation, reducing environmental externalities caused by agriculture. To be effective, water-saving policies focused on irrigation technology uptake should consider those factors that encourage perceptions of water scarcity, such as water prices and availability of water. An effective adoption of irrigation technology is unlikely when such factors are overlooked.

Some general results that have emerged from Spanish case studies can inform policy makers that aim to spread irrigation technologies in other water-scarce regions of the world. For example, we found that personal relationships between irrigation stakeholders are the key for the transmission of information, reducing perceived uncertainty about newly introduced technologies. We also identified the type of farmers that policies should target to ensure a successful early diffusion of technologies: younger and more educated farmers, with bigger farms, belonging to farmers' associations and with reduced credit constraints. Our results also suggest that the creation of demonstrative plots, where technology can be tested and tried by end-users, would foster the adoption process. Finally, we found that environmental restrictions, such as guaranteed water supply, would contribute to early adoption, as well as paying for the water.

In conclusion, our analysis of adoption and diffusion processes for drip irrigation technologies contributes a better understanding of how to successfully implement the adoption of water-saving technologies to mitigate the negative effects of climate change over water availability for irrigation in agriculture.

Acknowledgements This work was carried out under the AGRISERVI project: AGL2015-64411-R (MINECO/FEDER, UE). A.L.B. worked on this manuscript during an Experienced Researcher Fellowship from the Alexander von Humboldt Stiftung/Foundation (ARiD – Adaptive Resilience in Drylands).

References

Alcon, F., De Miguel, M. D., & Fernandez-Zamudio, M. A. (2006). Modelización de la difusión de la tecnología de riego localizado en el Campo de Cartagena. *Revista Española de Estudios Agrosociales y Pesqueros, 210*, 227–245.

Alcon, F., de Miguel, M. D., & Burton, M. (2011). Duration analysis of adoption of drip irrigation technology in southeastern Spain. *Technological Forecasting and Social Change, 78*(6), 991–1001.

Alcon, F., Egea, G., & Nortes, P. (2013). Financial feasibility of implementing regulated and sustained deficit irrigation in almond orchards. *Irrigation Science, 31*, 931–941.

Alcon, F., Tapsuwan, S., Brouwer, R., & de Miguel, M. D. (2014). Adoption of irrigation water policies to guarantee water supply: A choice experiment. *Environmental Science & Policy, 44*, 226–236.

Bass, F. (1969). A new product growth model for consumer durables. *Management Science, 15*, 215–227.

Banks. (1994). *Growth and diffusion phenomena: Mathematical frameworks and applications*. New York: Springer.

Burton, M., Rigby, D., & Young, T. (2003). Modelling the adoption of organic horticultural technology in the UK using duration analysis. *Australian Journal of Agricultural and Resource Economics, 47*(1), 29–54.

Carey, J. M., & Zilberman, D. (2002). A model of investment under uncertainty: Modern irrigation technology and emerging markets in water. *American Journal of Agricultural Economics, 84*(1), 171.

Cason, T. N., & Uhlaner, R. T. (1991). Agricultural productions impact on water and energy demand – A choice modeling approach. *Resources and Energy, 13*, 307–321.

Caswell, M., Lichtenberg, E., & Zilberman, D. (1990). The effects of pricing policies on water conservation and drainage. *American Journal of Agricultural Economics, 72*, 883.

Cramer, J. S. (2003). *Logit models from economics and other fields*. Cambridge: Cambridge University Press.

Dasberg, S., & Or, D. (1999). *Drip irrigation*. Berlin: Springer.

D'Emden, F. H., Llewellyn, R. S., & Burton, M. P. (2006). Adoption of conservation tillage in Australian cropping regions: An application of duration analysis. *Technological Forecasting and Social Change, 73*(6), 630–647.

Dieperink, C., Brand, I., & WJV, V. (2004). Diffusion of energy-saving innovations in industry and the built environment: Dutch studies as inputs for a more integrated analytical framework. *Energy Policy, 32*, 773–784.

Dinar, A., & Yaron, D. (1990). Influence of quality and scarcity of inputs on the adoption of modern irrigation technologies. *Western Journal of Agricultural Economics, 12*, 224–233.

Engler, A., Jara-Rojas, R., & Bopp, C. (2016). Efficient use of water resources in vineyards: A recursive joint estimation for the adoption of irrigation technology and scheduling. *Water Resources Management, 30*(14), 5369–5383.

Falkenmark, M. (2000). Competing freshwater and ecological services in the river basin perspective: An expanded conceptual framework. *Water International, 25*, 172–177.

Feder, G., & Umali, D. L. (1993). The adoption of agricultural innovations: A review. *Technological Forecasting and Social Change, 43*(3–4), 215–239.

Foltz, J. D. (2003). The economics of water-conserving technology adoption in Tunisia: An empirical estimation of farmer technology choice. *Economic Development and Cultural Change, 51*, 359–373.

Fourt, L. A., & Woodlock, J. W. (1960). Early prediction of market success for new grocery products. *Journal of Marketing, 25*, 31–38.

Freeman, C. (1995). The national system of innovation in historical-perspective. *Cambridge Journal of Economics, 19*, 5–24.

Gatignon, H., & Robertson, T. S. (1991). Innovative decision processes. In T. S. Robertson & H. H. Kassarjian (Eds.), *Handbook of consumer behaviour* (pp. 316–348). Englewood Cliffs, NJ: Prentice-Hall.

Green, G., & Sunding, D. (1997). Land allocation, soil quality, and the demand for irrigation technology. *Journal of Agricultural and Resource Economics, 22*(2), 367–375.

Green, G., Sunding, D., Zilberman, D., & Doug, P. (1996). Explaining irrigation technology choices: A microparameter approach. *American Journal of Agricultural Economics, 78*(4), 1064–1072.

IPCC. (2007). *Working Group 1 fourth assessment report 'the physical science basis'*. Geneva: IPCC.

IFC. (2014). *Impact of efficient irrigation technology on small farmers*. Washington, DC: International Finance Corporation, World Bank Group Retrieved December 19, 2016, from http://www.ifc.org.

Jenkin, S. P. (1995). Easy estimation methods for discrete-time duration models. *Oxford Bulletin of Economics and Statistics, 57*(1), 129–138.

Karlheinz, K., & Even, A. L. (2000). Diffusion theory and practice disseminating quality management and software process improvement innovations. *Information Technology & People, 13*, 11.

Lancaster, T. (1990). *The econometric análisis of transition data.* Cambridge: Cambridge University Press.

Lindner, R. (1987). Adoption and diffusion of technology: An overview. In B. R. Champ, E. Highly, & J. V. Remenyi (Eds.), *Technological change in porharvest handling and transportation of grain in the humid tropic* (pp. 144–151). Bangkok: Australian Centre for International Agricultural Research.

López-Becerra, E. I., Arcas-Lario, N., & Alcon, F. (2016). The websites adoption in the Spanish agrifood firms. *Spanish Journal of Agricultural Research, 14*(4), e0107. https://doi.org/10.5424/sjar/2016144-10113.

Mahajan, V., Muller, E., & Bass, F. N. (1990). New products diffusion model in marketing: A review and directions for research. *Journal of Marketing, 54*, 1–26.

Mansfield, E. (1961). Technical change and the rate of imitation. *Econometrica, 29*(4), 741–766.

Mohammadzadeh, S., Sadighi, H., & Pezeshki Rad, G. (2014). Modeling the process of drip irrigation system adoption by Apple Orchardists in the Barandooz River Basin of Urmia Lake Catchment, Iran. *Journal of Agricultural Science and Technology, 16*(6), 1253–1266.

Moreno, G., & Sunding, D. L. (2005). Joint estimation of technology adoption and land allocation with implications for the design of conservation policy. *American Journal of Agricultural Economics, 87*, 1009–1019.

Negri, D., & Brooks, D. (1990). Determinants of irrigation technology choice. *Western Journal of Agricultural Economics, 15*, 213–223.

OECD. (2010). *Sustainable management of water resources in agriculture.* Paris: OECD Publishing.

OECD. (2011). *A green growth strategy for food and agriculture.* Paris: OECD Publishing.

Pannell, D. J., Marshall, G. R., Barr, N., Curtis, A., Vanclay, F., & Wilkinson, R. (2006). Understanding and promoting adoption of conservation practices by rural landholders. *Australian Journal of Experimental Economics, 46*, 1407–1424.

Puy, A., García Avilés, J. M., Balbo, A. L., Keller, M., Riedesel, S., Blum, D., & Bubenzer, O. (2016). Drip irrigation uptake in traditional irrigated fields: The edaphological impact. *Journal of Environmental Management.* https://doi.org/10.1016/j.jenvman.2016.07.017.

Rogers, E. M. (2003). *Diffusion of innovations.* New York: The Free Press.

Shrestha, R. B., & Gopalakrishnan, C. (1993). Adoption and diffusion of drip irrigation technology: An econometric analysis. *Economic Development and Cultural Change, 41*(2), 407–418.

Sidibé, A. (2005). Farm-level adoption of soil and water conservation techniques in northern Burkina Faso. *Agricultural Water Management, 71*, 211–224.

Skaggs, R. K. (2001). Predicting drip irrigation use and adoption in a desert region. *Agricultural Water Management, 51*(2), 125–142.

Van den Bulte, C. (2000). New product diffusion acceleration: Measurement and analysis. *Marketing Science, 19*, 366–380.

WWAP (World Water Assessment Programme). (2012). *The United Nations world water development report 4: Managing water under uncertainty and risk.* Paris: UNESCO.

Chapter 15
Combating Climate Change Impacts for Shrimp Aquaculture Through Adaptations: Sri Lankan Perspective

J. M. P. K. Jayasinghe, D. G. N. D. Gamage, and J. M. H. A. Jayasinghe

Abstract Fisheries and aquaculture have been identified as important sources of food, nutrition, income and livelihoods for hundreds of millions of people around the world. World per capita fish supply has reached 20 kg in 2014. Aquaculture is one of the main contributors that provide a considerable percentage of fish for human consumption. By 2014, fish accounted for about 17% of the global population's intake of animal protein and 6.7% of all protein consumed. In addition, fish provided more than 3.1 billion people with almost 20% of their average per capita intake of animal protein. Global total capture fishery production in 2014 was 93.4 million metric tons (MT) while aquaculture production is estimated at 73.8 million MT, with a projected first-sale value of US$160.2 billion.

Global shrimp aquaculture production has reached 4.58 MMT in 2014 and may remain at the same level in near future. Shrimp culture makes vital contributions to national and global economies, poverty reduction and food security for the world's well-being and prosperity. Asia has always led the world production of cultivated shrimps. Expected changes in climate, extreme weather conditions and climatic events, sea level rise, ocean acidification and rise in temperature are expected to create significant impacts on coastal ecosystems and aquaculture in coastal areas. Adaptations for likely impacts of climate change are reachable through better management practices in site selection, pond construction and preparation, selection of post larvae for stocking, pond management, bottom sediment management and disease management together with reducing non-climate stressors such as pollution, conservation of sensitive ecosystems and adoption of dynamic management policies.

Keywords Shrimp aquaculture · Climate change adaptations · Better management practices · Food and nutritional security · Disease

---

J. M. P. K. Jayasinghe (✉) · D. G. N. D. Gamage
Faculty of Livestock, Fisheries and Nutrition, Department of Aquaculture & Fisheries, Wayamba University of Sri Lanka, Gonawila, Sri Lanka

J. M. H. A. Jayasinghe
Faculty of Engineering, University of Moratuwa, Moratuwa, Sri Lanka

© Springer Nature Switzerland AG 2019
A. Sarkar et al. (eds.), *Sustainable Solutions for Food Security*,
https://doi.org/10.1007/978-3-319-77878-5_15

15.1 Aquaculture and Fisheries

Global total capture fishery production in 2014 has been estimated at 93.4 million MT (Fig. 15.1), with a contribution of 81.5 million MT from marine waters and 11.9 million MT from inland waters. There is an increase in the global supply of fish for human consumption irrespective of population growth during past five decades, increasing at an average annual rate of 3.2% during the period 1961–2013, resulting in increased per capita availability (FAO 2016). According to FAO 2016, world per capita fish consumption increased from an average of 9.9 kg in the 1960s to 14.4 kg in the 1990s and to 19.7 kg in 2013, and is targeted to increase beyond 20 kg in the future. Increase in production together with reduction in postharvest loses, efficient fish distribution chains, and growing demand linked to population growth, rising incomes and urbanization, value addition, product development have contributed for the increased per capita consumption of fish and fishery products.

FAO (2016) has compiled the information available on present status of World Fisheries and Aquaculture. Aquaculture is the culture of aquatic animals and plants in a controlled aquatic environment, mainly for consumption. Exploiting the fish from natural water bodies is referred to as capture fisheries. According to them around 56.6 million people were engaged in the primary sector of capture fisheries and aquaculture in 2014, out of whom 36% were full-time fishermen, 23% part time, and the rest were mainly occasional fishers.

In 2014, total aquaculture production was 73.8 million MT (Fig. 15.1—World capture fisheries and aquaculture production), with an estimated first-sale value of US$160.2 billion and included 49.8 million MT of finfish (US$99.2 billion), 16.1 million MT of molluscs (US$19 billion), 6.9 million MT of crustaceans (US$36.2 billion) and 7.3 million MT of other aquatic animals worth of US$3.7 billion

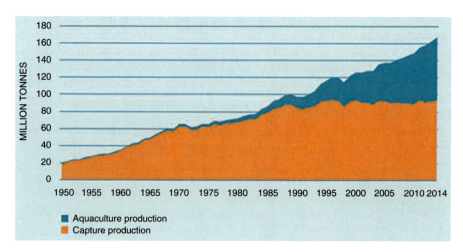

Fig. 15.1 Jayasinghe—World capture fisheries and aquaculture production (World Capture Fisheries and Aquaculture Production (FAO 2016)

(FAO 2016). Total aquaculture production contributes directly or in value added forms mainly for human consumption. Aquaculture has contributed 44.1% to the total production in 2014. All continents have shown the same general trend of an increasing share of aquaculture production to total fish production, except in Oceania.

15.2 Shrimp Aquaculture

15.2.1 Significance of Shrimp Aquaculture

According to (FAO/NACA 2001) Shrimp aquaculture has been identified as an industry that contributes significantly to national and global economies, poverty reduction and food security. Marine shrimp farming is a traditional aquaculture activity in many Asian counties. Extreme success of this aquaculture activity in Asia was mainly due to availability of coastal lands suitable for establishment of farms, successes in hatchery technology, and technological advances in the culture technology.

Traditional shrimp culture in South East Asia practices extensive type of culture in relatively large earthen ponds in low stocking densities using post larvae collected from wild. They mainly depended on natural productivity, tidal movements for water exchange and ponds were without mechanical aeration. Monoculture or polyculture with fish species was practised depending on the availability of fish fry in some countries.

Intensive shrimp farming became a reality with the success in hatchery technology where larval life cycles of shrimps were completed in hatcheries making available post larvae for stocking in ponds. Shrimp farming at commercial level got established during late 1960s and early 1970s. Technological advances and high profitability led to semi-intensive and supra-intensive forms of farming, and worldwide distribution of farms. Taiwan was one of the pioneering countries which became a leading producer in the 1980s and was then producing 21% of Asia's cultured shrimps, the highest output for the region. Taiwan market collapsed in 1988 due to unsustainable farming practices leading to frequent disease outbreaks and environmental problems. Later Thailand became the leading producer in black tiger shrimp. Ecuador and Brazil are the pioneers in shrimp farming in South America. At present, shrimp farms are distributed in over fifty countries including Middle East.

15.2.2 Farmed Shrimp Species

Out of the many species of shrimps, desirable characteristics for farming have been identified in a few species. Relatively larger size, established hatchery technology, disease resistance, adoptability to pond environment and climate, availability of market, relatively higher growth rate are some of the main criteria used in selecting shrimps for culture operations.

Pacific white shrimp (*Litopenaeus vannamei*, "white shrimp") native to the Pacific coast grows to a size of 23 cm. and accounts for 95% of the production in Latin America. It is easy to breed in captivity; this species has been now introduced to several Asian countries due to its resistance to white spot disease. But this species is susceptible to Taura syndrome virus (TSV). *Penaeus monodon,* ("black tiger shrimp") is a native to the Indian Ocean and to the Pacific Ocean. It can grow to a larger size and is widely farmed in Asia. Due to high susceptibility to white spot disease, it is gradually being replaced by *L. vannamei*. Western blue shrimp (*Penaeus stylirostris*) is farmed in the western hemisphere. Chinese white shrimp (*Penaeus chinensis*), occurs along the coast of China and the western coast of Korea and is farmed in China. Kuruma shrimp (*Penaeus japonicus*) is farmed mainly in Japan and Taiwan. Indian white shrimp (*Penaeus indicus*) is a native of the coasts of the Indian Ocean and is widely bred in India, Iran and Middle East and along the African shores. Banana shrimp (*Penaeus merguiensis*) is another cultured species from the coastal waters of the Indian Ocean.

15.2.3 *Farmed Shrimp Production and Trends*

Shrimp aquaculture, which only began to make significant contributions to global shrimp production in the 1980s, overtook wild harvest in 2007 and has continued to claim a growing share of the market in each subsequent year. As of 2014, farmed shrimp production amounted to 4.18 MMT. Global Aquaculture Alliance's (GOAL) 2016 survey of production trends in shrimp farming in main shrimp farming countries is presented in Fig. 15.2 (Shrimp aquaculture production and trends in major

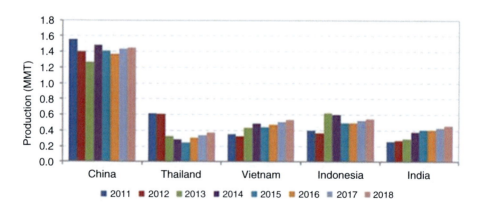

Fig. 15.2 Jayasinghe—Shrimp aquaculture production and trends in major farming nations in Asia (Shrimp aquaculture production and trends in major farming nations in Asia) (GOAL Shrimp Survey 2016)

farming nations in Asia). A general trend can be observed to increase the total production in year 2017 and 2018 in most of the main shrimp farming nations. Global production is expected to reach around 4.44 MMT in 2018. The Asian shrimp culture industry appears to be on the path of recovery after a substantial production drop in 2012 and 2013 in countries such as China and Thailand, caused by Early Mortality Syndrome (EMS). Contribution from Asia may reach 3.65 MMT by 2018, due to the expected increased inputs from Thailand, Indonesia, Vietnam and India. According to GOAL, 2016, China is expected to continue to remain as the leading producer.

15.2.3.1 Shrimp Aquaculture Industry in Sri Lanka

The fisheries industry including aquaculture plays an important role in the economy of Sri Lanka by providing livelihoods for around 2.6 million individuals including direct employment opportunities for 560,000 individuals. Others are involved in marketing, product development, value addition, processing for export, fish feed manufacturing, employment in fishery harbors and employment as technicians and laborers in fish farms and hatcheries. Generation of income, foreign exchange earnings and provision of reasonably priced protein for the rural and the urban masses in the country are other contributions. The industry contributes around 2% to the gross domestic production (GDP). The fisheries sector in Sri Lanka consists of coastal fisheries, offshore fisheries, and inland and coastal aquaculture. Coastal aquaculture has a great potential to diversify, secure income and food security among coastal communities in Sri Lanka. Shrimp farming industry provides the greatest contribution to the coastal aquaculture. The potential for this industry to provide a large number of jobs and export income makes its development very attractive to the Government of Sri Lanka. Figure 15.3 shows shrimp farming districts in Sri Lanka (Batticaloa and Puttalam).

Shrimp farming in Sri Lanka is presently restricted to North Western Province and Eastern Province. According to recent surveys; the total number of farms is around 950 with an estimated area of about 4500 ha. Sri Lanka produces 7000 MT of farmed shrimp annually and 30–40% of the production is exported. Nearly 2000 MT shrimps of shrimps including wild-captured are consumed locally. Penaeus monodon is the main species cultured. The majority of grow out shrimp farms in Sri Lanka follow semi-intensive culture practice. Farmed shrimp exports account for approximately 50% of the total export earnings from Sri Lankan fisheries sector.

15.3 Food and Nutritional Security and Aquaculture

Aquaculture has been identified as the fastest growing food production sector in the world. An increasing trend has been observed for aquaculture indicating a 12-fold increase during the last few decades. Industry has recorded average annual growth

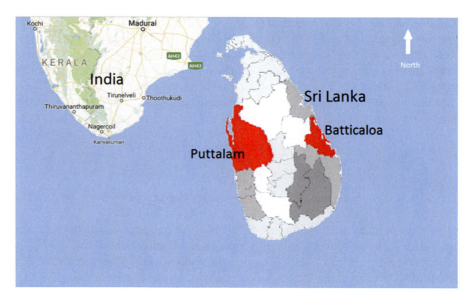

Fig. 15.3 Jayasinghe—Shrimp farming districts in Sri Lanka (Batticaloa and Puttalam)

of around 8% in recent past. World Bank (2013) has estimated that the demand for farmed aquaculture products will exceed 93 MMT by 2030. The issue is therefore to increase production while minimizing external environmental impacts and continue to lowering natural resource footprints of aquaculture. The nutritional value and health benefits of aquaculture products, the importance of aquaculture activities as a source of income and livelihoods and the efficiency of aquaculture products to produce/transform proteins have been identified as main contributory factors for human nutrition and food security through aquaculture.

According to HLPE (High Level Panel of Experts on food security) 2014, Fish contributes to food security and nutrition directly through availability of nutrient-rich food both at the household and at local, provincial and national market levels (Fig. 15.4—Conceptual representation of the different pathways between fish and food security and nutrition). Indirect pathways involve the trade of fish and generation of revenues, at household level or at higher (national) levels, including through income for crew-members and for those involved in fishery related activities such as fish processing factory workers. Income allows access to other food commodities. In Sri Lanka high valued aquaculture products are exported but cheaper fishery products such as canned and dried fish are imported maintaining a positive trade balance.

15.3.1 Fish as a Critical Food Source

According to HLPE 2014, capture fisheries and aquaculture provide 3.0 billion people with almost 20% of their average per capita intake of animal protein, and a further 1.3 billion people with about 15% of their per capita intake. This share can

15 Combating Climate Change Impacts for Shrimp Aquaculture Through Adaptations... 293

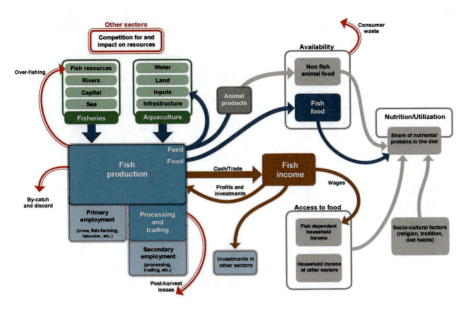

Fig. 15.4 Jayasinghe—Conceptual representation of the different pathways between fish and food security and nutrition (Different pathways by which fish contributes to food security and nutrition (HLPE 2014)

exceed 50% in some countries. In Sri Lanka 65% of the animal protein consumed comes from fish and fishery products. Fish provide a very significant proportion of protein in the human diet in most small island states like Maldives where 60% protein comes from fish. The continual growth in fish production, mostly from aquaculture since the 1990s has contributed significantly to increase the supply of fish during the last two decades.

Omega-3 (n-3) fatty acids are widely recognized as essential nutrients for the health and well-being of humans, particularly with respect to the n-3 long-chain polyunsaturated fatty acids (LC-PUFA), eicosapentaenoic (EPA; 20:5n-3) and docosahexaenoic (DHA; 22:6n-3) acids, that exert a range of health benefits through their molecular, cellular and physiological actions. Fish has been identified as a major dietary source of n-3 LC-PUFA for human. Recommended daily intake levels of EPA and DHA vary regionally, based on the scientific and health advisory information available. However there is a general agreement that population should consume at least two portions of fish per week, one of which should be oily. With the global population increase, decline in natural fishery resource and the population's demand for fish and fishery products, fatty acid requirement has to come from aquaculture. Fish is also an important provider of a range of micronutrients not widely available from other sources in the diets of the poor. More and more attention is being given to fish products as a source of vitamins and minerals which can be an excellent source of many essential minerals such as iodine, selenium, zinc, iron, calcium, phosphorus and potassium, but also vitamins such as A and D, and several vitamins from the B-group.

15.3.2 Relative Resource Efficiency of Aquaculture Production Systems

Aquaculture systems have been identified as efficient converters of feed into protein when compared to other animal husbandry systems such as poultry, piggery and cattle farming. It has been estimated that poultry converts about 18% of the food consumed and pigs about 13% as compared with 30% in the case of fish (Hasan and Halwart 2009).

HLPE (2014) indicates that the production of 1 kg of beef protein requires 61.1 kg of grain while 1 kg of pork protein requires 38 kg as compared with 13.5 kg in the case of fish.

Generally the species used in aquaculture are poikilothermic animals where their body temperature fluctuates with the environmental temperature and no energy is required to maintain a constant body temperature using energy as in the case of terrestrial farmed animals used in animal husbandry. For aquatic vertebrate animals the aquatic media they live in provide the physical support and they need only minimal skeletal support that they need not allocate large proportion of the food they eat to maintain the body.

Aquaculture production systems have a lower carbon footprint per kilogram of output when compared with terrestrial animal production systems. Nitrogen and phosphorous emissions from aquaculture production systems are much lower when compared with beef and pork production systems. It is an advantage that some of the aquaculture production systems such as bivalves can absorb nitrogen and phosphorous emissions from other systems (HLPE 2014). According to FAO 2016, climate change will affect food security in Asia by the middle of the twenty-first century. South Asia appears to be most severely affected.

15.3.3 Classification of Shrimp Aquaculture Systems

Shrimp culture systems are classified into three, namely, extensive, semi-intensive and intensive. The classification is based mainly on pond facilities, stocking density, food supply, water management, yield, technical know-how and skills of the farmer and other major inputs. While semi-intensive and intensive farming have gained some progress in recent years, the extensive system still remains the major practice possibly because of the relatively large landholdings (50–300 ha) per farmer. In order to shift to the semi-intensive or intensive system, farmers require greater amount of inputs, high-level technical know-how, and supervision (Apud 1985).

When considering water management, closed, open, semi-closed and complete recirculation are the available options. In open system water is extracted from the water source and effluents are discharged to the water source without treatment.

Most of the small-scale farmers adopt this system as they do not have enough farm area to treat water. In the semi-closed option incoming water to the farm is treated using a reservoir but effluents are discharged without treatment to the water source. Complete recirculation is the most sustainable option where water is completely circulated within the system. Physical, chemical and biological treatments are used to improve the water quality. The pressure on water source is minimal and risks of disease outbreaks are minimal. Most of the small-scale farmers adopt the closed system. They chemically treat the water in the culture pond itself before stocking. The culture period is less and the water quality is managed within the pond using probiotics. The size at harvest and period of culture are relatively low in this option.

15.4 Climate Change Impacts on Aquaculture

Several climate change impacts on shrimp culture have been identified in India (Muralidhar et al. 2010a, b). Those can be generalized to the region. Farmers have prioritized those impacts in the order of seasonal changes in weather, heavy rains, floods and cyclone in shrimp farming area and high temperature, floods, low/unseasonal rainfall, and rise in water level in coastal shrimp farming areas. These are the most felt effects by farmers. The seasonal changes considered were mainly temperature variations and delay in monsoon.

15.4.1 Expected Changes in Climate of Sri Lanka and Impacts on Shrimp Culture in Sri Lanka

15.4.1.1 Climate of Sri Lanka

Sri Lanka is an island in the Indian Ocean and located in southeast of India, between 5°55′ and 9°51′ N latitude, and 79°41′ and 81°53′ E longitude. There are three climatic zones, namely wet zone, dry zone and the intermediate zone. There are two seasonal wind regimes. South-west monsoon prevails from May to September and the north-east monsoon from December to February. These wind regimes together with topography are responsible for distribution and amount of rainfall. In Sri Lanka, climate change has been identified to increase the frequency and intensity of natural disasters, floods, droughts and cyclones. Storm surges are predicted to become more frequent and to be more devastating. Alterations in monsoonal patterns, deviations and fluctuations in annual rainfall are also predicted. Greater spatial variation between wet and dry areas in rainfall is expected and it may result in wet areas becoming more wetter and dry areas becoming drier. These identified shifts are likely to adversely affect aquaculture.

15.4.2 Effects of Climate Change on Shrimp Culture

Change in global climate can create significant impacts in the oceans and on the coastal environment where coastal aquaculture activities are flourishing. Climate change may result in new hazards and increase the frequency and intensity of existing hazards. Various processes related to climate change can cause significant changes particularly with respect to salinity and temperatures in brackish water bodies influencing aquaculture production, quality of the product and health of the aquaculture organisms.

Different elements of climatic change, can cause identifiable effects in the shrimp farming environment leading to health problems, decrease in production and quality of the product, premature harvests and economic loss or complete crop failure. The stress generated through changes related to climate change leading to incidences of disease can be identified as a main problem.

15.4.2.1 Precipitation

Sri Lanka is hot and humid with significant variation in climate over time and space. This variation is reinforced by the country's topography, which creates its unique rainfall pattern. The precipitation patterns have changed over the time. Historically, decreasing trend for precipitation has been observed over the past 30–40 years, although it is not statistically significant. The number of rainy days has decreased. There is an increasing trend for the frequency of extreme rainfall events, which would lead to more floods. Modifications in seasonal pattern in precipitation, changes in the extent, extreme heavy rainfall, prolong droughts are the most significant key factors of climate change that can affect shrimp aquaculture in coastal areas. Most of the shrimp farms are situated in the dry zones of the countries. Changes in rainfall patterns and more prolonged droughts in the dry zone areas would allow to greater evaporation which would negatively impact the shrimp aquaculture (Muralidhar et al. 2010a, b). When considering the situation in Sri Lanka evaporation is normally high in shrimp farming belt resulting in high salinities.

Timely onset and delay in onset of monsoon causes an increase in salinity of water especially in area where tidal influence is low. Culture ponds are to be diluted to reduce the salinity to acceptable levels. Larger volumes of fresh water have to be sourced exploiting the ground water resource which leads to conflicts with other users of fresh water and the depletion of the ground water resource. Prolonged droughts may also lead to increase in salinity beyond the acceptable and optimal levels of the cultured shrimp (Muralidhar et al. 2010a, b; Jayasinghe and Wijesekara 2014). In contrast, increased rainfall can result in sudden drop in salinity and other associated water-quality problems like high turbidity and changes in dissolved oxygen and nutrient levels. In areas where acid sulphate soils are converted to shrimp farming, draining of acidic water into ponds with flood water will result in sudden lowering of pH. The stress generated can lead to conditions which cause disease and related poor production and mass mortality of the farm crop. In periods of low salinity, water exchange gets restricted, leading to accumulation of pollutants and accumulation of toxic metabolic end products in the pond environment. Physical damages to the farm infrastructure such as pond dikes and sluice gates due to flooding as a result of heavy rain are evident. Floods allow shrimps to escape from ponds, and floods also contaminate pond water with poor quality water and wild organisms (Abery et al. 2009). Mass mortalities in shrimps are recorded during floods due to rapid oxygen depletions and sudden lowering of pH (Jayasinghe and Wijesekara 2014).

The impacts are more severe for small-scale farmers as they have located their farms in floodplains and in lagoon reservations. Shortened growing season and premature harvests are other factors that affect farmers in general (Muralidhar et al. 2010a, b). In Sri Lanka there is a crop calendar developed based on resource availability where farmers of different zones are given different periods of the year for their farming practices (Jayasinghe and Wijesekara 2014). Expected changes and deviations in rainfall pattern and intensity can severely disturb the crop calendar as well as changes occurring in sedimentation and nutrition loading in the ponds that can stimulate toxic algal blooms which are a threat to shrimps.

15.4.2.2 Temperature

Increase in temperature has significant consequences on coastal, marine and inland fisheries as well as on aquaculture. Water and air temperatures are expected to rise. Increased water temperatures lead to other associated physical changes in pond environment, such as shifts in dissolved oxygen levels. These have been linked to more frequent algal blooms and increase in the intensity and frequency of disease outbreaks. Changes in temperature of only a few degrees have significant negative impacts on aquaculture, and only a few positive impacts. The farm performance of tropical and sub-tropical species such as *P. monodon and P. merguiensis* can be increased with increase in water temperature up to a certain extent (Muralidhar et al. 2010a, b). Rise in temperature is a threat to cold-water species such as *M. japonicus* (Jayasinghe 1991). But in Sri Lanka, *Penaeus monodon* is the main culture shrimp species. Increased water temperatures especially during the dry season lead to greater evaporation in culture pond. Resultant increases in pond salinity stress the shrimp resulting in less appetite, high concentrations of toxic gases and overgrowth of algae. Low dissolved oxygen, delay in molting, slow growth, introduction of new predators and pathogens are other associated problems (Abery et al. 2009, Jayasinghe and Wijesekara 2014). Metabolic rates of shrimps will increase with increase in environmental temperature and this may result in higher food conversion ratios (FCR) affecting farm performance and profits. Increase in water temperatures has been related to increases in the intensity and frequency of disease outbreaks in shrimp culture systems in several countries.

15.4.2.3 Extreme Weather Condition and Climatic Events

When considering intensity of the extreme natural disasters and frequency of occurrence has been increased in recent years. Sri Lanka being a small island in the Indian Ocean in the path of two monsoons is likely to be affected by weather related hazards. Floods are mostly due to monsoonal rains or effects of low pressure systems developed in and around the country and droughts are due to failure of monsoonal rain which are the most common hazards experienced in Sri Lanka (Ministry of Disaster Management 2014). The intensity and frequency of extreme climate events, mainly the floods and droughts have increased in recent years as a consequence of global warming leading to an increase in natural disasters. The country has already experienced 2 years of serious drought and one major flood event within the first 5 years of the twenty-first century (Imbulana et al. 2006). The natural hazards such as droughts, floods, landslides, cyclones, and coastal erosion currently cause significant damage in Sri Lanka.

15.4.2.4 Cyclones

Since Sri Lanka is located outside the cyclone belt, severity of cyclone is lower in the Island when compared to Indian subcontinent. However, Sri Lanka has been influenced by cyclones occurring in the Bay of Bengal (Sri Lanka Disaster Knowledge Network 2009). It has been observed that intensity and severity of cyclones have been increased during the past few decades. But this situation may change in the future with climate change. Impacts of cyclones are mainly related to damages associated with infrastructure such as electricity lines, roads and farm buildings. Biological hazards of cyclones include contaminations across the ponds and spread of diseases.

15.4.2.5 Droughts

Dry and intermediate zones are vulnerable to droughts from February to April. Droughts occur in the south-eastern, north central and north-western areas of Sri Lanka due to low rainfall during monsoons. However if there is a substantial drought in the normal rainy season from May to June, it can extend into September. According to Global Facility for Disaster Reduction and Recovery (GFDRR) annual report for year 2011, in the past 30 years droughts have affected more than six million people (GFDRR 2011). Puttalam is the most vulnerable district to drought with respect to aquaculture. Frequency and intensity of droughts have resulted in reduction in the culture area due to the rise in salinity as well as lack of freshwater. Salinity can increase beyond the optimum level which is considered 10–25 ppt for black tiger shrimp within few days due to evaporation. It leads to poor survival and growth of shrimps as salinity plays an important role in the growth of culture organisms through osmoregulation of body minerals from that of the surrounding water.

However black tiger shrimp can tolerate a wide range of salinity. But optimum range is required for better survival and growth.

15.4.2.6 Floods

Climate risk and adaptation country profile study conducted by GFDRR 2011, has mentioned that floods are increasing from 1974 to 2004 and that floods have affected more than ten million people in the last 30 years. Floods are associated with extreme rainfall conditions. The south-west monsoons and the north-east monsoons cause flooding in different parts of the country. The main shrimp farming area, the Puttalam District is vulnerable to flooding. Floods have a higher frequency of occurrence in Sri Lanka than other natural disasters. However flood intensity and occurrence had been low in Puttalam District before shrimp farming was initiated. Construction of farms in floodplains and lagoon reservations has obstructed the receding floods and floods are more frequent now in this district. Flooding can affect inland aquaculture and capture fishery due to pollution, sedimentation and any adverse changes in water quality parameters of surface water bodies that sustain this fishery (Abery et al. 2009). Further it causes sudden reduction in salinity which reduces production in shrimp culture systems. Concentration of the metabolic end products increases within the ponds due to inability of farmers to exchange water from the lagoons and estuaries. Increasing flooding can cause inundation of the farms, erosion of bunds, and damage infrastructure facilities such as buildings, electrical installations, sluices, shutters and screens resulting in loss of crop and introduction of predators and pathogen (Muralidhar et al. 2010a, b). Sudden fall in pH has been recorded in water in coastal areas due to oxidation of pyrites in coastal sediments during heavy floods after prolonged dry spells (Jayasinghe 1991).

15.4.2.7 Sea Level Rises (SLR)

Sri Lanka is expected to be affected by global warming associated SLR. The increasing concentration of GHG is a key contributor to increasing atmospheric temperature leading to rise in sea level. In addition the thermal expansion of oceanic waters with increasing temperature, changing precipitation, melting of snow/ice and other factors contribute to partial inundation. SLR is one of the indicating factors of global warming. Geological subsidence, sedimentation are also important contributors to SLR. Global SLR is reported to be around 1.7 ± 0.5 mm per year from 1901 to 2010 during which sea level has increased by around 19 cm. In the case of Sri Lanka, SLR is estimated at 0.3 m by 2010 and 1.0 m by 2070. It is slightly higher than the global average predicted for SLR.

Coastal regions of Sri Lanka play a vital role in the economic growth of the country. SLR is a major challenge in the administration and management of coastal zones. Higher sea levels lead to inundation of coastal ecosystems such as mangroves and salt marshes, which play a major role in ecological functions. The goods and services provided by these ecosystems are essential to the sustainability of coastal aquaculture. Loss of land and availability of freshwater are other concerns. Water quality in the lagoons and estuaries is subjected to change due to increased salt water input and alterations in present water quality management practices are needed. A recent study conducted on hazard profile of Sri Lanka with sea level rise has identified that Puttalam District has the highest impact of sea level rise as it is located in lower elevation. Further the study reveals that the total available brackish water area by the year 2100 is estimated at 14800 ha in Puttalam and 2700 ha in Batticaloa due to the inundation as a result of sea level rise.

15.4.3 Health and Climate Change

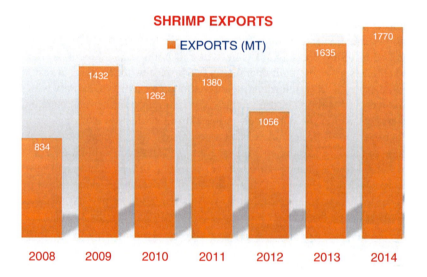

Diseases are the most feared threat, or one of the most feared threats to shrimp farming. Major viral, bacterial, fungal and other diseases recorded in shrimp culture systems are listed in Table 15.1. White spot disease in Indo-Pacific region, Taura syndrome virus (TSV) in South America and early mortality syndrome (EMS) have been identified as major threats in the recent history. Spread of pathogens and diseases is thought to be a major threat under climate change scenarios. The issue of WSV must be made a priority in the Indo-Pacific region as this is the main threat to the industry in that region in identifying adaptations. Better management practices (BMPs) including relevant biosecurity measures are considered as main adaptations.

Currently, WSV has no treatment. Avoidance is the best way to prevent disease outbreaks (Menasveta 2002 and Zhan et al. 1998). Implementation of proper management practices to reduce environmental stress is one of the best methods to reduce the risk of disease. (OIE 2011; Walker and Mohan 2009; Corsin et al. 2005;

Table 15.1 Common diseases/pathogens observed in the shrimp farming industry in the Indo-Pacific region

Viral diseases	Bacterial and fungal diseases	Other diseases
White spot syndrome virus (WSSV) Yellow head Virus group (YHV) Taura syndrome virus (TSV) Monodon baculovirus (MBV) group Infectious hypodermal and haematopoietic necrosis virus (IHHNV)	Vibriosis: – Hatchery vibriosis – Luminescent vibrio – White faecal disease – Early mortality syndrome (EMS/AHPNS) Other bacteria: – Rickettsia Fungal: – Larval mycosis – Fusariosis	Epicommensals and parasites: – *Leucothrix mucor* – Peritrich protozoans – Gregarines – Microsporidians Nutritional imbalances Toxic syndromes and environmental extremes

Lightner 2005). BMPs are the main tool for handling stress generated through climate change impacts.

Extensive investigations have been carried out in many countries to examine the effect of management practices on minimizing risks of disease outbreaks. A study conducted by MPEDA/NACA 2003, in India has identified several BMPs to reduce disease risks in small-scale shrimp farms.

15.4.4 National Climate Change Policy

Being an island nation subjected to tropical climate patterns, Sri Lanka is highly vulnerable to climate change impacts. Extreme weather events such as high intensity rainfall, flash floods and extended dry weather periods are now becoming frequent events in Sri Lanka. In addition to adaptive measures, Sri Lanka will fall in line with the global efforts to minimize the greenhouse gas emissions within the framework of sustainable development and principles enshrined in the United Nations Framework Convention on Climate Change (UNFCCC) and its Kyoto Protocol (KP).

The National Climate Change Policy of Sri Lanka provide guidance and directions to address the adverse impacts of climate change efficiently and effectively. The National Climate Change Policy includes a vision, mission, goal and a set of guiding principles followed by broad policy statements under Vulnerability, Adaptation, Mitigation, Sustainable Consumption and Production, Knowledge Management and General Statements. Collaborative action at all levels is necessary to transform this policy into a meaningful set of actions to meet the challenges of climate change. The Climate Change Secretariat (CCS) already plays a vital role in developing the institutional and regulatory mechanisms to address climate change. The National Climate Change Adaptation Strategy could provide a framework in which to develop this synergy further.

In coastal areas where aquaculture development takes place, Coast Conservation and Coastal Resources Management Department (CC&CRMD) is responsible for the implementation of the Coastal Zone Management plans. Department of Fisheries and Aquatic Resources (DFAR), Department of Wildlife Conservation (DWC), the Department of Forest Conservation (DFC), the Urban Development Authority (UDA) and the District Secretaries of coastal districts are cooperating with CC & CRMD in developing and implementing policy decisions for mitigating impacts of climate change and in planning processes. Farmer societies, Universities involved in aquaculture education and research are also consulted during policy development processes.

Necessity to include climate change in coastal planning and ecosystem restoration has been identified by CC&CRMD. Several policy decisions that have been implemented are directed towards conservation of remaining mangrove areas and restoration of degraded mangrove sites to enhance the capacities to minimize the impacts.

15.4.5 Farmer Conceptions

Shrimp farmers are aware of some aspects of climate change through several information sources such as radio, television, newspapers and interactions with extension and university students. They have experienced above-average rainfall, changes in rainfall pattern, unusual droughts and have initiated adaptation strategies. Sea level rise, water level changes, inundation of coastal areas, and water temperature rise have not been experienced yet by the farming community.

15.4.6 Better Management Practices as Adaptations for Climate Change

Better management practices (BMPs) are the key to the sustainability of the aquaculture industry. BMPs increase the total production, size at harvest, product quality, general health of shrimps and environmental sustainability. BMPs are developed in many areas from site selection to processing of end-products and includes site selection, pond preparation, water quality management, sediment management, selection of post larvae and health management. BMPs can also be used as adaptations to minimize the impacts of climate change. There are technical interventions and management interventions in better management practices that can be used in climate-compatible development.

Network of Aquaculture Centers in Asia-Pacific (NACA) has been involved in development and adoption of good conduct regulations, known as BMPs, since 2000 in a number of countries in the Asia-Pacific region (Mohan and DeSilva

(2010); Padiyar et al. 2008; NACA/FSPS/MOFI 2005). Many BMPs are derived from the Food and Agriculture Organization (FAO) Codes of Conduct for Responsible Fisheries (FAO/NACA/UNEP/WB/WWF 2006; Bene 2005; Barget et al. 2003). Several studies (FAO 2011; Umesh et al. 2009; ASEAN 2005; Engle and Valderrama 2004; MPEDA/NACA 2003) have revealed that recommended and developed BMPs have reduced the risk of disease outbreaks and improved social, ecological and economic sustainability and food safety. A programme was undertaken in Sri Lanka by IDRC, CIDA and Wayamba University of Sri Lanka in a project titled "Promoting Rural Income from Sustainable Aquaculture through Social Learning in Sri Lanka" in which the development of cluster-specific BMPs were identified for the sustainability of the industry in Sri Lanka.

15.4.6.1 Management and Technical Interventions

BMPs identified as management interventions useful in adapting to climate change impacts are given in Table 15.2. Shrimp aquaculture is increasingly constrained by changes in temperature, precipitation, drought, extreme weather conditions and storms/floods that affect farm performance, infrastructure and livelihoods of farming community. Ecological changes, inundation of low-lying lands and salt water intrusions may cause substantial dislocation of communities and disruption to farming systems. Management interventions identified under pond preparation, post larvae selection, pond management and biosecurity mainly focuses on adapting farming practices to minimize or to avoid climate change impacts that stress the shrimps, leading to disease outbreaks and to manage overall farm performance and the quality of the product.

Variation and extreme fluctuations in salinity, sudden changes in pH, lowering of oxygen levels, plankton blooms, and accumulation of metabolic toxic end products are some of the adverse changes expected in pond environment due to climate change. These factors can make shrimps more vulnerable to diseases. Alterations in crop calendar, slow growth, feeding problems, molting problems, poor product quality, brood stock collection problems, and poor farm performance are also identified to be impacts of climate change. Management adaptations listed under pond preparation, selection of post larvae for stocking, pond management during the culture cycle and implementation of the biosecurity (Table 15.2) can address the majority of the problems associated with climate change, improving the health and overall farm performance, ensuring the sustainability of shrimp farming.

Technical interventions useful in adapting to climate change are given in Table 15.3. These interventions will avoid the selection of shrimp farm sites in environmentally sensitive areas and provide possible interventions to reduce possible adverse impacts. Strategic Environment Assessment (SEA) is a systematic decision support process, aiming to ensure that environmental and possibly other sustainability aspects are considered effectively in policy, plan and programme making. Environment Impact Assessment (EIA) can look into all possible impacts of climate change and suggest mitigation and adaption measures. Recommendations of EIA

Table 15.2 List of BMPs identified as management interventions useful in adapting to climate change impacts

Area of activity	Intervention
Preparation of ponds for shrimp farming	Observing a 2-month fallow period after every harvest
	Drying and cleaning the pond bottom
	Removal of sediments accumulated during the previous culture cycle
	Ploughing and harrowing the pond
	Use of lime and dolomite to pond bottom to evaluate the pH of the soil and the lime requirement
	Use of stock tank or reservoir to store and treat water to be used for filling the pond
Selection of post larvae (PL) for stocking	PCR screening and other quality checks to select healthy larvae
	PL acclimatization to the pond environment before stocking
Pond management during the culture cycle	Fertilizer application to pond water to improve the natural productivity
	Dolomite and lime application to pond water to adjust the pH of the pond water
	Feed monitoring using feed trays
	Application of vitamins, minerals
	Water exchange during the crop cycle as and when necessary
	Regular monitoring of pond water for dissolved oxygen (DO), pH, salinity, and alkalinity of pond water
	Regular monitoring of ammonia, hydrogen sulphide
	Monitoring of Vibrio counts in pond water to assess pathogenic and beneficial bacterial populations
	PCR screening of shrimps after 1 month of stocking
	Conducting regular monitoring for growth and general health of shrimps
	Regular monitoring of bottom sediments of the pond
	Aeration of pond and proper positioning of aerators
	Use of only treated water to replenish the pond
Implementation of biosecurity measures	Use of bird nets to prevent the entry of birds
	Peripheral fence to prevent the entry of crabs as they can act as carriers for pathogens
	Filter water through twin net bag filter
	Peripheral fence to prevent the entry of dogs and other domestic animals
	Establish foot baths and disinfection of equipment
	Use of appropriate probiotics after monitoring the bacterial counts
	Allocation of separate sets of equipment for different ponds to avoid contamination

will prevent destruction of sensitive ecosystems such as mangroves, salt marshes and sea grass beds, and ensure their ecosystem services, reduce the possible adverse environmental impacts of bad site selection and reduce vulnerability to floods and heavy rainfall. Farmers will fall in line with conditions laid down in Environmental Protection Licence (EPL) to minimize the adverse impacts of shrimp farming operations. Acidification of water and sediments due to oxidation of pyrite in acid sul-

Table 15.3 List of BMPs identified as technical interventions useful in adapting to climate change impacts

Area of activity	Intervention
Strengthening of legal frame work	Establish coastal zone management plan incorporating shrimp farming zones Strategic Environmental Assessment (SEA) for potential shrimp farming zones Environmental Impact Assessment (EIA), Initial Environmental Assessment (IEA) for shrimp culture projects Environmental Protection Licence (EPL) for operation of shrimp farms Develop standards for shrimp farm effluents
Site selection for shrimp farms	Selection of supra tidal areas to establishing shrimp farms Avoidance of intertidal areas for establishing shrimp farms Avoidance of areas with pyritic soils that exhibit acid sulphate soil conditions Avoidance of environmentally sensitive areas such as mangroves, salt marshes and sea grasses Establish buffer zones, mangrove belts in-between farms Restoration/rehabilitation of degraded mangrove areas Avoid floodplains and lagoon reservations for establishment of farms
Infrastructure facility	Electricity supply for farms, use of renewable energy sources Desilting of lagoons/estuaries and water ways used as sources of water and sources for discharge of effluents Rehabilitation of water resources used to extract water Establish weather forecasting and early warning systems on natural hazards Address the problem of sand bar formation across the river outfalls Establish disease surveillance system to forecast disease/epidemic situations in shrimp farming areas Promote more environmentally friendly systems such as complete recirculation systems. Low intensity shrimp culture and organic shrimp culture Facilitate polyculture systems Develop an efficient information exchange system among stakeholders
Other interventions	Strengthening farmer organizations Establish crop calendar to minimize the pressure on water source and maximize the utilization of existing facility Organize administrative zones and sub-zones for shrimp farming Strengthening of farmers and awareness programmes for farmers

phate soil areas and other consequences of acidification such as increased solubility of iron, manganese and aluminium can be avoided by not setting up farms in acid sulphate soil areas. Restoration of mangroves, concept of shrimp farms in mangroves, establishment of buffer zones and green belts can improve resilience to rising sea levels. Figure 15.5 shows a shrimp farm constructed in a mangrove area. Mangrove areas can act as a protective belt (Fig. 15.6).

In addition, reduction of soil erosion pressure on environment and hydrology are other benefits. Mangroves have a capacity to absorb pollutants, purify water and to enhance the carrying capacity of the environment to sustain shrimp farming. Restoration/rehabilitation of mangroves with suitable species can create bio-shields

Fig. 15.5 Jayasinghe—Shrimp farm constructed in mangrove area

Fig. 15.6 Jayasinghe—Mangrove belt bordering a lagoon

to protect against cyclones, floods and extreme climatic events. Reducing the impact of floods can be facilitated through clearing of sand bars in river mouths. Effective communication and information exchange will enhance the adaptivity for climate change impacts by farmers.

Some of the BMPs discussed are also helpful in mitigation of climate change impacts. Reduction in energy utilization is possible through use of renewable energy, promotion of low intensity shrimp culture, biological treatment of water, and organic shrimp culture reducing CO_2 emissions. Carbon absorption and sequestration, protection of biodiversity are some of the important contributions of shrimp farming with mangroves.

Our experience in Sri Lanka indicates that majority of the farmers are very receptive for the adoption of BMPs. In addition to climate change adaptations, adoption of BMPs have contributed towards increased farm production, quality of the harvest, size at harvest, food conversion efficiency. BMPs have reduced the incidences of diseases, improving overall sustainability of the industry. Those who are not very receptive are small-scale cluster farmers with inadequate exposure to current issues and financial constraints.

Aquaculture extension services are to be strengthened to create more awareness on impacts of climate change on aquaculture industry and to enhance the adaptation response to possible impacts of climate change.

In summary it can be concluded that aquaculture contributes significantly towards national and global economies towards poverty reduction and food and nutrition security of communities, via a food production system with low carbon footprint and low nitrogen and phosphorous emissions. Combating climate change impacts on industry through adaptations will ensure the sustainability of the industry.

Acknowledgements The authors appreciate and would like to acknowledge all the support provided by the Wayamba University of Sri Lanka, Climate Change Secretariat, Ms. N. Balasubramaniam, and all other stakeholders with whom we have worked.

References

Abery, N. W., et al. (2009). *Perception of climate change impacts and adaptation of shrimp farming in Ca Mau and Bac Lieu, Vietnam: Farmer focus group discussions and stakeholder workshop report*. Bangkok: Network of Aquaculture Centers in Asia-Pacific. Retrieved December 20, 2016, from www.enaca.org/aquaclimate.

Apud, F. D. (1985). Extensive and semi-intensive culture of prawn and shrimp in the Philippines. In Y. Taki, J. H. Primavera, & J. A. Llobrera (Eds.), *Proceedings of the first international conference on the culture of penaeid prawns/shrimps, 4–7 December 1984* (pp. 105–113). Iloilo City: Southeast Asian Fisheries Development Center, Aquaculture Department.

ASEAN, 2005. Association of Southeast Asian Nations manual: ASEAN good shrimp farm management practice. *Fisheries publication series #1*. Retrieved December 17, 2016, from http://www.enaca.org/modules/wfdownloads/singlefile.php?cid=49&lid=122

Barget, U., Subasinghe, R., Willmann, R., Rana, K., & Martinez, M. (2003). *Towards sustainable shrimp culture development: Implementing the FAO code of conduct for responsible fisheries (CCRF)* (p. 35). Rome: Fisheries Department, Food and Agriculture Organization of the United Nations (FAO).

Bene, C. (2005). The good, the bad and the ugly: Discourse, policy controversies and the role of science in the politics of shrimp farming development. *Development Policy Review, 23*(5), 585–614.

Corsin, F., Turnbull, J. F., Mohan, C. V., Hao, N. V., & Morgan, K. L. (2005). Pond-level risk factors for white spot disease outbreaks. In P. Walker, R. Lester, & M. G. Bondad-Reantaso (Eds.), *Diseases in Asian aquaculture V* (pp. 75–92). Manila: Fish Health Section, Asian Fisheries Society.

Engle, C., & Valderrama, D. (2004). Economic effects of implementing selected components of best management practices (BMPs) for semi-intensive shrimp farms in Honduras. *Aquaculture Economics and Management, 8*(3-4), 157–177.

FAO. (2011). *Global capture production statistics dataset 1950–2009 and global aquaculture production statistics dataset (quantity and value) 1950–2009*. FAO: Rome. Retrieved December 13, 2016, from http://www.fao.org/fishery/statistics/software/fishstat/en.

FAO. (2016). *Climate change implications for fisheries and aquaculture: Summary of the findings of the Intergovernmental Panel on Climate Change Fifth Assessment Report, by Anika Seggel, Cassandra De Young and Doris Soto*. FAO fisheries and aquaculture circular no. 1122, Rome.

FAO/NACA. (2001). *Asia regional technical guidelines on health management for the responsible movement of live aquatic animals and the Beijing consensus and implementation strategy. FAO fisheries technical paper no. 402* (p. 53). Rome: FAO. Retrieved December 5, 2016, from ftp://ftp.fao.org/docrep/fao/005/x8485e/x8485e00.pdf.

FAO/NACA/UNEP/WB/WWF. (2006). *International principles for responsible shrimp farming* (p. 20). Bangkok: Network of Aquaculture Centers in Asia-Pacific (NACA). Retrieved December 17, 2016, from http://www.enaca.org/uploads/international-shrimp-principles-06.pdf.

GFDRR. (2011). *Climate risk and adaptation country profile 2012. Case study on conserving important mangrove ecosystem in Puttalam Lagoon. Sri Lanka: World Bank Group. Green Movement of Sri Lanka Inc.* Retrieved October 16, 2016, from http://gmsl.lk/mangrove.php.

Hasan, M. R., & Halwart, M. (2009). *Fish as feed inputs for aquaculture practices, sustainability and implications. FAO fisheries and aquaculture technical paper no. 518*. Rome: FAO. Retrieved December 22, 2016, from http://www.fao.org/docrep/012/i1140e/i1140e.pdf.

HLPE. (2014). *Sustainable fisheries and aquaculture for food security and nutrition*. Rome: A report by the High Level Panel of Experts on Food Security and Nutrition of the Committee on World Food Security.

Imbulana, K. A. U. S., Wijesekara, N. T. S., & Neupane, B. R. (2006). *Sri Lanka water development report 2010*. Colombo: UNESCO and Ministry of Irrigation and Water Resources Management.

Jayasinghe J. M. P. K. (1991). *The utilization of acid sulphate zone for shrimp culture on the West coast of Sri Lanka* (p. 210). Ph.D. thesis, University of Stirlin.

Jayasinghe, J. M. P. K., & Wijesekara, R. G. S. (2014). *Shrimp health management in Sri Lanka* (pp. 70–72). Gonawila: Wayamba University of Sri Lanka.

Lightner, D. V. (2005). Biosecurity in shrimp farming: Pathogen exclusion through use of SPF stock and routine surveillance. *Journal of World Aquaculture Society, 36*(3), 229–248.

Menasveta, P. (2002). Improved grow out systems for disease prevention and environmental sustainability in Asia. *Reviews in Fisheries Science, 10*(3–4), 391–402.

Ministry of Disaster Management. (2014). *Hazard profile of Sri Lanka*. Retrieved December 22, 2016, from http://www.disastermin.gov.lk

MPEDA/NACA, 2003. *Shrimp health management extension manual. Prepared by the Network of Aquaculture Centers in Asia-Pacific (NACA) and the Marine Products Export Development Authority (MPEDA), India, in cooperation with the aquatic animal health research institute, Bangkok, Thailand; Siam natural resources Ltd., Bangkok, Thailand; and Australian veterinary animal health services, Australia.* MPEDA, Cochin. 1–36. . Retrieved December 3, 2016, from http://library.enaca.org/Shrimp/manual/ShrimpHealthManual.pdf

Muralidhar, M., et al. (2010a). *Case study on the impacts of climate change on shrimp farming and developing adaptation measures for small-scale shrimp farmers in Krishna District, Andhra Pradesh, India*. Bhubaneswar: Network of Aquaculture Centers in Asia-Pacific. Retrieved December 27, 2016, from www.enaca.org/aqu.

Muralidhar, M., et al. (2010b). *Perception of climate change impacts and adaptation of shrimp farming in India: Farmer focus group discussions and stakeholder workshop report* (2nd ed.). Bhubaneswar: Network of Aquaculture Centers in Asia-Pacific. Retrieved December 15, 2016, from www.enaca.org/aquaclimate.

15 Combating Climate Change Impacts for Shrimp Aquaculture Through Adaptations... 309

Mohan, C.V., DeSilva, S. (2010). Better management practices (BMPs)-gateway to ensuring sustainability of small scale aquaculture and meeting modern day market challenges and opportunities. *Aquaculture Asia*, 15: 9–14.

NACA/FSPS/MOFI. 2005. *Reducing the risk of aquatic animal disease outbreaks and improving environmental management of coastal aquaculture in Viet Nam: final report of the NACA/SUMA projects FSPS1*. Retrieved January 1, 2017, from http://library.enaca.org/NACA-Publications/NACASUMA_Project_Completion_report.pdf

OIE. (2011). *Manual of diagnostic tests for aquatic animals* (pp. 121–131). Paris: Office International des Epizooties. Retrieved December 18, 2016, from http://www.oie.int/manual-of-diagnostic-tests-for-aquatic-animals/.

Padiyar, P. A., Phillips, M. J., Bhat, B. V., Mohan, C. V., Ravi, B. G., Mohan, A. B. C., & Sai, P. (2008). Cluster level adoption of better management practices in shrimp (P. monodon) farming: An experience from Andhra Pradesh, India. In M. G. Bondad-Reantaso, C. V. Mohan, M. Crumlish, & R. P. Subasinghe (Eds.), *Diseases in Asian aquaculture VI* (pp. 409–418). Manila: Fish Health Section, Asian Fisheries Society.

Sri Lanka Disaster Knowledge Network. (2009). *Major natural disasters in Sri Lanka*. Retrieved January 20, 2016, from http://www.saarcsadkn.org/countries/Srilanka/hazard_profile.aspx

Umesh, N. R., Chandra Mohan, A. B., Ravi Babu, G., Padiyar, P. A., Phillips, M. J., Mohan, C. V., & Bhat, B. V. (2009). Shrimp farmer in India: Empowering small scale farmer through a cluster-based approach. In S. S. De Silva & F. B. Davy (Eds.), *Success stories in Asian aquaculture* (pp. 43–68). Dordrecht: Springer.

Walker, P. J., Mohan, C. V. (2009). Viral disease emergence in shrimp aquaculture: origins, impact and the effectiveness of health management strategies. *Reviews in Aquaculture* 1(2):125–154.

World Bank. (2013). Fish to 2030: prospects for fisheries and aquaculture (English). Agriculture and environmental services discussion paper; no. 3. Washington DC; World Bank Group. http://documents.worldbank.org/curated/en/458631468152376668/Fish-to-2030-prospects-for-fisheries-andaquaculture

Zhan, W. B., Wang, Y. H., Fryer, J. L., Yu, K. K., Fukuda, H., & Meng, Q. X. (1998). White spot syndrome virus infection of cultured shrimp in China. *Aquatic Animal Health, 10*(4), 405–410.

Chapter 16
Application of the Terroir Concept on Traditional Tea Cultivation in Uji Area

Fitrio Ashardiono

Abstract Climatic changes and extreme temperature fluctuations have been occurring more frequently with evidences of continuous increase in its intensity. Direct effects of these changes are especially felt by the agricultural industries especially those which are utilizing the natural environmental elements, such as the tea cultivation in the Uji Area. As the oldest and most famous green tea producing region in Japan, teas from the Uji Area owes its reputation to the distinct quality characteristics created through the result of adopting traditional agriculture practices that hold a long heritage. Changes in the climatic conditions, especially temperature fluctuations, have directly affected the perceived quality of the harvested tea leaves in the region, caused by temperature variability during the growth of the first flush tea leaves. The problem is further exacerbated by frost events, drought, heavy rain and temperature extremes which have directly affect the yield of the tea production. Based on the analysis of climatic data and tea production statistics, lower mean air temperature can be correlated with low harvest yield. Observations and surveys conducted in the Uji Area have showed that these changes have not only affected the cultivation processes but also indirectly led to social and economic issues within the tea grower community. It is necessary to develop a new cultivation concept which takes account of the natural environmental elements and agriculture practices, such as the terroir concept which is utilized in French wine industry. By capitalizing on this new concept, the Uji Area tea growers would be able to proactively adapt to the ongoing changes, thus ensuring the sustainability of the Uji Area as a well-known tea growing region.

Keywords Climate change adaptation · Uji tea cultivation · Terroir concept · Traditional agriculture knowledge

---

F. Ashardiono (✉)
Asia-Japan Research Institute, Ritsumeikan University,
Ibaraki, Osaka, Japan
e-mail: fitrio-a@fc.ritsumei.ac.jp

© Springer Nature Switzerland AG 2019
A. Sarkar et al. (eds.), *Sustainable Solutions for Food Security*,
https://doi.org/10.1007/978-3-319-77878-5_16

16.1 Introduction

Although the name of Uji Tea might not be familiar for most people, among the Japanese, Uji Tea is very popular as a high-quality green tea used for tea ceremonies, and regarded as a delicacy among green tea enthusiasts. Although there are many types of green teas produced in Japan, what is commonly known outside Japan would be *matcha* (powdered green tea) or *sencha* (Japanese green tea). In comparison with wine products from France, where the information regarding vineyard location as well as the wine grape types are printed on the label of a wine bottle, unfortunately on most the tea product packagings which are sold in Japan, the information about the tea plantation origin is not printed. However, for Japanese green tea enthusiasts, the teas which are cultivated and produced in the Uji Area, especially matcha, is considered the benchmark for high-quality powdered green tea in Japan.

For generations, tea growers in the Uji Area (Fig. 16.1) have retained traditional cultivation methods which are relying heavily on manual labor for most of the

Fig. 16.1 Location of the Uji area

cultivation processes, where they manually hand-plucked the tea leaves during harvest. One of the unique cultivation methods utilized by the tea growers in the area is the use of traditional covering methods during the tea cultivation processes (Figs. 16.2, 16.3, 16.4, and 16.5). Knowledge of this practice has been continuously passed down through generations of tea growers, furthermore this method is unique to the area and especially essential to produce high-quality matcha and *gyokuro* (high-quality Japanese green tea). In this special method, as documented by Kimura and Kanda (2013), the tea bushes are covered with sunlight-blocking materials after the first leaf bud break occurred, lasting for approximately three to four weeks. In the first two to three weeks of the covering, light intensity is reduced by 95%, and in the last week prior to harvesting, it is reduced further to 98%. With the reduced sunlight condition, it is viewed that the tea plants would produce more chlorophyll in the tea leaves, thus increasing the amino acid contents and reducing the catechin level, which is believed to give more flavor and sweetness to the tea products.

Based on the surveys and observations, it is revealed that the tea growers have their own methods and system to achieve reductions in light intensity. The observed variation includes (1) selection of covering material, (2) layering methods and installation height, and (3) duration of covering period (Ashardiono and Cassim, 2015). Covering materials used in this method, traditionally are woven reeds and straws, although other artificial materials like black vinyl mesh sheets are currently also being used. This method contributes directly to the distinct flavor and taste of

Fig. 16.2 Covering method using black vinyl mesh sheet

Fig. 16.3 Covering Method Using Woven Reeds and Straws

Fig. 16.4 Tea leaf plucking under shade

16 Application of the Terroir Concept on Traditional Tea Cultivation in Uji Area

Fig. 16.5 Tea leaf plucking under shade

Uji Tea. Because of this complex and resource-consuming cultivation method, tea harvest in the Uji Area can only be conducted once a year during the spring season (April–May).

In general there are three types of tea which are produced in the Uji Area, namely *tencha*, gyokuro, and sencha. Most of these tea types are usually produced using tea leaves from the same tea cultivars (cultivated varieties), while the differences are mostly based on the cultivation processes. Sencha is the most popular green tea consumed in Japan, whereas this type of tea is fully grown under the sun. In contrast with sencha cultivation, tencha and gyokuro are grown under the shade or covered for about 21 days prior to harvesting. Depending on the length of the covering period, the harvested tea leaves can be processed into gyokuro (less than 21 days of covering) or tencha (more than 21 days of covering), where the latter is stone-ground to make matcha.

The Uji Area which is located in the south of Kyoto City, is one of the oldest and most famous tea growing regions in Japan, furthermore based on historical archives, the tea cultivation in the area started in the thirteenth century (Uji City, 2010). Originally Uji Tea refers to the tea products which are cultivated and processed inside the Uji Area, which is well known for its extraordinary quality, and was only used to cater the Japanese nobility. Traditionally, only a small quantity of tea is produced in the Uji Area because of its resource consuming cultivation methods. In order to comply with the continuous high demand from consumers, in 2006 *Kyoto-fu*

Chagyokumiai [Kyoto Tea Cooperative], a wholesaler collective, defines Uji Tea as tea products which are cultivated in four prefectures which are: Kyoto, Nara, Shiga, and Mie; and processed by a tea wholesaler based in Kyoto Prefecture using techniques derived from the Uji Area (Kyoto Tea Cooperative, 2013).

Although this approach might be beneficial to ensure the continuity of Uji Tea as a brand, in the long run it will have a negative impact on the traditional tea cultivation activities in the Uji Area, as the current definition does not consider the significance of the Uji Area's natural environmental characteristics. Their approach is the opposite of the terroir concept which can be observed in the French wine industry, where the combination of natural environmental characteristics of the region and the utilized winemaking techniques are considered crucial element in the quality characteristics of the wine products.

16.2 Changing Climatic Condition

Recorded climatic data obtained from long-term observations have generally indicated that the average temperature is increasing in many parts of the world. The increase in the average temperature might be beneficial for agriculture as colder climate regions would become warmer, allowing cultivation of crops which previously have been incompatible for colder climates becomes feasible while on the contrary, current, well-known cultivation regions might become unsuitable in the future as the temperature gradually becomes too warm (Jones, 2005; Holland and Smit, 2010; Jones and Webb, 2010).

In the wine grape cultivation, it is predicted that changing cultivation conditions might trigger a shift in suitable locations for cultivation of some varieties to obtain a high-quality harvest (Jones, 2007a, b). These changes in cultivation regions would lead to a loss of cultivation knowledge which has been accumulated in the original cultivation regions, especially for adapting and utilization of the natural environmental characteristics. The influence of higher temperature towards the tea cultivation is predicted to reduce tea production yield as stated by Wijeratne (1996).

Changes in the climatic conditions, especially air temperature will have direct effects to the growth cycle of the tea plant which are especially crucial during the bud break and flush harvest periods (De Costa et al., 2007; Duncan et al., 2016). Moreover, high-quality leaves can only be obtained from tea plants which are grown under stable climatic conditions, as the formation and growth of leaf fibers are heavily influenced by the climatic conditions (Watson, 2008a, b). The texture of tea leaf fibers plays a crucial role in deciding the types of tea produced from the harvested tea leaves. Leaves with smooth and soft fibers (fine leaves) are essential to produce high-quality teas while slightly fibrous leaf (medium leaves) and matured fibrous leaf (coarse leaves) are generally processed as lower-quality tea products (Konomoto et al., 2006; Watson, 2008a, b).

The general effects of climate change on tea cultivation, can be categorized into two types, which are: (1) average temperature increase (warming of the climate) and

16 Application of the Terroir Concept on Traditional Tea Cultivation in Uji Area 317

(2) increasing occurrences of extreme weather events (drought, high-intensity rainfall, and frost events) (Japan Meteorological Agency, 2018). Tea cultivation possesses similar traits as the wine grape cultivation, especially on its sensitivity towards the changes in microclimatic conditions. Based on this, the results derived from these observations are aimed to further understand how these changes affect the tea plants, thus becoming important references for other crops cultivation processes. Regarding the tea production, Wijeratne (1996) also mentioned that increasing temperature and changing rainfall pattern are very harmful to the growth of the tea plants, where these changes would ultimately affect the quality and quantity of the final tea products. Changes in the rainfall pattern could be observed as longer drought, increasing occurrences of erratic rainfall pattern, which accompanied by a sudden increase of rainfall volume and intensity.

The rapid transformation of the climatic conditions in the form of air temperature increase and the increasing occurrences of extreme weather events, is predicted to significantly affect the tea cultivation processes (De Costa, 2008; Wijeratne et al., 2007). Widespread observation have shown that wine grapes harvesting dates have advanced, especially in the last 10–30 years (de Orduna, 2010), while in the case of Japanese tea, the harvesting period has shown tendency to become late compared to the previous years (Ashardiono, 2014; Ashardiono and Cassim, 2014). These changes were also identified in the Uji Area, thus from the preliminary findings it is most likely that these abnormal climatic events will be occurring more frequently in the Uji Area. Based on the research findings, in the period of 2002–2012 the observed changes in the Uji Area: (1) sudden drop in temperature during the spring season (March–May); (2) higher temperature during fall season (October–November); (3) changes in quantity and period of rainy season (May–June); (4) longer periods of drought; (5) diminishing morning fog, all of which lead to the diminishing characteristics of high-quality Uji Tea (Ashardiono and Cassim, 2015).

To further understand the occurring changes, analysis is conducted on the climatic data (2002–2014) obtained from Kyoto Prefecture Tea Industry Research Institute, which consist hourly data of the first flush tea cultivation period, starting from early January to early May (1: January; 2: February; 3: March; 4: April; 5: May), and averaged into early [the first 10 days of the month] (a), mid [the 11th day until the 20th day] (b) and late [the 21st day onwards] (c). The average air temperature in the Uji Area from January to May (2002–2014) can be seen in Table 16.1—Average Mean Air Temperature in the Uji Area and Fig. 16.6—Average Mean Air Temperature in the Uji Area, from these data it can be observed how the fluctuations in the average mean air temperature during each period have become more apparent in the past years.

Further observation of the statistical data showed, in general the average temperature values during the first flush cultivation period have been declining over the years. Although it can also be observed that there were identifiable temperature fluctuations in late February (2c) and mid-March (3b) values. From these results, it can be interpreted, while the main trend of the average temperature values is on the decline, there are indications of temperature fluctuations towards higher temperatures especially in late February (2c) and mid-March (3b).

Table 16.1 Average mean air temperature in the Uji Area

	1a	1b	1c	2a	2b	2c	3a	3b	3c	4a	4b	4c	5a
2002	3.9	1.3	3.7	4	3.6	5.6	5.1	6.8	10.3	12.4	14.7	13.1	17.8
2003	3.5	7.1	4.3	5	3.6	8.3	6.7	10.7	10	14.2	14.6	17	18.8
2004	5.1	4	2.7	3.8	5.6	8.1	5.3	9.4	10.3	11.9	17.5	15.2	18.7
2005	3.7	4.1	4.2	4.3	5.4	3.5	6.2	6.7	8.7	13	14.1	17.3	18.1
2006	2.5	4.9	3.2	3.2	5.6	6.9	6.7	6.2	7.5	10.6	13	13.2	18.8
2007	4.9	4.6	4.9	6	6.2	7.4	8	4.6	10.3	10.4	13.1	16	18.5
2008	5.3	4	3.8	2.9	3.4	4.1	5.9	11.5	10.7	12	14.5	16.4	19.1
2009	5	3.6	5.9	5.1	7.1	7.5	7.9	9.7	8.3	11	17.5	14.1	18.1
2010	4.4	3.7	4.9	5	4.6	11.4	8.8	9.9	7.4	12.6	13	13.1	18.8
2011	2.3	1.5	1.7	3.9	3.2	8.5	4.4	6.9	6.2	10.1	11.2	12.3	17.5
2012	3.1	3.5	2.5	2	1.9	4.4	7.9	5.1	7.2	8.4	12.8	16.8	172
2013	1.8	2.7	3.3	4.4	2.3	3.6	6.8	8.6	8.8	11.8	11.9	11.9	13
2014	3.3	1.6	3.8	3.5	1.9	5.9	4.3	7.4	10.2	10.6	11.7	14.6	15.7
Mean	3.75	3.58	3.76	4.09	4.18	6.55	6.46	7.96	8.91	11.46	13.82	14.70	17.70
Median	3.70	3.70	3.80	4.00	3.60	6.90	6.70	7.36	8.80	11.80	13.10	14.65	18.10
Std. dev	1.15	1.59	1.12	1.06	1.70	2.32	1.43	2.17	1.49	1.49	1.98	1.88	1.68

Source: Kyoto Prefecture Tea Industry Research (2015)

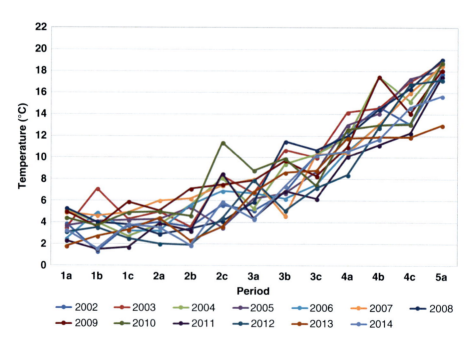

Fig. 16.6 Average mean air temperature in the Uji area

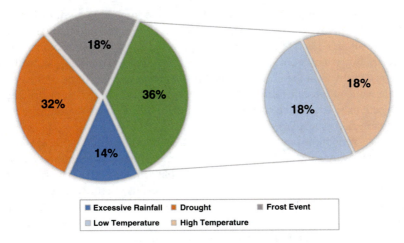

Fig. 16.7 Perceived changes of climatic factors in the Uji area

The ongoing changes in the climatic conditions of the Uji Area are evidence of increasing trends in extreme weather events. These events will undoubtedly affect the tea cultivation process cycle, most importantly because the tea plants are sensitive toward changes in the microclimatic conditions, especially temperature fluctuations. To achieve the desired tea products quality, young leaves need to be plucked before their fibers become too hard; observations have shown how exposure to extreme air temperature fluctuations would accelerate the hardening process of tea leaf fibers, hence diminishing the quality of the tea products. Moreover, because of its traits, Wijeratne (1996) mentioned that increasing temperature and changing rainfall pattern are very harmful to the growth of the tea plants, which ultimately affects the quality and quantity of the final tea products.

Social surveys which were conducted inside the Uji Area tea grower's community have also showed evidence on how the occurrences of extreme climatic events have become more frequent over the past ten years. Seen in Fig 16.7, from the social survey conducted with 15 tea growers which represent 15 tea plantations from the total 21 tea plantations in Uji Area, most of the tea growers interviewed have claimed that climate change has been affecting the tea cultivation processes, and attributed to temperature fluctuations (36%), drought (32%), frost events (18%), and excessive rainfall (14%) (Ashardiono and Cassim, 2015).

16.3 Burgundy Wine Grape Cultivation and Uji Tea Cultivation

Based on the current climatic conditions in the Uji Area, it is necessary to develop a new tea cultivation and production concept which could swiftly and effectively adapt to the ongoing climatic changes. This concept would not only be utilized to develop new adaptation methods, but also providing value additions to the tea products by employing new cultivation methods developed from the new concept. To achieve these objectives, the new concept would need to take account the unique characteristics which come from the Uji Area natural environment as well as the traditional cultivation methods practiced in the area. Deriving some of the concepts from the wine grape cultivation, the new tea cultivation concept is constructed with the objectives to ensure the sustainability of the tea cultivation region from a social and economic perspectives.

As observed, the tea growers in the Uji Area have been cultivating tea for many generations, where most of the tea grower families have been around for several hundred years. These tea growers have been able to fully utilize the characteristics of natural environmental factors in their tea plantations to cultivate desirable, high-quality tea products. This approach is similar to the wine grape cultivation in the France's Burgundy Area; in order to produce wines which could be associated with its origin, wine grape growers in each area utilized traditional cultivation methods which are unique to the areas where the wine grapes are being grown (Gade, 2004; van Leeuwen and Seguin, 2006; Gergaud and Ginsburgh, 2008).

Wine grape growers in the Burgundy Region are using traditional cultivation knowledge which have been continuously passed down from their predecessors, allowing them to select optimal cultivars which are suitable with the natural environmental conditions in their vineyards. The ripening condition of the wine grape plays a crucial role in determining the quality of wine products. The ripening condition of each vineyard differs from that of others, as it is highly dependent on the unique microclimatic conditions of each area (van Leeuwen and Seguin, 2006).

Winemaking requires a certain level of wine grape quality which is highly influenced by the natural environmental conditions of the vineyards. Wine grape growers take account of the natural environmental factors such as climatic conditions, soil types, and other exposures while cultivating the vines; thus, at the shortest, a typical old-world French vineyard would need 20–30 years before it is able to produce a good-quality wine grape harvest (Gergaud and Ginsburgh 2008). Tea cultivation possesses similar traits as wine grape cultivation as both plants are sensitive to the changes in the microclimatic condition. Through these observations it can be understood how natural environmental factors have been affecting the tea plants.

The factors above have been included in the *terroir* concept (van Leeuwen and Seguin, 2006), incorporating traditional cultivation practices which originated and were developed in the region. Together, these factors are important for defining the products' characteristics, quality, and quantity. Based on this concept, the terroir concept is employed to identify the changes that occurred in the region, especially concerning the links between climatic conditions and tea production characteristics.

Through the notion of terroir, it can be argued that the special quality of an agricultural product is determined by the character of the place where the product comes from (Gade, 2004). In growing grapes for wine, human factors such as history, socioeconomic conditions, as well as viticultural and oenological techniques, are also part of terroir (Seguin, 1986). In regard to the original concept and the scope of this research, the terroir concept is also linked to the unique biophysical properties of a particular area (Berard and Marchenay, 2006) which affects the quality of the resulting agricultural products.

Historically, the terroir concept in the wine industry is incorporated in the winemaking process as a means of showing the strong relationship between wine products of the area, and community where the product is produced. This inquiry aims to identify and implement the terroir concept for tea cultivation, especially in the Uji Area, where tea cultivation in the area has been conducted for more than 800 years. There are many indications of similar elements between traditional old-world style wine grape cultivation and the tea cultivation in Japan, especially in both cultivation processes the importance of natural environmental elements is highly regarded. Wine grape growers, as well as tea growers, conduct meticulously detailed processes throughout the cultivation methods, carefully utilizing the environmental elements to their advantage. With these similarities in the cultivation elements, utilization of the terroir concept in the Uji tea cultivation will add further value and optimize the cultivation process, thereby ensuring the sustainability of the area as a well-known tea cultivation region.

16.4 Terroir Concept for Uji Tea Cultivation

Implementation of the terroir concept in the Uji Area will lead to the identification of important terroir elements, furthermore helping to understand how these elements interacts as well as to comprehend the effect towards the characteristics of tea products. Based on the general terroir definition, these elements could be divided into two main parts, which are the natural environmental element and agriculture practices element. In this definition, the relationship between ongoing climatic changes can be seen as a factor that affects both of the terroir elements to a certain extent.

Based on these definitions, the terroir concept is defined as the relationship between the natural environment elemental and the agriculture practices element where these elements directly influencing the characteristics of an agricultural product (Ashardiono and Cassim 2015) (Fig. 16.8). These two elements represent the natural and human geographical characteristics of the Uji Area, which includes the accumulation of tea cultivation knowledge inside the tea grower community in the area.

The natural environmental element comprises of climate, soil, topography, and cultivars. Among these four environmental factors, this research highlights the climatic factors as the ongoing changes in the climatic conditions will have direct and indirect effects on the tea cultivation processes and its final products. The other important element in the terroir concept is the agriculture practices element which is the basis of tea cultivation knowledge in the Uji Area, where this knowledge is largely family-inherited knowledge, continuously passed down through successive generations of the tea grower's family.

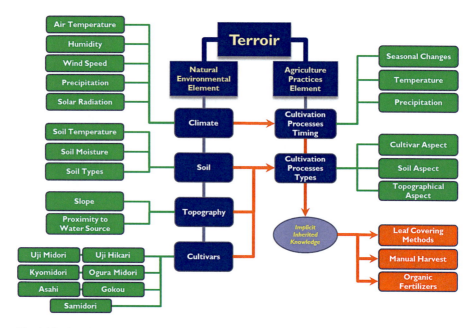

Fig. 16.8 Terroir concept in the Uji area tea cultivation

16.4.1 Natural Environmental Element

As one of the important elements in the terroir concept, the natural environmental element consists of factors which represent unique characteristics of environmental condition in a certain cultivation area. These important factors are climate, soil, topography, and cultivar (cultivated varieties), where the combination of these four main factors have shaped and created a certain natural environmental condition which is unique only to that area. Hence, agriculture plants such as tea, which grown under these specific conditions would produce agricultural products which have certain characteristics derived from these unique natural conditions. To fully understand these factors, measurements and analyses will be required to identify how these factors are interacting.

Climatic factor consists of several aspects which are directly related to the microclimate and mesoclimate condition of the agriculture area, furthermore other environmental related aspects are also directly influenced by the changes in the climatic conditions. With the indications of rapid changes in the climatic condition, these conditions will undoubtedly affect the characteristics of the other environmental aspects, where the five aspects considered crucial in the tea cultivation are: Air Temperature (°C), Humidity (%), Wind Speed (m/s), Precipitation (mm), and Solar Radiation (W/m^2) (Wijeratne, 1996; Konomoto et al., 2006; Watson, 2008a, b).

Other advanced aspects which are formed through the combination of several aspects will also be important for understanding the impact of changing climatic conditions, especially during important periods of the cultivation process. Air temperature is one of the most important aspects of the climate factor, where it can be further broken down into average temperature, minimum temperature, and maximum temperature. Generally tea plants are active under conditions where the air temperature is above 10 °C and would enter a dormant state when it is below 10 °C. It is clear that these internal mechanisms of the plants, which are crucial to the cultivation process, are largely dependent on the air temperature condition (Wijeratne, 1996; Konomoto et al., 2006; Watson, 2008a, b).

Humidity factor is largely affected by the air temperature and precipitation level; additionally occurrences of evapotranspiration from the soil and plant leaves would directly influence the humidity level. Wind speed is also indirectly affects not only the humidity factor but also the air temperature. The absence of wind can increase humidity level, whereas a strong wind can lead to a sudden gust, heavy downpour, drop in the air temperature, as well as causing frost events. In terms of solar radiation factors, the quantity of radiation received by the plants would become the signal for plant's metabolism to continue its phenological processes. Excessive solar radiation would lead to the increase of maximum daytime air temperature, thus, in extreme instances it will lead to serious issues such as drought, especially when accompanied by the absence of precipitation.

The Soil factor mainly consists of physical properties aspect and conditions of the soil, where the factors include Soil Temperature (°C), Soil Moisture (cb), and Soil Types. The conditions of these factors can be seen to be largely influenced by the changes in the air temperature and quantity of precipitation. The areas covered by the plants are also influencing these changes, acting as a buffer zone limiting penetration of sunlight, and thereby minimizing the temperature changes on the surface of the soil. In the Uji Area there are two types of soils, namely those where clay or sand predominate. In several tea plantations, tea growers mix these two soil types to obtain certain results. Based on anecdotal information obtained from tea growers, it was mentioned that the tea plants which are grown on sand type soils would produce teas with a better tea aroma and a deeper green color, whereas those which are grown on clay type of soil would produce teas which have a strong flavor.

Among these different soil types, sandy soil has large particles, and because there are spaces among the particles, it has low water retention, thus nutrient leaching is more frequent. Clay soil on the other hand has small particles with smaller spaces in between, resulting in good water retention properties. Through social surveys and observations, it can be seen how the soil factors are closely linked to the traditional cultivation practices, where the tea growers apply different types of organic fertilizers at different timings, and conduct deep soil maintenances periodically. These activities would ensure the soil health within their tea plantation, as well as ensuring enough nutrients are available for the tea plants to grow.

The topographical factor of the Uji Area is closely related with soil factors; where clay soil is found in the hillside, the sand soil is mostly found near the riverbed and flatlands. The topographical factors include Slope (degree), and Proximity to the water reservoir (m); which directly influencing the tea cultivation practices. Although the tea plantations which are located in the hillside are subject to lower winter temperature, they have fewer frost incidents. Furthermore as clay soil has better moisture retention, it has lower probability of experiencing droughts, thus because the surrounding areas are mostly high canopy forest, it shade the tea plantations and minimize water loss. Due to its location, that is, on slopes, cultivation practices are slightly more complex since the topographical gradient of the plot should be such that it achieves optimal water flow.

Tea plantations which are located around the riverbed and flatlands are mostly near the residential areas, thus more accessible compared to the ones on the hillside. Because of its location, these plantations are more vulnerable to extreme weather events such as high temperature and drought, as well as facing the risk of flooding during a heavy downpour. In parallel with its vulnerabilities, easy access to the water reservoir and proximity to the city drainage system would allow tea growers to manage the water flow in their tea plantations efficiently.

Cultivars are defined as cultivated varieties which are native to the area, or specially created through selective plant breeding to further incorporate the natural environmental characteristics in the area. Tea seeds first came to the Uji Area from China around 1191 AD, and soon after that because of its suitability, major tea cultivation began in the area . For more than 800 years of tea cultivation, the natural adaptation capabilities of the tea plant have made it suitable to optimally grow in the climatic conditions of the Uji Area. Based on these qualities, the tea plants have developed several characteristics which can only be attributable to those which are categorized as native plants. Apart from this, the varieties of tea plants in the Uji Area have also been developed from selective plant breeding which were conducted by the tea growers.

To achieve certain desired traits of the tea plant, especially to further optimize them to the micro climate condition, tea growers have carefully selected and breed tea plants based on their characteristics. Currently there are seven varieties of tea plants which have been developed in the Uji Area: *Uji Midori, Kyomidori, Samidori, Asahi, Uji Hikari, Ogura Midori*, and *Gokou*. Although these are the official varieties of tea plants which are originated and registered from the Uji Area, there are many other undocumented varieties which are exclusively bred and cultivated by many of the tea growing families in the Uji Area.

At the moment, samidori cultivar is the most widely cultivated variety by tea growers in the Uji Area, with the proportion in most of the tea plantations being around 70–80% with other tea cultivars making up the remaining numbers. Although samidori does not produce high-quality tea leaves, because of its long window of harvest period and its resistance to cold, it is widely selected to mitigate risk factors related to climatic and environmental conditions.

16.4.2 Agriculture Practices Element

The element describes the influence and the importance of human factors in an agriculture process, where in this inquiry it could be observed in Uji Tea cultivation. As one of the important elements of the terroir concept, the tea cultivation practices which are being used by the Uji Area tea growers are distinctive and irreplaceable. These practices utilize cultivation knowledge, which is accumulation of experiences of tea growers from the previous generations which is continuously passed down inside their families. This knowledge of tea cultivation is unique and native only to the Uji Area, as it has been developed based on the natural characteristics of the area's environmental element. The cultivation methods and special cultivation techniques developed by the tea growers are designed by taking the unique environmental characteristics into consideration.

Application of these cultivation methods is based on two deciding factors which are the timing on conducting cultivation processes and the types of cultivation processes applied. Both of these deciding factors are influenced by the four environmental factors, where the tea growers have taken advantages of these factors to obtain the desired level of tea production. These types of interactions between the environmental factors and the cultivation processes applied by the tea growers are unique and only exist in the Uji Tea cultivation. Each tea grower will need to fully comprehend the interactions between these terroir elements, before they could be acknowledged as a full-fledged tea grower in the Uji Area.

Through years of experience and observations, the tea growers have been able to understand the microclimatic condition in their tea plantation down to the smallest plot. By understanding these conditions, the tea growers have been able to identify the correlation between the microclimatic conditions and the phenological process of the tea plants. Based on this knowledge the Uji Area tea growers are able to conduct cultivation processes which are best suited to the current existing conditions as a means of adaptation (Berkes et al., 2000), especially in order to bring out the unique characteristics of the tea leaves. Based on preliminary observation, the timing of cultivation processes which are conducted by the tea growers mainly were based on the seasonal changes, air temperature, and precipitation level.

Some of the original cultivation processes, such as leaf covering, application of organic fertilizers, and soil maintenances are conducted depending on the seasonal changes in the climatic factors. The tea growers generally would start covering the tea leaves around mid-spring when the temperature is slowly becoming warm. Application of fertilizers and soil maintenances are also conducted depending on the season and precipitation level. This information shows how the changing climatic conditions have been directly affecting the tea grower's decision on the timing for conducting cultivation processes.

The second factor in the agriculture practices describes the types and kinds of cultivation processes which are utilized in the Uji Area. The types of cultivation processes selected by the tea growers are mostly motivated by each grower's objectives. They will utilize cultivation processes which are most suited to conditions of their plantation's natural environment. Based on this observation, most of the decisions for selecting the types of cultivation processes are based on the cultivar factor, soil factor, and topographical factor.

Depending on the combination of soil types and conditions, the slope of the tea plantation, and types of cultivars, the tea growers can select the most suitable cultivation processes to achieve the desired harvest quality and quantity, where in this

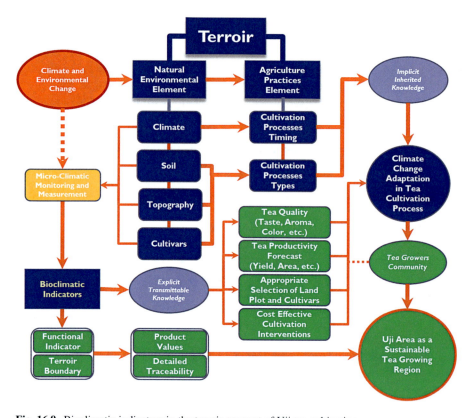

Fig. 16.9 Bioclimatic indicators in the terroir concept of Uji tea cultivation

regard, the combination of several cultivation processes are applied by the tea growers. Other cultivation processes which are being utilized to in response to the changing climatic conditions are watering system, frost protection fans, and other techniques.

Through utilization of the terroir concept for tea cultivation in the Uji Area (Fig. 16.9), there are possibilities for new wine grape cultivation derived bioclimatic indicators (Steadman, 1979; Tonietto and Carbonneau, 2004; O'Donnell and Ignizio, 2012) to be developed for the tea cultivation. The new indicators would function as a tool for the tea growers to further understand the phenological elements in the tea cultivation processes, as well as to provide precision bioclimatic information. In the long run, this tool will also become the foundation for developing new tea cultivation processes. Bioclimatic indicators will be the basis for new tea cultivation knowledge which is explicit and transmittable among the tea growers, complementing the implicit cultivation knowledge which they have already inherited from their families.

Application of the terroir concept and its bioclimatic indicators in the tea cultivation would actualize the development of tea cultivation processes which are more optimized and efficient, while at the same time preserving its traditional cultivation methods, which would further bring out the unique quality characteristic of the tea products. Through the application of these advanced cultivation methods in the Uji

Area, the tea growers would be able to cope with and swiftly adapt to the changing climatic conditions, thus allowing them to maintain the tea production yield and quality. Application of this concept will also ensure the sustainability of the area as a well-known tea growing region.

16.5 Conclusion

Currently the Uji Area tea growers have already understood the impact of rapid climatic changes to some extent, especially about its adverse effects towards the tea cultivation. Despite knowing the potential dangers of climatic impacts in the future, they are less worried about the worst possible climate scenario, even though the current indications have shown that the Uji Area is heading towards that direction. This is mainly because currently the tea growers have already facing several issues besides climatic changes. Although they have a well-known status, the tea growers are largely facing issues which are mostly economic in origin, of which declining sales value and increased expenses in the tea cultivation processes have slowly affect the tea grower's financial condition.

Additionally, this condition has made the younger generation of the tea grower's family to become less interested in continuing their family run business because of these economic uncertainties. Before they can get involved in the tea cultivation, young tea growers have to encounter the challenges in understanding the intricate and complex traditional tea cultivation and distribution processes, which further increases their reluctance to enter the industry. From the social survey result, it is evident that securing a successor has also become an important issue in the Uji Area, despite its reputation as a premier tea-producing region in Japan.

As explained in this study, from the new terroir concept new bioclimatic indicators as utilized in the wine grape cultivation can be derived, where the tea growers in the Uji Area will be able to utilize it for maintaining and improving the tea production yield. The combination of the tea growers inherited tea cultivation knowledge with the precision information from the bioclimatic indicator values will significantly increase the tea grower's capabilities to effectively select appropriate land plot and cultivars for tea cultivation. This will help the tea growers to make decisions in carrying out appropriate cultivation processes, which will further optimize and improve the current tea cultivation methods.

At the moment, most of the tea growers in the Uji Area obtain climatic information from external sources such as electronic media, newspapers, and the internet. These sources can only provide information on a region-wide mesoclimatic scale, and to a certain degree, it requires additional corrections to reflect the real microclimatic condition on each tea plantation. With the current rapid climatic changes, these open climatic data sources would not be able to provide climatic information which is required for the tea growers. Without this information, the tea growers would become less aware of the changes in their tea plantations and it might lead to cultivation processes decisions which are ineffective and resource consuming. From this analysis, it can be argued that it is critical for the local government and agricultural institutions to put more focus on assisting the tea growers in obtaining precise microclimatic information which can be utilized in their tea plantations.

To utilize the bioclimatic indicators, the tea growers would have to continuously monitor and record the microclimatic changes in their tea plantations, where from these monitoring activities the tea growers would be able to further understand the microclimatic conditions of their tea plantations and their impacts on the tea production quality. This, in turn, will make them more aware of the significances of changing climatic conditions toward their livelihood, and allowing them to adapt to any of the changes more swiftly.

From this inquiry it can be concluded, through the combined applications of traditional cultivation knowledge and terroir-derived bioclimatic indicators, the tea growers would be able to maintain the quantity and quality of the tea products cultivated in the Uji Area. Furthermore, it could lead to the creation of a new product value which is based on the characteristics of the natural environmental elements. The new product values would emerge in two forms, which are values internal to the tea plant and externally generated values. The new internal values would likely emerge as an improvement to the tea products' characteristics themselves, such as fragrance, color, taste, and health benefit, while the new external values would come from the development of detailed systematic records of the tea cultivation processes which can be translated into product traceability, safety, and quality assurance aspects.

The combination of these detailed records (external values) and the sensory attributes of the tea itself (internal values) can be utilized to create a strong brand value for the tea product, based on the rational evidence-based investigation. These new values of the tea products would create new market demands, leading to new economic benefits for the tea growers and their community in the Uji Area. This would, in turn, make it easier to find successors within tea-growing families and to lower barriers to new entrants which are attracted by the dynamism and profitability of tea cultivation. On this note, securing a successor in the tea cultivation industry would contribute to the socio-economic benefit of the Uji Area, which will ultimately ensure the sustainability of the Uji Area as a well-known tea-growing region.

References

Ashardiono, F. (2014). Climate change adaptation for agro-forestry: Sustainability and potentials in the tea industry. *Seisaku Kagaku, 21*(2), 99–113.

Ashardiono, F., & Cassim, M. (2014). Climate change adaptation for agro-forestry industries: Sustainability challenges in Uji Tea cultivation. *Procedia Environmental Sciences, 20*, 823–831.

Ashardiono, F., & Cassim, M. (2015). Adapting to climate changes: Challenges for Uji tea cultivation. *International Journal for Sustainable Future for Human Security J-SustaiN, 3*(1), 32–36.

Berard, L., & Marchenay, P. (2006). Local products and geographical indications: Taking account of local knowledge and biodiversity. *International Social Science Journal, 187*, 109–116.

Berkes, F., Colding, J., & Folke, C. (2000). Rediscovery of traditional ecological knowledge as adaptive management. *Ecological Applications, 10*(5), 1251–1262.

De Costa, W. J. (2008). Climate change in Sri Lanka: Myth or reality? Evidence from long term meteorogical data. *Journal of the National Science Foundation of Sri Lanka, 36*, 63–88.

De Costa, W. J., Mohotti, A. J., & Wijeratne, M. A. (2007). Ecophysiology of tea. *Brazilian Journal of Plant Physiology, 19*(4), 299–332.

Duncan, J., Saikia, S., Gupta, N., & Biggs, E. (2016). Observing climate impacts on tea yield in Assam, India. *Applied Geography, 77*, 64–71.

Gade, D. W. (2004). Tradition, territory, and terroir in French viniculture: Cassis, France, and Appelation Controlee. *Annals of the Association of American Geographers, 94*(4), 848–867.

Gergaud, O., & Ginsburgh, V. (2008). Natural endowments, production technologies and the quality of wines in Bordeaux: Does terroir matter? *The Economic Journal, 118*(June), F142–F157.

Holland, T., & Smit, B. (2010). Climate change and the wine industry: Current research themes and new directions. *Journal of Wine Research, 21*(2-3), 125–136.

Japan Meteorogical Agency. (2018). *Global warming projection and climate change monitoring.* Retrieved from Tokyo Climate Center WMO Regional Climate Center in RA II (Asia) http://ds.data.jma.go.jp/tcc/tcc/products/gwp/gwp.html

Jones, G. V. (2005). Climate change in the western United States grape growing regions. *Acta Holticulturae, 689*, 41–60.

Jones, G. V. (2007a). *Climate change and the global wine industry* (pp. 1–8). Thirteenth Australian wine Industry Technical Conference, Adelaide.

Jones, G. V. (2007b). *Climate change: Observations, projections and general implications for viticulture and wine production.* Zaragoza: Climate and Viticulture Congress.

Jones, G. V., & Webb, L. B. (2010). Climate change, viticulture, and wine: Challenges and opportunities. *Journal of Wine Research, 21*(2-3), 103–106.

Kimura, Y., & Kanda, M. (2013). Characteristic of the light spectrum environment of Honzu (ortodox shading) covering culture and influence of UV irradiation or shielding on the quality of new tea shoots. *Chakenhou, 116*, 1–13.

Konomoto, H., Morita, A., Kondou, S., Ozawa, A., & Nakamura, Y. (2006). *Zukai Chaseisan no saishin gijutsu – saibaihen [Illustration, Newest technology in tea production – cultivation edition-].* Shizuoka: Chamber of Tea Association of Shizuoka Prefecture.

Kyoto Prefecture. (2015). *Kyoto-fu Chagyoutoukei* [Kyoto Prefecture Tea Industry Statistic].

Kyoto Tea Cooperative. (2013). *We love Ujicha.* Retrieved from Kyoto-fu Chakyoudoukumiai http://www.kyocha.or.jp/we-love-ujicha/

van Leeuwen, C., & Seguin, G. (2006). The concept of terroir in viticulture. *Journal of Wine Research, 17*(1), 1–10.

O'Donnell, M. S., & Ignizio, D. A. (2012). *Bioclimatic predictors for supporting ecological applications in the conterminous.* Reston, VA: U.S. Geological Survey.

de Orduna, R. M. (2010). Climate change associated effects on grape and wine quality and production. *Food Research International, 43*, 1844–1855.

Seguin, G. (1986). Terroirs and pedology of vinegrowing. *Experientia, 42*, 861–873.

Steadman, R. G. (1979). The assessment of sultriness. Part I: a temperature-humidity index based on human physiology and clothing science. *Journal of Applied Meteorology, 18*, 861–873.

Tonietto, J., & Carbonneau, A. (2004). A multicriteria climatic classification system for grape-growing regions worldwide. *Agricultural and Forest Meteorology, 124*, 81–97.

Uji City. (2010). History of Uji tea. Retrieved from Kyoto Prefecture Uji City http://www.city.uji.kyoto.jp/en/tea/tea_01.html

Watson, M. (2008a). Climatic requirements and soil. In A. K. Zoysa (Ed.), *Handbook on tea* (pp. 10–15). Talawakelle: Tea Research Institute of Sri Lanka.

Watson, M. (2008b). Harvesting of tea. In A. K. Zoysa (Ed.), *Handbook of tea* (pp. 94–104). Talawakelle: Tea Research Institute of Sri Lanka.

Wijeratne, M. A. (1996). Vulnerability of Sri Lanka tea production to global climate change. *Water, Air, and Soil Pollution, 92*, 87–94.

Wijeratne, M., Anandacoomaraswamy, A., Amarathunga, M., Ratnasiri, J., Basnayake, B., & Kalra, N. (2007). Assessment of impact of climate change on productivity of tea (Camellia sinensis L.) plantations in Sri Lanka. *Journal of the National Science Foundation of Sri Lanka, 35*(2), 119–126.

Chapter 17
The Opportunity of Rural Space with Urban Relationships: Urban Agriculture as Contemporary Cultural Landscape for Resilience by Design

Luis Maldonado Rius

Difference which occurs across time is what we call "change".
Gregory Bateson.

Abstract Urban Agriculture has emerged as a key practice and interdisciplinary subject in an increasingly urban world. Its work ranges from climate and landscape dynamics to new infrastructures including changes in food system, public health, energy model, and urban planning and management. However, it still presents a contradiction between the healthy well-being of the rural realm identified with "nature" and cities as "parasites" and urban phenomena as an assault on nature. The conflict has given room to responses sticking on the division between rural and urban space.

What do we mean by urban agriculture? What is its space and how to work on it? What do we assess? The results of recent practice and research seem to show no relevant spatial features. What "makes the difference" would be the continuity of the resource, the spatial pattern or land mosaics, in time. We do not value the place as much as the relationships the place provides.

The chapter introduces the discussion on the future role of agrarian space in the conurbation of Barcelona to present and discuss urban agriculture as a key meaningful activity to reshape and rethink old land mosaics as new contemporary cultural landscapes: the opportunity of rural space with urban relationships. From a "city region perspective", the model could allow for important implications for resilience by design, combining landscape architecture, infrastructure planning, civil engineering, urban ecology, economics and the arts.

Keywords Urban agriculture · Land mosaic · Cultural landscape · Landscape architecture · Design · Chance · Adaptation · Relationships · Urban · Rural ·

L. Maldonado Rius (✉)
Universitat Politècnica de Catalunya · UPC Barcelona Tech, Barcelona, Spain

Department of Agri-Food Engineering and Biotechnology (DEAB-ESAB),
Castelldefels (Barcelona), Spain
e-mail: luis.maldonado@upc.edu

© Springer Nature Switzerland AG 2019
A. Sarkar et al. (eds.), *Sustainable Solutions for Food Security*,
https://doi.org/10.1007/978-3-319-77878-5_17

17.1 Introduction

Can we address human health from spatial or landscape design? Nothing can take place without space, but the obviousness of such a simple statement hides the fact that "space" and "place" are two great absentees of Western philosophy and thinking (Casey 1997). However, for those designing open space, such as urban or land planners, architects and landscape architects, "space" is our field of work. Indeed, "time" has also had a growing role in our thinking, research and practice after ecology and systemic thinking arose in the mid-twentieth century. In the same way that we learned from applied ecology to design processes and relationships instead of simple sets of elements or objects together devoid of context (Prominski 2005), we also came to know that human health cannot be understood in isolation from its natural and social environment (Hillman 2006). In a few decades, environmental problems have shifted from being local and isolated to global, territorially based and compelling.

We do know we cannot divide present problems into pieces, but how should we face the tangle of relationships involved? Orff (2016) has recently pointed out that our new situation is not exactly a problem to be solved, but a new mind-set. Hence, we do not speak about natural or human health, or about food system and diet, but about human behaviour at all scales. We confront the paradox of being the problem and the solution at the same time: if we humans are agents of change, how can we tackle the situation to provide a real change? If we designers are not scientists, what can our design practice and research offer? (Orff 2016; Stiphout 2014).

Cultural landscapes and cities once shaped our settlements across the world. Present land mosaics are the remains of those cultural landscapes. They are the result of a man-made balanced relationship with nature including the main principles of sustainability and resilience: environmental health, social justice and sustainable economic growth.

In Barcelona, as in other Western cities, urban agriculture is a subject of recent study and growing importance. Its main interest is its cross-disciplinary approach relating global topics with others of local or regional significance. The discussion on climate change, energy model or ecological sustainability fits with the particular implementation in the territory of food production and diet, landscape management and maintenance, ecosystem services, social values, local economy, access and exploitation of resources (Lohrberg et al. 2016). It has the goodness, therefore, of putting different policies, management and governance, literally, on the ground, at close and large scales. Another recognised value is its ability to work on the link of citizenship with the territory on which it is based: identity, leisure and gastronomy are part of the direct experience of the concepts mentioned.

The benefits of urban agriculture are greater in urban areas where it affects a greater number of people (Lohrberg et al. 2016).However, its spatial values or its placement in urban territories have not been systematically studied. The problem stems from the radical separation and enormous distance between production and consumption; between cities and the countryside, in current food system (Wiskerke 2016) and economic model.

In "The Opportunity of Rural Space with Urban Relationships: Urban Agriculture as Contemporary Cultural Landscape for Resilience by Design", I present and discuss urban agriculture as a key practice that could make landscape mosaics around cities meaningful again. I use the ongoing work discussing the future role of agrarian land in the set of open spaces in the studies for the new urban plan of the metropolitan area of Barcelona, to argue how these former cultural landscapes may allow us to rethink and reshape natural, social and economic relationships in an increasingly urban world. By the end of the chapter, "Scales Fit In", explain how new or renewed landscape mosaics hosting urban agriculture activities would not just be simply providing citizens of healthier food or diet or preserving a certain natural stage, but be the seed of a real change now. The opportunity of rural space under urban influence is to cut off the "distance dynamics" (Wiskerke 2016) that make urban food systems, and by extension the economic model, fossil-energy dependent. Finally, I briefly comment on two presentations at Barcelona in which the authors showed how research-by-design and landscape architecture widen the scope of the argument.

17.2 Agriculture and Cities: Urban Space vs. Rural Space

To speak of urban agriculture seems contradictory because rural and urban space and activities are perceived as opposing realities (Hillman 2006). Recent decoupling of cities from the soil that feeds its population and growing environmental problems due to their expansion, are leading many citizens to identify agriculture with nature. Consequently, urban development, its activities and necessities are perceived as an assault on the rural environment and, by extension, on nature. Ultimately, this recent perception comes from the changes in the supply of dairy products, due to the development of new and faster means of transportation since the Industrial Revolution (Steele 2008). These means gradually broke the spatial link between field and city, although it was not always like this.

In the case of Barcelona, we have abundant cartography to track the shift. What makes it such a well-studied city is that almost every moment in history has left a recognisable mark on the city's map (Solà-Morales 2008) and a catalogue of extensive documentation. Barcelona was founded 2000 years ago on a small hill surrounded by a fertile plain gently sloping to the sea. The successive city walls of the Roman and Middle Ages and the agrarian space around the city are easily visible in the set of plans drawn from the end of the seventeenth and mid-eighteenth century due to the successive sieges of the city. It is cartography, not yet topography, but it let us recognise the same agro-forestry mosaic that probably covered much of the inhabited territory in Europe at the time including close vegetable gardens and orchards that provided cities with fresh food.

Agriculture and city were born together at once. It was the ability to periodically produce enough food that gave rise to the first stable settlements (Steele 2008).

All great ancient cities were founded on or next to fertile soils that made them possible. Athens, Rome, Paris and London are outstanding examples in Europe. Still today, the word "agro" has a meaning as a separate word in Spanish and not only as a prefix. In addition to its first meaning "extension of arable land", it still means "territory of the city" (Coromines 1987). Agriculture or arable land and the territory of the city were the same because one was unthinkable without the other. In the same way, many of the myths and founding rites of ancient cities came from agriculture (Rykwert 1988).

Aerial photographs of the centre of Barcelona show what has happened today. As in most European cities, the city walls have disappeared and its perimeter can be identified in the layout of the streets and in a few short stretches of wall remaining. Fields and gardens have been totally occupied by dense blocks of buildings. The perimeter of the agro can still be recognised in Barcelona in the outline of the Cerda's extension plan. Water springs and channels watering the agricultural plain have been buried and forgotten. Their memory barely survives in the street index. In a city as dense as Barcelona, enclosed between the mountains and the sea, and in the flood area of two rivers, the complete construction of the central plain was taken for granted. But the perception changed when it overcame these natural barriers and extended the urban continuum beyond the central plain.

The interpretation of urban and rural relationship as conflict has given rise to two main answers sticking on the radical separation of city and agriculture. On one hand, technology as a hope that obviates climate and allows leaving soil, even farmers, out; on the other, land planning conceived as protection of rural space and manager of environmental resources against the voracity of the aggressor.

Technical innovation and agriculture have always been paired as the invention of the plough or the successive techniques of watering demonstrate. We cannot distinguish one from another. What is striking about current technical research is the tendency to obviate the territorial link, that is, the soil. As the last example of the MIT Open Agriculture Initiative shows (OpenAg 2015) the positive reading is the possible access of the entire population to fresh produce in situ, regardless of the conditions of the place. But the possibility makes place expendable, on account of the technical hope of obviating the ground. In the end, it implies that everything—everything—can be built. Hence, fertile soil as a finite and limiting resource ceases to be a problem.

On the other hand, urban planning, grounded on land zoning and management, has been clearly overcome by the urbanisation process. Agricultural land never appears by itself in regional or urban planning in Spain. It belongs to what is named "undeveloped land", as if its only purpose would be waiting to be built upon.

The territorial plan for the Metropolitan Region of Barcelona (PTMB 2010) increased the protected surface of natural and agricultural interest from 103,217 to 242,143 ha, a 74.8% of the plan area including 53,503 ha of agricultural land. The aim of the protection was to stop the steady decrease of agriculture at the expense of urban, infrastructural and forest land uses since mid-twentieth century: from 24,713 ha, a 39% of the Metropolitan Area in 1956 to 5,728 ha (9%) in 2009 (Fig. 17.1). Its model was the "Metropolitan Ecological Matrix" posed by Enric Batlle (2000). As the plan foresees, Batlle's hypothesis linked the protected forestry

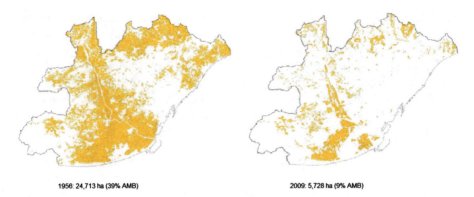

Fig. 17.1 Agricultural land decrease in the Metropolitan Area of Barcelona between 1956 and 2009 © Metropolitan Area of Barcelona (AMB)—Servei de redacció del PDU, first exhibited at "Barcelona Metropolis" (2015)

areas at the mountain ranges, the Green Ring of Barcelona, with a wide range of open spaces including rivers, beaches, farmland and marginal spaces along planned motorways to create a new urban counterpoint of the dense fabric of the conurbation. Barcelona followed the first and most important example in Europe of land planning by a city focusing on agriculture, the Parco Agricolo Sud Milano (PASM 1990) as a model for its own agrarian park (PABLL 2002). But land planning protects by resorting to prohibition: land and use planning keys say what and where anything can be done. Given the impossibility of foreseeing the future, the proposal blocks any initiative for adaptation or change.

17.3 What Do We Mean by Urban Agriculture?

A consequence of the separate vision of agriculture and city is the reduction of urban agriculture to what is informal, unprofessional, precarious and residual in the heart of the city, as a consequence of the economic crisis. It is confused with an outdoor activity for retirees, renting a small allotment in a community garden or occupying an abandoned plot to spend time growing something. Urban agriculture is confused with some of its social values. However, we are talking about a new situation.

As explained in the "World Urbanization Prospects" of the United Nations (2014), urban population outnumbered rural population for the first time in history in 2015. The situation varies from continent to continent. In Europe and North America, there is already a clear urban "majority". On the other hand, in Asia, and especially in Africa, it barely exceeds the threshold of 50%. Medium- and long-term forecasts predict a slight growth in total and urban population in the West, but a great urban development in South America and, above all, in Asia and Africa. The report predicts that two-thirds of the world's total population will be urban by 2050.

Data visualisation of the evolution and forecast of population growth for the major cities of the world between 1950 and 2030 (Smith 2017) show the "stagnation" of Western cities and the present great concentration in Asia (India, Japan, Korea and especially China). It also lets identify faster and younger growth areas in the future at Africa and the opposite: cities or countries with ageing pyramids of population as Europe and North America, excluding Mexico, and Japan. For Barcelona, the forecasts coincide with what is described for its region, with a slight increase of population until 2030 to close to six million.

This new urban horizon implies "various challenges" for cities as Wiskerke (2016) has recently explained. Challenges in development, governance and sustainability include food access and supply in the agenda for the first time in a significant way, in line with what Steele (2008) put forward in "Hungry City". When we talk about urban agriculture, we talk about how to feed a continent as deeply urbanised as Europe, or a world we foresee as urban, under reasonable and sustainable conditions. The question is whether urban agriculture can become a melting pot of the diverse scope of variables enumerated by Wiskerke.

If a few decades ago "city" was the problem, in the last United Nations Habitat (2016), "cities" are the solution (Kimmelman 2016). Globally, cities account for 70% of emissions and consume 75% of resources, hence the idea of cities as a problem and the image of urban aggression and plunder of nature. Cities and urban areas produce 80% of world GDP and, most importantly, they occupy just 2–3% of the world's surface (UN 2016), a 4.6% in Europe (Eurostat 2013). To the old idea of city as a space of prosperity, exchange, culture and knowledge, we could perhaps add that of protection of the land. The concentration of population in cities would "protect" soil, resources and life on the planet.

Nevertheless, "urban" or "city" may be too vague as it covers extremely diverse models. A clear and extreme example of contrasts representing the city model perspective would be the recurring comparison of emissions between the conurbations of Barcelona and Atlanta, due to their extremely different urban densities for the same population (Bertaud and Richardson 2004).

17.4 From Urban Agriculture to Land Mosaics: Rural Space with Urban Relationships

The aim of the European Cooperation in Science and Technology (COST) Action, Urban Agriculture Europe (2012–2016), was to answer all these issues from as many disciplines and interests as possible. The initiative highlighted the ecological, social, economic and strategic importance and benefits of agriculture near cities in Europe. The study was structured around four working groups—business, governance, metabolism and space—coordinated by a general one setting up conceptual definitions. Regular meetings were held in successive reference areas across Europe providing first-hand knowledge of diverse case cities led by local researchers.[1] Work conclusions were published in 2016 (Lohrberg et al. 2016).

---

[1] COST UAE meeting reports. Retrieved March, 2017, http://www.urbanagricultureeurope.la.rwth-aachen.de/action/events.html

17 The Opportunity of Rural Space with Urban Relationships

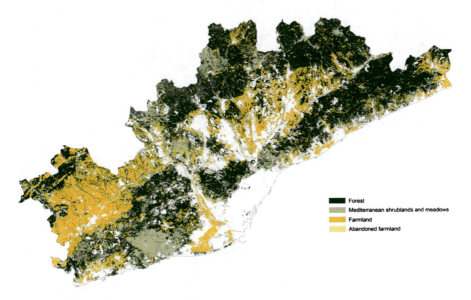

Fig. 17.2 Agriculture and Forestry at the Metropolitan Region of Barcelona by Luis Maldonado and Xavier Recasens; basis: © Institut Cartogràfic i Geològic de Catalunya (ICGC) 2016; data: Research, Innovation and Knowledge Transfer in Terrestrial Ecology (CREAF) 2007

Research, discussion and visits posed an extreme diversity of inputs. The disparity of situations corresponded to the disparity of: sizes—from 100 ha to less than 50 m²; character—urban, rural and industrial, private and public management, promotion or ownership; and purpose—local and global professional farming along with leisure, education, therapy and social inclusion (Fig. 17.2). In terms of space, given the diversity of situations, and since all they belonged to endorsed samples of what urban agriculture is, the questions were: What do we assess? Which is its space and how to work it? What do they have, spatially, in common?

At Barcelona, the initial goal was to draw what planning never draws, agriculture and forestry. We looked for common spatial characteristics that would allow posing productive open spaces as one more urban system in the metropolitan region through drawings simplifying the extensive record of land covers. The use of the graphical diagrams of M. Yokohari allowed distinguishing clear units by their position with respect to the urbanised environment, blank in the drawing, in a distant territorial scale (Maldonado et al. 2016). Yokohari's types of urban agriculture do not speak of space or any specific physical characteristic, but of the spatial relationship between city and agrarian space. The five models are named (see Fig. 17.3, from left to right) as Urban Agriculture on the fringe of the city; in/between; swallowed; emerging and remote from the city but in the same administration body of the city; never mind agricultural origin, history or age regarding the nearby city.

Models of the city-agriculture relationship did not allow to give particular guidelines about space, but to interpret agricultural patterns in relation to the urban process. In Barcelona, agricultural patterns are never determined by the city. They are conditioned firstly by topography, hydrology and soil properties, and secondly

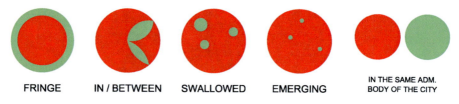

Fig. 17.3 Models of urban agriculture by Makoto Yokohari, printed with permission

by the infrastructures supporting production. Each particular shape is the result of the search looking for the optimal solution between the highest yield and the continuity of resources along time for each particular situation, even in the case of new squatter gardens along motorways and railways. Thus, the resource is not land or it is not just land, but the final pattern of organisation. The resource is not a space or a particular physical characteristic, but the arrangement that favours, and is a reflection at the same time, of a balanced relationship between the parties. From a spatial point of view, the first conclusion was the apparent obviousness of the spatial link. A second one was that urban agriculture lacks of a set of proper spatial characteristics (Lohrberg et al. 2016). The smaller size of the plots and a greater fragmentation of space are due to the pressure of urban activities and necessities. Finally, the importance of borders and paths was noted in the capacity of the place to adapt and mediate a wide range of ecosystem services. All these characteristics are not new. They have been studied in Barcelona (Gómez-Baggethun et al. 2013) and highlighted by the studies of landscape ecology (Forman 2004).

17.5 What Do We Appreciate?

After the Spanish Civil War, Barcelona and other economic areas to be rebuilt were widely photographed to cover the lack of mapping or planning. When we study these documents, prior to the expansion of the metropolitan cities in the 1960s, what we identify are not elements but sets of elements. We see "the interaction between environment and human activity" (Farina 2000). Figure 17.4 shows a fragment of the agrarian park (Parc Agrari del Baix Llobregat) south of Barcelona. At each side of the main draining channel, we can still see two of the patterns of organisation of the current agrarian space. "In/between" the metropolitan fabric, the case of the Parc Agrari del Baix Llobregat is especially useful as it is the only protected productive area of Barcelona.

The main resources of delta agriculture in Baix Llobregat are soil and water. However, it is the drainage that decisively determines the pattern. The delta plain is closed by a dune along the shoreline. The network of drainage channels allows overcoming this barrier, determining access, division, orientation, margins, building, even the main crops in the area. Comparing a current photograph with that of 70 years ago we would see that, fundamentally, its organisation has not changed

Fig. 17.4 Baix-Llobregat Agrarian Park 2016 © ICGC 2016

Fig. 17.5 Baix-Llobregat Delta 1956 © Ministerio de Defensa de España 1956 and ICGC 2016

very much. What we do appreciate would be the continuity of the resource—of the place—in time. We do not value the support as much as what the support provides. The key question is if the relationships that shaped it still remain (Fig. 17.5).

The opportunity of the open spaces with possible agricultural use in the metropolitan area of Barcelona resides in what is left of the agroforestry mosaic. Land mosaics covered a big part of this territory and of all Europe until the beginning of the past century, before mechanisation and, above all, before economic globalisation (Tello and Galán del Castillo 2013). In spite of their decline, they still are one of its hallmarks. The support us such understood, corresponds to what we know as a "cultural landscape" (Farina 2000). The characteristics that have made this a matter of study and interest are both its functional complexity and its ability to give continuity over time—resilience—to all kinds of ecological, human, social and economic relations, maintaining a certain structural and spatial continuity. A good way to envisage it is to understand the cultural landscape as the result of the interaction of the natural (N), the social (S) and economic (E) realms. The cultural landscape and the various forms of our land mosaics would appear around the centre of the network of relationships. Functional and structural complexity, their fragility and biodiversity, characterise these configurations. But what is most striking is their ability to adapt to successive changes in time without breaking the structure of the set—the land mosaic—or depleting system resources (Wrbka et al. 2004; Farina 2000).

Barcelona's main example, the Protection Act of the Parc Agrari del Baix Llobregat and its management (PABLL 2002), are largely responsible for the fact that we can speak of it as a contemporary cultural landscape. We can still call it "cultural" because it still integrates the diversified exploitation of the original natural, social and economic resources of the place through agriculture. We can rename it as "contemporary" because its management and exploitation have successfully adapted the place to global economy without destroying its landscape basis (Alfranca et al. 2013). Even so, the declaration, management and evolution of Barcelona's agrarian park are not exempt from contradictions: it was protected for environmental reasons (N), managed through agriculture and farmers interests (S), and finally criticised according to the market economy (E), without bearing in mind other costs and benefits for the rest the metropolitan area. Each initiative focuses in one realm and disregards the whole system.

Across Europe, the structure of landscape mosaics has undergone a number of disturbances. Land mosaics were simplified to give place to mechanised and large-scale management systems in the landscape of the global economy. Economic and social shifts have led to their degradation or disappearance, depending on the place (Tello and Galán del Castillo 2013). The resource—the place—has resisted or slowed down the process by its capacity for adaptation. Criticism based on market economics confuses value and price by reducing those spaces to a market commodity. The global economy assumes free social and environmental costs that do not have market value: one of the costs is erasing the structure, the land mosaic. Research on the valuation of ecosystem services is recurrent (see at Gómez-Baggethun et al. 2013). Given the impossibility of assessing, indications are nevertheless used to allow venturing a figure. The problem is that the loss of the resource has no price because there is no way back.

The renewed interest in landscape mosaics is not just as mere support of environmental and economic values because of its capacity for adaptation and regeneration by the balanced exploitation of resources (Farina 2000). It is also due to the fact that it is a man-made landscape: we shaped, or somehow "designed", them.

17.6 Why Are Landscape Mosaics Important?

The interest in landscape mosaic in the context of Barcelona is not new. Richard T. T. Forman gave the first vision of the city from an applied ecology viewpoint 10 years ago (Forman 2004), and has used the Barcelona case repeatedly (Forman

2008). Finally, he was invited to the initial workshops that started the drafting of the new planning for the metropolitan area (AMB 2014).

The urban plan for Barcelona and its area of immediate influence dates back to 1974. The Metropolitan Area of Barcelona began the study process for the definition of its new planning in 2013 (AMB 2014). The documentation and initial works were exhibited and published in 2015 (Crosas 2015). Subsequently, a series of sectoral round tables have been convened to discuss as many viewpoints as possible among planners, professionals, researchers, stakeholders, metropolitan or local administrators. The aim of the ongoing work is to define the guidelines for the new plan. The interest for urban agriculture and for land mosaic is therefore framed in its possible role as productive space and physical and environmental support of multiple services and bonds in the set of open spaces of the city (Gómez-Baggethun et al. 2013; Batlle 2000). The need to update planning after 35 years of intense changes now combines with global challenges and its spatial and social consequences. Expanded to the metropolitan scale, a significant trend is the continuous and reported reduction of agricultural land (Carreras et al. 2015) and mosaic structures (Marull et al. 2008).

For years, and Barcelona is a good example, main policies tried to solve complex problems using heavy grey engineering techniques as a way of simplifying nature by steps. Today, applied ecology gives us the conceptual basis to link the most recent changes and to improve new ones (Marull et al. 2010). We have learned that we can better grasp human settlement within nature if we understand it as ecology of disturbance (Tello and Galán del Castillo 2013).[2] In the discussion between land spacing and land sharing, land sharing has proved being more effective in heterogeneous metropolitan areas (Fisher et al. 2008). It has also been checked that certain levels of human activity in these areas can increase their biodiversity (Tello and Galán del Castillo 2013) showing that it is better to keep agricultural space active and occupied rather than protect it, because it preserves the dynamic balance, complexity level and connectivity of the natural system.

17.7 What Do We Learn from Land Planning?

Cities tend to homogenise their space, throwing entropy out to their surrounding territory. Planning as a protection of the rural space and resource manager is the main model in Barcelona as in Europe. However, the perception we have of the agrarian space near cities is that of disorder: a series of remnants fragmented by the expansion of the city or by the services or activities it demands, but needs periodically to move away from the centre. Contrary to what we expect, what we perceive is not an increasing order, but quite the opposite. Apparent disorder is seen as a failure of the planning system.

---

[2] As commented by Enric Tello from Ramón Margalef

The problem has much to do with the notion of "order" we apply. We identify planning with order, and order with classification, according to the principles of the classic scientific method, because planning, like science or the law, is intended to be objective, clear and not subject to interpretation, once approved. As when we classify, uniformity and homogeneity are ways of simplifying and making a complex reality feasible. Infrastructures relate urban and rural areas, solving the supply of energy, goods and food. Many uses or activities find their place according to what was foreseen for them. The problem is that neither nature nor humans necessarily always behave according to what planners or planning expect. As Yokohari's diagrams show, some agrarian activities persist or emerge in what was supposed to be urban space; and urban ones appear on what seemed undoubtedly rural. Like any living system, the ensemble tends towards increasing entropy. It tends to a "natural" balance including human behaviour (see Figs. 17.6 and 17.7). In the COST project, for instance, the effects of the extremely neat, instrumented and guarantee-based French planning in Toulouse and the absence of it at all levels in Warsaw gave exactly the same spatial results (see at Lohrberg et al. 2016).

The conclusions of checking the planning model let us infer important implications. Land planning is but a means that establishes a legal mechanism that temporarily solves the problem—the urban pressure on the land—but that, proposed in this way, is always surpassed by reality. It is the design of space as a support of the set of relationships that guarantees the continuity or creation of the spatial patterns and ecosystem services, along with the future of those who live, work it and the resources that make it a valuable asset.

Urban agriculture is not urban because the space it occupies or its activities are different or simply placed in urban areas. It is urban because its relationships are urban, even if its space is rural or the city seems far away. What makes the difference, the "urban approach" (Wiskerke 2016), is essential. The key to its design is not in the reproduction or continuity of a subsidised stage, but in giving the place the capacity to adapt to the expectations and needs generated by the city and those who maintain and live in it. Thus, planning should not preserve a particular configuration but drive the design of a new place.

17.8 From Landscape Mosaics to Cultural Landscapes: Scales Fit In

From the food system perspective, our planning model corresponds to that of the current global economy of stock. The current food system can be divided into separated production and consumption areas linked by transport, in the same way that urban and regional planning foresees wide and homogeneous areas linked by heavy transport and energy infrastructures. The following diagram explains what we call the food system through a series of boxes (Fig. 17.8).

Transport and energy exchange are continuous along the whole chain. It is almost a linear diagram, fully based on our common way of planning, because a large part

17 The Opportunity of Rural Space with Urban Relationships 343

Fig. 17.6 Land mosaics loss at the Metropolitan Region of Barcelona. (**a**) 1956; (**b**) 2016; (**c**) Current land cover © Ministerio de Defensa de España 1956 (**a**), ICGC 2016 (**b**) and CREAF 2007 (**c**)

Fig. 17.7 Urban sprawl and land mosaics at the Metropolitan Region of Barcelona. (**a**) 1956; (**b**) 2016; (**c**) Current land cover © Ministerio de Defensa de España 1956 (**a**), ICGC 2016 (**b**) and CREAF 2007 (**c**)

17 The Opportunity of Rural Space with Urban Relationships

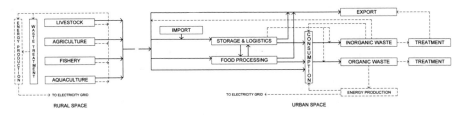

Fig. 17.8 Current food system: rural production and urban consumption

of the energy consumption or of its productive capacity has to do with the current organisation of the food system in space. In global economy of stock, the "agro" of the city is the entire planet. Its endurance is justified on the basis of economic development in spite of its energy inefficiency, environmental ravages and unhealthy consequences.

From a spatial viewpoint, the most surprising feature is the enormous physical distance between production and consumption or, spatially translated, between rural and urban spaces. As Wiskerke (2016) has pointed out, our current food system depends on "distance dynamics" and distances make it fuel dependent. The division of policies centred on production or consumption areas stick on the detachment.

The diagram can also correspond to those interpreting flows of energy, water, goods, food and waste treatment in present cities (Crosas 2015; EEA 2015; Tillie et al. 2014). Even so, there are no studies relating both sides, and no data or planning in this regard. The need for economic growth has historically justified the irreversibility of the consumption of fossil fuels, and the consequent emissions on which the current economic model and its implementation in the territory is based.

Recent models of urban metabolism interpreting city as a living organism and studies of circular economy seek to reverse the trend by local action and support the idea that a model developed at local or regional level would allow the energy transition, producing benefits to local economy and generating sustainable growth (Owen and Liddell 2016). The basis of all of them is working with shorter and more balanced loops linking all parts, processes and relationships (EEA 2015). Thus, the scale of the general scheme of the food system is transformed into a regional or urban one. The proximity of production and consumption would allow the recovery of local diets and food supply, reducing the dependence on fossil resources for transport and food preservation (Fig. 17.9).

From the scope of energy flows, the shorter length of the loops means that energy consumption and losses of the system are much smaller. The closer scale would make the gradual substitution of the type of energy consumed for renewable energies technologically possible and economically plausible, since it is not necessary to do everything at once. Current technology does not allow replacing completely the energy system on which planning or food systems are based if the model used is the existing one. As in the case of food, production and consumption are very far away and centralised in large stocks, or in few hands or large production plants, thus making the transition impossible.

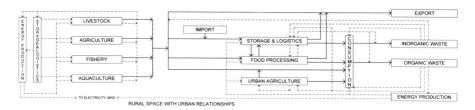

Fig. 17.9 Urban (or regional) food system

Spatial challenges due to energy transition are completely different in each model. An extreme way to grasp it is to think that much of the energy we have consumed in the last 150 years was underground (Sijmons 2014). Initial policies of energy transition focused on lower consumption, but are now infertile due to the massive incorporation of developing countries, such as China or India, to a development model based on "cheap" energy sources like oil and coal. The increasing difficulty and cost of obtaining them, pollution or the end of the technically available supply will mean that energy will have to be produced on the surface.

The output of current renewable systems allows making a forecast of the necessary surface we would need to dedicate to it. Again, its impact and implementation depend on the model. If, as happens now, we concentrate its production at a few centralised points, the impact of its size is enormous. However, recent research and existing pilot plants do not show this trend (Edwards et al. 2015). If, as Dirk Sijmons (2014) explains, energy has to be "harvested" continuously and very closely throughout the territory, then we are technologically ready and the existing grid of power transport, the increasing concentration of population and the heterogeneity of the territory are favourable for its gradual implementation. Likewise, it also fits in with those cases in which power production could be combined with the use of organic waste and the product of forest maintenance. Waste becomes a resource.

We can also speak about general transport in the same way. In Barcelona, the great concentration of population would make possible to rethink mobility at local and intermediate levels. The combination of short pedestrian and bicycle routes in loops between 5 and 10 km away, together with an efficient public transport network, would drastically reduce the number of vehicles on the roads (Crosas 2015).

In light of social and public health and well-being, the possible benefits of the system are not only those derived from food-related factors, such as a healthier or more suitable diet. We can easily appreciate the lower level of environmental pollution as well as the democratisation of food and energy production, and the consequent improvement of access and management of resources and the distribution of the resulting wealth.

Maintaining or creating new landscape mosaics in the Barcelona metropolitan area through urban agriculture allows for preserving the network of paths, borders and buffers that give ecological continuity, biodiversity and appropriate levels of complexity to the natural system (Marull et al. 2010). The point—and the difficulty—is that in highly urbanised environments these spaces are also a fundamental support of numerous flows and social processes. Leisure, sport and education are necessary activities for a greater well-being of the population, but also have, in turn, an impact on the environment. The use of open space around the city has grown. Growing expectations of contact with nature and the use of new technologies, such as GPS location, have facilitated access to a growing number of people. The activities have also changed: they cannot be located in a specific place. They move and change constantly: they are dynamic, too. It is an unstoppable process in which in many cases the city has created its own barriers.[3] In the same way that urban agriculture favours reinterpreting the production of the mosaic space and maintaining it, it could also encourage its reinterpretation and design as a support for these activities. It allows working in a punctual and networked way.

A big consequence of the set of possible changes described would be the need to adapt or invent their physical or spatial support. Out of the set of systems in which services, management and governance of highly populated areas are commonly organised, what is seen as a failure from the planning system—the heterogeneity to which the territory tends—or as an obsolete burden—the fragmentation inherited from the mosaic landscape—becomes an opportunity: all scales seemingly fit in.

Ensuring productive open space is the key to the future health of the city, its citizens and all its living systems, and an important resource for its adaptation to Climate Change. This can be achieved through regionally scaling the loops of production, transportation and consumption of food, energy, transport and social flows. Present land mosaics can be the basis to support, rethink and adapt the conurbation metabolism. We already have a lot of data, but future research must focus in the traceability of each system, its spatial translation, their links and scales to foster and boost local adaptation and regional integration.

17.9   Resilience by Design: "Back to Future"

The case of urban agriculture and what remains of the agro-forestry mosaic in Barcelona is only one aspect of an ongoing global task and it is, of course, not unique. The relationship between rural and urban space has led us to the link of citizens and food and, around this, given the foreseeable increase of the urban population, to global environmental health through land mosaics from an urban perspective as a basis for framing health and well-being of people as our future and, perhaps, that of life, on the planet.

I have argued from a spatial point of view, but the spatial and landscape effects of possible changes in the food system have no place in reports. A few cities as Rotterdam (Tillie et al. 2014) or Barcelona (Crosas 2015) newly began to do it.

At the 2016 International Biennial of Landscape Architecture in Barcelona, Dirk Sijmons, presented his ongoing research-by-design work relating landscape and energy change (Sijmons 2014; Sijmons and van Dorst 2012). His presentation and his books are an outstanding example of landscape spatial approach and are accompanied by magnificent data visualisation but, as he remarked with his closing

---

[3] I owe this comment to Carles Llop.

slide at a recent conference on city metabolism in Brussels: "... data are not yet information, and information is not yet knowledge, and knowledge is not yet wisdom, and wisdom doesn't give you an action perspective, yet..." and concluded: "...even if you have an action perspective you don't have a plan yet..." (Sijmons 2017). His intention was to warn the attendees of the danger of getting drunk on data or trying to imitate science in our way of proceeding.

Our landscapes are not alien to the laws of thermodynamics (Tello and Galán del Castillo 2013; Marull et al. 2010). Energy fluxes and balance are basic to understand the changes and benefits that would entail an adaptation of food system and the transition to renewable energies. Thermodynamics, ecology or science, in a generic way, are the basis for new planning and design strategies (Stremke and Koh 2011). But our contribution is design and research-by-design. As a designer and from a design perspective, Sijmons pointed out in Barcelona our biggest fear: he spoke of a "crisis of imagination", beyond the capacity that our professional skills give us to visualise certain changes or our proposals.

I have presented through Barcelona the landscape mosaic around the cities or rural space with urban relations as a chance for cities adaptation, but do not infer from the explanation that we can only work through existing spatial models. In the case of Barcelona, as in many others worldwide, it is an opportunity and it would be stupid and impoverishing to disregard it. Unfortunately, it is not always so simple. The point is that we can design; we can reinvent it, and not leave it randomly in the hands of nature.[4]

The ongoing work of the Metropolitan Area of Barcelona seeks an "action perspective" for a new plan (Crosas 2015; AMB 2014). First conclusions stick on Batlle's (2000) original "Metropolitan Ecological Matrix and the different scales of green infrastructures" (AMB 2014) as a design draft for Barcelona future outline (Fig. 17.10).

Barcelona still does not have an encompassing design of its open spaces including productive agriculture within the urban fabric. Nevertheless, I would like to finish this chapter by shortly mentioning an example of the imagination that Sijmons misses. A few hours after Sijmons contribution, Kate Orff's address closed the ninth Landscape Biennial in Barcelona. Like that of Sijmons, Orff's work is coherently displayed in her teaching and publications, design-by-research and professional practice (Orff 2016). Some of Orff's works with her office Scape, such as the celebrated Oyster-tecture proposal, "deny" Sijmons' fears.

Oyster-tecture proposes building a green/blue infrastructure to protect New York Harbour from sea level rise due to climate change. The proposal "aims to improve habitat and water quality, restore biodiversity to tidal marshes and encourage new relationships between New Yorkers and their harbor." I write "building" but the proper word would be "growing" their idea of "a living reef ... that supports marine growth, generates a 3D landscape mosaic that attenuates waves and cleans ... harbor water by harnessing the biotic filtration processes of oysters, mussels, and eelgrass."[5]

---

[4] I owe this comment to Xavier Mayor

[5] Quotation marks from Scape website. Retrieved March, 2017, from www.scapestudio.com/. Full description of the proposal at Kate Orff's, *Toward an Urban Ecology* (2016), pp. 89–116.

17 The Opportunity of Rural Space with Urban Relationships

Fig. 17.10 The "Metropolitan Ecological Matrix" for Barcelona by Enric Batlle. Reprinted with permission

Back and forward from research to design, from the previous geomorphology of the place and the history and former relationships of the settlement to the living cycle of marine life, and to community participation, Scape's work for adaptation to climate change is not just a challenge but a mindset driving and fostering the intervention.

The work has been widely explained, exhibited, published and praised. It is worthy to note that the proposal is based on science—ecology—as knowledge, not as a model, and does not expect to fight or solve a problem but adapt to it through understanding how a living system works, not by what we expect it to do. It is low technology based on relationships, not on isolated objects. The design is cheap, or at least divisible into multiple parts, making it affordable to cover all scales. As an idea, not a device, it involves behaviour dynamically, not a programme, growing and working with natural energy and community power including leisure, education, local economy and human activities to nurture other living systems and improving all habitat conditions to foster cohabitation (Fig. 17.11).

First Oyster-tecture presentations were always closed, with an image of the harbour community cheering at the "oyster parade". The image depicts a part of the aquaculture process—every spring and fall the barges would be loaded with oyster seed cages to plant oyster seed on the shell and rock structure breakwater[6]—and the community celebration of the periodical reset of the living cycle of the reef. Today we are used to gorgeous pictures showing our designs as finished projects

---
[6] I owe this remark to Kate Orff.

Fig. 17.11 "Oyster-tecture parade" by Kate Orff/Scape. Reprinted with permission

with people living in them, but to show a party is unusual. Like the image of any party, it makes the audience smile. Imagining a community celebration may seem to be a wink to the attendees, but it is extremely important as it represents the key of a new cultural landscape encompassing nature, society and economy—even food— or, to use Orff's own words, community, city and ecology through design.

Acknowledgements This chapter is the result of several works from 2013 to 2016. In chronological order, I first thank Frank Lohrberg of RWTH Aachen, for inviting me to join the COST Action UAE project, its visits and discussions; and Lilli Licka of BOKU Vienna, as chair of its Space working group, for her wisdom and confidence. I partly owe my critical thinking on Baix Llobregat Agricultural Park to Contxita Sánchez, president of the Centre d'Estudis Comarcals del Baix Llobregat, who repeatedly encouraged me to present my studies on the space from different viewpoints. I am grateful to Carles Riba of Universitat Politècnica de Catalunya (UPC) for inviting me to be part of UPC's commissions on "Diet and Energy" and "Habitability and Energy" and sharing his energy-flow understanding, and to Xavier Flotats of UPC for his comments and references on new-energies production conditions. I am very thankful to Antoni Farrero, from the Metropolitan Area of Barcelona, and Enric Batlle of UPC, who drove the round table "Metropolitan Landscape: Ecology, Leisure and Production", for inviting me to participate, present and discuss my ideas on the future role of agrarian spaces in the city. Finally, I am especially grateful to Makoto Yokohari of the University of Tokyo, to Enric Batlle of Batlle i Roig Arquitectes and to Kate Orff and her office Scape for their kindness and permission to use their inspiring images.

References

Alfranca, O., Maldonado, L., & Recasens, X. (2013). Cultural landscape in periurban territories: The case of Baix Llobregat (Spain). In O. Alfranca (Ed.), *Economics of culture: New aspects and new trends* (pp. 217–238). New York, NY: Nova Science.

AMB. (2014). *Quaderns-PDU metropolità, 1–8*. Retrieved March, 2017, from http://www.amb.cat/web/territori/urbanisme/pdu

Batlle, E. (2000). El anillo verde de Barcelona y la matriz ecológica metropolitana. *Arquitectos, 155*(3), 74–81 (In Spanish).

Bertaud, A., & Richardson, A. W. (2004). Transit and density: Atlanta, the United States and Western Europe. In A. W. Richardson & B. CHC (Eds.), *Urban Sprawl in Western Europe and the United States* (pp. 293–310). London: Routledge.

Carreras, J. M., Otero, M., & Ruiz, E. (2015). *Els usos del sòl a l'àrea i regió metropolitana de Barcelona, 1956–2006*. Barcelona: AMB Retrieved March, 2017, from http://www.amb.cat/web/territori/.

Casey, E. S. (1997). *The fate of place: A philosophical history*. Berkeley, CA: University of California Press.

Coromines, J. (1987). *Breve diccionario etimológico de la lengua castellana*. Madrid: Gredos.

Crosas, C. (Ed.). (2015). *Barcelona metropolis exhibition catalogue, volumes 1, 2 and Atlas*. Barcelona: AMB Retrieved March, 2017, from http://www.amb.cat/web/territori/urbanisme/pdu/.

Edwards, J., Maazuza, O., & Burn, S. (2015). A review of policy drivers and barriers for the use of anaerobic digestion in Europe, the United States and Australia. *Renewable and Sustainable Energy Reviews, 52*, 815–828.

EEA (2015) Urban sustainability issues – what is a resource-efficient city? EEA technical report, 23. Retrieved March, 2017, from www.eea.europa.eu/publications/

Eurostat (2013) Land cover statistics. Retrieved March, 2017, from http://ec.europa.eu/eurostat/statistics-explained/index.php/Land_cover_statistics

Farina, A. (2000). The cultural landscape as a model for the integration of ecology and economics. *Bioscience, 50*(5), 313–320.

Fisher, J., Brosi, B. J., Daily, G. C., & Tallis, H. (2008). Should agricultural policies encourage land sparing or wildlife-friendly farming. *Frontiers in Ecology and the Environment, 6*(7), 380–385.

Forman, R. T. T. (2008). The Barcelona Region's land mosaic. In F. RTT (Ed.), *Urban regions: Ecology and planning beyond the city* (pp. 243–274). Cambridge: Cambridge University Press.

Forman, R. T. T. (2004). *Mosaico territorial para la región metropolitana de Barcelona*. Barcelona: Gustavo Gili (In Spanish).

Gómez-Baggethun, E., Gren, S., Barton, D. N., Langemeyer, J., McPhearson, T., O'Farrell, P., Andersson, E., Hamstead, Z., & Kremer, P. (2013). Urban ecosystem services. In T. Elmqvist, M. Fragkias, J. Goodness, B. Güneralp, P. J. Marcotullio, R. I. McDonald, S. Parnell, M. Schewenius, M. Sendstad, K. C. Seto, & C. Wilkinson (Eds.), *Urbanization, biodiversity and ecosystem services: Challenges and opportunities* (pp. 175–251). Berlin: Springer.

Hillman, J. (2006). City and soul. In Uniform edition of the writings of James Hillman, vol. 2: *City & soul* (pp. 20–26). Washington, DC: Spring Publications.

Kimmelman M (2016) The kind of thinking cities need, The New York Times Sunday review, 28 Oct 2016. Retrieved March, 2017, from https://www.nytimes.com/2016/10/30/opinion/sunday/

Lohrberg, F., Lička, L., Scazzosi, L., & Timpe, A. (Eds.). (2016). *Urban agriculture Europe*. Berlin: Jovis.

Maldonado, L., Alfranca, O., Callau, S., Giacchè, G., Tóth, A., & Recasens, X. (2016). Barcelona: Outstanding agricultural diversity in a dense and small area. In F. Lohrberg et al. (Eds.), *Urban Agriculture Europe* (pp. 39–45). Berlin: Jovis.

Marull, J., Pino, J., & Tello, E. (2008). The loss of landscape efficiency: An ecological analysis of land-use changes in Western Mediterranean agriculture, global environment. *Journal of History and Natural and Social Sciences, 3*, 112–150.

Marull, J., Pino, J., Tello, E., & Cordobilla, M. J. (2010). Social metabolism, landscape change and land use planning. The Metropolitan Region of Barcelona as a referent. *Land Use Policy, 27*, 497–510.

OpenAg (2015) MIT media lab Website. Retrieved March, 2017, from http://openag.media.mit.edu/

Orff, K. (2016). *Toward an urban ecology: SCAPE/landscape architecture*. New York, NY: The Monacelli Press.

Owen, A. M., & Liddell, J. (2016). Implementing a circular economy at city scale – A challenge for data and decision making, not technology. *Unspecified International Seeds Conference 2016: Sustainable Ecological Engineering Design for Society,* Leeds, pp. 12–14. Extended version. Retrieved March, 2017, from http://eprints.whiterose.ac.uk/

PASM. (1990). Parco Agricolo Sud Milano website. Retrieved March, 2017, from http://parcosud.cittametropolitana.mi.it/

PABLL. (2002). Parc Agrari del Baix Llobregat website. Retrieved March, 2017, from http://parcs.diba.cat/web/

Prominski, M. (2005). Designing landscapes as evolutionary systems. *The Design Journal, 8*(3), 25–34 Retrieved March, 2017, from http://www.academia.edu/4363070/.

PTMB. (2010). *Pla Territorial Metropolità de Barcelona*. Barcelona: Generalitat de Catalunya. Retrieved March, 2017, from http://territori.gencat.cat/es/01_departament/05_plans/01_planificacio_territorial/

Rykwert, J. (1988). *The idea of a town: The anthropology of urban form in Rome, Italy and the ancient world*. Cambridge, MA: The MIT Press.

Sijmons, D. F. (2017) Where do we start? Masterclass 02.02.2017 in: *Masterclass series designing with flows*. Retrieved March, 2017, from https://www.youtube.com/

Sijmons, D. F. (2014). *Landscape and energy: Designing transition*. Rotterdam: NAi publ.

Sijmons, D. F., & van Dorst, M. J. (2012). Strong feelings: Emotional landscape of wind turbines. In S. Stremke & A. van Dobbelsteen (Eds.), *Sustainable energy landscapes, designing & development* (pp. 45–67). Boca Raton, FL: Taylor & Francis Group.

Smith, D. A. (2017). World city populations 1950–2030: Proportional circle time series map. *Environment and Planning A, 49*(1), 3–5. https://doi.org/10.1177/0308518X16641414.

Solà-Morales, M. (2008). *Ten lessons on Barcelona*. Barcelona: COAC.

Steele, C. (2008). *Hungry city: How food shapes our lives*. London: Random House.

Stiphout, M. V. (2014). *Samen – Together*. Amsterdam: Amsterdam Academy of Architecture and Architectura & Natura Publishers Retrieved March, 2017, from https://www.issuu.com/bouwkunst/docs/.

Stremke, S., & Koh, J. (2011). Integration of ecological and thermodynamic concepts in the design of sustainable energy landscapes. *Landscape Journal, 30*(2), 194–213.

Tello, E., & Galán del Castillo, E. (2013). Sustainable agrarian systems and transitions in the agrarian metabolism. Social Inequity, Institutional Changes and Landscape's Transformation in Catalonia (1850–2010). *HALAC, II*(2), 267–306.

Tillie, T. M. J. D., Klijn, O., Frijters, E., Borsboom, J., Looije, M., & Sijmons, D. F. (2014). *Urban metabolism, sustainable development of Rotterdam*. Rotterdam: IABR 2014 Urban by Nature. Retrieved March, 2017, from http://www.iabr.nl/en/publicatie/.

United Nations Habitat (2016). World cities report 2016. Retrieved March, 2017, from http://wcr.unhabitat.org/.

United Nations. (2014). World urbanization prospects: The 2014 revision. Retrieved March, 2017, from https://esa.un.org/unpd/wup/

Wiskerke, J. (2016). Urban food systems. In H. De Zeeuw & P. Drechsel (Eds.), *Cities and agriculture: Developing urban food systems* (pp. 1–25). London: Routledge.

Wrbka, T., Erb, K. H., Schulz, N. B., Peterseil, J., Hahn, C., & Haberl, H. (2004). Linking pattern and process in cultural landscapes. An empirical study based on spatially explicit indicators. *Land Use Policy, 21*, 289–306.

Part IV
Strategies for Access to Food, Technology, Knowledge and Equity, and Risk Governance

Chapter 18
Stakeholders' Perceptions on Effective Community Participation in Climate Change Adaptation

Subhajyoti Samaddar, Akudugu Jonas Ayaribilla, Martin Oteng-Ababio, Frederick Dayour, and Muneta Yokomatsu

Abstract Till date, successful community-based climate change adaptation projects and programs are rare; rather, the resentment and frustration among the local populace are ever increasing. Community-based climate change adaptation programs become nothing more than a trap to circumvent the local communities to get some plans sanctioned, encoded by the external agencies. The reason is that participation is not a simple, straightforward notion. In this chapter, it is argued that given manifold comprehension of participation, its unshackled, combative frameworks and numerous as well as dubious operation methods and techniques, the actual implementation of the participatory projects and programs is in the hand of implementation agencies. Their willingness, understanding, skills, and capacities determine to a great extent how successfully local communities can be engaged in the climate change adaptation programs. If the community's participation in climate change adaptation projects needs to be enhanced, it is critical to explore how stakeholders including government officials, technocrats, project managers, and donor agencies conceptualize and idealize community participation. But, in climate change adaptation studies, no such initiative has ever been made. This chapter aims to identify stakeholders' perspectives on effective ways, steps and factors for ensuring effective community participation in climate change adaptation programs and projects based on a case study in the Wa West district of Northern Ghana. We interviewed key stakeholders including government and non-government official involved in various climate change adaptation programs.

Keywords Community participation · Stakeholders' perspectives · Climate change adaptation · Ghana

---

S. Samaddar (✉) · M. Yokomatsu
Disaster Prevention Research Institute, Kyoto University, Kyoto, Japan
e-mail: samaddar@imdr.dpri.kyoto-u.ac.jp

A. J. Ayaribilla · F. Dayour
University for Development Studies, Tamale, Ghana

M. Oteng-Ababio
Department of Geography and Resource Development, University of Ghana, Accra, Ghana

© Springer Nature Switzerland AG 2019
A. Sarkar et al. (eds.), *Sustainable Solutions for Food Security*,
https://doi.org/10.1007/978-3-319-77878-5_18

18.1 Introduction

Community participation is a commonplace element in climate change adaptation (hereafter CCA) discourses (UN 1992; IPCC 2001) due to its potential to catalyze several positive benefits for successful project implementation, including enhancing community's awareness building, conflict resolution between stakeholders, building trust among project managers, ensuring more acceptable decision-making processes, and building community's self-reliance (Van Aalst et al. 2008; Sheppard et al. 2011; Reid et al. 2009; Samaddar et al. 2011). An additional motivation stems from the fact that community participation is also seen as the citizenry's democratic right to participate in consensus building particularly regarding policy decisions that directly and indirectly affect their life and livelihoods, though such a process is seldom effectively realized (Few et al. 2007; Hiwasaki et al. 2015; Prabhakar et al. 2009; Sheppard et al. 2011; Dodman and Mitlin 2013). Prior studies often attributed such failures to the intention, willingness, attitude, organizational capacity, managerial skills, and bureaucratic structure of the implementing agencies (Collins and Ison 2009; Hiwasaki et al. 2015; Cheng and Mattor 2006; Yang and Callahan 2007). These agencies include the project managers, public officials, field engineers, development partners, and NGO workers, among others from different institutions.

In most cases, ineffective community participation is seen either as a failure of the implementing agencies to motivate the local community to become willing partners in the CCA programs, or attributed to their an elitist, authoritarian approach that intimidates the poor, marginalized, and vulnerable local communities (Lorenzoni et al. 2007; Larsen and Gunnarsson-Östling 2009; Few et al. 2007; Hiwasaki et al. 2015; Reid et al. 2009; Dodman and Mitlin 2013). Our purpose is not to delve into the merit or otherwise of the debate but to formulate questions about the future of community participation in the climate change adaptation discourse and whether the beliefs and perspectives of the dominant stakeholders' matter for a meaningful CCA programs and projects. Currently, there appears to be fewer answers than questions, but we believe that clarifying and addressing such questions is an urgent task for the citizens, policymakers, civil servants, researchers, and civil society workers.

18.1.1 Do Stakeholders' Perspectives Matter in Climate Change Adaptation?

Principally, participation evinces a vague, contentious, and value-laden idea (Arnstein 1969; Rowe and Frewer 2000; Chess and Purcell 1999). It defies a universal definition, as for decades researchers battle over its true framework (Reed 2008; Blackstock et al. 2007; Blahna and Yonts-Shepard 1989). This lack of consensus is reflected as to who should represent the community. There is no consensus on who is to be involved, when and to what extent, what sociopolitical and cultural conditions are essential, what results to expect from participation. For some, community participation means only informing the community, for example providing information about what is climate change and how it can affect the community and how the potential impact on them can be managed (Patt et al. 2005; Roncoli et al. 2009). Such level of participation restricts community's involvement only as a passive recipient of information. Others also advocate for a level of participation where the community is involved in identifying risk and getting their candid counter-opinions about the plans prepared by the technical experts (Sheppard et al. 2011; Lorenzoni et al. 2007; Hung and Chen 2013; Galicia et al. 2015). Such a level of participation is seen as a top-down, autocratic model which must be replaced with a more proactive involvement of the community (Samaddar et al. 2015; Few et al. 2007). This new level of participation is envisioned as the one where the community has equal rights to open and table issues related to the project (Webler et al. 2001; Okada et al. 2013). In this instance, the community becomes an empowered collaborator through the preparatory stage to the implementation level (Samaddar et al. 2015). Although no consensus has been reached on the subject, we concur that in practice, the level of community involvement greatly depends on how those who command the administrative, legal, and financial authority of the project conceptualize and operationalize the idea of participation (Shaw 2006; Cheng and Mattor 2006; Yang and Callahan 2007).

18.1.2 The Yardstick of Community Participation: Controversies and Limitation

Generally, building a comprehensive participatory framework is seen as an antidote to reducing the acrimony over the concept of participation and project implementation. This has resulted in the development of a number of community participation frameworks in different academic domains, including natural resource management, environmental management, joint forest management, risk management (Dyer et al. 2014; Reed 2008; Blackstock et al. 2007; Blahna and Yonts-Shepard 1989; Carr et al. 2012; Samaddar et al. 2017) but quite limited in CCA studies (Sheppard et al. 2011; Collins and Ison 2009). In most cases, the proposed participatory frameworks provide a list of criteria, framed as a process (e.g., representation of all stakeholders, agreed objectives, fairness and trust, accountability, and good facilitation) (Rowe and Frewer 2000; Chess and Purcell 1999; Webler et al. 2001; Samaddar et al. 2017) and as an outcome (e.g., ownership, timely completion of project, conflict resolution, sustainability) (Chess and Purcell 1999; Teitelbaum 2014; Dyer et al. 2014; Reed 2008; Blackstock et al. 2007) as yardsticks for measuring effective community involvement. This initiative of developing participatory framework undoubtedly has helped to make progress towards better implementation of community-based programs and elevated the intellectual discourses on local participation. Nonetheless, the nature and success of community participation remains subjected to the wish, intention, attitude, and capacities of the dominant stakeholders or implementing agencies because of several reasons some of which are discussed below.

Abstract criteria: First, most of the derived criteria under community participation framework are considered to be very abstract or broad which defies uniform interpretation in the local context (Collins and Ison 2009; Dyer et al. 2014; Santos and Chess 2003; Dodman and Mitlin 2013). For example the process criterion "fairness" can be differently interpreted and given different operational meaning in two geographical locations of a same project (Blackstock et al. 2007). The decision is in the hands of local power holders or implementing agencies to define and decide the project framework and operational meaning. The universal framework can hardly pinpoint any implementable, universal matrix of participation (Dyer et al. 2014; Rydin and Pennington 2000).

No fixed criteria for participation: Second, flowing from the earlier observation, there is no fixed criteria for participation (Webler et al. 2001; Santos and Chess 2003). The types and number of criteria and their importance vary directly with the purpose, types, and location of the projects (Rosener 1981; Moore 1996; Webler et al. 2001). Scholars and practitioners parallel each other on the set of criteria for measuring effective community participation. Thus, the success of the community-based projects depends largely on how stakeholders including government officials, field engineers, and NGO staff interpret the guidelines of participation (Few et al. 2007).

Authority and ownership of participatory discourse: Third, the very root of the derived framework of community participation is under the dominance of powerful stakeholders. The frameworks for effective participation have evolved from different scholarly foundations—ranging from theoretically derived frameworks (Renn et al. 1995), to case study analysis (Chess and Purcell 1999; Rowe and Frewer 2000; Reed 2008) and summaries from real-life project experiences (Dyer et al. 2014; Samaddar et al. 2017). In all cases, the held criteria for participation are determined by the relatively powerful, privileged stakeholders including researchers, project managers, government officials, and donor agencies (Rosener 1981; Moore 1996). The local communities have hardly had the opportunity to define what participation ought to be. Consequently, the derived criteria for participation only reflect the belief and intention of the powerful stakeholders (Samaddar et al. 2015b). There is hardly any method or approach adopted to form the participatory criteria based on inductive inquiry or emic perspective, that is, the local community's perspectives on ideal community participation. It is critical, therefore, to examine how the implementing agencies and their collaborators perceive community participation with the intent of rectifying, revising and influencing community-based projects through meaningful local community engagement.

18.1.3 Tools and Methods of Community Participation: The Control of Stakeholders

Along with the definition and framework of participation, the methods and tools are considered important as they are the mechanisms that translate the participatory idea and frameworks into practice. But the adoption of participatory methods or tools is value-laden too (Rowe and Frewer 2005). This is due to the plethora of equally promising participatory tools and techniques, all proclaiming to engage community meaningfully in the decision-making process (Toth and Hizsnyik 2008; Roncoli et al. 2009; Maraseni 2012; Samaddar et al. 2011, 2015). Nevertheless, the differences between these tools and techniques are vivid in terms of structure, functions, steps and procedures (Van Aalst et al. 2008; Toth and Hizsnyik 2008; Rowe and Frewer 2005). It is the implementing agencies from public or private sectors who are solely responsible for selecting the tools and techniques in a project (Na et al. 2009; Chambers 1997; Okada et al. 2013). The adoption of tools and methods eventually decides the direction, speed, intensity and outcomes of the participation (Rowe and Frewer 2005). For example, some methods are structurally rigid and proffer a maximum amount of power to the facilitators to decide who is to be involved, how and when and to what extent. Some studies argued that when the tools are so highly facilitator-controlled, actual participation is compromised (Yamori 2009; Chambers 1994). It is also argued that most participatory tools in disaster management and climate change adaptation focus only on identifying problems and issues (Na et al. 2009), but do not offer the future course of actions, such as what can be done and how. Hence, the implementing agencies tend to restrict community's involvement only at the risk identification level. The entire planning machinery remains in the hands of stakeholders (Chambers 1997).

Given the valued laden perspectives of the participatory idea, the powerful stakeholders influence and control the implementation of community-based projects greatly. Consequently, the community's engagement in climate change adaptation is truly intangible till date (Few et al. 2007; Larsen and Gunnarsson-Östling 2009; Allen 2006). Field reports show that for the community, community participation appears to be a powerful tool for local disaster management authorities to legitimatize their plans which they wanted to pursue for a long but failed (Cheng and Mattor 2006). The level, type and objective of community-based projects are designed and implemented by the project managers, donor agencies or government officials. If the community's participation in climate change adaptation programs and projects needs to be enhanced, it is critical to explore how stakeholders including government official, technocrats, project managers and donor agencies conceptualize and idealize community participation. What are the key factors for effective community participation? This chapter aims to identify stakeholders' perspectives on effective ways, steps and factors for ensuring effective community participation in climate change adaptation programs and projects. This study is based on the Wa West district, Northern Ghana. We interviewed key stakeholders including government and non-government officials involved in various climate change adaptation programs.

18.2 Climate Change, Vulnerability and Wa West District in Northern Ghana

The study area Wa West district (see Map 18.1), located in the western part of Upper West Region of Ghana, is one of the most susceptible locations to climate change in the West African sub-region (Samaddar et al. 2014; Kusakari et al. 2014). The district is surrounded by Nadowli district in North, Sawla-Tuna-Kaba district in the south and Burkina Faso in the east.

The Upper West Region is divided into three administrative units, i.e., Wa Municipality, Wa East District, and Wa West District. According to the National Population and Housing Census, the Wa West District's population in 2010 was 81,348 (i.e., Male: 40,227; Female: 41,121). The average temperature in the district varies from 25 to 36 °C, but local observers claim that due to climate change, the area now often experiences high temperature close to 40 °C during the months of February to April (Kusakari et al. 2014). Additionally, the district is said to be experiencing increased evaporation, decreased and highly variable rainfall pattern, as well as frequent and pronounced drought spells (Laux et al. 2008). The district's climatic dynamics resonate with the national figures as earlier studies in Ghana show a temperature increase of 1 °C in the last three decades, whereas average

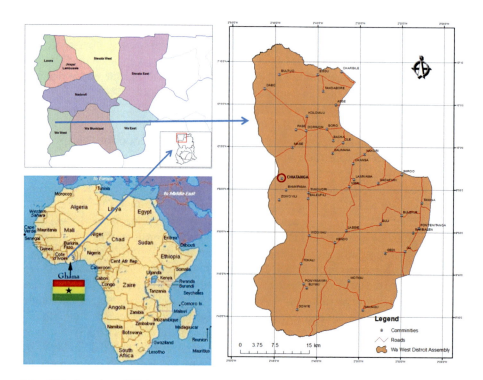

Map 18.1 Location of Wa West District, Ghana

rainfall has reduced by 20% (Yaro 2013). Indeed it is predicted that by 2050, the country's temperature will increase 2.0 °C and by 2080 the temperature rise will be 3.9 °C (Ghana Environmental Protection Agency 2007). Regarding the rainfall pattern, normally April used to be the month for the onset of the rain, but studies show that that has now shifted to May and it is predicted that in future, by 2040 the onset will be shifted further to June or even later in Northern Ghana (Jung and Kunstman 2007). Given this changing climate scenario, scientists predict that Northern Ghana in future will experience extreme drought and floods (Armah et al. 2010), perhaps, reminiscent of flooding incidences experienced in 2007 and 2008 (Samaddar et al. 2014). The climate change is likely to impact most on the poverty-stricken population because their livelihood is based on subsistence, rain-fed agriculture economy which is predicted to be very severely affected by flood and drought (Kusakari et al. 2014). The impacts of climate change are expected to manifest in the area of food security, out-migration, and hunger months in the region (Akudugu et al. 2012). There is, therefore, the need to take up programs and projects focusing on enhancing the coping capacities of the local communities through alternative livelihood systems, innovative technology adoption by rural households, mobilizing resources at the grassroot level (Samaddar et al. 2018).

In fact, several climate change adaptation projects and programs are initiated in recent times by the national and local government, and by international donor and aid agencies to create alternative livelihood systems, augment local coping capacities of vulnerable communities, linking poverty alleviation with climate change adaptation (Laube et al. 2012; Schraven 2010). The more these programs have been initiated, the more it becomes obvious that for a successful implementation of these programs, the participation of the local people is indispensable (Westerhoff and Smit 2009; Laube et al. 2012). The conundrum, however, relates to how project managers, public officials, NGO workers, development officers and other stakeholders conceptualize how to effectively involve the local community in climate change adaptation programs (Samaddar et al. 2015b, 2018). Some yardstick or baseline criteria should be drawn to actualize local community participation in climate change adaptation programs in Northern Ghana.

18.3   Methods

We interviewed officials and representatives of several government and nongovernmental organizations who are directly involved in climate change adaptation and disaster management programs in the Wa West District of the Upper West Region of Ghana. From the government side, we interviewed ten respondents from different sectors and institutions including National Disaster Management Authority (NADMO), Ministry of Forest and Agriculture (MoFA), Environmental Protection Agency (EPA), Ghana Irrigation Development Authority, Ministry of Fisheries and Aquaculture Development, Forest Service Department, Ghana National Fire Service, Ministry of Local Government and Rural Development, Department of

Community Development, and District Town Planning Department. We also interviewed five locally functioning NGOs working on climate change and environmental issues and two FM Community Radio Stations conducting various awareness building and educational programs on climate change, environmental issues, and agriculture-related issues in the region.

To obtain stakeholders' perception on effective community participation, we primarily targeted the heads of the institutions or organizations, or the manager of any program (on climate change and disaster risks) for the interview, however, in cases where the heads recommended other officials or individuals who handle climate change adaptation issues, we interview the suggested individuals. Consequently, we interviewed one person from each institute. However, in case the respondent wanted other office colleagues or coworkers to join him/her in the interview, we allowed them and welcome their comments and suggestions as well. In such cases, the interviews were more like small group discussions. We experienced four such cases. Before the interview, we intimated to the respondents that the goal of the survey was not to get their official perspectives based on institutional affiliations, but rather, more personal perspectives as people who have much working experience regarding climate change adaptation and make decisions for the successful accomplishment of project objectives. Therefore, the respondents' opinions and views on participation did not require any endorsement by their representative organizations or institutions. Respondents were assured that no personal information will be disclosed in our report. We observed that the announcement helped to get candid and honest answers from the respondents.

The interview was conducted in two phases, first in December 2015 and then in July 2016. First, we prepared a list of prospective stakeholders or institutions for the interview and then consulted with the local researchers and experts (from University for Development Studies, Wa) to delete or add relevant institutions for the interview. To get affirmation from the respondents, 2 weeks before the interview, we sent request letters and emails to the prospective institutions and organizations for an interview, giving the background of the project and purpose of the study. In case, we received no feedback, we visited the office and met the designated head personally to get a consent and fix a date for interview. The first and second authors of the chapter mainly conducted field surveys and took face-to-face, open-ended interviews. On an average, an interview took one and half hours, but some interviews took much longer time. There was no fixed set of questions or questioners, but keeping in mind open-ended interview, we prepared a set of questions as shown in Table 18.1 to carry out the interview in a more organized and efficient manner. The research method was qualitative in nature because it intended to obtain data in an inductive fashion, that is, to get a bottom-up perspectives. Qualitative research methods are reflexive and inductive that encourage respondents to express their own views and help to obtain multiple realities of the problem (Denzin and Lincoln 2011). These attributes match well with the present research objectives since it aims to get stakeholders' own perspectives on community participation based on their situation and experience.

18 Stakeholders' Perceptions on Effective Community Participation in Climate...

Table 18.1 Preface and question guidelines for interview

Preface and introductory questions	– We want to know from you how we can successfully engage and collaborate with local communities in climate change adaptation programs – We want to know your experience, suggestions, opinions, and views on criteria for effective community participation in climate change adaptation? Why do people want to participate and why not? What are their stakes? – We want you to share details of some of your success and failed cases? – We want you to list your successful projects and tell us why in those projects local communities were better involved? – Give us a comparative picture between projects and tell us in which projects local community's involvement was better seen and why?
Process of community participation	– Can you outline the ideal process for effective involvement of local communities in the climate change adaptation projects project? – What have you learned from your projects or from the field regarding involved community in CCA projects? – What are the Dos and Don'ts you will suggest for successful community participation? – If you again like to start a CCA project and want full participation of the local community—then what procedure will you follow? – In which project were you able to successfully involve the community and why do you think it was successfully done? – What do you think about how local communities prefer to get involved in climate change adaptation project and why?
Outcomes of community participation	– If you are given the role of evaluation of community-based CCA projects, on what basis can you say the local communities successfully got involved in a project? – Several organizations have been claiming that they have successfully implemented CCA projects with the meaningful participation of local communities in Wa region. Do you think these claims are true? On what basis, can you judge their claims – What are the outcome-based criteria for community participation in climate change adaptation, and why are these factors critical?

For the data analysis, we followed the content analysis method (Graneheim and Lundman 2004) by using coding process and category dispositions. First, the recorded data were transcribed verbatim. In the next stage, we read the transcribed data and assigned codes to them. First the codes were assigned with two broad categories—process and outcomes of participation and then these codes were assigned with sub-codes to develop subcategories.

18.4 Results

The derived criteria for effective community participation from the stakeholder survey do not actually fall into process and outcome-based broad categories. Rather, the emerging factors of effective participation can be divided more into "participatory condition" and "outcomes" as shown in Fig. 18.1. Participatory condition includes both steps or procedures and certain prerequisite conditions or

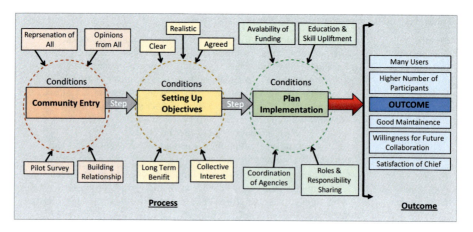

Fig. 18.1 Process conditions and outcome for community participation

environment necessary for the successful execution of the participatory programs. The "outcomes" are projects results indicating how successfully communities are involved in the project. The respondents expressed concern that "outcome" based parameters are very shaky and weak as the criteria may always alter based on the place and community, types of programs and projects. Respondents mentioned that outcome-based criteria do not have any direct impacts on enhancing community's involvement in CCA project, but can serve as evaluation parameters to understand the degree of community involvement.

18.4.1 Participatory Process and Conditions

Respondents asserted that community entry, setting good objectives and plan implementation are three inevitable steps for ensuring effective community participation in CCA projects. The three critical steps of participation are elaborated below:

18.4.1.1 Community Entry and Village Governance

Successful community engagement is strongly associated with the nature of external actors' entry into the community as depicted in Fig. 18.2 and Table 18.2. A good community entry is critical in getting the traditional authorities and local communities to give legitimacy to the project. It also facilitates effective community and resource mobilization. In Northern Ghana, social and local political issues revolve around the chieftaincy institution. The chieftaincy institution is a time tested traditional political system of the people. This traditional political system is more pronounced in the rural communities, which often form the target sites for climate interventions.

18 Stakeholders' Perceptions on Effective Community Participation in Climate...

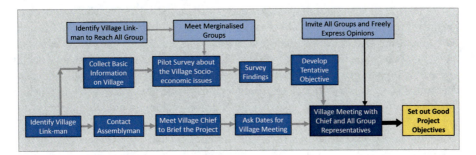

Fig. 18.2 Steps of community entry for effective local community participation

Table 18.2 Descriptions of "community entry" for effective community participation

Subcategories of community entry	Description
Find link-man/source man	– Not the chief or traditional leaders – Village assemblyman, young leader or school teacher or someone with whom you have worked before – Preferably young and not a part of the village governance system – They will take you to most vulnerable, poor families remotely located – Conduct a pilot survey. Visit 2–3 settlements in a day before meeting with the chief
Informal survey/pilot survey	– Small informal group meeting in each cluster – Specifically, meet women. They work hardest but have no power – Women give you more useful information because they take up the family responsibilities – Formal survey is not always necessary. Take notes – Know the community—culture, group dynamics, and climate change impacts on the villagers – Inform chief of your survey findings and what you want to do with your projects – You can always refine your plans through discussions with the people
Relationship with the chief	– Chiefs are generally receptive but sometimes ignorant about many issues. So internal knowledge, expert inputs, and scientific information are necessary – Meet the chief frequently to develop rapport – Respect indigenous values – Do not hide your true intentions – The Chief will cooperate with you if you are honest and the project will benefit everybody – Always wait for the chief's consent. If you do not agree with his decision, meet him again and again. Ask for his help – Without the chief's consent, no local resident can join the project
Village meeting	– It is more of a formality – Ensure the presence of all groups. Encourage them to express their opinions – Village meeting means giving endorsement to your work or plan. This is an evidence of your presence – Present survey findings in village meetings for validation

At the top of the chieftaincy system is a head called the chief who inherits this position from his royal forefathers and is the sole or ultimate decision makers on village politics and collective decision (Owusu-Mensah 2014). There also exists an advisory body consisting of the village elders, linguist, village priest and queen mother (Magazia) to assist the king in various matters (Mahama 2009; Owusu-Mensah 2014). This traditional and so-called autocratic governance system reached its peak, enjoying supremacy over village decision-making in precolonial and colonial era. But, political reforms over the years resulted in the setting up of decentralized and democratic structures with political and legislative power such as District Assemblies and their sub-structures. This new local government system provided for the election of Assemblymen as representatives of the local people to the District Assemblies. The Assemblymen are elected at the community or village level (Owusu-Mensah 2014; Crook 2005; Guri 2006). Although chiefs lost their formal judicial, administrative, and military power in modern democratic Ghana, the chieftaincy institution is still culturally embedded in the social and political life of Ghana. The role of chiefs in village politics and governance as well as day-to-day decision-making are still pivotal and final (Knierzinger 2011; Owusu-Mensah 2014; Crook 2005). In areas where the village land still belongs to the chief, he decides the access, use and management of the land. The chief still plays critical roles in initiating and monitoring community development projects and mobilizes resources including communal labor or village lands and forests for the village development in this part of the world (Guri 2006; Mahama 2009).

Given the importance of the village chief and the traditional governance system, the stakeholders advocated the building of strong rapport with the chiefs and other members of the village governance system. Without acceptance from the village chief, no plan can be legitimized. The rapport building with the chief is not only necessary for legitimizing the plan, it also facilitates resource mobilization, including land, forest timbers, and stones available in the village. The chieftaincy institution is instrumental in mobilizing for collective action and communal labors for village development. Without a consent from the chief, no one will be willing to participate in a project.

However, community entry through the chief could be problematic. In communities that have autocratic chiefs, little or no space is often provided to other marginalized groups including women, politically rival groups or lineages, new settlers to express their interests and rights. Thus, it is advisable to explore other options of making a good survey in the community to know the needs, visions and perspectives of the common citizens. An official from the Irrigation Development Authority of the Wa West District explained: "Develop your own source man. Assemblyman will be okay to understand the villagers' sentiments. Assemblyman can also help you to reach some unknown (poor or marginalized) groups of individuals who will be rarely seen in public meetings. But remember that assemblymen are often politically motivated. If you have a village link-man, he can help you to understand the ongoing village politics, and group conflict. Through him, you can reach different sections of villagers – keeping different views and beliefs. Then having in-depth grassroots level understanding, you call a village meeting headed by the chief to reach a consensus. Yes, but note that the chief's approval is a must. Keep him in confidence. Otherwise you can never pursue the project successfully".

Based on the respondents' views, the ideal process of community entry is illustrated as shown in the self- explanatory Fig. 18.2.

18.4.1.2 Setting Up Objectives

The stakeholders argued that setting up good objective is a prerequisite for meaningful community participation in climate change adaptation programs. It clearly gives the participants an indication of expected outcomes. A good objective should carry the following features: it should (1) be clear, (2) be realistic, (3) be agreed upon by all groups and stakeholders, (4) safeguard collective interests, and (5) have long-term benefits. The features of a good project objective are interlinked and generally affect each other as shown in Fig. 18.1. For example, to serve short-term gains, projects often include objectives that partially favor a particular group. Similarly, an unrealistic, over-ambitious project often fails to set clear objectives.

But the ability to set good objectives is often affected by (Table 18.3):

1. One of such local dynamics is poverty. The region is poverty-stricken which creates challenges for designing projects with long-term suitable goals. In view of the poverty situation, the local communities want projects that meet their immediate needs and provide material gains including financial support, relief assistance or inputs for agriculture. Thus, projects offering relief materials after flood and drought or direct cash or agricultural supplies can easily receive attention as well as commitment from the villagers. Conversely, projects working towards capacity building or self-reliance of the community are getting unpopular among the people. Thus, CCA projects should be linked with poverty alleviation and income generation to attract local community's willingness to participate.
2. The second local dynamics affecting the setting of good objectives is the strong desire for material gains and the high rate of illiteracy among the people. Respondents from the public sector complained that NGOs in the region have created a culture of "getting relief". As most of the agencies are greatly funded by the foreign donors, they launch projects that offer lots of wealth and financial benefits to the local people. Consequently, local communities are not interested in working with government agencies as their projects are small, do not offer any material gains or free materials, but rather demand mutual support and contribution from the communities. An official of the Forestry Commission explained: "NGOs have money from outside, mostly from foreign countries. They offer shelters, food or sometimes basic infrastructure directly. Villagers are very happy to work with them. But we (Government organizations) cannot offer same. We use their (villagers) resources because we want to make them independent in future. But, it takes time and costs lots of labor. People then don't agree to whatever we propose. Since from your entry into the village, they (villagers) murmur – "what we can get from the project". Illiteracy drives them to see only

Table 18.3 Setting good objective: challenges and characteristics

Code	Descriptions
Challenges to set good objectives	
Poverty and climate change adaptation	– Poverty in the region is high. This is the biggest concern for the development of the region. Poverty induces people's vulnerability – People forget flood immediately after few months. Because they live in a hand to mouth situation. Their daily concern is whether they have enough food or not – When people are starving, you cannot ask them to join in CCA project Getting stomach full food is their daily concern – Poor people are greedy because the hunger forces them to be unashamed People want cash, crops, and other family assistance from the project
Short term objectives and material gain and illiteracy	– People now believe in quick fix. People do not mind depending on NGOs, foreign aid agencies for their own wellbeing. They become very enthusiastic if they know they will relief materials – Always ask what they can get in free and happier if you give them cash – Entering community becomes difficult because of NGOs. People enquire from day one—what they can get free? – No commitments. NGOs are encouraging this attitude by giving free houses, seeds, crops, and fertilizers – What happens when the project is withdrawn—frustration, no real development – Less interested in village well-being. More focus on getting free stuffs for own—petty politics. No collective interest – Lack of education, illiteracy—people become ignorant until they face the threat. Without education if they join in program, they cannot contribute anything – Illiteracy is a bigger issue than poverty. People are ignorant about climate change impacts and the need for sustainable development
Local conflicts and power structure	– There are conflicts among groups based on religion or lineage. Villages do not always live in peace and harmony – Village politics stop good projects. Very difficult to unite them. Each one has their own vested interests – Dominant groups or chief's own kin enjoy the power to influence decisions – Women are more practical and committed. But women have no power
Broad and abstract objective	– Regional headquarters set the project objectives, but no idea about real issues or local problems – The project should be tailor made—set area-specific project objectives – Broad objectives create confusion and conflicts among stakeholders. You can give any meaning you want – Donors, especially foreign aid agencies are sometimes very stubborn. They have fixed objectives and project framework – Donor agencies pressurize local NGOs to follow a set of objectives without considering local situation and culture

(continued)

18 Stakeholders' Perceptions on Effective Community Participation in Climate... 369

Table 18.3 (continued)

Code	Descriptions
Misconduct and lies	– NGOs or many other organizations tell lies. Do not disclose the project objectives because they are worried about the outcome in case the local community does not like the objectives – Keeping the projects vague or obscure helps an organization to be on the safe side – You are never committed for something very concrete. No one can raise questions
Characteristics of good objectives	
Clear	– Detail out each objective as much as possible – Match possible results with objectives – Attach concrete actions with objectives. Then attach roles and responsibilities of different individuals and groups – Relate objectives with local issues and culture – Facilitate visual and oral communication. Avoid texts and write-ups. Illiteracy is very high. People do not understand the project contents
Realistic	– Display your objectives to the community. Ask their opinion—achievable or not? If not, why not? If yes how? – Check resource, time, money, and organizational capacity. Do not be a true romantic – Ask contributions from the communities—labor, land, sand, and timber – Realistic objectives will make objectives clear. People want to participate if they see that the project objectives are feasible
Agreed by all groups	– Do survey, informal meeting. Collect people's need – Check your objectives during survey. Delete or add new objectives – Put weak stakeholders' voice in your agenda. Try to reduce conflicts among the groups
Collective interest	– The project should be a help for the improvement of the entire village—like afforestation, water conservation, or irrigation facility – Stop giving relief – Awareness and educational enlightenment are important
Long term perspective	– Capacity building – Sustainable, self-dependent – No matter if the donor leaves after 3–5 years, the local people themselves can run the project – Focus on rehabilitation, not relief. Upgrade people's knowledge and skills – Integrate poverty alleviation or income generation aspects in climate change adaptation projects – Educate people

short term benefits and personal material gains. Awareness building and improving general literacy profile of the communities are essential in changing their mindset and encouraging them to participate in projects that offer sustainable growth and future resilience of the community.

3. The third factor affecting the formulation of good objectives is local conflicts and power structure. It will be a mistake to consider that communities are

homogeneous entities with people always living in harmony. Differences and conflicts among the villagers along the lines of lineage, gender, kinship, and settlers (conflicts between early and late settlers) are normal occurrences. These are stumbling blocks to reaching consensus on objectives within the community. One of the Agricultural Officers in the district explained: "To you, it may appear as a consensus when they declare a MoU (memorandum of understanding) through village chief (in a village meeting). But there are many undercurrents and tensions within the community. If you don't address them and prepare your plan only on what the chief says, then I tell you it is granted that people will show less interest to join you…if not they are getting something free. And to reach an agreed objective becomes more difficult when your project involves more than one village. I think working with each other and collaborating with neighboring villages is something that is historically absent in this region." The respondents advocated that the collective interest of the community should reflect in the project objectives and for that effective community entry as mentioned in the earlier section is critical.

4. The fourth factor is the use of broad and abstract objectives. Climate change adaptation related projects, whether coming from government sector or non-government sector, are mostly designed at the national and regional levels and by outsiders, such as foreign donor agencies or national level experts. As a result, project objectives and framework are often very general and give little clarity and scope as to how these can be applied at the local level. A staff of one of the local nongovernmental organizations explained: "The project frameworks often do not match with the local demands and situations. I remember we were arguing with a European agency about why their sanitary project cannot be successfully implemented if they do not revise the framework. But the agency was adamant believing that as it was successful in Asia, it will also work well here as well. Who will convince them?" As one of the district planning officers notes: "Many NGOs are worried about losing funds if they oppose the top management's or donor's suggestion. Hence, they compromise with local requirements. Result? If local people need boreholes, you are giving people live-stocks".

5. The fifth factor affecting the formulation of good objectives is the incidence of misrepresentation. Many stakeholders, particularly NGOs (nongovernmental organizations) often hide their true intentions whenever they feel there could be disagreement regarding what the local community wants and what the projects intend to achieve. The NGOs fear that by knowing the expected outcomes of the project, the local communities may show no interest. Because of this fear, many organizations keep the objectives very obscure and abstract. Allegedly, some even give false impression of the projects to the local people. When the local communities finally get to know the real intentions of the projects and the NGO actors, they become demoralized and eventually refuse to participate.

In our attempt to explore good practices in formulating good and shared project objectives, the following measures are being suggested (Table 18.3):

1. Community should be clearly told the possible results or benefits the project can bring and how that could be realized. There is also the need to educate and sensitize community about the long-term benefits of such projects to the communities to enable them to appreciate the importance of long-term interventions in building self-reliant communities. This will help the local communities move away from quick interventions and focus on long-term sustainable projects. To this end, improving community's general literacy status will also contribute towards this objective.
2. Given the prevalence of internal conflicts and politics in local communities, external actors should reach out to marginalized individuals or groups including women, poor peasants, new sellers, and ethnic minority groups, who have little voice in local decision-making, before setting up project objectives.
3. CCA project should also be linked with poverty alleviation and income generation goals. The promise of livelihood will solicit strong local participation. Without income generation potentiality of the project, the local communities would have very little interest to join in any climate change projects.
4. To reduce the obscurity of the project objectives, the agency should clearly write the project objective based on local context and requirement. Each project objective should be linked with series of consequent actions so that all stakeholders can visualize the project.
5. Given the prevailing illiteracy and ignorance, the project managers should try to facilitate visual and oral communication and avoid written documents or texts during village meeting for the project brief.

18.4.1.3 Plan Implementation

Respondents believe that local community turn away from CCA projects, especially from infrastructure improvement or heavily engineering based projects, because such are seldom successful implemented. The actual implementation of the project is often hampered and consequently local community's participation significantly drops. Several reasons account for this trend:

1. *Lack of funding*: Steady funding is more often an issue because active engagement of the community in a project demands more budget than usual. The capacity building and community empowerment programs often cost more money and time than usual.
2. *Apathy in taking roles and responsibilities*: Respondents claimed local people keep distance from the projects when the time demands contribution from them. When local communities fail to take responsibility of projects they have committed themselves to, implementation of such projects becomes problematic.
3. *Illiteracy, lack of scientific education and skills:* The literacy rate in Wa West district is very low. Even those who are educated have no scientific knowledge and understanding on climate change and environmental issues. Their technical

knowledge and skills are also poor. Therefore, it is difficult to involve communities in plan implementation process of scientific projects.

4. *Lack of coordination between projects, departments, and organizations:* There is a lack of coordination between different projects and departments as well as between organizations. For the holistic and sustainable development of the local community, integration of different projects and mutual collaboration between organizations are encouraged to reduce project cost, enhance synergy, and tackle the climate change impacts in a more integrated manner. An official of one the NGOs noted: "Several organizations go to the same community with same or similar types of projects. They visit again and again. Whereas many other issues of climate change are being left untouched. These matters are in community's priority list for long. It means you did not solve the problems. Villagers are also confused as to who come from which organizations and how many times they should support them for same purpose". He further noted: "If you give water, it could support farming activities. With smaller budget, we can improve many areas of the village. People's participation will be automatically high." According to an official of the Forestry Commission, "Mutual understanding and coordination between various departments and even organizations is so very important. And when many organizations work together you can solve many problems and meet many demands of the community. Also, for the growth of one sector of the community, for example forestry, you need to improve other areas, here it could be water, roads. There is less repetitiveness of projects and works."

18.4.2 Outcome Based Factors

The respondents believe that the participatory condition or process is the most critical to ensure local community's meaningful involvement in the project. However, adhering to a preferred process do not always produce the desired results. The project types, objectives, the capacity of the organization, the cultural and social environment of the site, as well as the size of the community play invisible roles for successful engagement of communities in climate change adaptation.

Therefore, respondents suggested that during and after the projects, the project implementing agencies should evaluate the actual progress of community participation in a project. The following measures are suggested to check the outcomes of community participation:

1. Survey: The implementing agencies should conduct survey among the beneficiaries to know how happy or satisfied they are with the project and with their involvement. During the survey, the suggestions from the beneficiaries should be also collected to improve the project implementation. The target should be to interview people as many as possible.

2. Hire evaluator: The project can also higher an outside evaluator to check the success of the project. It would provide much better accountability and neutrality of

the project. The report of the evaluator should be displayed to the village communities.

Apart from these two evaluation measures, the respondents also suggested certain parameters that can help to understand the extent of community participation in CCA projects:

1. The number of people using the products of the project—such as water supply system, mini-dams, or sanitary facilities.
2. The maintenance of the project outcomes. If it is maintained well means people are happy with their involvement.
3. The number of people that participated in a project. If more than 60% of all households of the village participated in a project, the project should consider that they have successfully involved the local community.
4. The wish of the village chief to collaborate with the same external agencies in future is also an indication of how well the community was involved in a past project. A respondent however is afraid that: "Remember the authoritative nature of this position may mean that the wish to collaborate with project in future is for personal gains. Remember they receive money and other donations from foreign NGOs on behalf of the community and could be benefiting greatly from these to the neglect of their subjects. But not because all groups and people across the community were properly involved. Hence using the willingness of the chief to support future projects as a criterion for proper involvement can be call to question."

18.5 Conclusions

Broadly, climate change adaptation discourses in northern Ghana in general has relied on scientific expertise as a framework for decision-making for a considerable time. Attempts to introduce local community involvement however are more recent and address two key issues. The first issue is the fact that the increase in the number of climatic variability events has led to a very distinct increase in community mistrust as regards scientific approaches to environmental problems (Samaddar et al. 2018). Secondly, rising community exposure to the vagaries of the climate has led to a growing demand for community participation in climate change adaptation, but its successful implementation remains uncertain and illusionary (Samaddar et al. 2015b). Thus, the introduction of community involvement in this context appears as a concept which responds by giving a defined set of answers to these emerging challenges. Often there are several claims from government and nongovernmental organizations on successful execution of community-based or community-led CCA projects and programs. But these claims are seldom verified. The reality on the "successes or otherwise" manifest when the project agencies fail to produce any scientific evidence (published data corroborating their claims) or rarely replicate the successes in other areas.

More than the generalization challenge or modelling questions, the resentment and frustration among the local populace are ever increasing over their involvement in such ostensible community-based climate change adaptation projects. The impression of community based climate change adaption programs to the community becomes nothing more than a trap to circumvent the local communities to get plans legitimized, actually encoded by the external agencies. Often, local communities feel marginalized when they realize project decisions and plans have already been decided without their involvement and consent. Given the complex nature of participation, its unshackled framework and numerous operation methods and techniques, its actual implementation rests with the implementing agencies—their willingness, understanding, skills, and capacities, and these greatly influence how successfully local communities can be engaged in CCA programs. In most cases, these implementing agencies include a wide range of government and nongovernmental organizations from regional, national and international level and their individual perception plays a critical role in realizing the community-based CCA projects in practice.

These notwithstanding, in most cases climate change adaptation studies rarely institute appropriate initiatives geared towards understanding the perception of stakeholders about effective community participation. Our study contributes to current literature by investigating what stakeholders think about the ideal process and outcomes of meaningful community participation in CCA. We took Wa West district, a severely climate challenged region in West Africa as our case study area and interviewed various governmental and nongovernmental organizations' representatives, employees, and executives to investigate what they perceive to be effective community participation in climate change adaptation.

The survey results showed that stakeholders believe the participatory environment or process condition is the most critical for effective community participation. The participatory environment includes a three steps process (including community entry, setting good objectives, and plan implementation) and certain situational conditions or sociopolitical environment that are influencing the execution of participatory programs. Some respondents from different implementing agencies attributed the difficulties and glitches in community participation mostly to the community's own internal problem and their general unwillingness to contribute to the CCA projects. The factors cited as influencing this development include the community's lack of motivation, illiteracy, poor knowledge and skills, ignorance, internal conflicts and lobbying, and inaccessibility to local resources. According to these stakeholders, the most important step towards ensuring successful community-based CCA project is to set up a clear objective, that is, to have a realistic goal with long-term development perspectives. It was revealed that the local community's internal group conflicts and petty politics do hinder the process to reach agreed objectives. Similarly, illiteracy and ignorance and above all growing poverty are responsible for local communities' inclination towards short term, material gains. Hence, even after a long effort from the implementing agencies, community participation in CCA project has dropped significantly in recent times. Community pays little heed when the project demands their sincere and intensive involvement for effective CCA through capacity building and local resource management.

Further the implementing agencies conceded that though a good implementation plan is equally an important factor for effective community-based CCA projects, in most cases lack of commitment in terms of roles and responsibilities from the local communities spells unintended failure. Further, it was opined that the community's lack of skills and knowledge often impede their meaningful involvement in any technical and systemic discussion which are critical in plan implementation. Rather local community's participation adds more budget constraints over the successful implementation of the CCA project. In nutshell, making CCA projects more people-centric according to our respondents requires debugging the pitfalls prevailing within the communities. Surprisingly, the key stakeholders did not mention a single factor that calls for improving the attitude and organizational setups of implementing agencies for improved community participation. Whereas, several field-based studies have reported that communities are ever willing to participate in a joint project, the opportunity for equal and fair participation by the community or marginalized groups counts further towards the effective involvement of the communities in a project (Webler et al. 2001; Moore 1996).

Several studies have amply indicated the need to factor in the marginalized, poor and vulnerable communities for a fairer, accountable, and trustworthy governance system and effective local community involvement (Samaddar et al. 2017; Tam 2006; Moote et al. 1997). Failure to institute these measures only results in local community's participation getting stuck at the level of physical participation without any decision-making authority. It also then defeats the much touted benefits of local participation such as community's ownership and self-reliance often associated successful community based project (Reed 2008; Shaw 2006; Dyer et al. 2014; Blackstock et al. 2007; Okada et al. 2013).

The present study showed that dominant stockholders still hold the same banal perspective of participation where the local communities are nothing more than passive recipient of information. The role of the community is picturized as nothing more than to assist the government and the development partners to implement their visions and plans. The government and other agencies are presumed to be taking the leading roles because they are often the main resource providers, particularly giving financial back-ups, and having more technical expertise and experience. Therefore, it is clear from the present study, there is no much progression from the top-down communication and collaboration approach in CCA which often proved to be an inefficient mechanism to establish resilient communities against climate change induced risks. The question however is how to reduce the gap between the two perspectives of community participation, one that is held by the dominant stockholders and the other by the marginalized group that constantly demands fairness, equality, accountability, and trust in the participation process (Renn et al. 1995; Rowe and Frewer 2000). At present there is no study available in climate change domain addressing these issues, and therefore, it could be a future research task to strengthen local communities' participation in the planning process.

References

Akudugu, M. A., Dittoh, S., & Mahama, E. S. (2012). The implications of climate change on food security and rural livelihoods: Experiences from Northern Ghana. *Journal of Environment and Earth Science, 2*(3), 21–29.

Allen, K. M. (2006). Community-based disaster preparedness and climate adaptation: Local capacity-building in the Philippines. *Disasters, 30*(1), 81–101.

Armah, F. A., Yawson, D. O., Yengoh, G. T., Odoi, J. O., & Afrifa, E. K. (2010). Impact of floods on livelihoods and vulnerability of natural resource dependent communities in Northern Ghana. *Water, 2*(2), 120–139.

Arnstein, S. R. (1969). A ladder of citizen participation. *Journal of the American Institute of Planners, 35*(4), 216–224.

Blackstock, K. L., Kelly, G. J., & Horsey, B. L. (2007). Developing and applying a framework to evaluate participatory research for sustainability. *Ecological Economics, 60*(4), 726–742.

Blahna, D. J., & Yonts-Shepard, S. (1989). Public involvement in resource planning: Toward bridging the gap between policy and implementation. *Society & Natural Resources, 2*(1), 209–227.

Carr, G., Blöschl, G., & Loucks, D. P. (2012). Evaluating participation in water resource management: A review. *Water Resources Research, 48*(11), W11401.

Chambers, R. (1994). The origins and practice of participatory rural appraisal. *World Development, 22*(7), 953–969.

Chambers, R. (1997). *Whose reality counts? Putting the first last.* London: Intermediate Technology Publications Ltd (ITP).

Cheng, A. S., & Mattor, K. M. (2006). Why won't they come? Stakeholder perspectives on collaborative national forest planning by participation level. *Environmental Management, 38*(4), 545–561.

Chess, C., & Purcell, K. (1999). Public participation and the environment: Do we know what works? *Environmental Science & Technology, 33*(16), 2685–2692.

Collins, K., & Ison, R. (2009). Jumping off Arnstein's ladder: Social learning as a new policy paradigm for climate change adaptation. *Environmental Policy and Governance, 19*(6), 358–373.

Crook, R. (2005). The role of traditional institutions in political change and development. *CDD/ODI Policy Brief, 4*, 1–5.

Denzin, N. K., & Lincoln, Y. S. (Eds.). (2011). *The Sage handbook of qualitative research.* Thousand Oaks, CA: Sage.

Dodman, D., & Mitlin, D. (2013). Challenges for community-based adaptation: Discovering the potential for transformation. *Journal of International Development, 25*(5), 640–659.

Dyer, J., Stringer, L. C., Dougill, A. J., Leventon, J., Nshimbi, M., Chama, F., & Syampungani, S. (2014). Assessing participatory practices in community-based natural resource management: Experiences in community engagement from southern Africa. *Journal of Environmental Management, 137*, 137–145.

Few, R., Brown, K., & Tompkins, E. L. (2007). Public participation and climate change adaptation: Avoiding the illusion of inclusion. *Climate Policy, 7*(1), 46–59.

Galicia, L., Gómez-Mendoza, L., & Magaña, V. (2015). Climate change impacts and adaptation strategies in temperate forests in Central Mexico: A participatory approach. *Mitigation and Adaptation Strategies for Global Change, 20*(1), 21–42.

Ghana Environmental Protection Agency. (2007). Climate change and the Ghanaian economy. In *Policy advice series* (Vol. 1). Accra: Ghana Government Policy Document.

Guri, B. (2006). *Traditional authorities, decentralization and development: A concept paper for strengthening the capacity of traditional authorities for good governance and development at the local level.* Madina: Center for Indigenous Knowledge and Organizational Development (CIKOD).

Graneheim, U. H., & Lundman, B. (2004). Qualitative content analysis in nursing research: concepts, procedures and measures to achieve trustworthiness. *Nurse education today, 24*(2), 105–112.

18 Stakeholders' Perceptions on Effective Community Participation in Climate... 377

Hiwasaki, L., Luna, E., & Marçal, J. A. (2015). Local and indigenous knowledge on climate-related hazards of coastal and small island communities in Southeast Asia. *Climatic Change, 128*(1–2), 35–56.

Hung, H. C., & Chen, L. Y. (2013). Incorporating stakeholders' knowledge into assessing vulnerability to climatic hazards: Application to the river basin management in Taiwan. *Climatic Change, 120*(1–2), 491–507.

Intergovernmental Panel on Climate Change (IPCC). (2001). *Climate change 2001: Synthesis report.* Cambridge: Cambridge University.

Jung, G., & Kunstman, H. (2007). High Resolution regional climate modelling for the Volta Basin of West Africa. *Journal of Geophysical Research, 112*, 17.

Knierzinger, J. (2011). *Chieftaincy and development in Ghana: From political intermediaries to neotraditional development brokers.* Institut für Ethnologie und Afrikastudien, Johannes Gutenberg-Universität: Mainz.

Kusakari, Y., Asubonteng, K. O., Jasaw, G. S., Dayour, F., Dzivenu, T., Lolig, V., & Kranjac-Berisavljevic, G. (2014). Farmer-perceived effects of climate change on livelihoods in Wa West District, Upper West region of Ghana. *Journal of Disaster Research, 9*(4), 516–528.

Larsen, K., & Gunnarsson-Östling, U. (2009). Climate change scenarios and citizen-participation: Mitigation and adaptation perspectives in constructing sustainable futures. *Habitat International, 33*(3), 260–266.

Laube, W., Schraven, B., & Awo, M. (2012). Smallholder adaptation to climate change: Dynamics and limits in Northern Ghana. *Climatic Change, 111*(3), 753–774.

Laux, P., Kunstmann, H., & Bárdossy, A. (2008). Predicting the regional onset of the rainy season in West Africa. *International Journal of Climatology, 28*(3), 329–342.

Lorenzoni, I., Nicholson-Cole, S., & Whitmarsh, L. (2007). Barriers perceived to engaging with climate change among the UK public and their policy implications. *Global Environmental Change, 17*(3), 445–459.

Mahama, C. (2009). Local government and traditional authorities in concert: Towards a more productive relationship. *Commonwealth Journal of Local Governance, 4*, 7–25.

Maraseni, T. N. (2012). Climate change, poverty and livelihoods: Adaptation practices by rural mountain communities in Nepal. *Environmental Science & Policy, 21*, 24–34.

Moore, S. A. (1996). Defining "successful" environmental dispute resolution: Case studies from public land planning in the United States and Australia. *Environmental Impact Assessment Review, 16*(3), 151–169.

Moote, M. A., McClaran, M. P., & Chickering, D. K. (1997). Theory in practice: Applying participatory democracy theory to public land planning. *Environmental Management, 21*(6), 877–889.

Na, J., Okada, N., & Fang, L. (2009). A collaborative action development approach to improving community disaster reduction by the Yonmenkaigi workshop method. *Journal of Natural Disaster Science, 30*, 57–69.

Okada, N., Fang, L., & Kilgour, D. M. (2013). Community-based decision making in Japan. *Group Decision and Negotiation, 22*, 45–52.

Owusu-Mensah, I. (2014). Politics, chieftaincy and customary law in Ghana's Fourth Republic. *Journal of Pan African Studies, 6*(7), 261.

Patt, A., Suarez, P., & Gwata, C. (2005). Effects of seasonal climate forecasts and participatory workshops among subsistence farmers in Zimbabwe. *Proceedings of the National Academy of Sciences of the United States of America, 102*(35), 12623–12628.

Prabhakar, S. V. R. K., Srinivasan, A., & Shaw, R. (2009). Climate change and local level disaster risk reduction planning: Need, opportunities and challenges. *Mitigation and Adaptation Strategies for Global Change, 14*(1), 7–33.

Reed, M. S. (2008). Stakeholder participation for environmental management: A literature review. *Biological Conservation, 141*(10), 2417–2431.

Reid, H., Alam, M., Berger, R., Cannon, T., Huq, S., & Milligan, A. (2009). Community-based adaptation to climate change: An overview. *Participatory Learning and Action, 60*(1), 11–33

Retrieved June 11, 2017, from http://www.indiaenvironmentportal.org.in/files/Community-based%20adaptation%20to%20climate%20change.pdf#page=13.

Renn, O., Webler, T., & Wiedemann, P. M. (Eds.). (1995). *Fairness and competence in citizen participation: Evaluating models for environmental discourse* (Vol. 10). New York, NY: Springer.

Roncoli, C., Jost, C., Kirshen, P., Sanon, M., Ingram, K. T., Woodin, M., et al. (2009). From accessing to assessing forecasts: An end-to-end study of participatory climate forecast dissemination in Burkina Faso (West Africa). *Climatic Change, 92*(3–4), 433.

Rowe, G., & Frewer, L. J. (2000). Public participation methods: A framework for evaluation. *Science, Technology & Human Values, 25*(1), 3–29.

Rowe, G., & Frewer, L. J. (2005). A typology of public engagement mechanisms. *Science, Technology & Human Values, 30*(2), 251–290.

Rydin, Y., & Pennington, M. (2000). Public participation and local environmental planning: The collective action problem and the potential of social capital. *Local Environment, 5*(2), 153–169.

Rosener, J. B. (1981). User-oriented evaluation: A new way to view citizen participation. *The Journal of Applied Behavioral Science, 17*(4), 583–596.

Samaddar, S., Chatterjee, R., Misra, B. A., & Tatano, H. (2011). *Participatory risk mapping for identifying spatial risks in flood prone slum areas, Mumbai* (pp. 137–146). Kyoto: Annuals of Disaster Prevention Research Institute, Kyoto University.

Samaddar, S., Yokomatsu, M., Dzivenu, T., Oteng-Ababio, M., Adams, M. R., Dayour, F., & Ishikawa, H. (2014). Assessing rural communities concerns for improved climate change adaptation strategies in northern Ghana. *Journal of Disaster Research, 9*(4), 529–541.

Samaddar, S., Choi, J., Misra, B. A., & Tatano, H. (2015). Insights on social learning and collaborative action plan development for disaster risk reduction: Practicing Yonmenkaigi System Method (YSM) in flood-prone Mumbai. *Natural Hazards, 75*(2), 1531–1554.

Samaddar, S., Yokomatsu, M., Dayour, F., Oteng-Ababio, M., Dzivenu, T., Adams, M., & Ishikawa, H. (2015b). Evaluating effective public participation in disaster management and climate change adaptation: Insights from Northern Ghana through a user-based approach. *Risk, Hazards & Crisis in Public Policy, 6*(1), 117–143.

Samaddar, S., Okada, N., Choi, J., & Tatano, H. (2017). What constitutes successful participatory disaster risk management? Insights from post-earthquake reconstruction work in rural Gujarat, India. *Natural Hazards, 85*(1), 111–138.

Samaddar, S., Yokomatsu, M., Dayour, F., Oteng-Ababio, M., Dzivenu, T., & Ishikawa, H. (2018). Exploring the role of trust in risk communication among climate-induced vulnerable rural communities in Wa West District, Ghana. In O. Saito, G. Kranjac-Berisavljevic, K. A. Takeuchi, & E. Gyasi (Eds.), *Strategies for building resilience against climate and ecosystem changes in Sub-Saharan Africa. Science for Sustainable Societies* (pp. 247–264). Singapore: Springer.

Santos, S. L., & Chess, C. (2003). Evaluating citizen advisory boards: The importance of theory and participant-based criteria and practical implications. *Risk Analysis, 23*(2), 269–279.

Schraven, B. (2010). Irrigate or migrate? Local livelihood adaptation in Northern Ghana in response to ecological changes and economic challenges. *Unpublished Ph.D. thesis, University of Bonn, ZEF.*

Shaw, R. (2006). Critical issues of community based flood mitigation: Examples from Bangladesh and Vietnam. *Science and Culture, 72*(1/2), 62.

Sheppard, S. R., Shaw, A., Flanders, D., Burch, S., Wiek, A., Carmichael, J., & Cohen, S. (2011). Future visioning of local climate change: A framework for community engagement and planning with scenarios and visualisation. *Futures, 43*(4), 400–412.

Tam, C. L. (2006). Harmony hurts: Participation and silent conflict at an Indonesian fish pond. *Environmental Management, 38*(1), 1–15.

Teitelbaum, S. (2014). Criteria and indicators for the assessment of community forestry outcomes: A comparative analysis from Canada. *Journal of Environmental Management, 132*, 257–267.

Toth, F. L., & Hizsnyik, E. (2008). Managing the inconceivable: Participatory assessments of impacts and responses to extreme climate change. *Climatic Change, 91*(1), 81–101.

United Nations (UN). (1992). *United Nations framework convention on climate change*. Rio De Janerio: United Nations Conference on Environment and Development (UNCED).

Van Aalst, M. K., Cannon, T., & Burton, I. (2008). Community level adaptation to climate change: The potential role of participatory community risk assessment. *Global Environmental Change, 18*(1), 165–179.

Webler, T., Tuler, S., & Krueger, R. O. B. (2001). What is a good public participation process? Five perspectives from the public. *Environmental Management, 27*(3), 435–450.

Westerhoff, L., & Smit, B. (2009). The rains are disappointing us: Dynamic vulnerability and adaptation to multiple stressors in the Afram Plains, Ghana. *Mitigation and Adaptation Strategies for Global Change, 14*(4), 317–337.

Yamori, K. (2009). Action research on disaster reduction education: Building a "community of practice" through a gaming approach. *Journal of Natural Disaster Science, 30*, 83–96.

Yang, K., & Callahan, K. (2007). Citizen involvement efforts and bureaucratic responsiveness: Participatory values, stakeholder pressures, and administrative practicality. *Public Administration Review, 67*(2), 249–264.

Yaro, J. A. (2013). The perception of and adaptation to climate variability/change in Ghana by small-scale and commercial farmers. *Regional Environmental Change, 13*(6), 1259–1272.

Chapter 19
Disadvantaged Communities in Indonesian Semi-Arid Regions: An Investigation of Food Security Issues in Selected Subsistence Communities in West Timor

Yenny Tjoe, Paulus Adrianus Ratumakin, Moazzem Hossain, and Peter Davey

Abstract Traditional subsistence farming is an important part of rural society, the yield is a measure of the main source of food to maintain health and livelihoods of rural households. This chapter chiefly investigated the food security issues in AtoinMeto, a subsistence community in semi-arid West Timor, Indonesia. It discusses the concept of subsistence living from the perspective of food sovereignty and food security. Data were collected in Kupang and Timor Tengah Selatan Regencies in West Timor, via mixed-methods of participant observations, and both quantitative household surveys, and in-depth key informant interviews..

This study found that local knowledge and values of AtoinMeto is founded on their existing clan regime and emotionally bonded moral values. This community maintains food sovereignty without overly using the local resources: following seasonal cycles to grow staple food (self-sufficient) and earn cash income via multiple activities within and outside the community to offset declining food stock. However, the system has weaknesses and to support their adaptation to climate change, this chapter suggests three solutions to enhance their food production, improve nutritional value of local diets and develop their ability to market produce.

Electronic supplementary material: The online version of this chapter (https://doi.org/10.1007/978-3-319-77878-5_19) contains supplementary material, which is available to authorized users.

Y. Tjoe (✉) · M. Hossain
Department of International Business and Asian Studies, Griffith University,
Nathan, QLD, Australia
e-mail: y.tjoe@griffith.edu.au

P. A. Ratumakin
Faculty of Social and Political Science, Catholic University of Widya Mandira,
Kupang, Nusa Tenggara Timur, Indonesia

P. Davey
School of Environment, Griffith University, Nathan, QLD, Australia

© Springer Nature Switzerland AG 2019
A. Sarkar et al. (eds.), *Sustainable Solutions for Food Security*,
https://doi.org/10.1007/978-3-319-77878-5_19

The findings of this study imply that, in order to attain sustainable food security for the disadvantaged subsistence community, it is vital that any solutions link to the existing community's knowledge of and values within the cycle of food production and resource use. International organisations and governments must consider this important point and answer the question: How to apply collaborations between technology and local knowledge to the development process?

Keywords Subsistence community · Semi-arid · Local knowledge · Community values · Ancestral territories · Food security · Sustainable livelihoods · AtoinMeto · Tribal community · West Timor

19.1 Introduction

In recent years, there is increasing emphasis on the climate change literature in regards to the importance of community knowledge and values for future adaptation strategies, arguing that the "purpose to adapt" differs greatly between advanced societies and low technology based communities (Berkes 2009; Comenetz and Caviedes 2002; Fraser 2003; Leonard et al. 2013; Petheram et al. 2010; Riedlinger 1999; Wolf and Mosser 2011). The question to be asked is: Why local knowledge and values are important?

Food and Agricultural Organization (2010) and the World Bank (2015a) acknowledge the role of local knowledge and values as human and social capitals which provide problem-solving and survival strategies for rural communities. It should be noted that although food is one of the main priorities for survival in disadvantaged rural areas, it is not the absolute amount of food that determines one's ability to sustain and survive; rather the capacity to produce food locally with an aim to benefit the local members without causing significant harm to the environment. This point is central to local knowledge and values in sustainable food production and is often misconceived or overlooked in the development process.

This chapter investigates the local knowledge and values of a subsistence community in West Timor, the *AtoinMeto*. Individual community members use horticultural practices to grow corn for food security. The way of life is focused on a clan system and customary laws. As part of the national development and regional poverty elimination programs, this community have had reasonable interactions with public health services, market systems and formal education. Conflicting perspective occurs where some external policy makers, practitioners and scholars view the customary laws and beliefs that govern the conduct of this community as not being conducive to development. This chapter presents a clearer portrait of this conflicting view and recommends a better solution to assist this community in securing their food needs and future survival in the era of climate change adaptation.

Given the importance of this issue, there are two purposes of this chapter. First, is to provide a snapshot of the AtoinMeto institutional regime (i.e. the clan system and customary laws) and the local concept of sustainability. The AtoinMeto Clan regime plays a key role in granting individuals access to and control of ancestral lands, while their local concept of sustainability provides moral values to individual members in their sovereignty to attain food needs and their own survival. The second purpose of this chapter is to highlight some of the actions being taken in responding to the food crisis in the region and to consider the integration between various actors. Increasing climatic hazards such as long-term droughts and high-intensity rainfall during the main corn sowing period have had a decisive impact on local food production which relies mostly on rainwater. Over the past decade, food productions in almost all districts in West Timor Island have been impacted by drought and rainfall variability (BNPB 2009; Muslimatun and Fanggidae 2009; World Food Programme Indonesia 2016). With climate change likely to exacerbate extreme weather, crop yields are expected to decline drastically and the local government has allocated budget to combat future food insecurity issues in the region (Tempo.co 2014, 2016). This chapter argues that it is vital that solutions to food security for subsistence rural communities consider established local knowledge and values. Within West Timor, many non-governmental organisations (NGOs) have been undertaking substantial fieldwork and working together with the rural communities to help improve local capacity to produce food and to remain food-independent. While there are many positive examples of infrastructure development and food relief by the national and regional governments, it is not clear that the integration of purpose and action exists among different actors in addressing priority food security issues. How this integration will be achieved is not the main focus of this chapter. However, to ensure that all people have *sufficient, safe and nutritious food* in this semi-arid region of West Timor will require integration among and between actors including multiple professions and their joint expertise.

This chapter includes an overview of the subsistence community from the perspective of food security and food sovereignty; and a discussion of the study area and the local knowledge and values of the AtoinMeto in sustaining livelihoods in West Timor. The weaknesses of the local food system in terms of nutrition and economic access and the impacts of drought and heavy rainfall anomalies on local production will be discussed, including the actions to tackle food crises by the national and regional governments and NGOs. Finally, three solutions to food security for this particular group are proposed.

19.2 Subsistence Community: Food Security or Food Sovereignty?

The term "subsistence" refers to a way of life that is led by minimalist low levels of living, less productive techniques, less input purchased, and strong influence of social-cultural considerations in decision-making processes (Wharton 1969). Pursuing subsistence goals, rather than commercial profit have been identified as the feature of rural communal societies throughout the world. From an economic perspective, Abele and Frohberg (2003) comment that the main purpose rural communities engage in subsistence production is for self-sufficiency and social-cultural goals. The excess production in good times can be sold for additional cash money, but the production is not controlled by profit and competition, rather it is for the sustenance of individuals, families and the community (Goldsmith 1998; Seavoy 2000). This lifestyle is often found in indigenous and tribal communities that have close attachment to ancestral lands with established customary systems (FAO 2010; World Bank 2003).

Studies have found that subsistence production is effective for rural poor to prevent food insecurity (Davidova et al. 2012), to provide additional income (Larsen 2009) and to reduce households' exposure to price fluctuations (Kuokkanen 2011). By following the seasonal cycle of available resources, subsistence communities are able to produce well-adapted and diversified crops, while integrating a combination of activities (farming, hunting, fishing, gathering and other activities) to obtain nutrition and secure food; examples include crops, medicinal plants, heat, clothing and shelter (Conklin 1961; Fisk 1975; Kuokkanen 2011; Manner 1981). These features are indeed in line with the concept of food sovereignty which emphasises the capability to ensure self-sufficiency; where individuals are able to look after their food needs and survival, without being disadvantaged by climate, regulatory bodies, markets or other instruments.

In the sixth Session of the UN Forum for Indigenous Peoples (UNPFII 2007), acknowledging the Declaration of Atitlán in 2002 which defines Food Sovereignty as:

> The right of Peoples to define their own policies and strategies for the sustainable production, distribution, and consumption of food, with respect for their own cultures and their own systems of managing natural resources and rural areas, and is considered to be a precondition for Food Security.

However, on the other side of the argument living conditions in these communities are far from desirable to achieve local food security. According to the Food and Agriculture Organization,[1] "food security" is achieved when all individuals have *sufficient, safe and nutritious food* at all times. This is assessed from four independent dimensions, the four pillars on food security, namely availability, access, utilisation and stability (FAO 2009). In the following sections, it will be shown in the case of *AtoinMeto*'s food sovereignty these four dimensions are not independent of each other. This community's food security is a developing process that follows the stages from availability to utilisation and economic access, while the stability of their food system is maintained by clan and its *Adat*[2] laws, as well as local values when using the available resources within ancestral lands and forests.

---

[1] As detailed in the *Rome Principles for Sustainable Global Food Security,* agreed to at the FAO Declaration of the World Food Summit on Food Security, in Rome from 16 to 18 November 2009.

[2] In Indonesian term, *Adat* means custom of a political-social institution, in which the associated mores and laws were derived from way of life of a particular tribe in the Dutch East Indies (now Indonesia) which has been established far before the European occupation in the seventeenth century. In post-independence, following the passing of Law No. 1/ 1957, on Basic Law on Local Administration (and then replaced by Law No. 18/ 1965), *Adat* as a political-social institution at

19.3 Study Area and Methods

The study was conducted in three *AtoinMeto* communities in West Timor Island of Indonesia: one in Kupang Regency (B) and the other two in TTS Regency (A and C, see Fig. 19.1). This study was part of a doctoral research program between 2012 and 2016 at Griffith University, investigating rural livelihood sustainability and drought adaptation by the research team. Fieldwork was conducted from June to November 2013, comprised of mixed-method data collections including quantitative household surveys ($n = 627$), qualitative in-depth interviews with customary elders, community figure heads and individual households and further participant observations.

For quantitative household surveys, a series of questions were developed to reflect the five assets of Sustainable Rural Livelihoods (see Chambers and Conway 1992; DFID 1999), i.e. natural resources, human knowledge, social capital, financial

Fig. 19.1 Location of the three communities in West Timor: Regions A and C in Timor Tengah Selatan Regency, and region B in Kupang Regency. Map adopted from: Maximilian Dörrbecker (Chumwa) [CC BY-SA 2.0 (http://creativecommons.org/licenses/by-sa/2.0)], via Wikimedia Commons

---

hamlet and village level was gradually replaced by *Desa* (the new administrative village unit). In the modern Indonesia, *Adat* still plays an important role in governing the use of natural resources in some of the tribal communities in remote areas. These communities are recognised as 'geographically isolated customary law communities' (*Komunitas Adat Terpencil*). Such recognition tends to indicate the poverty and primitive aspects of this community (Duncan 2004).

sources and physical infrastructure. The quantitative method adapted the livelihood vulnerability index (LVI) measure from Hahn et al. (2009) and constructed a composite index to measure the livelihood vulnerability of dryland communities in West Timor.[3] While qualitative in-depth interviews and participant observations together with aspects of local knowledge and values were explored including:

- The history of the community.
- Daily routine of a typical village family.
- Local corn growing process and corn preservation techniques.
- Things that they do when facing problems related to dryness and drought.
- Aspects of life that they perceive as important, meaningful and valuable.

Qualitative data were analysed using grounded theory coding based on the guidance of Strauss and Corbin (1990), Glaser (1978) and Flick (2009).

This research was granted approval by the East Nusa Tenggara Local Government for conducting research and data collection in the three regions under study. Throughout all interviews, the researcher was accompanied by a local translator to ensure correct interpretation of the terms and meanings. Interviews were conducted in Bahasa Indonesia, with a few combinations of *Meto* language. A digital recorder was used to record all interviews. Further to this, this survey was conducted with approval of the Griffith University Human Research Ethics Committee (HREC).

West Timor is a semi-arid island, located in East Nusa Tenggara (NTT), the third poorest[4] province of Indonesia. The island is characterised by unfavourable climate for growing crops, limited amount of natural resources and a dry and hot climate. A region which is often described as prone to drought and food shortages. Corn is the staple food of the region. Agriculture remains an important sector in the regional economy, in 2013 it accounted for nearly 35% of NTT regional-GDP (BPS NTT 2014) and provides livelihoods for over 61% of the region's working population (BPS NTT 2015).

About two-thirds of the West Timor Island consists of the *AtoinMeto* communities (Fox 1999). The land area is 14,200 km². In 2016, the total population was over 1.8 million and the population growth rate has remained constant at 2% annually over the last three decades (see Table 19.1). The majority of the population live in the rural Kupang and TTS Regencies (19% and 25% respectively) and in the urban Kupang City (22% of total population). In Kupang and TTS Regencies, subsistence farming is commonly found where people grow corn during the rainy season (December to March) and raise animals for domestic consumption. Population growth in TTS has decreased to 0.7% annually from 1.7% in the 1980s, with a higher percentage of women (0.97 males/females) compared to Kupang (1.05 males/females) (Fig. 19.2). The less number of men in TTS may be due to the higher

---

[3] The methods and results of the LVI for the three *AtoinMeto* communities in West Timor have been published (see Tjoe 2016a).

[4] In 2016, NTT is ranked the third poorest province in the country, with 22% of its population living on less than Rp 11,550 a day, an amount equivalent to AUD1.15 (BPS Indonesia 2016). Besides West Timor Island, NTT Province has three other major islands, namely Flores, Sumba, and Alor.

Table 19.1 Total population and population growth rate (%), 1980–2016, West Timor Island, Kupang and TTS Regencies

Year	West Timor Island	Kupang	TTS
1980	1,007,987	403,167	289,655
1990	1,250,123 (2.0)	522,944 (2.4)	348,067 (1.7)
2000[a]	1,542,850 (1.9)	444,800 (−1.5)	404,700 (1.4)
2010[a]	1,664,042 (0.7)	304,548 (−3.4)	441,155 (0.8)
2016	1,865,105 (1.6)	360,228 (2.4)	461,681 (0.7)

BPS NTT (2017)
[a]Secession of new regencies/city from Kupang Regency in post-decentralisation since 1999

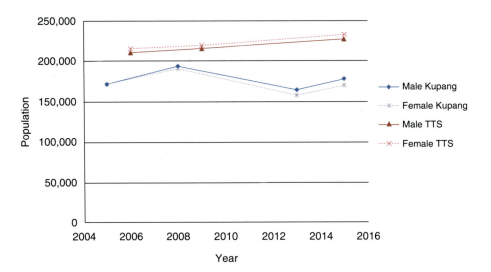

Fig. 19.2 Population by gender in Kupang and TTS Regencies, 2005–2015. Source: BPS NTT 2017

male fetal or infant mortality and the gradual out-migration of the adult males. This concludes that there is a lack of male labour to work the land for food production.

19.4 *AtoinMeto* Food System: A Bottom-Up Approach to Food Security

Traditionally, *AtoinMeto* People cope with droughts through a range of strategies that ultimately aim to secure the availability of corn (stored under their kitchen roof) and water within their ancestral territories. This study found that clan regime and local value play a key role in preserving these local assets (land, forests and water) for the members to cultivate and gather food, water and other resources for their own subsistence.

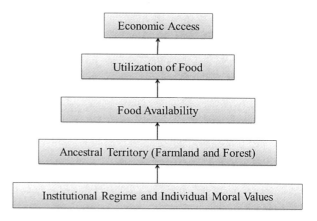

Fig. 19.3 Elements of food system in AtoinMeto

Clan-based land and water management has been found to be effective in the rural society of West Timor because of the "ikatan emosional" (emotional bonding) rooted in the value of users who identify themselves as sharing a common ancestor and history of the clan (Ratumakin et al. 2016). Additionally, in the three communities under study, there is a local concept of sustainability where *one should limit his own desire for the common good, to be self-sufficient and be grateful for what he has owned* (Tjoe 2016b). According to the local people in these communities, sustaining livelihoods in the region is not difficult if one (1) exercises self-control in using the available resources; (2) is self-sufficient in meeting food needs; and (3) achieves self-content when having low harvest. As these peoples are clan-based community, the "emotional bonding" also leads to individual moral duty to not overly exploit the natural assets, so that the younger generations can continue to have the necessary resources to live in and enjoy this semi-arid land.

The clan regime and local value become the foundation of their food production system which historically has allowed the community to have a stable food system throughout the years. The food system in AtoinMeto community resembles the following stages (Fig. 19.3):

19.5 Clan System, *Adat* Laws and Ancestral Territory

In rural West Timor, a number of clans may live within one village administrative system. In the context of *AtoinMeto*, individual clans have their customary laws, known as *Adat* and are independent of the village administration. Also, each clan has its ancestral territories, consisting of farmlands and forests and the resources within, while clan members are grouped into four *Ume* (house) based on their status as descendants of the same ancestor (Tjoe 2016b). Depending on the status, members inherit a certain block of farmland and are granted access to forest as their rights for livelihood. Their emotional bonding is reflected in their belief that ancestral territories are sacred because the ancestors who had passed away continue to protect the resources (land, water and forestry) and the humans in these territories. Hence, any social occasion that affects land status (such as marriage between two clans or territorial disputes) will require the holding of community meetings with the presence of *Tua Adat* (clan customary elders) to clarify issues of territory, origin of ancestors and location of their burial areas, referred to as boneyards.

Clan regime continues to function because *AtoinMeto* People benefit from such membership, especially in getting the assurance that food, water and other resources within their ancestral territories are protected by *Adat* laws. Some members are landless due to decades of out-migration and decisions to return to the community, while some others have smaller farmlands due to their lower status in clan system (referred to as the lower rung member). These members can regain their right to subsistence after consulting with *Tua Adat*, by offering labour to work on a piece of land owned by their kin or negotiate for assistance with those members with multiple farmlands. This type of share-cropping subsistence, known as *garap*, is commonly practiced in West Timor.

Clan and *Adat* laws also play an important role in the control of water resources, which is derived from the local knowledge of community-based resource management. In studying the management of water sources in selected communities in Kupang, TTS and Malaka Regencies, Ratumakin et al. (2016) found that community-based management of water sources is more sustainable and these communities exhibited seven of the Ostrom's eight principles of managing a commons, plus two additional principles that reflect the local knowledge of West Timorese society (Table 19.2).

In dryland West Timor, water is a scarce resource and an essential source of livelihood for rural communities—"*where there is a water source, there is life*" (Tjoe 2016b)—communities always established their settlements around the water sources. Water sources are used and managed as a common-pool resource, and the usage and maintenance are strictly tied to local customary structures, usually the custom of a clan

Table 19.2 Determinants of sustainable management of water sources in West Timor

Ostrom's eight principles of managing a commons:
1. Clearly defined boundaries
2. Proportional equivalence between benefits and costs
3. Collective-choice arrangements for most individuals
4. Monitoring users and resources
5. Graduated sanctions
6. Conflict-resolution mechanisms that are rapidly accessible
7. Minimal recognition of rights to organise
8. Nested enterprises, in case of resources that are parts of larger systems[a]
Two principles in local context:
1. Collective memory (how the water source was discovered and by who) and emotional ties between members of community and the water source
2. Adoption of local *Adat* (customary) structure in the control of water source

Source: Ratumakin et al. (2016)

[a]The study did not find this principle to occur in the studied communities

that the community believed was the water discoverer, hence the right to control this water source is granted to the discoverer. Today, local folklores about founders of water sources and ancestral territory are still recited in clan rituals to enforce members' moral values. This ensures accountability in using natural resources within the territory, and maintaining connections with the ancestors, this process safeguards the water sources and protected water resources for keeping the local community alive in the future.

19.6 Food Production and Food Availability

We harvest once and eat for a year. (Mama Yun, female farmer, Kupang Regency)

The food (corn) production process takes place on individual farmlands, often called "*Bikin Kebun*" (meaning to make garden). It begins with the clearing of farmland by individual farmers in the hottest season (usually in September and October) to utilise the heat and high temperature which dries and reduces the weeds. It then ends after the sowing of seeds during rainy season (December to March) where farmers utilise the combination of weather, temperature and rainwater to grow the crops. For *AtoinMeto* People, *bikin kebun* is regarded as a sign of industriousness and self-sufficiency. The locals commonly used a phrase: *today's harvest equals one year of the household's food stock*. Figure 19.4 below illustrates the process of their food production and food availability.

AtoinMeto People employ *Tumpang Sari* (mixed cropping) in their sowing stage, which allows N-fixing legumes to coexist along with corn planting. Four types of seeds are sown into each planting hole: three corn, one pumpkin, one Chinese long bean (Vigna unguiculata subsp. sesquipedalis) and one pigeon pea (Cajanus cajan). Throughout January and February, rainwater continues falling and brings water and humus from the hills down to the valley bypassing their farms, where the corn kernels have developed and have grown large. Individual households have to remove weeds in between corn stalks to prevent them from hindering the growth of corn or becoming pests.

From March to April, households which have taken care of their "garden" attentively will have plenty of crops ready to be harvested. Corn and pumpkins are harvested within 3 months, while Chinese long beans and pigeon peas are normally harvested in the fifth month after sowing and may continue until the seventh or ninth month. Entering the dry season (April–November), these people allow their farmland to lie fallow from April to September, while continue to visit this farmland to harvest pumpkin, pigeon pea and Chinese long bean.

AtoinMeto are independent and have full control over the food system, yet they actually have a period of "food stock finishing" before entering the period of "full stock of food", particularly their stock of corn as staple food. In rural West Timor, the availability of food in each household can be categorised into two types (Boli Sura et al. 2010): first type is called *amsenat*, a situation where households have full stock of food (corn, pumpkin, and various types of beans); and second type is called *fun amnahas*, meaning normal hungry, a situation that often takes place between

19 Food Security and Subsistence Community in West Timor

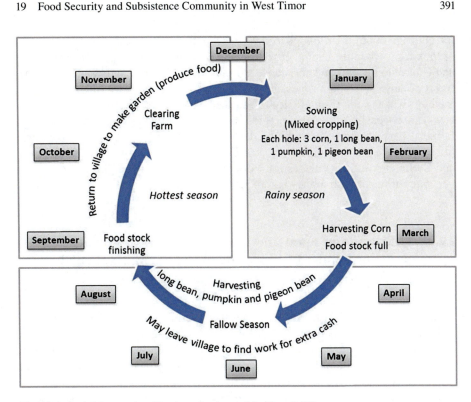

Fig. 19.4 AtoinMeto cycle of food production and food availability

September[5] and March, where households experience insufficient food because the 1-year stock in their kitchen is finished. To overcome this period of "normal hungry", the three communities under study carried out a combination of activities to support their food needs, including:

1. Gathering produces from trees cultivated in the community forest.
2. Generating cash through the sale of animals and produces from the community forest such as *kemiri* (candlenut), tamarind (*Tamarindus indica*), and timber products for building materials.
3. Temporarily departing the community (mainly males) to work for extra income, particularly during fallow season (May to August). The males go to towns to work as paid labour in construction projects, while some females (in Kupang Regency) go to another village to work in irrigated paddy fields.

This study found that local people in Region B combined subsistence agriculture and cash income activities as their strategies to cope and adapt to the impacts of changing climate. The accessible public transport has allowed the locals to easily

---

[5] Those who have smaller farm and production may experience early fun amnahas, starting June to March.

travel in and out of their village to sell their produce in several farmer markets and/or look for work in the capital city (Kupang). In contrast, local people in Region A and C have to depend on subsistence agriculture in their farmland and community forest because of the poor roads and the distance to market places (taking 6–10 h of travel to Kupang the Capital City to sell their produce for cash). Cattle (pigs and cows) are not easily sold in these regions unless there is a *papa lele* (agent who buys farmers' harvest and sells to the market) who visits to look for good breeds and prices. Most of the time produce is stored in the houses. Local people in Region A and C often leave their villages for 2 days to sell their produce, or for 3 months to generate sufficient cash money before returning home.

19.7 Utilisation of Food and Economic Access to Foods and Non-foods: Produces from the Farm, Backyard and Community Forest

From the farm, the newly harvested corn is then preserved using a traditional storing technique of hanging bunches of cobs under the even temperature of the kitchen-roof; and allowing the heat and smoke of cooking to continuously dry the kernels. This traditional technique effectively preserves and protects corn from "fufuk" (corn flea beetle). This corn is edible for a year or more, ensuring the availability of food for individual households until the next harvest. In addition to subsistence crops in their farmland, the *AtoinMeto* People also breeds a number of animals in their backyard and cultivates a range of crops for various purposes.

From the list of produces shown in Table 19.3, the diet of the *AtoinMeto* People is composed mostly of carbohydrate foods (corn, rice and cassava), with plant proteins (mainly from beans). Animals are not consumed on a regular basis, because animals are used as a reciprocal exchange to support members in the core-family who are holding a wedding or funeral. In addition, animals (especially beef cattle) are bred for paying educational expenses as many parents have become aware of the benefit of education for their children in obtaining modern employment.

It was found from the survey that households do not spend much income on food items, rather a large sum of expenditures is used for children's educational fees and school uniforms and ingredients for cooking (such as cooking oil, sugar, salt and coffee). Over half of the total respondents (57.7%) had expended between Rp 101,000 and Rp 500,000 (USD$10 and USD$50) and 9% over Rp 1,000,000 (USD $100) in the past month—these households have at least three children to support, at high school or university level. 32.2% of total respondents spent between Rp 15,000 and Rp 100,000 (USD$1.50 and USD$10) in the past month for basic cooking and cleaning needs. Six respondents (1%) reported spending Rp0 because they rarely buy these cooking items listed above.

Figure 19.5 below summarises the *AtoinMeto* People's food sovereignty and food security system, where individual pillar of food security is built up from the

Table 19.3 Utilisation of food in *AtoinMeto* Food System

Utilisation of food	Location of production	Variety
Domestic needs	Farmland	Corn Pumpkin *Kacang turis* or pigeon pea (*Cajanus cajan*) Chinese long bean (*Vigna unguiculata subsp. sesquipedalis*) Chilli paddys Tomatoes
Educational expenses, social-cultural occasions and as cash crops	Backyard, fallowed lands, forest	Small animals: chicken, dog, pig, goat Large animals: beef cattle

Alternative foods, shelter needs and cash crops	Community forest	Alternative food and cash crops:
		  Lima beans (*Phaseolus lunatus*)
		  Rice beans (*Vigna umbellata*)
		  Cassava
		  Pumpkin
		  Papaya
		  Banana
		  Mango
		  Apple guava
		  Tamarind (*Tamarindus indica*)
		  Candlenut
		  Betel leaf
		Shelter materials and cash crops:
		  Various palm trees (gebang palm, coconut, betel nut)
		  Lamtoro (*Leucaena leucocephala*)
		  Mahoni (broad-leaf mahogany)
		  Jati (teak or tropical hardwood)
		  Gamal (*Gliricidia*)

Source: Tjoe (2016b)

pillar underneath with local customary laws and values as the foundation that maintains stability of the system. Nevertheless, this system has weaknesses and with increasing climate anomalies in the region, these communities need support to adapt and to improve the food security system. The next section discusses these issues.

19.8 Nutritious and Economic Values of Local Production

For many decades, poverty issues in West Timor and other islands within NTT province have been inseparable from continuing child malnutrition. Prevalence for underweight children and malnutrition in NTT is among the highest in Indonesia. Over the years between 2007 and 2013, this rate remained twice the national

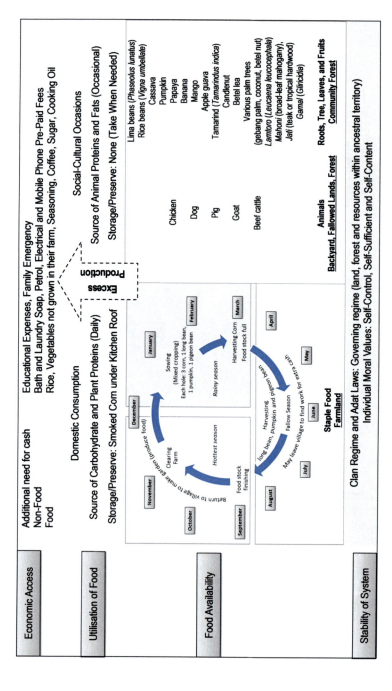

Fig. 19.5 The four pillars and bottom-up process of food security in AtoinMeto, West Timor

19 Food Security and Subsistence Community in West Timor

Table 19.4 Nutrition security in NTT

Nutrition security indicators	2007				2013			
	Kupang	TTS	NTT	National	Kupang	TTS	NTT	National
% Children underweight (5–12 years old)	21.6 (M) 15.3 (F)	29 (M) 23 (F)	23 (M) 19 (F)	12.1	20.4	19.1	20.4 (M) 18.3 (F)	11.2
% Children undernourishment (<5 years old)	37.9	40.2	33.6	18.4	33.4	46.5	33	19.6
Mortality rate/1000 (<5 years old)	N/A	N/A	80	44	N/A	N/A	58[a]	40[a]
Ratio of Puskesmas/1000 people	6.74	5.56	5.69	3.65	7.73	5.98	7.28	3.89

Source: BPPK (2008, 2013), Ompusunggu et al. (2013), Pusdatin (2008, 2014)
[a]Data for year 2012

average (BPPK 2008, 2013). However, there has been major improvement in the provision of health services for the poor, through the increasing number of Puskesmas (community health centres) to treat both the pregnant mothers and their children (Table 19.4). The ratio of community health centres per 1000 population in 2013 was 7.28 in NTT (higher than the national average of 3.89). In the regencies of the three communities under study (Kupang and TTS in West Timor), the figures are even higher; for every 1000 people, there are eight puskesmas in Kupang, and 6 puskesmas in TTS.

In Indonesia, Puskesmas provides basic healthcare to local communities at districts level, under the supervision of the Minitry of Health. Its main services include maternal care, prenatal and postnatal care, immunisation, and communicable disease control (especially malaria prevention programs in NTT). At the village level, Pustu (supporting units of Puskesmas) and Posyandu (integrated service post) deliver the healthcare to the villagers. In the effort to meet the earlier eight Millenium Development Goals (now SDGs), the Indonesian Government set several health development targets in its National Medium-Term Development Plan (RPJMN) 2010–2014, one of the targets was to reduce the prevalence of child malnutrition to 15% by 2015 (Kemenkes 2011). These targets have not been met even with increased access to community health centres.

Since 2011, the national government has allocated budget via the Ministry of Health *Dana BOK* (*Bantuan Operasional Kesehatan*, or Health Operational Assistance) to implement *Program Penambahan Makanan Tambahan—Pemulihan* (PMT-P, or Supplementary Food-Recovery Program) to regions of high prevalence of child malnutrition. The program was carried out at local Posyandu with the help of local women's group (PKK Desa). Each day for 3 months, mothers would bring their toddler (children aged 5 or below) to the Posyandu in order to receive training and counseling, as well as a set of nutritional recovery foods (predetermined menu by the department of health) for stunted toddlers.

From the fieldwork, this study found that services provided in Pustu and Posyandu are very limited and mostly focus on immunisation and providing vitamins to infants and pregnant mothers; and there was no PMT-P activities. According to the Regulation of the Minister of Health No. 11/2015 (on Technical Guidance of BOK), elimitating child malnutrition is still the top priority of MoH and PMT-P Programs remain one of the activities financed by BOK.

In addition, the government also allows various aid agencies to finance the PMT-P Program some additional funding coming from the community, businesses and APBD (local government's budget). A local newspaper reported that PMT-P program in NTT Province had not resumed since 2014 due to the complicated procedure, including submitting proposals and bank details for fund disbursement (Radio Republic Indonesia 2015). In other regions of Indonesia, this program also has not been optimally implemented for a number of reasons,[6] including limited budget, lack of health staff and lack of timely reporting systems.

Providing counselling and distributing supplementary food from health centres only provided temporary support and recovery for the mothers and children; but it did not effectively lead the households to healthy dietary habits. It was found that in general, pregnant mothers and mothers prefer traditional methods taught by their mother or mother-in-law for pre and postnatal care. Similar patterns are observed in their diets, where they follow the recipes of their grandmother or mother in their daily activities. What is more interesting, is that in selling produces to market, such as produce from their farm, backyard and community forest, the local people tend to select the ones that are in best quality, leaving the poor quality produce (overly ripen) for domestic consumption. Rather than consuming eggs laid by the chickens in their backyard, eggs are regularly sold at the market. In summary further education and behaviour change are needed to support these communities to achieve a balance of their utilisation and sale of local food.

19.9 Impact of Rainfall Anomalies on Local Productions

Rainfall during December to March was reported by the local community as "very little" or "it stopped after we sow the seeds", unlike the normal rainy season, where the distribution of rainwater was more even over the three or four months period. Figure 19.6 below shows the annual rainfall in Kupang City, West Timor from 1986 to 2013. According to Lasiana Kupang Climatological Station and a local NGO (PIKUL), since 2010 there has been an uneven distribution of rainfall across the island of West Timor during the rainy season. The island experienced climate anomalies, in which during the dry season, the incidence of rain increased, and in some areas rain occurred more during the hottest season (September and October) which coincided with the garden cleaning period.

---

[6]This was found in villages of Gunungkidul Regency (see Pambudi 2015) and Surabaya Municipality (see Tyas Arum Sari Dewi and Rahaju 2015).

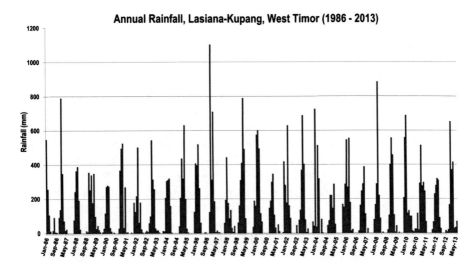

Fig. 19.6 Annual Rainfall in Kupang, West Timor. Source: Fieldwork, secondary data, Lasiana Kupang Climatological Station

On the contrary, during the rainy season, there is a decreased incidence of rain but an increased intensity of rainfall where big drops of rain fall with more force on the soil surface and increase the surface run-off (Fanggidae and Ratumakin 2014), thus resulting in loss of yields and crop failures.

The three communities under study also experienced a reduction in food production, and thus now need more cash to buy food from outside sources. Animals that are mainly sold to fund child registration to school or university intake or a family emergency are now being sold in order to buy more food from the market. Similarly, the resources in the forest are now being sold to buy other foods. Some households were willing to try planting even with this climate disruption but many others were pessimistic whether they would have a positive and fruitful result.

> The parents and elderly want to return to subsistence farming and make *kebun* [garden], but it might be the current daily needs [for cash] which have caused difficulties in allocating time and energy. If only there is enough energy, they might want to do both [subsistence farming and cash generating work], it all depends on how they can meet the daily needs. (Mama Yayu, Housewife, TTS Regency).

Households are gradually shifting to "cash" work, such as transmigrating short-term to work as labourers in the nearby town or migrating overseas. For those close to the coastline and beaches time can be spent in an alternate livelihood collecting *batu warna* (beach pebbles), earning approximately Rp 8000–Rp 30,000 per bag[7] (equivalent to approximately US$ 0.80–US$ 3.00).

The inability to secure stable food production has social and economic implications on these communities. Individual self-worth and self-esteem has been

---

[7] Each bag may weigh 20 kg to 50 kg, depending on the size of stones.

impacted. Economically, these peoples are now not only vulnerable to the impact of climate change, but also likely to be disadvantaged by the fluctuation of prices in the market, both as a supplier of low-skilled labour as well as a consumer who buys food from the market.

NTT Provincial government has committed to improve water resources infrastructure. Between 2014 and 2018, the regional development plan included the building of one dam, 27 large reservoirs for irrigation, 801 small farm reservoirs and 850 drilled wells (DPU NTT 2014). However, by conserving water in reservoirs, the role of these mega infrastructures tended to tackle drought issues on farmlands that require large supply of water, such as paddy productions. In fact, from 1986 to 2014, there have been over 530 reservoirs built in this region (BWS NTII 2016). Several initiatives funded by national government and International organisations have also addressed water scarcity related issues; Table 19.5 highlights three of the existing programs in NTT: Pamsimas (*Penyediaan Air Minum dan Sanitasi Berbasis Masyarakat*), P3A (*Perkumpulan Petani Pemakai Air*), and SPARC. Each initiative has delivered considerable amounts of outputs that are community-based, including provision of drinking water, construction of sanitation facilities, irrigation systems, and climate risk information system. However, many of infrastructures and systems which have been built as part of Pamsimas and P3A programs, have not functioned optimally, some are even left stalled, due to mismanagement and conflicts among local people (Ratumakin et al. 2016). There are a number of sustainability challenges, in which this study identified:

- Lack of awareness of the need of post-program infrastructure maintenance.
- Lack of effectiveness in post-program management, including limited sanctions against violators of rules and agreements, and no viable mechanism to resolve conflicts and grievances of communities.
- Lack of awareness of the need to protect all water sources.

In contrast to sustainable management of water sources in West Timor shown in Table 19.2, the above-highlighted three programs have overlooked the need for clearly defined boundaries of water resources, which can be problematic as water users do not know which are of watershed they should protect and maintain. Also, the two local context principles are commonly missing in implementation of Pamsimas and P3A. The adoption of Adat structure and acknowledgement of history of local water discoverer are essential for establishment of emotional bonding between users and water sources, enforcing the moral duty of individual users to protect water sources.[8]

---

[8] P3A group in Noelbaki village, Kupang Regency is well-known for its best irrigation model in NTT. The group acknowledged the Oematan family as descendant of the water discoverer and adopted local Adat structure in monitoring, control and protection of watershed and source water (Ratumakin et al. 2016).

Table 19.5 Initiatives and strategies to address issues of drought, low rainfall and climate change

Initiatives/strategies	PAMSIMAS Community-based water supply and sanitation program	P3A Water User Farmer Association	SPARC in NTT Strategic Planning and Action to strengthen climate resilience of Rural Communities in NTT
Program year, implementing partners, and management	Nationwide 2008–2012: Phase I 2013–2016: Phase II Implementing partners: – Ministry of Public Works – Ministry of Health – Ministry of Home Affairs PAMSIMAS is managed by an independent body BPSPAMS, comprises: – Chairman – Secretary – Treasurer – Various sections (technical, health, etc.) – Members Village head and religious leaders acting as protectors and advisors of program	Nationwide Since 1984 (Presidential Instruction No. 2/1984) Current status is regulated in the Government Decree No. 20/2006 P3A development is supported by agencies under: – Ministry of Public Works, – Ministry of Agriculture P3A is managed by a group of farmers, comprises: – Members meeting – Chairman and vice chairman – Secretary – Treasurer – Production input section – Technical executor – Business section – Block leaders and members	NTT: East Sumba, Manggarai and Sabu Raijua Feb 2013–Dec 2016 Implementing partners: – Ministry of Environment (MoE) – Bappeda NTT (NTT Provincial Development Planning Agency) Key stakeholders consulted include: – UN SDGs, National level: Ministries of Environment, Agriculture, Public Works, Bappenas (National Development and Planning Agency), BMKG (Meteorological, Climatological and Geophysical Agency), BNPB (National Board for Disaster Management), National Climate Change Council – Development partners: World Food Programme, FAO, and AusAID – Local level: Bappeda, government agencies – Private sectors: bank, academics, agribusiness, local NGOs and civil society organisations

(continued)

Table 19.5 (continued)

Initiatives/ strategies	PAMSIMAS Community-based water supply and sanitation program	P3A Water User Farmer Association	SPARC in NTT Strategic Planning and Action to strengthen climate resilience of Rural Communities in NTT
Major funding source	World Bank, Government of Indonesia, and Government of Australia. Local communities raise funds to finance the maintenance of water facilities	Local government funded the construction of capturing and irrigation infrastructure (sharing between provincial and regency governments based on area coverage). Individual groups raise funds to finance the maintenance, including irrigation system, building and networks among users	UNDP (United Nations Development Programme), SCCF (the Special Climate Change Fund, operated by the Global Environmental Facility), and NTT Provincial Government
Objectives	To increase the number of low-income rural and peri-urban populations accessing safer water and improved sanitation facilities, and enhanced hygiene behaviour	To accommodate water-related problems and assist farmers' needs for irrigation, agribusiness and marketing	To develop climate resilient institutions and strengthen rural communities' livelihoods, food and water security in anticipation of the impacts of climate change
Outputs in NTT	Between 2008 and 2015, construction of Pamsimas drinking water and sanitation facilities reached 819 villages across 19 regencies and 1 city of NTT	– 27 large reservoirs for irrigation – 611 farmer associations (P3A): – 71 well developed – 174 developing – 366 undeveloped	Expected outputs include: – Developing a community-based climate risk information system (covering 120 communities) – Supporting 40 communities to switch from subsistence farming to more flexible practices – Assisting 40 communities to diversify income sources to be less sensitive to climate change – Improving water resources infrastructure and management in 40 communities to anticipate changes in rainfall patterns

| Sustainability challenges | Performance of Pamsimas in NTT is still below the national average, and many of the water and sanitation facilities are now not functioning due to several issues:
– Lack of awareness of the need of post-program maintenance. Almost 50% of coverage regencies did not raise funds to finance the maintenance of the facilities
– The host (water source owner) initially agreed to the construction of water facilities, but then expects compensation (monthly payment from users), otherwise forbids people from using the water
– Lack of effectiveness in BPSPAMS in post-program period, especially resolving conflicts and grievances of communities regarding the costs and benefits of program
– Lack of community participation and transparency in reporting | In NTT, 60% of the 611 farmer associations are undeveloped and damages to irrigation infrastructure have been left unattended. Several local issues include:
– Lack of firm sanctions and conflict resolution mechanisms in the P3A when users violate rules or agreements
– Lack of awareness of the need of source water and watershed protection
– Lack of capacity and human resources in the management of farmer associations | Program evaluation and monitoring results are not yet published, but several risks may potentially impede program implementation and undermine the objectives to counter effects of climate change on rural communities, for examples:
– Regional governments (province or regency) fail to allocate funding due to competing interests
– Ineffective coordination among various agencies and inconsistency in reporting as the program involves large number of government and non-governmental institutions
– Bureaucratic processes or red tape |

Source: Pamsimas (2017); UNDP (2017); Ratumakin et al. (2016); World Bank (2015b)

19.10 Concluding Remarks: Solutions to Food Security

So, *Why is local knowledge and values important?*

This chapter provided a snapshot of food security system in *AtoinMeto*, a subsistence community in semi-arid West Timor, and identified that local food production system is sustainable as a result of the local effective clan regime and the emotionally bonded moral values in attaining food sovereignty in disadvantaged rural areas.

Subsistence living is not all about living standards, backwardness and traditions that inhibit development. In the context of disadvantaged areas such as the semi-arid rural West Timor, the collective conscience to live a subsistence way of life allows the limited resources to regenerate and replenish; so the members and importantly children of these communities can continue a sustainable livelihood in the region. In the three communities under study, the clan customary laws and individual values effectively keep this collective conscience intact and govern the use of natural resources within ancestral territories. In this regard, International Organisations and the national governments in the development process must consider this important point and how it applies to local communities.

From the above discussions, it shows that *AtoinMeto* People have a well-established food system that has given them the sovereignty in securing and sustaining food needs for many generations. However, the weaknesses are in the utilisation of the produce i.e. their diets which lack nutrition and their produce that lack market value. Increasing climate anomalies have impacted on the local production. The loss of yields and also crop failures are not merely due to longer droughts and poor rainfall, but also due to the high intensity of the rainfall, over shorter periods of time, during the farmers' sowing period.

The government's programs and infrastructure developments have focused more on giving food and water to the people and their farms. While missing out the fact that food sovereignty has been well established at the community level, government interventions can potentially harm the unique local knowledge for the problem-solving. In the era of global warming induced climate change, without a proper solution to sustain their food security, communities are likely to become increasingly dependent on government and market, and more farmers and young adults will migrate out of the village to find a paid job opportunity such as working as migrant workers to sustain their livelihoods.

Based on these issues identified by this study, this chapter suggests that the local knowledge and values in *AtoinMeto* remain as the foundation of their food system. Three solutions are proposed to improve the utilisation of local produces and to support this community in adaptation to climate change (Fig. 19.7). These solutions are to be implemented as collaborations between technology and local knowledge, through partnerships between the community and actors from multiple agencies, professions and joint expertise, in the following areas:

1. Innovation of the appropriate technology to tackle local issues of climate anomalies (drought, low incidence of rainfall and high intensity of rainfall) during the

19 Food Security and Subsistence Community in West Timor

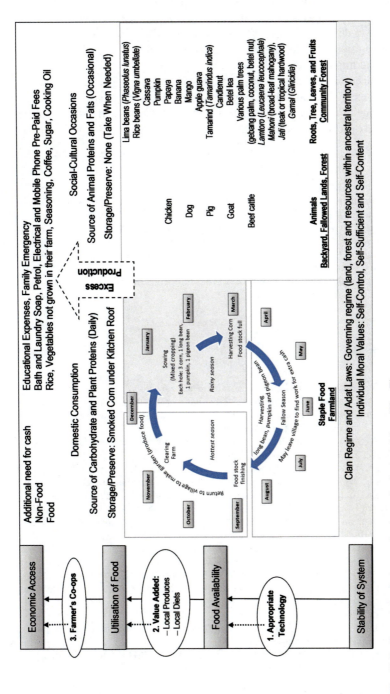

Fig. 19.7 Grounded solutions to food security in AtoinMeto, West Timor

sowing period, adapting to these conditions so the people can continue to perform subsistence farming and be self-sufficient.

2. Enhancement of knowledge in the utilisation of local produce by involving the people to develop local nutritious diets from their farm and forest products, so communities learn to prepare more nutritious diets for their family and create value-adding to local produces.

3. Revitalisation of farmers' cooperatives (Co-ops) to organise gatherings and the sale of excess production to larger markets in other towns and cities, improving the bargaining power of the farmers, generating community's fund through farmers' Co-ops, so individual members can have access to fund when needed.

The paper is explained in the video titled "Subsistence community in West Timor".

Acknowledgements This paper is based on part of the results of an approved Ph.D. research and supported fully by Griffith University. The authors would like to acknowledge local academics and practitioners in Kupang City, IRGSC (Institute of Resource, Governance and Social Change), FAN (Forum Academia NTT), and Perkumpulan PIKUL, for providing their valuable insights about rural development in NTT and their assistance on the procedure for conducting fieldwork in the region. Also, the authors would like to express their gratitude to the local hosts, local guide and the AtoinMeto communities in the three research sites for their participation and hospitality during the data collection period in West Timor.

References

Abele, S., & Frohberg, K. (Eds.). (2003). *Subsistence agriculture in Central and Eastern Europe: how to break the vicious cycle? Introductory note, IAMO studies on the Agricultural and Food Sector in Central and Eastern Europe* (Vol. 22, pp. I–VI). Halle: Institute of Agricultural Development in Central and Eastern Europe.

Berkes, F. (2009). Indigenous way of knowing and the study of environmental change. *Journal of the Royal Society of New Zealand, 39*, 151–156.

BNPB. (2009). *Peta Kejadian Bencana di Propinsi Nusa Tenggara Timur Tahun 2008 (East Nusa Tenggara 2008 Natural Disaster Incidence Map)*. Jakarta: Badan Nasional Penanggulangan Bencana (National Bureau for Disaster Management of the Republic of Indonesia).

Boli Sura, Y., Fanggidae, V., Medah, A., Nggili, S., & Simanjuntak, L. (2010). Keamanan pangan, pembangunan dan pemiskinan di Nusa Tenggara Timur: Studi kasus Desa Oelnasi, Kabupaten Kupang (Food security, development and impoverishment in East Nusa Tenggara: A case study of Oelnasi Village, Kupang Regency). *Journal of NTT Studies, 2*(1), 8–30 Retrieved from http://ntt-academia.org/nttstudies/Bolidkk2010.

BPPK. (2008). *Riset Kesehatan Dasar 2008 (Indonesia Basic Health Research 2008)*. Jakarta: Badan Penelitian dan Pengembangan Kesehatan (National Institute of Health Research and Development), Kementerian Kesehatan Republik Indonesia (Ministri of Health of the Republic of Indonesia).

BPPK. (2013). *Riset Kesehatan Dasar 2013 (Indonesia Basic Health Research 2013)*. Jakarta: Badan Penelitian dan Pengembangan Kesehatan (National Institute of Health Research and Development), Kementerian Kesehatan Republik Indonesia (Ministri of Health of the Republic of Indonesia).

BPS Indonesia. (2016). *Badan Pusat Statistik* (Central Bureau of Statistics Indonesia). Retrieved February 20, 2016, from https://www.bps.go.id/linkTableDinamis/view/id/1219

BPS NTT. (2014). *NTT in figures year 2013*. Kupang City: Badan Pusat Statistik NTT (NTT Provincial Central Bureau of Statistics), Nusa Tenggara Timur (NTT).

BPS NTT. (2015). *Produk Domestik Regional Bruto Menurut Lapangan Usaha, 2004–2013 (Gross Regional Domestic Product by Industrial Origin, 2004–2013)*. Kupang: Badan Pusat Statistik NTT (NTT Provincial Central Bureau of Statistics) Retrieved March 20, 2015, from http://ntt.bps.go.id/linkTableDinamis/view/id/41.

BPS NTT. (2017). *Jumlah Penduduk per Kabupaten/Kota Tahun 1980, 1990, 2000, 2010, 2011–2016 (Number of Population by Regency/City 1980, 1990, 2000, 2010, 2011–2016)*. Kupang: Badan Pusat Statistik NTT (NTT Provincial Central Bureau of Statistics) Retrieved August 17, 2017, from http://ntt.bps.go.id/linkTableDinamis/view/id/360.

BWS NTII. (2016). *Balai Wilayah Sungai Nusa Tenggara II* (Office of River Region Nusa Tenggara II). Retrieved from http://bwsnt2.org/web/

Chambers, R., & Conway, G. (1992). *Sustainable rural livelihoods: Practical concepts for the 21st century'. Discussion paper 296*. Institute of Development Studies, University of Sussex: Brighton.

Comenetz, J., & Caviedes, C. (2002). Climate variability, political crises, and historical population displacements in Ethiopia. *Global Environmental Change Part B: Environmental Hazards, 4*, 113–127. https://doi.org/10.1016/j.hazards.2003.08.001.

Conklin, H. C. (1961). The study of shifting cultivation. *Current Anthropology, 2*(1), 27–61.

Davidova, S., Fredriksson, L., Gorton, M., Mishev, P., & Petrovici, D. (2012). Subsistence farming, incomes, and agricultural livelihoods in the new member states of the European Union. *Environment and Planning C: Government and Policy, 30*, 209–227. https://doi.org/10.1068/c1195r.

Department for International Development (DFID). (1999). *Sustainable livelihoods guidance sheets* (p. 10). London: DFID Retrieved November 3, 2012, from http://files.ennonline.net/attachments/871/dfid-sustainablelivelihoods-guidance-sheet-section1.pdf.

DPU NTT. (2014). *Rencana strategis 2014–2018 (Strategic Plan 2014–2108)*. Kupang City: Dinas Pekerjaan Umum Provinsi NTT (Department of Public Works NTT Province), Nusa Tenggara Timur (NTT).

Duncan, C. R. (2004). From development to empowerment: Changing Indonesian government policies towards Indigenous minorities. In C. R. Duncan (Ed.), *Civilizing the margins: Southeast Asian Government Policies for the Development of Minorities* (pp. 86–111). Ithaca, NY: Cornell University Press.

Fanggidae, S., Ratumakin, P. A. (2014). *Nelayan dan petani membaca cuaca dan musim: Sebuah kajian tentang pengetahuan nelayan dan petani atas informasi cuaca dan musim (Fisherman and farmer in reading the climate and season: An analysis of the knowledge of fisherman and farmer on the climatic and seasonal information)*. Jakarta: Indonesia Climate Change Trust Fund and Perkumpulan, PIKUL.

FAO. (2009). *Rome declaration of the world food summit on food security, world summit on food security, 16–18 November 2009*. Rome: Food and Agriculture Organization of the United Nations.

FAO. (2010). *FAO policy on indigenous and tribal peoples*. Rome: FAO Retrieved October 15, 2016, from http://www.fao.org/docrep/013/i1857e/i1857e00.pdf.

Fisk, E. K. (1975). The neglect of traditional food production in pacific countries. *Australian Outlook, 29*(2), 149–160.

Flick, U. (2009). *An introduction to qualitative research* (4th ed.). London: SAGE.

Fox, J. J. (1999). Precedence in practice among the Atoni Pah Meto of Timor. In L. V. Aragon & S. D. Russell (Eds.), *Structuralism's transformations: order and revisions in Indonesia and Malaysia* (pp. 1–36). Tucson, AZ: Center for Southeast Asian Studies, Arizona State University.

Fraser, E. D. G. (2003). Social vulnerability and ecological fragility: Building bridges between social and natural sciences using the Irish Potato Famine as a case study. *Conservation Ecology, 7*(2), Article 9 Retrieved from http://www.consecol.org/vol7/iss2/art9.

Glaser, B. G. (1978). *Theoretical sensitivity*. Mill Valley, CA: University of California Press.

Goldsmith, E. (1998). *The way: An ecological world-view (revised and enlarged edition)*. Athens, GA: University of Georgia Press ISBN: 0820320307.

Hahn, M. B., Riederer, A. M., & Foster, S. O. (2009). The Livelihood Vulnerability Index: A pragmatic approach to assessing risks from climate variability and change— A case study in Mozambique. *Global Environmental Change, 19*(1), 74–88.

Kemenkes. (2011). *Panduan Penyelenggaraan Pemberian Makanan Tambahan Pemulihan Bagi Balita Gizi Kurang (Bantuan Operasional) (Guidebook for Implementation of Supplementary Food – Recovery Programme for Children Malnutrition (Operational Assistance))*. Jakarta: Ministry of Health of the Republic of Indonesia Retrieved October 20, 2016, from http://gizi.depkes.go.id/wp-content/uploads/2011/11/Panduan-PMT-BOK.pdf.

Kuokkanen, R. (2011). Indigenous economies, theories of subsistence, and women: Exploring the social economy model for indigenous governance. *American Indian Quarterly, 35*(2), 215–241.

Larsen, A. F. (2009). Semi-subsistence producers and biosecurity in the Slovenian Alps. *Sociologia Ruralis, 49*, 330–343.

Law No. 1/1957. Undang – Undang Nomor 1 Tahun 1957. Tentang Pokok – Pokok Pemerintahan Daerah (Basic Law on Local Administration).

Law No. 18/1965. Undang – Undang Nomor 18 Tahun 1965, Tentang Pokok – Pokok Pemerintahan Daerah (Basic Law on Local Administration).

Leonard, S., Parson, M., Olawsky, K., & Kofod, F. (2013). The role of culture and traditional knowledge in climate change adaptation: Insights from East Kimberley, Australia. *Global Environmental Change, 23*, 623–632.

Manner, H. I. (1981). Ecological succession in new and old swiddens of Montane Papua New Guinea. *Human Ecology, 9*, 359–360.

Minister of Health. (2015). Regulation of the Minister of Health No. 11/2015, Peraturan Menteri Kesehatan No. 11 Tahun 2015, Tentang Petunjuk Teknis Bantuan Operasional Kesehatan (on Technical Guidance of the Health Operational Assistance).

Muslimatun, S., & Fanggidae, S. (2009). *A brief review on the persistent of food insecurity and malnutrition problems in East Nusa Tenggara Province, Indonesia, Working Paper 12*. Kupang: Institute of Indonesia Tenggara Timur Studies (NTT Studies), IITTS Publications (Open Sources).

Ompusunggu, S., Syachroni, S. U., Yulianto, A., & Kulla, R. K. (2013). *Riset Kesehatan Dasar Dalam Angka Provinsi Nusa Tenggara Timur 2013 (NTT Province Basic Health Research in Figures 2013)*. Jakarta: Badan Penelitian dan Pengembangan Kesehatan (National Institute of Health Research and Development), Kementerian Kesehatan Republik Indonesia (Ministri of Health of the Republic of Indonesia) ISBN: 978-602-235-569-4.

Pambudi, R. S. (2015). *Evaluasi Program Pemberian Makanan Tambahan Pemulihan (PMT P) untuk Balita Gizi Buruk di Kabupaten Gunungkidul (Evaluation of Supplementary Food – Recovery Programme for Children Malnutrition in Gunungkidul)*, unpublished dissertation, Universitas Gadjah Mada, Yogyakarta. Retrieved October 26, 2016, from http://etd.repository.ugm.ac.id/index.php?act=view&buku_id=78141&mod=penelitian_detail&sub=PenelitianDetail&typ=html.

Pamsimas (2017) *Penyediaan Air Minum dan Sanitasi Berbasis Masyarakat* (Community-Based Water Supply and Sanitation Program).. Retrieved July 10, 2016, from http://www.pamsimas.org

Petheram, L., Zander, K. K., Campbell, B. M., High, C., & Stacey, N. (2010). Strange changes: Indigenous perspectives on climate change and adaptation in NE Arnhem Land (Australia). *Global Environmental Change, 20*(4), 681–692.

Pusdatin. (2008). *Peta Kesehatan Indonesia 2007 (Indonesia Health Map 2007)*. Jakarta: Pusat Data dan Informasi Departemen Kesehatan (Centre for Data and Information), Ministry of Health of the Republic of Indonesia.

Pusdatin. (2014). *Peta Kesehatan Indonesia 2013 (Indonesia Health Map 2013)*. Jakarta: Pusat Data dan Informasi Departemen Kesehatan (Centre for Data and Information), Ministry of Health of the Republic of Indonesia.

Radio Republik Indonesia. (2015). *Program PMT bagi balita gizi buruk di NTT terhambat prosedur dana Bansos* (PMT Programme for children malnutrition in NTT hampered by social grants procedures). Retrieved September 23, 2016, from http://www.rri.co.id/post/berita/177606/daerah/program_pmt_bagi_balita_gizi_burub_di_ntt_terhambat_prosedur_dana_bansos.html

Ratumakin, P. A., Kuswardono, P. T., Heo, M. J., & Weo, Y. U. P. (2016). *Pengetahuan Lokal Dalam Keberlanjutan Pengelolaan air (local knowledge in sustainability water management)*. Kupang: Perkumpulan PIKUL ISBN: 978-602-74097-0-5.

Riedlinger, D. (1999). Climate change and the Inuvialuit of Banks Island, NWT: Using traditional environmental knowledge to complement Western science. *InfoNorth (Arctic), 52*(4), 430–432.

Seavoy, R. E. (2000). *Subsistence and economic development*. Westport, CT: Greenwood Publishing Group ISBN 9780275967819.

Strauss, A. L., & Corbin, J. M. (1990). *Basics of qualitative research: Grounded theory procedures and techniques*. Newbury Park, CA: SAGE Publications ISBN: 0-8039-3250-2.

Tempo.co. (2014) *NTT farmers threathened by drought, famine.* Retrieved September 10, 2016, from http://en.tempo.co/read/news/2014/03/08/055560495/NTT-Farmers-Threatened-by-Draught-Famine

Tempo.co. (2016) *Kekeringan, NTT Terancam Kelaparan (Droughts, Famine Threatens East Nusa Tenggara (NTT))*. Retrieved September 11, 2016, from http://nasional.tempo.co/read/news/2016/01/11/058734836/kekeringan-ntt-terancam-kelaparan

Tjoe, Y. (2016a). Measuring the livelihood vulnerability index of a dry region in Indonesia. *World Journal of Science, Technology and Sustainable Development, 13*(4), 250–274. https://doi.org/10.1108/WJSTSD-01-2016-0013.

Tjoe, Y (2016b) *Sustaining livelihoods: An analysis of dryland communities in West Timor, Indonesia*, Ph.D. thesis, Griffith University, Australia.

Tyas Arum Sari Dewi B, & Rahaju, T. (2015). *Evaluasi Program Pemberian Makanan Tambahan Bagi Balita di Posyandu Melati V RW V di Kelurahan Lontar, Kecamatan Sambikerep, Kota Surabaya (Evaluation of Supplementary Food Programme for Children Under Five in the Integrated Service Post (Posyandu) Melati V RW V in Lontar Village, Sambikerep District, Surabaya City)*, unpublished dissertation, Universitas Negeri Surabaya, Surabaya. Retrieved October 26, 2016, from http://id.portalgaruda.org/index.php?page=19&ipp=10&ref=browse&mod=viewjournal&journal=4761.

UNDP. (2017). *Strategic planning and action to strengthen climate resilience of communities in Nusa Tenggara Timor Province (SPARC), United Nations Development Program*. Retrieved July 10, 2016, from http://adaptation-undp.org/projects/sccf-indonesia.

UNPFII. (2007). *United Nations permanent forum on indigenous issues special theme: Territories, lands and natural resources, sixth session, 14–25 May 2007, United Nations Headquarter.* New York, NY: United Nations.

Wharton, C. R. (1969). Subsistence agriculture: Concepts and scope. In C. R. Wharton (Ed.), *Subsistence agriculture and economic development* (pp. 12–20). New Brunswick, NJ: Transaction Publishers.

Wolf, J., & Mosser, S. (2011). Individual understandings, perceptions, and engagement with climate change: Insights from in-depth studies across the worlds. *WIREs Climate Change, 2*(4), 547–569.

World Bank. (2003). *Implementation of operational directive 4.20 on indigenous peoples: an independent desk review*. Washington, DC: World Bank.

World Bank. (2015a). *Indigenous Latin America in the twenty-first century*. Washington DC: World Bank Group Retrieved June 8, 2016, from https://openknowledge.worldbank.org/bitstream/handle/10986/23751/Indigenous0Lat0y000the0first0decade.pdf.

World Bank. (2015b). *PAMSIMAS – Responding to the water and sanitation challenges in rural Indonesia*. Washington DC: World Bank Group Retrieved August 19, 2016, from http://documents.worldbank.org/curated/en/938961468195535278/PAMSIMAS-Responding-to-the-water-and-sanitation-challenges-in-Rural-Indonesia.

World Food Programme Indonesia. (2016). *Indonesia food security monitoring bulleting, special focus: Impact of El Nino* (Vol. 2). Jakarta: World Food Programme Indonesia and Government of Indonesia Retrieved September 23, 2016, from http://www.wfp.org/content/indonesia-food-security-monitoring-2015.

Chapter 20
Enhancing Food Security in Subarctic Canada in the Context of Climate Change: The Harmonization of Indigenous Harvesting Pursuits and Agroforestry Activities to Form a Sustainable Import-Substitution Strategy

Leonard J. S. Tsuji, Meaghan Wilton, Nicole F. Spiegelaar, Maren Oelbermann, Christine D. Barbeau, Andrew Solomon, Christopher J. D. Tsuji, Eric N. Liberda, Richard Meldrum, and Jim D. Karagatzides

Abstract Lesser snow geese (*Chen caerulescens caerulescens*) that breed in subarctic and arctic Canada are overabundant due to a number of anthropogenic influences, including climate change. In 2011, First Nations Cree from subarctic Ontario, Canada, participated in a regional spring harvest of overabundant snow geese—these geese have caused desertification of the environment—using only steel shotshell, as part of a food security initiative. Benefits included the procurement and sharing of an uncontaminated, nutritious source of game birds, while also helping to protect the environment for future generations. However, in 2012, the spring was unusually warm which disrupted the timing and routes of waterfowl migration. Thus, we adapted our intervention to be locally focused, allowing for quick response

---

L. J. S. Tsuji (✉) · M. Wilton · N. F. Spiegelaar
Department of Physical and Environmental Sciences, University of Toronto Scarborough, Toronto, ON, Canada
e-mail: leonard.tsuji@utoronto.ca

M. Oelbermann · C. D. Barbeau
School of Environment, Resource, and Sustainability, University of Waterloo, Waterloo, ON, Canada

A. Solomon
Fort Albany First Nation, Fort Albany, ON, Canada

C. J. D. Tsuji
Department of Anthropology, McMaster University, Hamilton, ON, Canada

E. N. Liberda · R. Meldrum
School of Occupational and Public Health, Ryerson University, Toronto, ON, Canada

J. D. Karagatzides
Engineering & Environmental Technologies, Georgian College, Barrie, ON, Canada

© Springer Nature Switzerland AG 2019
A. Sarkar et al. (eds.), *Sustainable Solutions for Food Security*,
https://doi.org/10.1007/978-3-319-77878-5_20

to changes in timing and migration routes of waterfowl due to climate change. Although global warming can present a challenge to food security, the warming climate also offers opportunities for the growth of vegetables and fruits in subarctic Canada, previously beyond their distributional range. In 2012, we initiated a pilot, agroforestry intervention in subarctic Ontario, to foster adaptive capacity and resilience. Agroforestry is a stewardship practice that combines woody perennials with crops in beneficial arrangements and can increase food security. At a site that met agricultural-soil-contamination guidelines, an agroforestry plot was established between rows of willows, while a non-treed site acted as a control. Findings indicate that the agroforestry site is more productive with respect to potato and bean yields. In the present initiative, the subarctic food system is now viewed as a whole with our main objective being the harmonization of traditional harvesting and agroforestry activities into one locally, sustainable food system by utilizing by-products of the game bird harvest to nutrient enrich the soil. Adaptation to climate change and import substitution of fruits and vegetables are the end goals.

Keywords Climate change · Challenges and opportunities · Indigenous Canadians · Subarctic · Food security · Agroforestry · Waterfowl harvesting · Import substitution

20.1 Introduction

Food security can be defined as a state where people always have access to safe, nutritionally adequate, and culturally appropriate food required for a healthy life (FAO 1996). By contrast, food insecurity is said to exist when the above conditions are not fully realized. Food insecurity is disproportionately prevalent in northern (i.e., subarctic and arctic) Canada (CA) compared to the rest of the country (Council of Canadian Academics 2014); the unique food security challenges faced by Indigenous (First Nations, Inuit, and Metis) people have often been described in the academic literature (Council of Canadian Academics 2014)—and in particular—the recurring bouts of food insecurity in the First Nation Cree of subarctic Ontario (ON) have been well documented in the historical record (Tsuji 2017). Taking into account that food security has been identified as a determinant of health (McIntyre and Tarasuk 2002), food security interventions, especially in northern CA, have the potential to decrease the well-known health disparities gap between Indigenous and non-Indigenous people in CA.

Major barriers to the consumption of healthy foods that have consistently been identified through semi-directed interviews with adult First Nations Cree of subarctic ON are the expenses associated with traditional (subsistence) harvesting activities (Skinner et al. 2013), and the availability, quality, and expense of fruits and vegetables (Gates et al. 2012). Indeed, results from a school-based, nutrition survey in subarctic ON indicated that 34% of grade 6–12 students would eat more traditional (wild game and fish) food if "it was more available at home" (Hlimi et al. 2012). Further, it has been reported that even though school-based nutritional educational programs in subarctic ON Cree First Nations (Gates et al. 2011) can significantly increase the vegetables and fruit tried and liked with respect to grades 6–8 schoolchildren, significant changes in actual diets remains elusive. In other words, Cree schoolchildren would eat more fruits and vegetables if produce were available, but access to fruits and vegetables is limited by the availability at the community store and/or the expense of the produce (i.e., parents may not be able to afford the fruits and vegetables, even if available). Government of Canada northern initiatives have been mainly based on an import-subsidization strategies, and results have not been encouraging (Stanton 2011).

Other types of food security interventions in northern CA have been more conventional (e.g., snack and breakfast programs, and greenhouses). Greenhouses come in many forms, and there is no one best solution for northern communities, with greenhouse initiatives being scalable either up or down (Exner-Pirot 2012). Two main noncommercial options have been described for northern greenhouses (Exner-Pirot 2012): (1) Backyard/home/individual greenhouses. These greenhouses are the simplest in construction and the most inexpensive way to lengthen the short growing season in the north (Exner-Pirot 2012), other than using black tarp and row covers (Chesney 2015). These types of individual greenhouses typically use no power—and can be simple in structure consisting of only a wood box with a clear or translucent lid (e.g., plastic tarp covering) on hinges (i.e., a "cold frame"; Jabbour 2011)—to larger more complex structures with tubular frames (e.g., geodesic domes) covered in commercially fabricated fabrics (e.g., UV-protected woven plastic; HarvestRight 2017). Similar in function to the individual greenhouses, high tunnels (also known as polytunnels) consist of tunnel-shaped frames covered with polyethylene (plastic); they require no power and are becoming more popular in the Canadian north (Exner-Pirot 2012). (2) Community greenhouses. Scaling up to community-run greenhouses requires substantial capital costs and long-term funding for operating (e.g., power) and maintenance costs, even with volunteer labor (Exner-Pirot 2012). Community greenhouses can be relatively simple, such as the large polytunnel found in Wabowden, Manitoba, CA (Exner-Pirot 2012), or more complex, such as the Inuvik Community Greenhouse (Iunvik, Northwest Territories, CA) which is a converted hockey arena, where the roof was removed and replaced with polycarbonate plastic (Council of Canadian Academics 2014; Lees and Redman 2009). Thus, the use of greenhouses of all types in subarctic and arctic CA, offer the people the opportunity to improve their food security, by actively increasing the length of the growing season and allowing the production of locally grown fruits and vegetables.

Recently, interventions have been more innovative in the context of climate change. In late 2010, the regional intervention entitled "Sharing-the-Harvest" was initiated with participants coming from all the western James Bay Cree First Nations. This initiative addressed many of the identified barriers to subsistence hunting by providing transportation, equipment, and consumables to facilitate the harvesting of lesser snow geese (*Chen caerulescens caerulescens*) at Cape Henrietta Maria, on the western coast of Hudson Bay, ON (Fig. 20.1). The populations of snow geese breeding in subarctic and arctic Canada have increased geometrically

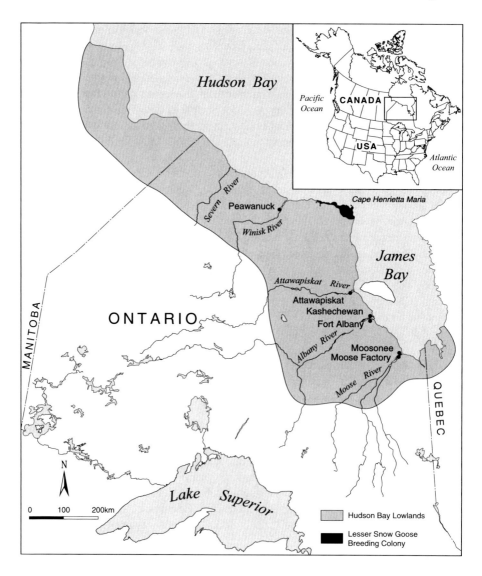

Fig. 20.1 Geographic location of Fort Albany First Nation, Hudson Bay Lowlands, and breeding colony for lesser snow geese in the Hudson Bay Lowlands

since the late 1960s—due to anthropogenic influences, including climate change—reaching an estimated population of seven million birds (Leafloor et al. 2012). The detrimental effects (e.g., desertification of the environment) associated with lesser snow geese foraging activities are well documented (Abraham et al. 2005; Jefferies et al. 2006). The harvesters used only steel shotshell, so as not to contaminate the snow goose meat with lead pellets and/or fragments during harvesting. In this way, the Cree people procured and shared a relatively uncontaminated source of game bird meat (Tsuji 2017). Indeed, organohalogen concentrations in spring-harvested snow geese have been previously shown to pose little threat with respect to consumption for the subarctic Cree (Tsuji et al. 2007), and this is also true for polycyclic aromatic hydrocarbons (Tsuji 2017), as well as for a suite of toxic metals (Braune et al. 1999). In addition, the Cree harvesters through this food security intervention—or mitigation strategy—have also assisted in the North American-wide intervention to decrease the number of snow geese, to allow the northern ecosystem time to recover (Abraham et al. 2005; Jefferies et al. 2006). Thus, the Cree have helped to address their food insecurity issue, while also protecting the environment for future generations (Tsuji 2017).

Another innovative food security intervention recognized the changing climate as an opportunity to grow fruits and vegetables under ambient conditions—that have been historically stunted in growth or outside the distributional range of the subarctic—and increased the availability and quality of vegetables through agroforestry stewardship practices. Agroforestry is a more sustainable agricultural strategy that has been utilized in impoverished areas of the world to combat food insecurity (Huai and Hamilton 2009; Vogl et al. 2004). By definition, agroforestry is a land-use system that combines woody perennials with crops in arrangements—spatial and/or temporal—that enhance beneficial biological interactions (Gordon and Newman 1997; Young 1997). Being more biological diverse than conventional agriculture, agroforestry stewardship has the potential to act as a more reliable local food system (Ninez 1987), in a changing climate, while also providing other ecosystems services, such as habitat for species, erosion and flood control, and carbon sequestration by woody perennials (i.e., climate change mitigation; Spiegelaar et al. 2013).

Although the Sharing-the-Harvest and the agroforestry food security interventions were successful in combatting food insecurity, by responding to challenges and opportunities related to climate change—both interventions have been somewhat reductionist in their vision, as the subarctic food system was not considered or addressed as a whole—but rather as separate pieces (i.e., Cree traditional foods, and fruits and vegetables) of a complex food security puzzle. Potential complementary interactions were not explored. Here, we describe how the above-stated shortcoming is being addressed through the harmonization of traditional Cree harvesting practices with agroforestry gardening activities, in an effort to form a more sustainable import-substitution strategy for enhancing food security in subarctic ON, in the context of climate change.

20.2 Study Area

20.2.1 Cultural Context

The remote western James Bay region of subarctic ON, CA, is populated by approximately 10,000 Cree who inhabit four First Nations and one town (Fig. 20.1). Historically, the spring harvest (typically April to May) of waterfowl, in particular the Canada goose, *Branta canadensis interior*, was a celebratory event for two reasons: the survival of the family unit through the challenging winter months

Table 20.1 Physical description of Fort Albany, Ontario, Canada (52°21′N, 81°69′W)

Physical characteristic	Description
Elevation	11 m above sea level
Mean annual ambient temperature (°C[a])	−2[b]
Mean (± sd) 2014 growing season temperature (°C)	13.5 ± 5.16 (May 24–October 6, 2014)
Maximum temperature (°C)	33.3 (occurred on June 29)
Minimum temperature (°C)	−2.1 (occurred on September 30)
First frost (°C)	−1.9 (occurred on September 17)
Annual rainfall (mm)	700–800 mm[a]
Growing season rainfall (mm)	323.1 mm
2014 growing season duration	94 days (June 25–September 25)
Soil	
Order	Organic Mesisols
Type	Silt loam
Moisture (%)	37–50
pH	7.7–7.8
Bulk density (g/m^3)	0.7
Organic matter (S value)	14.1–21.6
Carbon–nitrogen	14.5–15.0
Total N (µg/g)	5767–74,000
Total P (µg/g)	8.2–15.6
Total K (µg/g)	32.7–40.1

[a]Source: MSSC (1996)
[b]Monthly average for 2013–2016 was −0.5 °C and −2.1 °C for June 2013–June 2014

(Thompson and Hutchison 1989); and an important nutritional change from a very limited winter diet. Of all subsistence activities of the subarctic Cree, waterfowl harvesting in the spring and fall (September to October) are of primary importance (Thompson and Hutchison 1989). Geese are the more valued species with dabbling ducks harvested opportunistically (Thompson and Hutchison 1989). Fort Albany First Nation was selected as the pilot community for the present study for two reasons: engagement of the community in the previous regional harvesting initiative (Tsuji 2017); and the area's proven capacity to support the agroforestry initiative (Barbeau et al. 2015; see Table 20.1 for a physical description of Fort Albany).

20.2.2 Climate Change Context

Climate-change projections from the Fifth Assessment Report of the Intergovernmental Panel on Climate Change (IPCC) predicts the global-average, surface-air temperature will increase by 4.8 °C relative to 1986–2005, by 2100 (IPCC 2014). Moreover, disproportionate increases in surface-air temperatures at high latitudes are predicted worldwide (IPCC 2014), and specifically, in arctic CA, a projected rise in the range of 3.5–12.5 °C has been suggested for the 2071-2100 period (Furgal and Prowse 2008).

Further, there is increasing evidence that subarctic and arctic CA have undergone major climatic change (Anisimov et al. 2007; Hori et al. 2012; Tam et al. 2011). Indeed, the global-average, surface-air temperature has risen by 0.6 ± 0.2 °C over the past century—while the subarctic and arctic have experienced an increase of up to 5 °C (Anisimov et al. 2007; IPCC 2001a, b, c)—the most rapid rates of increasing average surface-air temperatures worldwide (Anisimov et al. 2007). Thus, the Hudson and James Bay regions of subarctic and arctic CA have been disproportionately affected by such rising temperatures (IPCC 2001a, b). For example, the duration of sea-ice cover in Hudson and James Bay has recently been decreasing due to earlier break-up and later freeze-up (Gagnon and Gough 2005; Gough et al. 2004), which is a result of climate change (Gough and Houser 2005). However, the trends in river-ice break-up dates in the western James Bay region are not as consistent, as there are many confounding variables, although the average, surface-air temperatures in spring and winter have increased in the region (Ho et al. 2005). In addition, Hori (2010) has employed climate projection models (using an ensemble approach) to reveal that surface-air temperatures will continue to increase in the subarctic James Bay region based on several climate models (and emission scenarios). Increasing air temperatures were predicted for all seasons (time periods: 2011–2100), and the results for seasonal total precipitation were variable and dependent on the season, time period, and emission scenario (Hori 2010). Overall, the subarctic James Bay region and other high latitude regions of the world are becoming more conducive to growing fruits and vegetables under ambient conditions—from an increasing surface-air temperature and growing-season length perspective—but other factors may be limiting (e.g., soil moisture; Chen et al. 2016).

20.3 Sharing-the-Harvest

20.3.1 Traditional Food Intake

Although subsistence harvesting for James Bay Cree remains an integral part of their culture—and provides many benefits such as nutrient-rich wild game, a relatively inexpensive source of protein (store-bought food is expensive), and promotes social and community cohesiveness—access to and availability of traditional foods has declined in many of the communities for a variety of reasons (Gates et al. 2016). Barriers to subsistence harvesting activities that have been consistently identified by the Cree themselves include the cost of transportation, hunting equipment, and consumables. It should be emphasized that the most culturally appropriate interventions must be informed by the people themselves (Skinner et al. 2013). Example interventions that were suggested by the Cree include:

[There should be] scheduled hunting trips where gas and supplies are paid [by the Band, the locally-elected First Nation government] and traditional food/meat caught given to lower income families...Lots of great hunters and trappers in this community – utilize them (Skinner et al. 2013:8)

Get Band Council to get some hunters to go hunting for spring and fall. Supply the hunters with guns, shells, gas for their trip. Whatever game [meat] is killed, it should be shared within the community (Skinner et al. 2013:8).

However, there is anxiety among people of James Bay (Hlimi et al. 2012) and throughout CA (van Oostdam et al. 1999) that contaminants are impairing human health and interfering with traditional cultural practices.

20.3.2 Potential Lead Contamination

Lead is a well characterized nonessential, toxic metal (Tsuji et al. 2008). In CA, even though lead shotshell was banned for most migratory bird hunting in 1999, lead shotshell can still legally be used for the hunting of upland game (birds and small mammals), because there are no federal laws restricting its use for this purpose (Tsuji 2017). Consequently, lead shotshell is still available and being used for waterfowl harvesting in remote areas. The food security issue relates to the consumption of wild game harvested with lead shotshell. In Cree, the radiographic presence of lead pellets/fragments in the digestive system is common (Tsuji and Nieboer 1997)—because approximately 10% of game birds harvested with lead ammunition becomes contaminated with shot-in lead pellets/fragments—exceeding the Canadian consumption guideline for lead in protein (0.5 $\mu g/g$ wet weight) with levels up to 19,900 $\mu g/g$ wet weight (Tsuji et al. 1999). Of importance, lead pellets/fragments located in the digestive system are not inert and become a source of lead exposure (Gustavsson and Gerhardson 2005). Thus, elevated tissue lead levels have been reported for people who consume wild game hunted with lead shotshell, and stable lead isotopes have been used to definitively identify lead shotshell as a major source of lead for subsistence harvesting groups (Tsuji et al. 2008). This is why only steel shotshell was used in our Sharing-the-Harvest initiative.

20.3.3 Lesser Snow Goose Management

Since the 1960s, changes in the agricultural landscape and refuge provisions in North America, as well as global warming, have led to a population explosion of lesser snow geese nesting in the James and Hudson Bay area (Abraham et al. 2005; Jefferies et al. 2006; Leafloor et al. 2012). By the 1990s, it became evident that the overpopulation of snow geese had to be managed, because their "foraging destabilizes the thin arctic soil...plants die, ponds dry up, and the parched earth cracks" (Rockwell et al. 1996:21). A radical departure from regular hunting practices was needed to protect the environment and save the snow geese that breed in northern CA; thus, a joint effort by the Canadian and American governments led to a special spring hunt and increased bird limits for non-Indigenous people (Abraham et al. 2005). Thus, the Cree have helped in part with the effort to preserve the snow goose population, as well as protect the environment (Tsuji 2017).

20.3.4 The Intervention

During the first year (i.e., 2011), ~3700 snow geese were harvested regionally by the Cree and shared, because the snow goose migration routes and timing of migration followed historical patterns. However, during the second year (i.e., 2012) only ~550 snow geese were harvested, due to an unusually warm spring and changes in the timing and migration routes. The snow geese did not behave as in the past. With input from the Cree harvesters, an adjustment was made to our initiative, whereby the regional focus was replaced by a more local one. Historically, the sit-and-wait at a specific place strategy worked well for waterfowl harvesting, but with climate change, the timing and routes of migration change quickly within a year (and/or yearly). In addition, the high cost of air transportation in the subarctic made the regional strategy unsustainable. Switching to a local focus allowed for increased surveillance of goose behavior (i.e., the use of the most up-to-date Indigenous knowledge) and better use of resources. Thinking locally, harvesters also suggested that molt-migrant giant Canada geese (*Branta canadensis maxima*) be harvested. Historically, this subspecies was not harvested by the Cree as they were almost extinct before they were reintroduced; these geese arrive in June, and like the snow geese due to anthropogenic reasons, are overabundant and need to be harvested. Local harvests of ~500 giant Canada geese per year from this initiative have been recorded, which is substantial for food security, because giant Canada geese weigh 6.8–9.1 kg (Tsuji 2017).

20.4 Import Substitution

20.4.1 Food Localization

The industrialized, global-food system is not sustainable, especially in remote northern Canadian communities who depend on air transportation as their main source of goods and foodstuff. One alternative is for these communities to build capacity and resilience by establishing locally based food security. Since the 1980s, the concept of a more local and ecologically sustainable food system as an alternative to a globalized food system has gained popularity (Polack et al. 2008). Import substitution can be defined as a process where food security activities lead to local production of food to replace food that has been previously imported, and can result in increased food freshness and variety (Bellows and Hamm 2001). Further, the role of food extends beyond nutritional acquisition, affecting environmental and spiritual wellbeing (Feenstra 1997).

20.4.2 Agroforestry

It has been suggested that climate change will affect the distribution and productivity of crops in North America, and vulnerable communities will experience disproportionately more food security challenges (Catford 2008). Current food localization strategies in First Nations are exploring community and home/yard gardens to achieve simultaneous benefits of food security and community cohesiveness; our intervention extends First Nations' community and home/yard garden initiatives in temperate CA to subarctic CA utilizing agroforestry stewardship practices. Our gardening initiative was also community-informed:

> You have to start first. You have to teach people by having something in common, how to care for the community garden and then … it will spread. You're just planting a seed … It's also a good way, I think, for developing a community, a cohesive community, a sociable community, from talking about something in common: our garden. (Spiegelaar and Tsuji 2013:12)

> Start a garden. You could grow things…We used to have a garden and we grew potatoes. We used the potatoes at the goose camp. We know that you can grow things here. (Skinner et al. 2013:8)

Our subarctic agroforestry intervention in Fort Albany First Nation.

I. Annuals. Based on community input, three potential agroforestry sites were chosen for biophysical assessments that included soil nutrient (Spiegelaar et al. 2013) and soil contamination analyses (Reyes et al. 2015). Of the three sites, only two sites met agricultural-soil-contamination guidelines. Two sites were used for gardening purposes (Table 20.1): one became a tree-lined (maturing *Salix* spp. (willow) trees) agroforestry community garden (Fig. 20.2); and the other a non-treed site (a control with no trees or shrubs growing within a 58-m radius) (Fig. 20.2). Tree-based "backyard" gardens were also established in the community (Fig. 20.2), as well as non-treed home/yard gardens (Fig. 20.2). All home/yard garden soils and any other soils used to augment garden soils were tested for organochlorines and metal/metalloids; only soils that were analyzed and found to meet contamination soil guidelines were used in our intervention. The agroforestry community gardens started with potatoes (*Solanum tuberosum* L.) and bush beans (*Phaseolus vulgaris* L.). Potatoes were commonly eaten in Fort Albany and very expensive to buy at the local store; and beans had high nutritional value, could be harvested at least twice within a growing season, and fixed nitrogen (N). Results from the agroforestry community garden showed that potatoes and bush beans could be grown under ambient conditions in the subarctic with yields similar to those for conventional agricultural

Fig. 20.2 Representative photographs of non-treed community garden (**a**), agroforestry (treed) community garden (**b**), and home/yard (**c, d**) gardens

endeavors in warmer climates (Barbeau et al. 2015). After successes with these crops–carrots (Umbelliferae) and kale (*Brassica oleracea*) were grown successfully—contributing important nutrients (e.g., beta-carotene, iron) to the Cree diet. In addition, a variety of vegetables have been grown under ambient condition in the home/yard gardens (Table 20.2). Also, cover crops were added to the agroforestry community garden plots to help maintain the quality of the soil by suppressing weeds, improving nutrient availability, adding organic matter to the soil, and protecting the soil from erosion and compaction. *II. Perennials*. Since edible perennials grow naturally within the western James Bay region such as raspberries (*Rubus* spp.), gooseberries (*Ribes* spp.), and cranberries (*Vaccinium* spp.), perennials have also been incorporated into our initiative. Moreover, berry bushes are becoming increasingly rare around the community due to environmental change and development. For people who continue to gather berries, they are required to travel further from Fort Albany and it is becoming expensive to do so. Thus, perennials have been planted at both the agroforestry community garden, non-treed site, and home/yard gardens. Since 2014, 20 gooseberry plants and one raspberry plant have been transplanted from two Cree hunting camps on James Bay. Further, we have added some perennials from a commercial farm to augment our perennials

Table 20.2 The variety of crops grown in community and home/yard gardens at Fort Albany, Ontario (Canada)

Common name	Plant family	Variety	Growth characteristics	Measurements[a] (mean ± SE)	Study site[a]	Management
Good growth						
Annual ryegrass	Poaceae	Gulf	Variable growth	$n = 12$ Ht 60.5 ± 4.8 cm	ACG	Ambient
Arugula	Brassicaceae	Sky Rocket	Yielding; good vegetative growth; reached maturity	$n = 6$ Ht 11.8 ± 0.5 cm	HG	Ambient
Organic buckwheat	Polygonaceae	N/A	Grows well as a cover crop; reaches maturity	$n = 12$ Ht 79.0 ± 2.4	ACG	Ambient
Bush beans	Fabaceae	Provider	Yielding; reached maturity; allow for two harvests in a season	$n = 48$ FW 69.9 ± 11.1 g Ht 32.6 ± 0.2 cm # trifoliates/plant 4 ± 0 total leaf area/plant 277 ± 25 cm^2 # beans/plant 5.0 ± 0.4 Total bean FW/plant 22.4 ± 2.8 g	ACG+HG	Ambient
Carrots	Apiaceae	Amsterdam Maxi	Yielding; edible root	$n = 100$ leaf ht 23.6 ± 0.5 cm root length 9.8 ± 0.3 cm root FW 13.4 ± 1.0 g root volume 14.1 ± 0.7 cm^3	ACG+HG	Black tarp and low tunnel
Chard	Amaranthaceae	Bright Lights	Yielding; good vegetative growth; reached maturity	$n = 6$ FW 56.0 ± 5.7 g # leaves/plant 6.8 ± 0.3 Ht 39.7 ± 1.5 cm	HG	Ambient
Crimson Clover	Fabaceae	Dixie	Grows well as an understory to other cover crops	$n = 12$ Ht 23.1 ± 1.3 cm	ACG	Ambient
Green onion	Amaryllidaceae	Guardsman	Adequate vegetative growth		HG	Ambient

Kale	Brassicaceae	Red Russian	Relatively high yielding; good vegetative growth; multiple harvest	$n = 12$ Ht 32.8 ± 0.9 cm	ACG[a] + HG	Ambient
Lettuce	Asteraceae	Black Seeded Simpson	Good vegetative growth	$n = 8$ FW 9.9 ± 0.6 g # leaves/plant 9 ± 0 total leaf area/plant 321.9 ± 12.19 cm^2	ACG + HG	Ambient
Onion Bulb	Amaryllidaceae	Yellow Dutch	Good top vegetative growth; limited bulb growth	$n = 6$ FW 56.8 ± 4.5 g # leaves/plant 7 ± 1 plant height 54.0 ± 2.0 cm bulb volume 16.4 ± 2.2 cm^3	HG	Ambient
Peas	Fabaceae	Strike	Yielding; reached maturity; allow for two harvests	$n = 6$ FW 29.3 ± 2.4 g peas/plant 4.6 ± 0.3 pea FW/plant 12.3 ± 0.8 g	HG	Ambient
Potatoes	Solanaceae	AC Chaleur	Yielding; reached maturity	$n = 40$ above ground FW 139.9 ± 14.9 g Ht 39.5 ± 0.3 cm total tuber FW 329.9 ± 32.3 # potatoes/plant 5.0 ± 0.0	ACG + HG	Ambient
Radish	Brassicaceae	Sora	Relatively high yielding; adequate bulb growth; reached maturity	$n = 6$ FW 121.2 ± 7.3 g # leaf/plant 9 ± 1 Ht 51.0 ± 6.1 cm bulb volume 87.9 ± 10.1 cm^3	HG	Ambient
Rapini	Brassicaceae	Sorrento	Yielding; good vegetative growth; reached maturity	$n = 6$ Ht 46.0 ± 4.0 cm	HG	Ambient

(continued)

Table 20.2 (continued)

Common name	Plant family	Variety	Growth characteristics	Measurements[a] (mean ± SE)	Study site[a]	Management
Daikon Tillage Radish	Brassicaceae	CCS-779	Good aboveground and belowground growth; reaches maturity	$n = 6$ Ht 95.0 ± 3.2 cm Tap root length 30.3 ± 1.9 cm	ACG	Ambient
Turnip	Brassicaceae	White Lady Hybrid	Relatively high yielding; adequate bulb growth; reached maturity	$n = 12$ FW 75.4 ± 4.0 g Ht 36.9 ± 0.6 cm # leaves/plant 12 ± 0 bulb FW 38.4 ± 2.1 g bulb volume 52.4 ± 3.4 cm^3	HG	Ambient
Zucchini	Cucurbitaceae	Spineless Beauty	Provided adequate size fruits	$n = 1$ (largest fruit) fruit volume = 1172.3 cm^3 fruit length 36.1 cm	HG	Start indoors and black tarp
Poor growth						
Khorobi	Brassicaceae	Kongo Hybrid	Yielded small bulbs; stunted growth	$n = 6$ Ht 17.6 ± 2.1 cm	HG	Ambient
Pumpkin	Cucurbitaceae	Aspen Hybrid	Provided a few immature fruits		HG	Start indoors and black tarp
Spinach	Amaranthaceae	Speedy Hybrid	Poor yields; stunted growth	$n = 6$ FW 5.8 ± 0.6 g Ht 24.2 ± 1.3 cm	HG[a]	Ambient
Sweet Corn	Poaceae	Vitality	Stunted growth; limited development		HG	Ambient

[a]*FW* plant fresh weight, *Ht* plant height, *HG* home/yard gardens, *ACG* agroforestry community garden

from the hunting camps to increase accessibility and availability of berries in the community. The perennials are not currently being harvested as they require several years to establish.

20.5 Harmonization of Indigenous Harvesting Pursuits and Agroforestry Activities

In general, land managed for food production requires nutrient replenishment after crop harvest and other outputs, such as runoff and leaching. Nitrogen (N), phosphorus (P), and potassium (K) are the three primary nutrients that are essential for crop growth and survival. To maintain soil fertility and more optimal food production, soils require inputs of fertilizers (i.e., material that supplies essential elements for plant growth) that are either chemically produced or natural/organic in origin (Brady and Weil 1999). When fertilizer (organic or inorganic) is applied, the aim is for the substance to be efficiently stored in the soil and used by the plants.

With respect to crop production in the subarctic, for practicality, cost-effectiveness, and sustainability reasons, materials that are locally available should be primarily used. Composting is a viable option that mixes and decomposes various locally sourced materials that contain essential plant nutrients. Table 20.3 provides a list of potential local sources of compostable materials and their estimated nutrient content. Composting is a natural-aerobic-biochemical process where microorganisms convert organic material into stable soil-like material (Schaub and Leonard 1996). Microbial activity determines the rate of decomposition and quality of compost. Factors that influence compost microbial activity are temperature, oxygen concentration, pH, moisture content, particle size, and carbon to N (C/N) ratio (College of ACES 2013). Carbon and N are the main substrates that provide energy and regulate functions of microorganisms that decompose composted material. A mixture of materials that produce a C/N of 25–30 is optimal for microbial populations in composting systems (Schaub and Leonard 1996). Composts with C/N < 25 results in excess N that converts to nitrate or ammonia, creating bad odors, and loss of N via leaching or volatilization (Stevenson and Cole 1999). By contrast, C/N ratios > 30 causes N to be immobilized impeding decomposition (Stevenson and Cole 1999).

Meat producers in CA use composting as an efficient option to dispose of their animal by-products. They mix the animal by-products with high C substrates, such as straw, wood chips, or sawdust (Fleming and MacAlpine 2006; Koebel et al. 2003). It is imperative to allow the internal compost temperature to reach 55 °C for several days or weeks (depending on the size of the pile) to destroy pathogens and weed seeds (Fleming and MacAlpine 2006; Fonstad et al. 2003). The compost then can be used to replenish nutrients and organic matter in the soil.

It should be emphasized that when Cree smoke geese, the feathers, innards, and bones are typically removed and not consumed, while wood ash is produced as a

Table 20.3 Potential nutrient content from locally sourced materials for compost at Fort Albany, Ontario, Canada

Local nutrient sources	C:N	Nitrogen (%)	Phosphorus (%)	Potassium (%)
Animal by-product				
Blood	3–3.5	12.5–15.8	0.6–1.5	0.6–1
Feathers	4–10	14–15	0.3	0.2
Innards	4–10	7	10	0.5
Bones	4	2–8	5.8–27	0.36
Fish emulsion	6.5–14.2	2.6–12	1–12	0.5–1.5
Plant residues				
Sawdust	43–750	0.06–1.98	0.05–0.16	0.22–4
Leaves	40–80	0.5–1.84	0.08–0.11	0.27–0.8
Grasses (green)	9–25	2–6	0.19	0.71
Grasses (dry)	48–150	0.3–1.1	0.4	1.55
Shrub trimmings	53	1.0		
Peat	52–89	1.5–3	0.25–5	0.5–1
Fire by-product				
Wood ash	–	–	0.5–5	2.3–7

References: (Carpenter and Beecher 1997; Fisher 2005; Gagnon and Berrouard 1994; Gale et al. 2006; Jeng et al. 2006; Koebel et al. 2003; Kurhy and Vitt 1996; NRAES 1992; Penhallegon 2003; Rynk 1992)

by-product of the smoking of the geese. Viewing the northern food system as a whole, these by-products have been utilized as a local source of compostable materials for our agroforestry initiative. The by-products from the game bird harvesting activities are concentrated with N, and contain P and K making the materials excellent compostable materials.

Composting waterfowl by-products with various high C plant residues have the potential to replenish Fort Albany soils with N and improve P; however, K levels in soils are more difficult to increase even though K is present in bird by-products. Wood is often used in Cree traditional activities such as heating and cooking in bush camps, and for smoking geese; the combustion creates wood ash as a by-product. Wood ash has been applied as a K fertilizer and to increase soil pH (Bougnom et al. 2009). Composts that incorporated wood ash have lower bulk densities, reduction in odors, increase in friability, and improvements with oxygen transfer (Carpenter and Beecher 1997). Kuba et al. (2008) found that applying 8% and 16% ash to compost improved plant performance. However, soils with high pH and high cation exchange capacity limit availability of cationic elements, such as calcium, magnesium, K, and sodium (Bougnom et al. 2009). The soils in Fort Albany have a pH of 7.7–7.8, which is higher than the reported optimum for food production (pH range 5.5–7.0; Gardiner and Miller 2008; Islam et al. 1980; Mosaic 2016). To our knowledge there are no studies dealing with the fertilization of subarctic peat soils with wood ash; thus, caution is warranted when incorporating materials that may increase soil pH in the composting process.

20.5.1 Composting in Subarctic CA

Overabundant giant Canada geese were harvested and shared throughout Fort Albany First Nation. Nonedible portions of geese—feathers, bones, and innards—produced as by-products during the above described harvesting were put into a continuous flow batch composter (Actium, Seaforth, ON; Fig. 20.3). This composter was chosen to withstand potential nuisance carnivores. The metal drum of the composter was 1.6 m^3 and could hold 682 kg of material. A manual crank was used to turn the composter to mix materials. High density fiberglass with an ultraviolet-resistant jacket insulated the drum. The location of the composter was in a semi-enclosed forested area to reduce temperature losses from cool winds. Composting activities ran from 24 June 2015 to 4 October 2015. The composter was turned at least once a week and after any material was added. Locally sourced materials were mostly used. For example, grass clippings were obtained from maintenance of the agroforestry gardens, untreated-wood shavings were used from the small community sawmill, and wood ash from heating and smoking activities. Imported materials were used when available and contained spoiled and/or scraps of fruit and vegetables provided by the local grocery store, community-food market, and the local restaurant. Fish by-products are another excellent composting material and available from Cree traditional activities, but were not used; overabundant waterfowl were used for the added benefit of helping to protect the environment.

To ensure the resultant compost was safe to use as a soil nutrient amendment, compost samples were analyzed for contaminants, such as organochlorines (i.e., pesticides and polychlorinated biphenyls; Table 20.4) and toxic metals/metalloids

Fig. 20.3 Continuous flow batch composter (Actium, Seaforth, ON, Canada) with a manual crank and capacity of 1.6 m^3 or 682 kg of material. High density fiberglass insulated with an ultraviolet resistant jacket

Table 20.4 Concentrations of organochlorines in locally sourced compost (Fort Albany, Ontario, Canada) as determined by gas chromatography using electron capture detector (Analytical Services Unit, Queen's University, ON, Canada)

	Blank	Control	Control target	Compost[a]
Organochlorines (ng/g)				
Alpha-BHC	<2.0	23.7	20.0	<2.0
Beta-BHC	<2.0	25.2	20.0	<2.0
Gamma-BHC	<2.0	23.9	20.0	<2.0
Delta-BHC	<2.0	24.4	20.0	<2.0
Heptachlor	<2.0	26.6	20.0	<2.0
Aldrin	<2.0	22.9	20.0	<2.0
Heptachlor Epox iso B	<2.0	24.0	20.0	<2.0
Endosulfan I	<2.0	23.7	20.0	<2.0
Dieldrin	<2.0	23.6	20.0	<2.0
4,4-DDE	<1.0	22.7	20.0	6.2
Endrin	<2.0	22.7	20.0	<2.0
Endosulfan II	<2.0	25.7	20.0	<2.0
Endrin Aldehyde	<5.0	24.5	20.0	<5.0
4,4-DDD	<1.0	22.5	20.0	<1.0
Endosulfan sulfate	<2.0	25.8	20.0	<2.0
4,4-DDT	<1.0	22.6	20.0	<1.0
Endrin Ketone	<2.0	–	–	<2.0
Methoxychlor	<2.0	25.1	20.0	<2.0
2,4-DDE	<1.0	–	–	<1.0
2,4-DDD	<1.0	–	–	<1.0
2,4-DDT	<1.0	–	–	<1.0
Polychlorinated Biphenyls (µg/g)				
Aroclor 1016	<0.10	–	–	<0.10
Aroclor 1221	<0.10	–	–	<0.10
Aroclor 1232	<0.10	–	–	<0.10
Aroclor 1242	<0.10	–	–	<0.10
Aroclor 1248	<0.10	–	–	<0.10
Aroclor 1254	<0.10	–	–	<0.10
Aroclor 1260[b]	<0.10	5.0	5.0	<0.10

[a]Average result of duplicate analysis
[b]This mixture has been previously been found at elevated levels in soil, plants and organisms in Fort Albany, ON, Canada

(Table 20.5). All concentrations were either below the detection limit or within acceptable levels. Microbiological concentrations of the compost were also found to be at acceptable levels (Table 20.6). To determine the quality of the compost produced, pH, electric conductivity, organic matter, total carbon, total N, ammonium, nitrate, K, and P were analyzed (Table 20.7). One element in the compost that will require further enrichment in future endeavors is P. Phosphorus can be increased if bones are pulverized prior to composting, because the reactive surface area would be increased. Field data of compost augmentation at the agroforestry site is in the process of being analyzed.

20 Enhancing Food Security in Subarctic Canada

Table 20.5 Metals/metalloids (µg/g) in locally sourced compost (Fort Albany, Ontario, Canada) as determined by inductively coupled plasma optical emission spectrometry and inductively coupled plasma mass spectrometry (Analytical Services Unit, Queen's University, ON, Canada)

Sample	Blank	MESS-3[a] found	MESS-3 expected	Sample 1	Sample 2[b]
Ag	<2.0	<2.0	<2.0	<2.0	<2.0
Al[c]	<250	16,600	20,000	7300	7600
As	<1.0	17	18	2.0	2.4
B[c]	<50	–	–	<50	<50
Ba	<5.0	303	350	36	40
Be	<4.0	<4.0	<4.0	<4.0	<4.0
Ca[c]	<500	12,800	14,000	76,000	62,000
Cd	<1.0	<1.0	<1.0	<1.0	<1.0
Co	<5.0	11	12	5.9	6.1
Cr	<20	28	36	<20	<20
Cu	<5.0	30	31	10	8.9
Fe	<50	32,700	35,000	14,000	15,000
K	<20	4220	4900	1100	1100
Mg	<20	12,500	13,000	24,000	24,000
Mn	<1.0	280	300	420	460
Mo	<2.0	<2.0	2.1	<2.0	<2.0
Na	<75	10,800	11,000	180	170
Ni	<5.0	34	37	11	13
P	<20	1030	1000	580	660
Pb	<10	21	19	<10	<10
S	<200	1590	1700	930	1100
Sb	<1.0	<10	1.1	<10	<10
Se	<10	<10	<10	<10	<10
Sn	<2.0	<2.0	<2.0	<2.0	<2.0
Sr	<5.0	57.4	64	43	40
Ti	<10	–	–	680	580
Tl	<1.0	<1.0	<1.0	<1.0	<1.0
U	<10	<10	<10	<10	<10
V	<10	69	84	23	23
Zn	<15	125	130	30	31

[a]Lab certified reference material
[b]Average result of duplicate analysis
[c]Reporting levels rose due to interferences

Table 20.6 Concentrations of microorganisms in locally sourced compost (Fort Albany, Ontario, Canada) as determined by several methods (A & L Laboratories Inc., ON, Canada; Maxxam Analytics, ON, Canada)

Microorganism	Sample 1	Sample 2	Unit	Method
Fecal coliform	>1000[a]	>1000	MPN[b]/g dry	TMECC 07.01
Campylobacter spp.	Negative		P-A[c]/25 g (ml)	MFLP-46
Escherichia coli	>1000	>1000	MPN/g dry	TMECC 07.01
Escherichia coli 0157:H7	Negative	Negative	P-A	MFLP-30
Listeria spp.	Negative	Negative	P-A/25 g (ml)	MFHPB-30
Listeria monocytogenes	Negative	Negative	P-A/25 g (ml)	MFHPB-30
Salmonella spp.	Negative	Negative	P-A/25 g (ml)	MFLP-75

[a]Compost made from plant residues (such as wood chips) sometimes tests "false positive" for fecal coliforms that are, in fact, just ubiquitous *Klebsiella* bacteria
[b]Most probable unit
[c]Presence–absence

Table 20.7 Chemical content (dry weight) of locally sourced compost at Fort Albany, Ontario, Canada (Agricultural and Food Laboratory, University of Guelph, ON, Canada)

Parameter	Units	Sample 1	Sample 2	Optimal range
Dry Matter	%	25.82	22.52	25–40
Ammonium	mg/kg	146	170	<100
Nitrate	mg/kg	434	518	~1000
pH	–	8.0	8.1	5.5–8.5
Electrical conductivity	mS/cm	1.95	2.02	0.75–2
Mg extractable	mg/kg	1260	1530	30–40
K extractable	mg/kg	9280	10,800	600–1700
Na extractable	mg/kg	841	968	<2000
Ca extractable	mg/kg	5240	6550	600–1000
Total C	%	38.8	44.7	~50
Inorganic C	%	2.28	0.572	~1–2
Organic C	%	36.5	44.1	~50
Ca total	%	2.37	1.84	1–4
Mg total	%	0.399	0.395	0.2–0.4
P total	%	0.205	0.224	0.25–1.1
K total	%	1.01	1.17	0.2–1.7
Na total	%	0.091	0.104	<2
Total Kjeldahl N	%	1.34	1.36	0.75–2.5

20.6 Concluding Remarks

The implementation of our northern food system intervention that harmonized traditional Cree harvesting pursuits and agroforestry activities has the potential to form a sustainable import-substitution strategy and increase adaptive capacity to climate change, in the pilot community. The harvesting of overabundant waterfowl and the sharing of wild meat, and fruits and vegetables will have obvious dietary benefits at the individual, household, and community levels. In addition, individuals participating in the harvesting and/or agroforestry activity itself will accrue important social and cultural benefits from interacting with fellow harvesters and from being on the "land," as well as be empowered by contributing to the wellbeing of themselves and others. The community and academic researchers together are breaking down barriers to food security in subarctic Canada. Our efforts create social capital, so that the reservoir of community resources available for future efforts will always be growing. Our import-substitution strategy will continue through other community-based programs.

Nonetheless, it should be recognized that although climate was the most important regional barrier to our agroforestry initiative in subarctic CA—and climate change projections predict an even more hospitable climate for the growing of fruits and vegetables in high latitude regions of the world in the future—other challenges, such as moisture, are more locally specific. Soil moisture can be a limiting factor—not enough, there are limited nutrients available for proper plant functioning, and too much, the plants become waterlogged, and disease occurs. Fort Albany soil is composed primarily of peat and clay, with some sand, which is conducive to high water retention, especially accounting for the low lying elevation of the region. Thus, to improve drainage, drainage ditches were dug around the agroforestry sites (we have added another site), and backyard gardens have utilized raised beds to reduce water logging. A nearby beaver dam has also impacted local water levels, but that is a separate issue that will be addressed by local trappers. On the other hand, it is relatively dry during the beginning of the growing season, where seedlings require more water so that their roots can properly establish. During dry periods, soil moisture is regulated through the use of rain barrels and/or water pumping from the nearby lake. Thus, temperature and nutrient parameters are not the only factors of importance.

The protocol (in full or in part) that we have described for subarctic ON (Fig. 20.4) can be easily adapted for other high-latitude communities in CA and around the world that still partake in traditional pursuits (e.g., hunting, fishing, trapping) or the protocol can be appropriately modified for any individual (or community) wishing to become less reliant on imported fruits and vegetables. In our intervention, we used several goose species that were overabundant. Other animal species are also overabundant around the world, such as deer, and could be used instead of geese. The use of fish would also be a good choice, even invasive ones that are edible. In this way, participating communities would receive not only a fresh and nutritious source of wild game meat (or fish), but also contribute to the health of the environment (Tsuji 2017); while also receiving fresh fruits and vegetables, and other benefits associated with being on the land.

Acknowledgements We thank the harvesters, community-based research assistants and coordinators, elders, Chiefs and Councils, and the rest of the research team for their participation in these interventions. The pilot work was supported by the Social Sciences and Humanities Research Council of Canada (856-2009-0003) and the interventions by the Canadian Institutes of Health

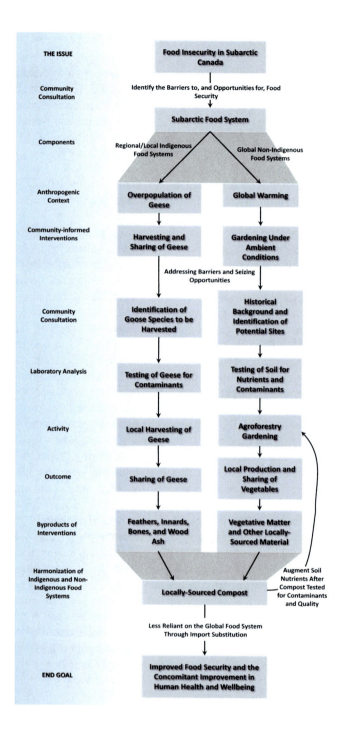

Fig. 20.4 Protocol used to harmonize the Cree traditional food system and the agroforestry food security initiative into a subarctic food system intervention with Fort Albany First Nation (Ontario, Canada). This import-substitution program has the potential to be adapted to communities across the globe with different locally sourced resources, both for wildlife (and/or fish) and vegetation. Note that community engagement and knowledge translation activities (e.g., workshops, community harvests, and youth camps) occurred throughout the agroforestry intervention, but are not represented in the figure

Research (Institute of Aboriginals Peoples' Health, and the Institute of Nutrition, Metabolism, and Diabetes; grants AHI-105525, AHI-120536, and MOP-133395).

Ethical approval: All procedures performed in studies involving human participants were in accordance with the ethical standards of the University of Waterloo, Waterloo, ON, CA, and the University of Toronto, Toronto, ON, CA. All applicable institutional and Government of Canada regulations concerning the harvesting of waterfowl by Indigenous peoples were followed. Informed consent was obtained from all individual participants included in the study.

References

Abraham, K. F., Jefferies, R. L., & Alisaukas, R. T. (2005). The dynamics of landscape change and snow geese in mid-continent North America. *Global Change Biology, 11*, 841–855.

Anisimov, O. A., Vaughan, D. G., Callaghan, T. V., et al. (2007). Polar regions (Arctic and Antarctic). In M. L. Parry, O. F. Canziani, J. P. Palutikof, P. J. van der Linden, & C. E. Hanson (Eds.), *Climate change 2007: Impacts, adaptation and vulnerability. Contribution of working group II to the fourth assessment report of the intergovernmental panel on climate change* (pp. 653–685). Cambridge: Cambridge University Press. Retrieved March 11, 2009, from http://www.ipcc.ch/pdf/assessment-report/ar4/wg2/ar4-wg2-chapter15.pdf.

Barbeau, C. D., Oelbermann, M., Karagatzides, J. D., & Tsuji, L. J. S. (2015). Sustainable agriculture and climate change: Producing potatoes (*Solanum tuberosum* L.) and bush beans (*Phaseolus vulgaris* L.) for improved food security and resilience in a Canadian sub-Arctic First Nation community. *Sustainability, 7*, 5664–5681. https://doi.org/10.3390/su7055664.

Bellows, A., & Hamm, M. (2001). Local autonomy and sustainable development: Testing import substitution in localizing food systems. *Agriculture and Human Values, 18*, 271–284.

Bougnom, B. P., Mair, J., Etoa, F. X., & Insam, H. (2009). Composts with wood ash addition: A risk or a chance for ameliorating acid tropical soils? *Geoderma, 153*, 402–407.

Brady, N. C., & Weil, R. R. (1999). *The nature and properties of soils* (12th ed.). Upper Saddle River, NJ: Prentice Hall.

Braune, B. M., Malone, B. J., Burgess, N. M., et al. (1999). Chemical residues in waterfowl and gamebirds harvested in Canada, 1987-95. Technical report series no. 326.

Carpenter, A., & Beecher, N. (1997). Wood ash finds niche in biosolids composting. *BioCycle, 38*, 37–39.

Catford, J. (2008). Food security, climate change and health promotion: Opening up the streams not just helping out downstream. *Health Promotion International, 23*, 105–108.

Chen, W., Zorn, P., White, L., Olthof, I., Zhang, Y., Fraser, R., & Leblanc, S. (2016). Decoupling between plant productivity and growing season length under a warming climate in Canada's arctic. *American Journal of Climate Change, 5*, 344–359. https://doi.org/10.4236/ajcc.2016.53026.

Chesney, A. (2015). *Increasing growing degree days in subarctic conditions: Food soverereignity in remote northern Aboriginal communities.* Undergraduate thesis. Waterloo, ON: University of Waterloo.

College of ACES. (2013). Composting for the homeowner: Science of composting. University of Ilinois Extension. Retrieved August 17, 2013, from http://web.extension.illinois.edu/home-compost/science.cfm

Council of Canadian Academics (2014) Aboriginal food security in northern Canada: An assessment of the state of knowledge. Ottawa: Council of Canadian Academics

Exner-Pirot, H. (2012). Guidelines for establishing a northern greenhouse project. Saskatoon, Saskatchewan, Canada: International Centre for Northern Governance and Development, University of Saskatchewan.

FAO. (1996). *Rome declaration on world food security and world food summit plan of action. World Food Summit 13-17 November 1996.* Italy: Rome.

Feenstra, G. W. (1997). Local food systems and sustainable communities. *American Journal of Alternative Agriculture, 12*, 28–36.

Fisher, B. (2005). Composting blood (slaughterhouse waste) mixed with various substrates. Final report. Ridgetown College-University of Guelph. Retrieved August 17, 2013, from http://www.ridgetownc.uoguelph.ca/research/documents/fleming_Final_Report_-_Compost_Blood.pdf

Fleming, R., MacAlpine, M. (2006). On-farm composting of cattle mortalities. University of Guelph Ridgetown Campus. Retrieved August 17, 2013, from http://www.ridgetownc.uoguelph.ca/research/documents/fleming_on_farm_carcass_composting.pdf

Fonstad, T. A., Meier, D. E., Ingram, L. J., & Leonard, J. (2003). Evaluation and demonstration of composting as an option for dead animal management in Saskatchewan. *Canadian Biosystems Engineering, 45*, 619–625.

Furgal, C., & Prowse, T. D. (2008). Northern Canada. In D. S. Lemmen, F. J. Warren, J. Lacroix, & E. Bush (Eds.), *From impacts to adaptation: Canada in a changing climate* (pp. 57–118). Ottawa: Government of Canada.

Gagnon, B., & Berrouard, S. (1994). Effects of several organic fertilizers on growth of greenhouse tomato transplants. *Canadian Journal of Plant Sciences, 74*, 167–168.

Gagnon, A. S., & Gough, W. A. (2005). Climate change scenarios for the Hudson Bay region: An intermodel comparison. *Climatic Change, 69*, 269–297. https://doi.org/10.1007/s10584-005-1815-8.

Gale, E. S., Sullivan, D. M., Cogger, C. G., Bary, A. I., Hemphill, D. D., & Myhre, E. A. (2006). Estimating plant-available nitrogen release from manures, composts, and specialty products. *Journal of Environmental Quality, 35*, 2321–2332.

Gardiner, D., & Miller, R. (2008). *Soils in our environment*. Upper Saddle River, NJ: Pearson Education.

Gates, A., Hanning, R. M., Gates, M., Isogai, A. D., Metatawabin, J., & Tsuji, L. J. S. (2011). A school nutrition program improves vegetable and fruit knowledge, preferences, and exposure in First Nation youth. *The Open Nutrition Journal, 5*, 22–27.

Gates, A., Hanning, R. M., Gates, M., Skinner, K., Martin, I. D., & Tsuji, L. J. S. (2012). Vegetable and fruit intakes of on-reserve First Nations schoolchildren compared to Canadian averages and current recommendations. *International Journal of Environmental Research and Public Health, 9*, 1379–1397.

Gates, A., Hanning, R. M., Gates, M., & Tsuji, L. J. S. (2016). The food and nutrient intakes of First Nations youth living in northern Ontario, Canada: Evaluation of a harvest sharing program. *Journal of Hunger and Environmental Nutrition, 11*. Retrieved from http://www.tandfonline.com/doi/full/10.1080/19320248.2016.1157552.

Gordon, A., & Newman, S. (1997). *Temperate agroforestry systems*. Wallingford: CAB International.

Gough, W. A., & Houser, C. (2005). Climate memory and long-range forecasting of sea ice conditions in Hudson Strait. *Polar Geography, 29*(1), 17–26. https://doi.org/10.1080/789610163.

Gough, W. A., Cornwell, A. R., & Tsuji, L. J. S. (2004). Trends in seasonal sea ice duration in southwestern Hudson Bay. *Arctic, 57*(3), 299–305.

Gustavsson, P., & Gerhardson, L. (2005). Intoxication from an accidently ingested lead shot retained in the gastro-intestinal tract. *Environmental Health Perspectives, 113*, 491–493.

HarvestRight. (2017). Greenhouses. Retrieved July 18, 2017, from https://harvestright.com/greenhouse-why-its-better/

Hlimi, T., Skinner, K., Hanning, R. M., Martin, I. D., & Tsuji, L. J. S. (2012). Traditional food consumption behaviour and concern with environmental contaminants among Cree Schoolchildren of the Mushkegowuk Territory. *International Journal of Circumpolar Health, 71*, 17344 (online). https://doi.org/10.3402/ijch.v7.17344.

Ho, E., Tsuji, L. J. S., & Gough, W. A. (2005). Trends in river-ice break-up data for the western James Bay region of Canada. *Polar Geography, 29*(4), 291–299. https://doi.org/10.1080/789610144.

Hori, Y. (2010). *The use of traditional environmental knowledge to assess the impact of climate change on subsistence fishing in the James Bay region, Ontario, Canada*. Master's thesis. Waterloo, ON: University of Waterloo.

Hori, Y., Tam, B., Gough, W. A., Ho-Foong, E., Karagatzides, J. D., & Tsuji, L. J. S. (2012). Use of traditional environmental knowledge to assess the impact of climate change on subsistence fishing in the James Bay region, Ontario, Canada. *Rural and Remote Health, 12*, 1878. Retrieved from http://www.rrh.org.au/articles/subviewnew.asp?ArticleID=1878.

Huai, H., & Hamilton, A. (2009). Characteristics and functions of traditional homegardens: A review. *Frontiers in Biology China, 4*, 151–157.

Intergovernmental Panel on Climate Change (IPCC). (2001a). Chapter 10: Regional climate information—evaluation and projections. In J. T. Houghton, Y. Ding, D. J. Griggs, M. Noguer, P. J. van der Linden, X. Dai, K. Maskell, & C. A. Johnson (Eds.), *Climate change 2001: The scientific basis. Contribution of Working Group I to the Third Assessment Report of the Intergovernmental Panel on Climate Change* (pp. 585–683). Cambridge: Cambridge University Press. Retrieved June 2, 2009, from http://www.ipcc.ch/ipccreports/tar/wg1/pdf/TAR-10.PDF.

Intergovernmental Panel on Climate Change (IPCC). (2001b). Chapter 13: Climate scenario development. In J. T. Houghton, Y. Ding, D. J. Griggs, M. Noguer, P. J. van der Linden, X. Dai, K. Maskell, & C. A. Johnson (Eds.), *Climate change 2001: The scientific basis. Contribution of working group I to the third assessment report of the Intergovernmental Panel on Climate Change* (pp. 741–768). Cambridge: Cambridge University Press. Retrieved May 12, 2009, from http://www.ipcc.ch/ipccreports/tar/wg1/pdf/TAR-13.PDF.

Intergovernmental Panel on Climate Change (IPCC). (2001c). Summary for policymakers. In J. J. McCarthy, O. F. Canziani, N. A. Leary, D. J. Dokken, & K. S. White (Eds.), *Climate change 2001: Impacts, adaptation and vulnerability. A report of working group II of the Intergovernmental Panel on Climate Change* (pp. 1–18). Cambridge: Cambridge University Press. Retrieved April 23, 2009 from, http://www.ipcc.ch/ipccreports/tar/wg2/pdf/wg2TARspm.pdf.

IPCC. (2014). Climate change 2014: Synthesis report. In The Core Writing Team, R. K. Pachauri, & L. A. Meyer (Eds.), *Contribution of working groups I, II and III to the fifth assessment report of the Intergovernmental Panel on Climate Change* (pp. 1–151). Geneva: IPCC.

Islam, A. K. M. S., Edwards, D. G., & Asher, C. J. (1980). pH optima for crop growth. *Plant and Soil, 54*(3), 339–357. https://doi.org/10.1007/BF02181830.

Jabbour, Niki. 2011. *The year-round vegetable gardener*. North Adams, MA: Storey Publishing.

Jefferies, R., Jano, A. P., & Abraham, K. F. (2006). A biotic agent promotes large-scale catastrophic change in coastal marshes on Hudson Bay. *Journal of Ecology, 94*, 234–242.

Jeng, A. S., Haraldsen, T. K., Grønlund, A., & Pedersen, P. A. (2006). Meat and bone meal as nitrogen and phosphorus fertilizer to cereals and rye grass. *Nutrient Cycling in Agroecosystems*. https://doi.org/10.1007/s10705-005-5170-y.

Koebel, G., Rafail, A., & Morris, J. (2003). On-farm composting of livestock and poultry mortalities. Ontario fact sheet, Ministry of Agriculture and Food. Retrieved August 8, 2013 from http://www.agbiosecurity.ca/healthybirds/Factsheets/Management/Supplementary/OMAFRAcompostingFactsheet.pdf

Kuba, T., Tscho, A., Partl, C., Meyer, K., & Insam, H. (2008). Wood ash admixture to organic wastes improves compost and its performance. *Agriculture Ecosystems and Environment, 127*, 43–49. https://doi.org/10.1016/j.agee.2008.02.012.

Kurhy, P., & Vitt, D. H. (1996). Fossil carbon/nitrogen ratios as a measure of peat decomposition. *Ecology, 77*, 271–275.

Leafloor, J. O., Moser, T. J., & Batt, B. D. J. (2012). Evaluation of special management measures for midcontinent lesser snow geese and Ross's geese. Arctic Goose Joint Venture Special Publication. U.S. Fish and Wildlife Service, Washington, DC and Canadian Wildlife Service, Ottawa, ON.

Lees, E. & Redman, H. (2009). Northwest Territories: Inuvik community greenhouse—Building a strong sense of community through recreational gardening, food production, knowledge sharing, and volunteer support. In *Bringing health to the planning table: A profile of promising practices in Canada and abroad.* (pp. 58–62). Ottawa: Public Health Agency of Canada.

McIntyre L, Tarasuk V (2002) Food security as a determinant of health. Public Health Agency of Canada. Retrieved February 22, 2008 from http://www.phac-aspc.gc.ca/ph-sp/phdd/pdf/overview_implications/08_food_e.pdf

Minister of Supply and Services Canada (MSSC). (1996). *A national ecological framework for Canada: Hudson plains ecozone and James Bay Lowland*. Ottawa: Minister of Supply and Services Canada.

Mosaic. (2016). Crop nutrition. Soil pH. . Retrieved July 21, 2017, from http://www.cropnutrition.com/efu-soil-ph

Natural Resource, Agriculture, and Engineering Service (NRAES). (1992). *On-farm composting handbook*. App. A characteristics of raw materials. Table A.1 Plant and life science publishing. Retrieved August 8, 2013, from http://compost.css.cornell.edu/OnFarmHandbook/apa.taba1.html.

Ninez, V. (1987). Household gardens: Theoretical and policy considerations. *Agricultural Systems, 23*, 167–186.

Penhallegon, R. (2003). Nitrogen - phosphorus - potassium values of organic fertilizers. Oregon State University Extension Service. Retrieved August 8, 2013, from http://extension.oregonstate.edu/douglas/sites/default/files/documents/lf/orgfertval.pdf

Polack, R., Wood, S., & Bradley, E. (2008). Fossil fuels and food security: Analysis and recommendations for community organizers. *Journal of Community Practice, 16*, 359–374.

Reyes, E. S., Liberda, E. N., & Tsuji, L. J. S. (2015). Human exposure to soil contamination in subarctic Ontario, Canada. *International Journal of Circumpolar Health, 74*, 27357. https://doi.org/10.3402/ijch.v74.27357.

Rockwell, R., Abraham, K., & Jefferies, R. (1996). Tundra under siege. *Natural History, 105*, 20–21.

Rynk, R. (1992). *On-farm composting handbook. Publication NRAES-54*. Ithaca, NY: Natural Resource, Agriculture and Engineering Service.

Schaub, S. M., & Leonard, J. J. (1996). Composting: An alternative waste management option for food processing industries. *Trends in Food Science and Technology, 7*, 263–268.

Skinner, K., Hanning, R. M., Desjardins, E., & Tsuji, L. J. S. (2013). Giving voice to food insecurity in a remote indigenous community in sub-arctic Ontario, Canada: Traditional ways, ways to cope, ways forward. *BMC Public Health, 13*, 427 (online). https://doi.org/10.1186/1471-2458-13-427.

Spiegelaar, N., & Tsuji, L. J. S. (2013). The impact of Euro-Canadian agrarian practices: In search of sustainable import-substitution strategies to enhance food security in subarctic Ontario, Canada. *Rural and Remote Health, 13*, 2211 online.

Spiegelaar, N., Tsuji, L. J. S., & Oelbermann, M. (2013). The potential use of agroforestry community gardens as a sustainable import-substitution strategy for enhancing food security in subarctic Ontario, Canada. *Sustainability*. https://doi.org/10.3390/su50x000x.

Stanton, B. (2011). *From food mail to nutrition North Canada*. Retrieved from http://publications.gc.ca/collections/collection_2011/parl/XC35-403-1-1-02-eng.pdf

Stevenson, F. J., & Cole, M. A. (1999). The internal cycle of nitrogen in soil. In F. J. Stevenson & M. A. Cole (Eds.), *Cycles of soil: Carbon, nitrogen, phosphorus, sulfur, micronutrients* (2nd ed.). New York, NY: John Wiley & Sons.

Tam, B., Gough, W. A., & Tsuji, L. J. S. (2011). The impact of warming on the appearance of furunculosis in fish of the James Bay region, Quebec, Canada. *Regional Environmental Change, 11*, 123–132. https://doi.org/10.1007/s10113-010-0122-8.

Thompson, J. E., & Hutchison, W. A. (1989). *Resource use by native and non-native hunters of the Ontario Hudson Bay lowland*. Toronto: Ontario Ministry of Natural Resources.

Tsuji, L. J. S. (2017). Sharing-the-harvest: Addressing food security and sustainability issues in First Nations of northern Ontario, Canada. Manuscript.

Tsuji, L. J. S., & Nieboer, E. (1997). Lead pellet ingestion in First Nation Cree of the western James Bay region of northern Ontario, Canada: Implications for a nontoxic shot alternative. *Ecosystem Health, 3*, 54–61.

Tsuji, L. J. S., Nieboer, E., Karagatzides, J. D., Hanning, R. M., & Katapatuk, B. (1999). Lead shot contamination in edible portions of game birds and its dietary implications. *Ecosystem Health, 5*, 183–192.

20 Enhancing Food Security in Subarctic Canada

Tsuji, L. J. S., Martin, I. D., Martin, E. S., LeBlanc, A., & Dumas, P. (2007). Spring-harvested game birds from the western James Bay region of northern Ontario, Canada: Organochlorine concentrations in breast muscle. *Science of the Total Environment*. https://doi.org/10.1016/j.scitotenv.2007.06039.

Tsuji, L. J. S., Wainman, B. C., Martin, I. D., Sutherland, C., Weber, J.-P., Dumas, P., & Nieboer, E. (2008). The identification of lead ammunition as a source of lead exposure in First Nations: The use of lead isotope ratios. *Science of the Total Environment*. https://doi.org/10.1016/j.scitotenv.2008.01.022.

Van Oostdam, J., Gilman, A., Dewailly, E., Usher, P., Wheatley, B., Kuhnlein, H., Neve, S., Walker, J., Tracy, B., Feeley, M., Jerome, V., & Kwavnick, B. (1999). Human health implications of environmental contaminants in arctic Canada: A review. *Science of the Total Environment, 230*, 1–82.

Vogl, C., Vogl-Lukasser, B., & Puri, R. (2004). Tools and methods for data collection in ethnobotanical studies of homegardens. *Field Methods, 16*, 285–306.

Young, A. (1997). *Agroforestry for soil management*. Wallingford: CAB International.

Chapter 21
Vulnerability Amidst Plenty? Food Security and Climate Change in Australia

Ruth Beilin, Michael Santhanam-Martin, and Tamara Sysak

Abstract The reality of climate change, and the expectation that agricultural production systems will need to adapt in response to it are now largely accepted by the Australian agricultural policy community. However, the effect of Australia's market-oriented agricultural policy is to delink the matter of adaptation from questions of Australian food security: national food security is considered assured by national income and global trade and food security is framed as the contribution that Australian farmers and agricultural technologists can make to the food security of others elsewhere in the world. Adaptation in this view is the process of farming system innovation, undertaken at individual enterprise scale, that allows Australian farm enterprises to remain profitable and globally competitive, even as environmental conditions change and on-farm vulnerabilities increase.

In this chapter, we argue that Australia's export-focused agricultural policy and more general assumptions of domestic food security result in the framing of adaptation as a technical process located at the scale of the farming enterprise and that this framing ignores important threats to Australian food security, in particular at the household scale. In fact, food producers and the rural communities where they live are themselves amongst the most vulnerable. In our reading, climate change is just one of the conditions creating this vulnerability for Australian rural communities, and adaptation must be understood in the context of multiple pressures that threaten the ability of farmers to carry on farming. This includes the volatility and

---

R. Beilin (✉)
Faculty of Science, University of Melbourne, Melbourne, VIC, Australia

Environmental Social Sciences, School of Ecosystem and Forest Sciences,
Faculty of Science, Baldwin Spencer West Annex, G55, University of Melbourne,
Melbourne, VIC, Australia
e-mail: rbeilin@unimelb.edu.au

M. Santhanam-Martin
Faculty of Veterinary and Agricultural Sciences, University of Melbourne,
Melbourne, VIC, Australia

T. Sysak
University of the Sunshine Coast, Sippy Downs, QLD, Australia

© Springer Nature Switzerland AG 2019
A. Sarkar et al. (eds.), *Sustainable Solutions for Food Security*,
https://doi.org/10.1007/978-3-319-77878-5_21

competitiveness of both domestic and export markets and the concentration of market power at key stages of agricultural value chains.

Keywords Food security · Lock-in · Resilience · Vulnerability · Climate change · Adaptation

21.1 Introduction

From 1983–1985 Ethiopia experienced a terrible famine. The world press was filled with excruciating images of dying children in the arms of emaciated mothers. At the time of the famine, military expenditure accounted for nearly half of Ethiopia's GDP, as the regime of President Mengistu attempted to contain and defeat liberation movements in Eritrea and Tigray (Webb et al. 1992). Spending on defence still appropriated 37% of GNP in 1991 (Tekabe 1998 in Devereaux 2000). But the advent and magnitude of the 1983–1984 drought allowed this and many other disastrous political and policy decisions that had contributed to the famine to recede from focus. The narrative of causation that came to prevail was an environmental discourse focused on land degradation, cyclic weather patterns that included drought, and questions about "modern" and appropriate land management.

In the last 35 years we have come to understand the physical conditions behind the Ethiopian drought as part of a larger, global weather pattern—anthropogenic climate change—that is, or should be, a significant driver of policy change for governments across the globe. The Ethiopian crisis is thus a salient reminder that, as our climate changes, the vulnerability of citizens is a function not of ecological and climate conditions alone, but rather of the way the economic and social policies of governments interact with them. This is the nexus where food security under climate change resides. Two anomalies of the Ethiopian crisis remain to mention: Firstly, Ethiopia continued to export some food products internationally during the height of the famine (Bello 1993). Secondly, in this famine, as in many others, food producers were amongst those who became dependent on food aid (Siyoum et al. 2012). Both these features are surprisingly relevant to the current discussion of food security in Australia.

21.1.1 Environmental Change as Security Crisis

One framing of the challenge that environmental change poses for the nation state is that it constitutes a threat to security: food security, water security and national security, for example. Walker and Cooper (2011):152) argue that the emergence of "resilience" as a way that policy-makers think about responding to security threats finds its genesis in American homeland policy in the context of the "War on Terror", and as being operationalised through a twofold approach. These are "critical infrastructures" and "operational resilience" of emergency responders, governments and enterprise. As they go on to note (p.154) "...The strategy of resilience replaces the short term relief effort—with its aim of restoring the *status quo* ante through post catastrophe reconstruction—with a call to *permanent adaptability in and through crisis* [our emphasis]." Adaptation for resilience and to avoid vulnerability have come to be part of the state and academic discourse, with socio-ecological resilience (SER) emphasising the biophysical environment and the role of society as parts of an interconnected system. Disaster and risk management—associated with engineered resilience and "bouncing back" after an event, attempts to resume the pre-shock stability settings. This tends to focus on infrastructure and societal order. Maintaining stability can be interpreted as protecting the norm, and in many situations, to do so leads to "potentially depoliticising" (Welsh 2014:2) the resilience discourse across government, industry and community. SER however, theoretically at least, acknowledges a dynamic experience, including predicting how to maintain a sense of stability even as things change and at least until thresholds are reached and new emergent systems or their properties are recognised (Walker and Salt 2006).

As the environmental discourses of climate change and resource management have come to acknowledge the likelihood of socio-ecological uncertainty, and increasing vulnerability in the face of maladaptation (Barnett and O'Neil 2010), Australian government policies associated with agriculture, water, and food urge adaptive and innovative responses, and a devolved, decentred and shared responsibility for resilience (as an endpoint or a process). The historical determination to use science and technology to respond to land degradation or overcoming exacerbated climate conditions, focuses attention on infrastructure and adaptive capacity (Hajkowicz and Eady 2015). In doing so, the focus is on predictable socio-ecological vulnerabilities (e.g. threats to farm businesses, changing seasons, water shortages) to maintain "business as usual" for the nation, and apparently ensure its food and water security.

Less noticeable from an aggregate level but seen at a local scale, the neoliberal economic paradigm emphasising trade liberalisation and free markets offers little or no protection or support for producers (through subsidies or incentives) facing unexpected shocks e.g. climate and/or markets. In effect, shocks are perceived as business risks. Diverse landscapes and weather create a panoply of possible outcomes within a huge continent like Australia. Responsibility for mitigating shocks must be taken on site, and in response to existing conditions. As Paschen and Beilin (2016) and Beilin et al. (2012) have argued, maladaptation occurs in part because the devolution of responsibility to the individual generally means that citizens' can only act within existing already institutionalised frameworks.

21.1.2 Food Insecurity in Affluent Nations: The Australian Case

Australia is an affluent, industrialised country with GDP per capita safely in the OECD Top 10. Australia is a significant agricultural exporter, exporting products worth some A$51 billion in 2013-14 (approximately 2% of GDP) (Commonwealth of Australia (CoA) 2015:4). In the Australian Government's *Agricultural Competitiveness White Paper* (CoA 2015), food security is referred to in a global rather than domestic sense even though it is estimated that the domestic demand for food will be about 90% more than what it was in 2000 (Michael and Crossley 2012). The largest export products (by value) are wheat, beef and dairy, but Australia also exports pulses, oilseeds, fruits and vegetables. The claim is routinely made that Australian agriculture, by virtue of exporting 65% of production (CoA 2015:4), feeds 60 million people elsewhere in the world (CoA 2015: 118)—currently near 1% of the world's population. In addition, providing more than 90% of the fresh produce consumed by its own population (DAFF 2014), though these data do not indicate who is financially able to buy fresh produce. Our export market is also targeted to middle class consumption, so not intended to alleviate food security in low-income countries, but to focus disproportionately on middle-class consumption in the South and South-East Asian and Oceania regions (Qureshi et al. 2013:136).

Australia is geographically isolated, but economically is tightly connected to international networks and flows of goods and capital, by virtue of its longstanding commitment to the global trade liberalisation agenda. Even though a large proportion of many agricultural products are sold in domestic markets, prices are set by global markets. The people who grow Australia's food make decisions about what to grow, and how much, in response to price signals influenced by these global markets (Campbell 2009; McMichael 2005).

We argue that Australia presents a cautionary tale to the affluent nations of the world. In accepting how industrial agriculture's successes have masked the ecological consequences of production systems that are not based on agro-ecological thinking (IPES-Food 2016) and coupled with exacerbated climate change, Australian producers face many challenges. These physical realities increase their risk and diminish their resilience, even as producers must reconcile physical shocks with dynamic domestic and international market dictates. Anthropogenic climate change is disrupting Australian agriculture (Stokes and Howden 2010; Hughes et al. 2015). Modelled scenarios predict potential yield reductions in cereals, beef and dairy (Hughes et al. 2015). Qureshi et al. (2013) modelled expected water scarcity scenarios in the southern Murray-Darling Basin and more generally across Australia based on drought experiences in the first decade of the twentieth century. They predicted "major declines in irrigated production in rice (43%), cereals (42%) and pasture related commodities (including dairy, beef and sheep) (23%)" (p139). Australian food production is significantly irrigation-dependent—especially fruit, vegetables and dairy, with irrigation water accounting for 50–70% of water withdrawn from aquifers for human consumption (NPIS 2012). Increasing temperatures

21 Vulnerability Amidst Plenty

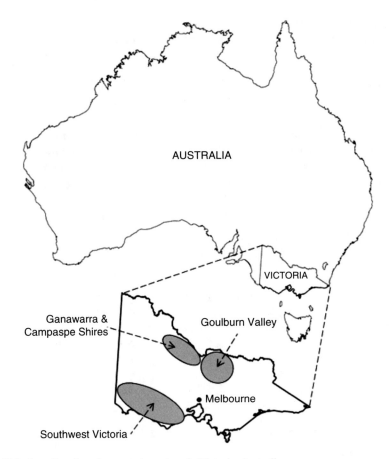

Fig. 21.1 Locality plan of case study regions in Victoria, Australia

and reduced rainfall are expected to deleteriously affect Australian production of household food items (MSSI 2015).

In this chapter we examine three case studies of agricultural regions in Australia (Fig. 21.1) where the food security of food producers or the longer-term provision for the domestic market is being compromised. Producers are locked-in to the technological and mechanistic responses that, in many cases, exacerbate their debts and diminish their chances of building more general resilience and adaptive capacity (Holling and Meffe 1996). Further, in terms of policy settings, the effect of food security being caught up and defined by privatised, global, corporate free trade (McMichael 2005:279) is to effectively hide the way in which local biophysical resources like water systems, and local producers and individual households are put at risk. Citizens become vulnerable to food insecurities, even while, as farmers, they continue to produce, because of the challenges of declining terms of trade and increasing "shocks" to the production systems in which they are working. Trade emphasizes output, not costs to the environment or family well-being, and as Clapp (2014:18) argues, such externalities have important implications for food security.

The United Nations currently defines food security as "a situation that exists when all people, at all times, have physical, social and economic access to sufficient, safe and nutritious food that meets their dietary needs and food preferences for an active and healthy life" (FAO et al. 2015:53). There are three aspects of this definition that we wish to highlight as context for the three vignettes that follows. Firstly, as Clapp (2014:23) notes, providing excess production for export does not guarantee that everyone will be well fed within a country. Secondly, as noted by Dixon (2016), in nations where local food systems are highly commodified and households are likely to be dependent on external sources for food access, declining incomes often herald a decline in fresh food purchasing and an increase in processed food consumption. Therefore, access to sufficient *nutritious* food is linked to incomes. Thirdly, and here we draw on the critique of food security that has been mounted by food sovereignty advocates (see e.g. Wittman et al. 2010), we argue that food security cannot be considered as existing outside of the democratic process—which is expected to be inclusive—and that this process depends on multiple policies that for example, link local natural resources to production. Therefore, an agricultural policy setting that ignores the well-being of certain strata of domestic society, or the certainty of increasing climate extremes, and the care of limited water and soil resources—and ignores these to ensure the overall commitment to export goals, is essentially forfeiting the food justice rights of current citizens and the prospects of future generations.

21.1.3 Food Insecurity and Access in a Landscape of Productivity

Southwest Victoria is known as one of the most productive regions of Australia. It has been predominantly a landscape of continuously cleared fields since the 1850s. These lie ploughed and ready for the industrial-scale production of animal feeds, crops cut for silage or feed, and the growing of pasture for dairy and beef. Within the immediate region around the major town of Warrnambool, there are four water supply protected areas associated with the 2015 Southwest Limestone Local Management Plan for water regulation (Southern Rural Water 2016). This plan indicates that there is approximately 85,000 ML/year available for consumptive use which includes farming, potable and industry supply. Licenses for access to this supply, beyond that associated with individual properties rights to stock and domestic water are fully taken up. This means that water trading can occur between those with licenses and those seeking more water, as regulated by the Southern Rural Water Board but no new licenses are available. In this region, irrigation is mainly for pasture and feed, and for managing dairies.. As testament to its productivity, there are fifteen dairy factories in the region represented by five major companies (Murray

Goulburn Cooperative, Fonterra Milk Australia, Saputo, Dairy Farmers and National Foods), with dairy farms varying in size and commonly around 150 hectares and 220 cows (WestVic Dairy 2014).

This vignette, undertaken in 2014–2015 originated from semi-structured interviews with 16 participants working in the regional food supply chain, from producers through to distributors and including local government, governance and agencies (Hewson 2015). It focused on the paradox of a region that is extremely productive in terms of efficiencies of scale and technologies, also being identified as a site of increasing food insecurity for food producers. The historical context for the vignette is the repeated crises that have affected the Australian dairy industry over the last decade including a milk price crash during the 2008 global financial crisis (Santhanam-Martin 2015), consistent downward pressure on milk prices resulting from "milk wars" between Australia's two dominant supermarket chains (Lawrence and Campbell 2014: 267), and another sudden milk price collapse occasioned by movements in the global market in 2016 (Beilharz 2016). These repeated shocks felt at the farm gate are indicative of the way Australia's integration into the global dairy market, mediated by the large milk processing companies, transfers large risks to those who are perhaps the most vulnerable participants in the value chain (Australian Dairy Farmers 2017). The Hewson (2015) study, coupled with ongoing research in the region, emphasises the vulnerability of "first world" farmers living in countries like Australia where trade liberalisation has exposed producers to difficult and even unfair, international economic conditions (see Clapp 2014).

Dixon (2016:382 citing Foodbanks Australia 2013), states that "one in ten Australians rely on food relief and half of these are children"; and this juxtaposition of plenty and access is cautionary. Clapp (2014:15) points out that there are rarely social safety nets for those who are compromised by a nation's ardent engagement with liberalisation initiatives. The majority of fresh food and vegetables produced, meat grown and fish harvested in this region goes elsewhere for sorting, processing and distribution. Some of it returns to the region after a round trip of over 660 km to Melbourne. The processing that does occur in the region is predominantly for export.

Vulnerability and food insecurity have many forms. In Australia, vulnerability is hidden in at least two significant ways. One way is structural adjustment. It is "disappeared" by government promoting continuing productivity, continuing export and trade liberalisation rules, with the mantra of export hiding the inequitable distribution and access of some foods that are critical to well-being. The second vulnerability is individual and the example is the consequences of structural adjustment on households and individuals. The Australian farming ethos has traditionally been for farm production to be unsubsidised, though historically federal governments of all persuasions supported a social-agrarian economy until the late 1970s (Watson 1979). No longer represented for example, by grain or wool boards, and subject to export market variations where they are generally "price takers" rather than "price setters"; domestic prices are determined on export markets creating intense pressure for farmers to be efficient producers (Watson 2012). As Beilin et al. (2012) indicate, path dependency through investment and limited response ability due to infrastructure and economic lock-in, means cascading pressures like drought, feed loss, petroleum price increases and similar shocks, create individual vulnerabilities that effect family well-being and longer term farming viability. Watson (2012) notes that food security associated with ability to produce is not an Australian issue. We argue however, that the concept of food security needs to be disaggregated to expose what foods are available. For example, there is evidence that fresh food access and affordability has become an indicator of well-being and its lack, of food insecurity (Hewson 2015; Foodbank Victoria Annual Report 2016). The loss of access to fresh food is one that reflects a common nutritional finding within socio-economically disadvantaged households. The nutritional indicator of whether families are eating fresh food is generally the measure of obesity. This region has a higher than national average rating for obesity (Australian Institute of Health and Welfare 2012) with data from the AIHW Healthy Communities Report (2016:5) indicating approximately 35% of the population is affected. (AIHW analysis is based on the Australian Bureau of Statistics National Health Survey 2014–2015). Clarifying the consequences of nutritional vulnerability through anthropometric indicators is an important area for future research.. Further, we argue that the shrouding of food access issues by producers and governments alike in this narrative, combines to preclude effective acknowledgment of need and solutions that are more than a "bandaid".

The complexity in finding official recognition of these producers' plights is concerning. It makes their difficulties and humiliation hard to document except through in-depth interviews about their experiences. Rural Finance Officers, who visit farm properties to assist with farm finance decisions, are frequently the ones who contact local food aid distribution centres. In southwest Victoria, the Foodbank Manager has arranged for these officers to be official distributors of food aid hampers. Farmers also spoke of being means tested to receive the national Farm Household Allowance. This welfare net is typically an acknowledgment of economic need but many farmers are not comfortable accepting that they need it because the process of declaring insufficient funds is perceived as humiliating. In a survey of Foodbank recipients, many indicated their emotional difficulties in declaring need (Foodbank Victoria Annual Report 2016:27).

In Australia, dairy farmers take up individual contracts with milk processing companies at the start of each year. The contracts are weighted to the discretion of the processors so that if prices are changed, farmers must accept those conditions. In the ideal contract, the negotiated quota for their production results in knowing and agreeing on an equitable price for the year. The farmers then make their annual plans based on that guaranteed price. In recent years, however, processors have significantly altered the contracted price during the year, effectively undermining the guarantee. Farmers are then indebted to the processing company for any advances and generally locked-in to supplying companies at the lower rate. Lack of choice about where to sell milk, and lack of confidence in the milk processing companies retaining prices and honouring farm commitments, weighs heavily on these farmers. The changing seasons mean different pasture mixes, the changing markets mean different cow numbers and there are increasing attempts to grow more of their feed to limit debt. As Murphy (2010) notes in a developing country context, and we hear in this case, the conversation among many farmers is not to end export trade but for governments to provide a place for their scale of production, and with protection from or incentives to support small-scale farming survival.

21.1.4 Incremental Adaptation in Australia's Fruit Bowl

The Goulburn Valley is situated some 200 km north of Melbourne. Its landform is in fact more of a riverine plain than a valley, and it is this flat to gently sloping topography, combined with generally fertile and well-drained soils and a reliable Mediterranean climate, that underpin its suitability for agricultural production. Its agricultural potential has been dramatically enhanced by the construction of irrigation infrastructure. Construction of dams and channels began at the end of the nineteenth century and by the beginning of the twenty-first century the region contained some 300,000 hectares of irrigable land. Major industries include dairy, temperate tree fruits including pears, apples, stone fruits and cherries, and processing tomatoes.

Our discussion focuses on the Goulburn Valley's fruit industry. We draw on formal research interviews with four Goulburn Valley orchardists conducted as part of a larger study in 2016 (Santhanam-Martin and Stevens 2017). Fruit is of particular interest in discussion of food security because fruit production in Australia has historically been much more focussed on meeting domestic demand only. Temperate tree fruit is also particularly vulnerable to an array of climate change impacts across the annual growth cycle (Webb and Whetton 2010).

Fruits grown in the Goulburn Valley include pome fruits (apples, pears and nashi) and stone fruits (peaches, nectarines, plums, apricots and cherries). In 2013 there were 306 commercial-scale fruit farms in the region, with a total area planted of some 11,500 hectares (RMCG and GVFGSWG 2013: 24–25). The vast majority of these farms are family-owned and operated. The Goulburn Valley fruit industry is of national significance, producing 10–85% by volume of the national crop for each of the fruits listed above (RMCG and GVFGSWG 2013: 16).

In response to a declining market for processing fruit (see Hattersley et al. 2012), most Goulburn Valley growers have recently reoriented their production to fresh fruit markets. Large additional volumes of fruit from the Goulburn Valley have further exacerbated the difficulties fruit growers throughout Australia face in achieving viable returns in an oversupplied and highly concentrated domestic fresh fruit market (Dixon and Isaacs 2012). One response from growers, actively encouraged by governments, is to focus on expanding exports. The hope is that, building on Australia's reputation for "clean and green" and high-quality produce, and with its geographic proximity to growing middle-class markets in south-east and east Asia, fresh fruit can be a new export industry.

This is the domain of pressure and opportunity within which Goulburn Valley orchardists are contemplating the potential impacts of climate change. Indeed, our interviewees were often unable to differentiate management actions they take in response to climate change from actions associated with adaptation to market demands: the two are intricately entwined. The technological responses to adaptive orchard management are many, and largely fall within the realm of "command and control" (Holling and Meffe 1996) within Walker and Cooper's (2011) "operational resilience" mode. These adaptations include chemical sprays available that can help trees come out of dormancy in spring, even if winter chill has been sub-optimal, and protective orchard covers (netting) to protect from hail and extreme heat damage. Most orchardists have already adopted highly efficient drip irrigation systems, to minimise water demand. All these technological adaptations entail costs. Installing orchard netting, for example, doubles the capital cost of a new orchard block. Every increment in costs demands an increment in revenue if profits are to be maintained. Thus this technological response to climate change keeps Goulburn Valley orchardists within the attractor of the industrial agriculture "lock-ins" identified by IPES-Food (2016).

There is one climate change threat that Goulburn Valley orchardists told us they will not be able to respond to through technological adaptation: lack of water. While irrigators of annual crops and pastures can drastically reduce their water use in drought years when irrigation water is unavailable, orchardists require at least a minimum irrigation application in all years, if tree health and productivity are to be maintained. To produce abundant crops of fruit that meets retailers' stringent size and quality specifications, it is critical that full water applications be maintained. Market-based reforms to water resource management in Australia mean that each fruit grower has access to a capped volume of water (for which they pay a volumetric charge) from the irrigation system in any given year, with the quantity available varying according to the quantity of water stored in the regional irrigation system's storage dams. Any additional water that is required must be purchased from other irrigators who are willing to sell their allocation. In a drought year water prices can escalate rapidly and dramatically, and can reach levels at which orchardists must pay more for the water than the crop is worth, just to keep their trees alive. This policy framework puts fruit growers in market-based competition with other water users, including environmental water use. One orchardist viewed this policy context as an existential threat to the Goulburn Valley fruit industry.

21.2 Locked in to Irrigation

The last vignette is drawn from research conducted between 2009 and 2013 in two towns in northern Victoria, Cohuna (Ganawarra Shire[1]) and Rochester (Campaspe Shire), during a time of severe drought (Sysak 2013). This region is situated to the northwest of the Goulburn Valley orchardists, is predominantly irrigated, and mostly dairy farming where the dairies account for approximately 25% of national milk production and 28% of Australia's dairy export industry (DEDJTR 2015). This

---

[1] A Shire is a local government area in rural Australia.

region had been in a 10-year drought, which was the longest drought in colonial memory, until it broke in 2011 with extreme floods.

Northern Victoria has the least rainfall of the three case studies (BOM 2013) and as with the orchardists, the dairy farmers are heavily reliant on irrigation water with the majority (83% in 2016) surface water from the Murray River (the longest river in Australia), its tributaries and dams and the remaining 17% groundwater (BOM 2017). Irrigation was the first technical response (introduced in 1879) to the dry climate and marginal land. There have always been regulations that govern the use of the irrigation water, but since the 1990s, the fluctuating arrangements between the State and Federal governments and ad hoc changes to policies (Appendix) have added an extra layer of complexity for farmers to negotiate when trying to adapt to climate change.

Australia's irrigation systems are governed by corporations set up by statutory authorities i.e. bodies set up under legislation, with the responsibility of the management, delivery and allocation of water for their district. Water markets began in Australia in 1984, and as any market they are used as an instrument to reallocate water for competing uses where prices reflect availability. Irrigators are able to trade their water entitlement (their right to a water allocation for each season subject to availability) and their allocations (their right to water for the current season—determined at the start of the season subject to change each month dependent on climate and storage levels). In northern Victoria, irrigation water is managed by Goulburn Murray Water, Australia's largest rural water corporation. Goulburn Murray Water also administers and enforces the Groundwater Management Plan for this area. They manage licence entitlements (the number of groundwater bores), restrict groundwater extraction when deemed necessary and limit the concentration of groundwater pumping (Department of Sustainability and Environment 2012). Since 2008 the Commonwealth Government does have the power to override allocations for a season.

During the drought, some farmers adopted highly resource intensive on-farm strategies, such as double cropping (two successive crops in a year), associated with depleting the soil over time, and making it an unsustainable longer-term strategy for this region. Some tried different feedlot systems that after two seasons became prohibitively expensive where they could not recover costs through production. Others used water trading as their primary adaptation strategy, particularly when allocations were at their lowest. Many farmers were excluded from these transactions because of either not enough water to trade or insufficient funds to buy it when available. As a consequence, some producers left farming. Those that stayed had increasing levels of debt and when water unbundling was introduced, many used this as a strategy for short-term cash flow and sold their water rights permanently. The Victorian government argued that the changes would "provide greater flexibility to irrigators and diverters to manage their water use and their risks, through easier water trade and increased entitlement certainty" (State of Victoria 2008). Most farmers were unable to then buy their rights back. Those in Campaspe Shire had the added stress of (in)voluntarily selling their water rights to the Federal Government as part of the modernisation project, which significantly redesigned the irrigation spine, reducing delivery to smaller channels. With much of the work for the modernisation project being outsourced to private companies, some farmers had to wait over twelve months for an on-farm consultation (Sysak 2013). The government had changed the way it offered on-farm extension services with the number of extension workers reduced and the remaining ones situated in larger centres (DPI 2009) meaning that farmers had to travel to access their services. Even though the dairy corporations had field officers, their own internal staffing cuts meant that there were less people available to visit farms.

It was the additional water stress as well as high profile media focusing on drought and farmer suicide, which led to a social welfare initiative, Farm Gate, subsidised by national drought relief emergency funding[2]. Even though farmers were at the whim of government water policy and international markets, the initiative was framed solely as drought relief and emergency social welfare. The social workers estimated that over 50% of farmers were in severe financial hardship, with many requiring immediate food assistance.

An evaluation report of the Farm Gate initiative (Tauridsky and Young 2009) described how many farmers felt abandoned by government but were still quite reluctant about accepting "goods" (food and other groceries). They were concerned they would be perceived as needy and judged as poor farm business managers. Importantly, this is where food insecurity among some of those producing for the nation and export, was acknowledged, but only in these Farm Gate conversations. The reality that these producers could not afford to feed themselves or their families properly was hardly known if we consider the national debate and national media.

By 2010 the national farm debt had doubled over 10 years, and for dairy the average debt increased by 20% in 2 years. Northern Victoria farmers were the worst off in the nation and could anticipate a farm cash income of minus $27,300 (cf. Western Australia farmers average of $240 500) (Dairy Australia 2010). In this same period, the number of registered dairy farmers dropped by a further 18% (Dairy Australia 2010).

In this vignette, vulnerability was reframed by local financial and social counsellors as a consequence of climate and markets colliding, creating "tipping points" for the social ecological resilience of the region. In resilience thinking, these tipping points generally lead to regime change, most often associated with incremental changes that undermine the core (in this case the family farmers) and which create a new set of feedbacks that trigger reorganisation (Folke et al. 2004). Regime change was underway, begun with the unbundling of water, and this program would have assisted many farmers to accept (no matter how bitterly) the new thresholds heralded by the unswerving commitment to the terms of market and technological drivers.

---

[2]Australia has had several incarnations of its drought policy since 1992. At the time of this case study, Exceptional Circumstances were in place where assistance was based on drought severity in a particular area. Farmers in these areas with low net cash income, off-farm income and high debt were eligible to receive income assistance (Productivity Commission 2009). In 2014 this was replaced by the Farm Household Allowance Scheme which is an income safety net for farmers in times of financial hardship, regardless of the source of the hardship (Department of Agriculture and Water Resources 2017).

21.3 Discussion

These three vignettes combine to present part of the narrative of production, insecurity and climate change among agricultural producers on the driest continent in the world. We argue that resilience of the current agricultural system is being stabilised primarily as the foundation of export growth; and that this normalising of trade as the core of the food system creates social and ecological vulnerabilities that are evident at the local level. At the local level it is only those who can mobilise resources quickly—usually not the most vulnerable—who are able to adapt to changing conditions to reduce their vulnerability (Adger 2006). In the first vignette, dairy producers in an historically important production area are unable to provide fresh fruit and vegetables for their families as a consequence of the repeated shocks that have affected their industry, often with their origins in the global dairy trade. In the second vignette we visit orchardists noted by their industry as being "top of the game" with regard to markets and climate awareness, who know that climate and policy will unite to decide their futures and, by extension, the availability of domestic fruit and vegetables. In the third, we turn to the irrigators, who are adapting to constantly changing water policy while losing flexibility for future adaptation options. In this vignette we see how water policy motivated by efficiency, removes options, dries up towns and locks-in those still in the irrigation networks—to farm layouts, equipment, technology and debt—that make alternative options (such as agro-ecological systems and the continuing of smaller-scale family farms (Magnan 2015)) untenable.

Australia undoubtedly, is subject to extreme climate variability (IPCC 2014), though Australian Federal Government attention to climate change has been highly variable. The Agriculture Competitiveness White Paper (CoA 2015) acknowledges climate change as impacting water and land management and promises to invest in on-farm and regional adaptation programs. In the case of water modernisation, it is argued that this program is part of the response to climate. It is apparent that in the quest for more export products, efficiency drivers take precedence over climate variables. We argue that while there are multiple pressures on producers as decision-makers, the deployment of climate change as a significant driver of immediate or possible food insecurity, and therefore, warranting action now to prevent future loss of capacity, is not found within government policy responses, and only sometimes referred to among producers. Rather, climate change (or "weather") is found in our case studies/vignettes, to be part of everyday responses among producers—and is treated as another variable in their production systems. It is "in the mix" that is officially centred on trade liberalisation and export policies.

As Bové and DuFour, 2001 in McMichael (2005:287) indicate, the world food production networks effectively create "food from nowhere". These supply chains efficiently obscure the social and ecological, geographic and technical realities of local conditions, and we can add their social-ecological resilience and vulnerability. This homogenisation into commodities allows global corporates and markets to externalise the real costs in soil and water, land and people that are known at the local site of production. It could be expected that the Australian nation-state would resist the loss of fragile ecological systems and scarce water as part of a national incentive to sustain food security in the face of predicted, exacerbated climate change and to ensure food justice for all its citizens; but to do so, is possibly cast as too radical, demanding a rethink of international trade agreements and the myriad political stratagems that currently sustain documents like the Agricultural Competitiveness White Paper that Federal Government endorses as the way forward.

This brings us back to food security. Clapp (2014:27) notes that "in market-oriented agricultural systems that rely on efficiency criteria, purchasing power is a primary determinant of food distribution rather than need…and, careful consideration of multiple social goals is required in the formulation of agricultural and food security policies, including policies guiding agricultural trade". We have not focused on food sovereignty arguments in this chapter, but rather on the issues that create vulnerabilities for producers, arguing that their vulnerability bodes ill for domestic food production into the future.

The Goulburn Valley fruit industry offers an example of how climate change interacts with policy and market context to create increased economic vulnerability for food producers. It can be argued that this is immaterial to Australians' food security. If changing policy priorities around water allocation and/or competition from lower-cost producers elsewhere in the world render it unprofitable to continue producing fruit in the Goulburn Valley then Australia will be able to use wealth generated from other activities to import fruit from these other producers. But the question of whether or not Australia retains a domestic fruit industry seems to us to be a pertinent and highly political one that deserves public discussion and debate. Perhaps Australians do value the opportunity to eat fresh fruit of known provenance and quality, and conceivably they also value production landscapes and communities such as exist in the Goulburn Valley. Arguably they would prefer not to be dependent on distant producers, distant governments and global value chains for something as basic as an apple for a school lunch box. This is especially as we head into a future where other countries' production is also liable to destabilisation by climate change, and where refrigerated freight to remote corners of the globe like Australia, seems destined to become much more expensive. If food security entails some democratic involvement in decisions about where our food comes from, then fresh produce including fruit is a point of vulnerability in the Australian food system that is worthy of vigilant attention.

For the farmers in northern Victoria, their vulnerability stems more from changes to water policy than climate extremes. The short time frame between allocation announcements and water supply means longer-tern planning is not possible. De Haan (2000) argues adaptation can only be considered in the global context of increasing interdependence. This is the situation in northern Victoria where international and domestic markets determine allocation as much as availability as water moves to its highest value use based on markets (Productivity Commission 2003) and not local conditions. In these instances, climate change adaptation is not about learning, it is reactive where farmers' vulnerability is amplified (Adger et al. 2005), locking the farmers further into an unsustainable path each time there is a policy change (Anderies et al. 2006). It also locks them into short-term thinking where they need to deliver short-term results (IPES 2016), often requiring significant economic outlays perpetuating debt and hardship at the household level, and where they do not always have economic access to sufficient food. Despite repeated high level advice over a number of years, pointing to food system vulnerabilities that will be exacerbated by climate change, the unrelenting commitment to an export driven focus, and manipulation of the water markets obfuscates the unevenness of food security for Australians.

In the three vignettes, climate change as part of the environmental discourse, stresses individual adaptation on farm, upgrading of technological innovation and a focus on individual ability to manage in conditions that are constantly being reassembled to benefit global markets. SER is currently based on a continuing and heightened investment in technology, which locks producers in to same-same systems that have very little flexibility; while all around them the adaptation mantra is for flexible decision making in the face of uncertainty. Clearly there are those who are sometimes able to take advantage of international markets to benefit in an immediate harvest season or over the short term—for example in the 2013 orange crop harvest (Saefong 2013). However, the diminishing number of family farm (non-corporate) smaller enterprises and their increasing debt, suggests that in the current regime, they are vulnerable to markets, climate and unaligned national policies that sees some of our producers some of the most food insecure in the nation.

21.4   Conclusion

The social and economic costs of climate change, coupled with Australia's history of participating in food systems that are structured as global supply chains (ostensibly to provide cheap sources of some foods) are made explicit in this chapter. The chapter has articulated the complexity of issues that food producers encounter in this global and national governance system. Food producers are not exempt from narratives of food insecurity and food welfare even as the national policy settings promote efficient production regimes, oriented to increasing technological adaptation. For the producers represented in our three vignettes, the availability of food and the quality of available food were not the principle restraints on their access. From Ethiopia in the 1980s the world eventually learned that the social and political policies of that government, took their farmers and their people to the brink of starvation; and it was not climate *alone* that led to the horrific famine. We wonder if nations, immersed in global production and markets can ignore such lessons from history.

In Australia, the shame of food poverty among producers contributes to hiding their plight, as does the uncertainty around water access. The disconnect at a national level between policies that are associated with long-term food security, such as climate, land, water and agriculture, intersect with what farmers and families experience as the irony of food injustice in a wealthy, export-oriented nation. Producers know that markets, climates and access to water pose immediate risks to their business and to their futures. Siloed government policies exacerbate the impossibility of producers being able to surmount the physical and social complexity; and of Australians as a population being able to adapt to the changes ahead.

Government policies and governance systems continue to deal with these immediate risks through policy responses where longer-term infrastructure, like dam building, is intended to mitigate that risk. The "technological-fix" locks the nation into path dependency and the adaptive practices that emerge, are themselves, constrained and even myopic. Attention on site goes to promoting the changes on-ground that are most responsive to the immediate driver, whereas attention needs to be focused on the government policies that reinforce the feedbacks underpinning these conditions. Food security in a changing climate requires a national and integrated approach to land and water management, and Australia's historical and current commitment to being a major agricultural exporter must be revisited in this context.

This is necessarily a democratic process. Australian food security, based on not externalising the environment and seeking social and environmental justice for all, may require a significant national regime change that defines the SER of the future. IPES (2016:60–61) indicates that Brazil, Cuba, Thailand and various EU nations have initiatives that encourage agro-ecological farming systems, and argue that these first steps to diversity will reframe the land and water system to reflect limitations and potential more acutely than current industrial agriculture can do. This, in combination with Clapp's (2014:30–31) call to re-examine trade policies to actively consider nation-state social goals; and, a commitment by all Australian governments to incorporate climate change indicators as part of internalising the market's current externalities, seem to this chapter's authors to be a way in which Australia can walk the talk into the twenty-first century as a nation of adaptive, innovative, equitable and food secure citizens.

We conclude this chapter with a follow up to the food aid situation described earlier. The FoodShare coordinator in the southwest region received a phone call from another NGO, Foodbank Victoria (see the supplied images) in mid-December. It distributes and supplies food for FoodShare. Foodbank undertook a donor campaign in response to farmer relief agencies, noting the increasing signs of food insecurity among dairy families across the State. Foodbank received funds for 1000 Christmas hampers to go to dairy farmer relief. No government funds were directly involved—all donations were from private citizens. It is charity in a failed and failing system. It is food aid in the place of adjustment and integrated policy responses.

Appendix: State and Federal Government Water Policy

State	Federal
Water Act 1989 (Vic.)—governs the amount of water irrigators can access each year (dependent on water levels in storage).	
1991—under the *Water Act 1989 (Vic.)* water trading allowed where water can be exchanged and redistributed.	
2004—water cap of 4% introduced where no more than 4% of water can be traded out of the area.	
2007—water bundling introduced where water is no longer attached to land title.	*Water Act 2007 (Cth)*—established the need for a Basin Plan for the management of the Basin water resources
	Water Amendment Act 2008 (Cth)—establishment of Murray Darling Basin Authority. Has the power to override state water allocations.
Northern Victoria Irrigation Renewal Project (2008)—to upgrade existing infrastructure and to eliminate some of the smaller channels—modifications made to the 4% cap.	
	Water Amendment (Long-term Average Sustainable Diversion Limit Adjustment) Act 2012—sets out the maximum amount of water that can be taken for consumptive use—transition from cap to SDL which will commence in 2019.
2011 Victorian Food Bowl Modernisation Project—agreement between Commonwealth and Victorian Governments. Established stage 2 of NVIRP. 50% share of water savings previously intended to be shared amongst irrigators in the Goulburn Murray Irrigation District (the "second 50% share") will be retained by the Victorian Government and sold	
	Water Amendment Act 2015—increased flexibility in the recovery of 450 Gl of water through efficiency measures where a statutory limit of 1500 Gl was set on Commonwealth purchases of surface water.
Water for Victoria 2016 (discussion paper)	Water Amendment (Review Implementation and Other Measures) Act 2016

References

Adger, W. N. (2006). Vulnerability. *Global Environmental Change, 16*(3), 268–281.

Adger, W. N., Arnell, N. W., & Tompkins, E. L. (2005). Successful adaptation to climate change across scales. *Global Environmental Change Part A, 15*(2), 77–86. https://doi.org/10.1016/j.glovencha.2004.12.005.

Anderies, J. M., Ryan, P., & Walker, B. H. (2006). Loss of resilience, crisis and institutional change: Lessons from an intensive agricultural system in southeastern Australia. *Ecosystems, 9*(6), 865–878.

Australian Dairy Farmers. (2017). Lessons learnt from the senate inquiry. *Australian Dairy Farmers Newsroom*. Retrived February 14, 2017, from http://www.australiandairyfarmers.com.au/media-corner/lessons-learnt-from-the-senate-inquiry

Australian Institute of Health and Welfare. (2012). *Australia's food & nutrition 2012. Cat. no. PHE 163*. Canberra: AIHW. Retrieved December 9, 2016, from http://www.aihw.gov.au/WorkArea/DownloadAsset.aspx?id=10737422837.

Australian Institute of Health and Welfare. (2016). *Healthy communities: Overweight and obesity rates across Australia, 2014–15 (In Focus). Cat. no. HPF 2*. Canberra: AIHW. Retrieved July 20, 2017, from http://myhealthycommunities.gov.au/Content/publications/downloads/AIHW_HC_Report_Overweight_and_Obesity_Report_December_2016.pdf?t=1500599329258.

Barnett, J., & O'Neil, S. J. (2010). Maladaptation. *Global Environmental Change., 20*, 211–213.

Beilharz, N. (2016). Dairy co-op Murray Goulburn cuts milk prices, MD Gary Helou departs. *ABC News*. Retrieved February 14, 2017, from http://www.abc.net.au/news/2016-04-27/murray-goulburn-price-drop-helou-departure/7361654

Beilin, R., Sysak, T., & Hill, S. (2012). Farmers and perverse outcomes: The quest for food and energy security,emissions reduction and climate adaptation. *Global Environmental Change., 22*, 463–471.

Bello, W. (1993). Population control: The real culprits and victims. *Thirld World Resurgence, 33*, 11–14.

Bureau of Meteorology. (2013). Map of average conditions. Australian Government. Retrieved February 4, 2017, from http://www.bom.gov.au/climate/averages/maps.shtml

Bureau of Meteorology. (2017). National water account 2016, Australian Government. Retrieved July 25, 2017, from http://www.bom.gov.au/water/nwa/2016/mdb/index.shtml

Campbell, H. (2009). Breaking new ground in food regimes theory: Corporate environmentalism, ecological feedbacks, and the 'food from somewhere' regime. *Agriculture and Human Values, 26*(4), 309–319.

Clapp, J. (2014). Trade liberalization and food security: Examining the linkages. Quaker United Nations Office, Geneva. Retrieved December 10, 2016, from http://quno.org/sites/default/files/resources/QUNO_Food%20Security_Clapp.pdf

Commonwealth of Australia (CoA). (2015). *Agricultural competitiveness white paper: Stronger farmers, stronger economy*. Canberra.

DAFF. (2014). *Australian food statistics 2012–13*. Department of Agriculture, Fisheries and Forestry. Retrieved March 24, 2017, from http://www.agriculture.gov.au/SiteCollectionDocuments/ag-food/publications/food-stats/australian-food-statistics-2012-13.pdf

Dairy Australia. (2010). *Dairy 2010: Situation and outlook, summary report*. Retrieved August 20, 2011, from http://www.dairyaustralia.com.au/Statistics-and-markets/~/media/Documents/Statistics-and-markets/Situation%20and%20outlook/Summary%20Report%20Dairy%20Situation%20and%20Outlook%202010.ashx

Department of Agriculture and Water. (2017). *Farm household allowance*. Commonwealth Government, Canberra. Retrieved July 29, 2017, from http://www.agriculture.gov.au/ag-farm-food/drought/assistance/farm-household-allowance

Department of Economic Development, Jobs, Transport and Resources. (2015). *Dairy industry profile*: December 2014. Retrived February 1, 2017, from http://agriculture.vic.gov.au/__data/assets/pdf_file/0006/292182/6-Dairy-Industry-Profile_December-2014-Update_MASTER.pdf

Department of Primary Industries. (2009). *Better services to farmers strategy*. State Government of Victoria, Melbourne.

Department of Sustainability and Environment. (2012). *Lower campaspe valley water supply protection area groundwater management plan, October 2012*. Victorian Government. Retrived July 25, 2017, from http://www.g-mwater.com.au/downloads/gmw/Groundwater/Lower_Campaspe_Valley_WSPA/Nov_2013_-_Lower_Campaspe_Valley_WSPA_Plan_A4_FINAL-fixed_for_web.pdf

Devereaux, S. (2000). *Food insecurity in Ethiopia: A discussion paper for DFID*. University of Sussex, International Development School. Sussex, UK. Retrived December 14, 2016, from https://www.ids.ac.uk/files/FoodSecEthiopia4.pdf

Dixon, J. (2016). The socio-economic and socio-cultural determinants of food and nutrition security in developed countries. In B. Pritchard, R. Ortiz, & M. Shekar (Eds.), *The Routledge handbook of food and nutrition security* (pp. 378-392). Oxford: Routledge Press.

Dixon, J., & Isaacs, B. (2012). There's certainly a lot of hurting out there: Navigating the trolley of progress down the supermarket aisle. *Agriculture and Human Values, 30*(2), 283–297. https://doi.org/10.1007/s10460-012-9409-3 (Online 12 November, 2012).

FAO, IFAD and WFP. (2015). *The state of food insecurity in the World 2015. Meeting the 2015 international hunger targets: Taking stock of uneven progress*. Rome: FAO.

Folke, C., Carpenter, S., Walker, B., & Holling, C. (2004). Regime shifts, resilience and biodiversity in ecosystem management. *Annual Review of Ecology., 35*, 557–581.

Foodbank-End-Hunger in Australia Report (2013). https://www.foodbank.org.au/wpcontent/uploads/2013/10/Foodbank-End-Hunger-Report-2013.pdf

Foodbank Victoria Annual Report. (2016). Retrieved July 20, 2017, from https://www.foodbankvictoria.org.au/wp-content/blogs.dir/18/files/2011/08/Foodbank-Hunger-Report-2016.pdf

de Haan, L. (2000). Globalization, localization and sustainable livelihood. *Sociologia Ruralis, 40*(3), 339–365.

Hajkowicz, S., & Eady, S. (2015). Rural industry futures: Megatrends impacting Australian agriculture over the coming twenty years, publication no. 15/065. Rural Industries Research and Development Corporatio, Commonwealth of Australia. Retrieved January 27, 2017, from https://rirdc.infoservices.com.au/downloads/15-065

Hattersley, L., Isaacs, B., & Burch, D. (2012). Supermarket power, own-labels, and manufacturer counterstrategies: International relations of cooperation and competition in the fruit canning industry. *Agriculture and Human Values, 30*(2), 225–233. https://doi.org/10.1007/s10460-012-9407-5 (Online November 22, 2012).

Hewson, E. (2015). *Sustainability, resilience and equity: An exploratory study of food system reform in Southwest Victoria. Masters of Environment Research Project*. Parkville: University of Melbourne.

Holling, C. S., & Meffe, G. K. (1996). Command and control and the pathology of natural resource management. *Conservation Biology, 10*(2), 328–337.

Hughes, L., Steffen, W., Rice, M., & Pearce, A. (2015). *Feeding a hungry nation: Climate change, food and farming in Australia*. Climate Council of Australia. Retrieved February 1, 2016, from https://www.climatecouncil.org.au/uploads/7579c324216d1e76e8a50095aac45d66.pdf

IPCC. (2014). *Climate change 2014: Impacts, adaptation, and vulnerability. Part A: Global and sectoral aspects*. In: Field CB, VR Barros, DJ Dokken, KJ Mach, MD Mastrandrea, TE Bilir, M Chetterjee, KL Ebi, YO Estrada, RC Genova, B. Girma, ES Kissel, AN Levy, S MacCracken, PR Mastrandea, and LL White (eds) Contribution of working group II to the fifth assessment report of the intergovernmental panel on climate change. Cambridge University Press, Cambridge, 1132 pp.

IPES-Food. (2016). *From uniformity to diversity: A paradigm shift from industrial agriculture to diversified agroecological systems*. International Panel of Experts on Sustainable

Food systems. Retrieved January 20, 2017, from http://www.ipesfood.org/images/Reports/UniformityToDiversity_FullReport.pdf

Lawrence, G., & Campbell, H. (2014). Neoliberalism in the Antipodes: Understanding the influence and limits of the neoliberal political project. In S. A. Wolf & A. Bonanno (Eds.), *The neoliberal regime in the agri-food sector: Crisis, resilience, and restructuring* (pp. 263–283). Abingdon: Routledge.

Magnan, A. (2015). The financialisation of agri-food in Canada and Australia: Corporate farmland and farm ownership in the grains and oilseed sector. *Journal of Rural Studies, 41*, 1–12.

McMichael, P. (2005). Global development and the corporate food regime. *Rural Sociology and Development, 11*, 269–303.

Michael, D. T., & Crossley, R. L. (2012). *Food security, risk management and climate change.* Gold Coast: National Climate Change Adaptation Research Facility.

MSSI. (2015). *Apetite for change: Global warming impacts on food and farming in Australia*, WWF Australia, Sydney. Retrived February 10, 2017, from http://sustainable.unimelb.edu.au/sites/default/files/MSSI_AppetiteForChange_Report_2015.pdf

Murphy, S. (2010). Changing perspectives: Small scale farmers, markets and globalisation. IIED. http://www.hivos.net/Hivos-Knowledge-Programme/

NPIS. (2012). Irrigation in Australia: Facts and figures. Cotton Research and Development Corporation, Narrabri, NSW. Retrived February 3, 2017, from http://www.nswic.org.au/pdf/irrigation_statistics/Facts%20Figures.pdf

Paschen, J.-A., & Beilin, R. (2016). Resilience multiple—Sounding a call for responsible practice. *Dialogues in Human Geography, 6*(1), 41–44.

Productivity Commission. (2003). *Water rights arrangements in Australia and Overseas.* Commonwealth Government, Commission Research Paper, Melbourne. Retrieved November 27, 2012, from http://www.pc.gov.au/research/commission/water-rights

Productivity Commission. (2009). *Government drought support, report no. 46.* Final inquiry.

Qureshi, M., Hanjra, M., & Ward, J. (2013). Impact of water scarcity in Australia on global food security in an era of climate change. *Food Policy, 38*, 136–145.

RMCG & GVFGSWG. (2013). *Goulburn valley fruit growing industry roadmap—Final.* RMCG in conjuction with the Goulburn Valley Fruit Growers Strategic Stakeholder Group. Retrieved February 6, 2017, from http://agriculture.vic.gov.au/__data/assets/word_doc/0019/254332/Goulburn-Valley-Fruit-Growing-Industry-Roadmap-2014.docx

Saefong, M. (2013). Orange juice outshines gold in futures markets, *Marketwatch*, June 7, 2013. Retrieved February 14, 2017, from http://www.marketwatch.com/story/orange-juice-outshines-gold-in-futures-markets-2013-06-07

Santhanam-Martin, M. (2015). Governing agriculture for rural community sustainability: A case study in the Australian dairy industry. Unpublished PhD thesis submitted to the Faculty of Veterinary and Agricultural Sciences. Melbourne: University of Melbourne

Santhanam-Martin, M, Stevens, L (2017) *Understanding apple and pear growers' climate adaptation decision-making.* Rural Innovation Research Group, University of Melbourne, Melbourne.

Siyoum, A.D., Hilhorst, D., & van Uffelen, G.-J. (2012). Food aid and dependency syndrome in Ethiopia: Local perceptions. *The Journal of Humanitarian Assistance.* Retrieved December 20, 2016, from http://sites.tufts.edu/jha/archives/1754

Southern Rural Water. (2016). Southwest limestone local management plan. Retrieved July 20, 2017, from http://www.srw.com.au/wp-content/uploads/2016/05/South-West-Limestone-LMP-May-2016.pdf

State of Victoria. (2008). *Future farming: Productive, competitive and sustainable.* Melbourne: Department of Primary Industries.

Stokes, C. J., & Howden, S. M. (Eds.). (2010). *Adapting agriculture to climate change: Preparing australian agriculture, forestry and fisheries for the future.* Collingwood, VIC: CSIRO Publishing.

21 Vulnerability Amidst Plenty

Sysak, T. (2013). *Drought, power and change: Using bourdieu to explore resilience and networks in two northern Victoria farming communities*. Thesis, Department of Resource Management and Geography, Melbourne School of Land and Environment, The University of Melbourne.

Tauridsky E, Young, J (2009) *Farm gate cold calling: An open or shut case? An evaluation of the practices of farm gate cold-calling in the campaspe shire January 2007—September 2008*. Campaspe Primary Care Partnership. Echuca, VIC. Retrieved from https://www.bouverie.org.au/images/uploads/Farm_Gate_Evaluation_Report.pdf

Walker, J., & Cooper, M. (2011). Genealogies of resilience: From systems ecology to the political economy of crisis adaptation. *Security Dialogue, 42*(2), 143–160.

Walker, B. H., & Salt, D. (2006). *Resilience thinking: Sustaining ecosystems and people in a changing world*. Washington: Island Press.

Watson, A. (1979). Rural policies. In A. Patience & B. Head (Eds.), *From Whitlam to Fraser* (pp. 157–172). Melbourne: Oxford University Press.

Watson, A. (2012). Food security in Australia: Fallacies, fantasies, fancies, foibles and furphies: Inquiry into the Development of Northern Australia Submission 4—Attachment 1. Retrieved from http://www.aph.gov.au/DocumentStore.ashx?id=ca7353ef-12df-4e8b-92fb-078f8199bae6&subId=206491

Webb, L., & Whetton, P. H. (2010). Horticulture. In C. J. Stokes & S. M. Howden (Eds.), *Adapting agriculture to climate change: Preparing Australian agriculture, forestry and fisheries for the future* (pp. 119–136). Collingwood, VIC: CSIRO Publishing.

Webb, P., Braun, J., & Yohannes, Y. (1992). *Famine in Ethiopia: Policy implications of coping failure at National and Household Levels*. Research Report 92, IFPRI.

Welsh, M. (2014). Resilience and responsibility: Governing uncertainty in a complex world. *The Geographical Journal, 180*, 15–26. https://doi.org/10.1111/geoj.12012.

WestVic Dairy. (2014). *Dairy in South West Victoria: Production data*, Dairy Australia. Retrieved February 12, 2017, from http://www.westvicdairy.com.au/AboutWestVicDairy/DairyinSouthWestVictoria/Productiondata.aspx

Wittman, H., Desmarais, A. A., & Wiebe, N. (2010). *Food sovereignty: Reconnecting food, nature & community*. Halifax: Fernwood.

Chapter 22
Agricultural Decision Support Tools: A Comparative Perspective on These Climate Services

Jonathan Lambert, Nagothu Udaya Sekhar, Allison Chatrchyan, and Art DeGaetano

Abstract Climate services such as agricultural decision support tools provide a link between climate information and agricultural practices for farmers, with a goal of improving best management practices and agricultural sustainability through the useful presentation of climate variability and change. Independent organizations throughout the world have developed tools to meet their region's specific needs, and these tools are generally commodity or issue specific. The Cornell Climate Smart Farming Program, an interdisciplinary program of the Cornell Institute for Climate Smart Solutions (CICSS), has developed a website and suite of climate-based agricultural decision support tools aimed at helping farmers make more informed decisions in the face of increasing climate uncertainty. Specific tools were developed based on the major climate impacts to Northeastern US agriculture and through a collaborative development process with stakeholders, researchers, and the Northeast Regional Climate Center. Through this process, CICSS performed a review of decision tools on a national and international scale, and in this text the role and impact of decision support tools are examined, along with the ability of researchers and tool developers to learn from stakeholders and share information via extension specialists. The need for monitoring, evaluation, and coordination among regional programs and organizations is also discussed.

Electronic supplementary material: The online version of this chapter (https://doi.org/10.1007/978-3-319-77878-5_22) contains supplementary material, which is available to authorized users.

J. Lambert · A. Chatrchyan (✉)
Department of Earth and Atmospheric Sciences, Cornell University, Ithaca, NY, USA

Cornell Institute for Climate Smart Solutions, Cornell University, Ithaca, NY, USA
e-mail: amc256@cornell.edu

N. U. Sekhar
Norwegian Institute for Bioeconomy Research, Ås, Norway

A. DeGaetano
Department of Earth and Atmospheric Sciences, Cornell University, Ithaca, NY, USA

Northeast Regional Climate Center, Cornell University, Ithaca, NY, USA

© Springer Nature Switzerland AG 2019
A. Sarkar et al. (eds.), *Sustainable Solutions for Food Security*,
https://doi.org/10.1007/978-3-319-77878-5_22

Keywords Climate services · Decision support tools · Climate change · Agriculture · Farming · Risk

22.1 Introduction

Climate change is imperiling agriculture and food security worldwide, with higher temperatures, changes in precipitation patterns, shifting pest and weed pressures, and changing daily and interannual variability (IPCC 2014). These impacts can lead to climate-related disasters and exacerbate extreme events, directly influencing agricultural productivity, livelihoods, food and water security, land use, market prices, trade policies, economic strategies, and much more (WMO 2014). Because climate variability and change are linked so closely to agriculture, food security, and supply chains, the Intergovernmental Panel on Climate Change (IPCC) recognizes the possible impacts and notes that meteorological information has the potential to improve early warning and preparedness in order to reduce the severity of these climate risks (IPCC 2014). Particularly in agricultural systems, climate forecasting and use of climate information has the ability to increase preparedness and promote better social, economic, and environmental outcomes (Meinke and Stone 2005). Therefore, through the creation and utilization of climate services, including agricultural decision support tools, there exist opportunities to bring together climate information and agricultural development processes to optimize best management practices and policies in developed and developing countries (WMO 2014).

Climate services are defined by the American Meteorological Society (2015) as "providing scientifically based information and products that enhance users' knowledge and understanding about the impacts of climate on their decisions and actions." Climate services are produced by government organizations, universities, nongovernmental organizations (NGOs), and private industry, and include many different types of services, tools, and support systems under their umbrella. For example, the Cornell Climate Smart Farming (CSF) Program through the Cornell Institute for Climate Smart Solutions (CICSS) focuses specifically on developing agricultural decision support tools (DSTs) such as growing degree day calculators, water deficit calculators, freeze risk tools, and more. The decision tool website serves as an interactive platform that integrates climate information in order to support decision-making at a farm or agricultural-system scale within a specific region. Other entities in the USA such as the Southeast Climate Consortium (SECC) and the Useful to Usable (U2U) Team based at Purdue University have also developed comparable regionally based tools, and share a similar scope and objective. Other organizations globally, including the National Institute for Bioeconomy Research (NIBIO) in Norway, the World Food Programme in Africa, and the National Institute for Agricultural Research (INIA) in Uruguay, focus on developing broad-based tools for entire countries or continental regions (Hansen and Coffey 2011). This chapter shares insights from programs and projects such as these to illustrate the range of scales and emphases of agricultural DSTs.

22.2 Decision Support Tools, From Conception to Evaluation

Farmers have been watching and recording the weather for generations with the hopes of determining useful patterns and outlooks to help plan their planting dates, crop varieties, and timing of inputs. Prior to the advent of seasonal weather forecasting in the late twentieth century, this type of climate-based knowledge was passed down through families and communities. However, with the growing impacts of climate change, including increasing extreme weather events and shifting climate variability, solely relying on weather forecasts and historical climatology may not provide enough sophisticated information to help farmers respond to climate change. A lack of timely and targeted responses to the agricultural impacts and stresses of climate change endangers farmers' livelihoods, production, and ultimately the global food supply.

Climate change will ultimately necessitate that agricultural, ecological, and social systems respond across multiple scales (Adger et al. 2005). The use of DSTs is a valuable strategy for adapting to climate change, as these tools promote active participation and engagement with climate data and trends, as well as with the providers of applied climate information. However, developers of DSTs are realizing it is not sufficient to make tools available to farmers without proper training and education on how to use them successfully. Engaging stakeholders from the beginning to gauge their needs and acquire input on usefulness and utility of these tools is critical. In this chapter, a stakeholder is defined as any citizen or group potentially affected by, or having a vested interest in an issue, program, action, or decision leading to an action (Decker et al. 1996). By fully engaging stakeholders in the development and use of DSTs from conception to monitoring and evaluation, developers can better ensure their success and sustainability.

From the first conception of an idea to the development of a DST, the goal in the initial design should be to provide a useful service that is not already available to a specific audience. A demonstrated need or benefit from a DST should catalyze the design, as the value of climate information comes from improved decisions, not solely presentation or availability of information (Hammer 2000). Specifically, with climate-based agricultural DSTs, simply showing the effects of climate variability and providing better climate forecasts is insufficient to raise the adaptive capacity of stakeholders through decision support tools (Jones et al. 2000) because this does not provide them the information or context to make informed decisions. In order to avoid this, Cox (1996) details the benefits of an analytical phase during the development process to deconstruct the models in use, the resolution, validation, and ultimately the appropriateness of the tool as it relates to the intended purpose. This analytical phase is important in quality control for the data input into the tools, as well as for stakeholder engagement so the tool can be accessible enough to be effectively used by stakeholders and affect their behavior change (Breuer et al. 2007).

Researchers have found that rapid adoption of new practices and tools will be most successful if those stakeholders who are most directly affected are included in the decision-making process (Howden et al. 2007; Meinke et al. 2006; Prokopy et al. 2015). Cox (1996) in particular warns against the lack of direct communication between tool developers and stakeholders, as the tool should not serve as just an intermediary, but a segue to conversations between all stakeholders and tool experts. By including decision-makers and stakeholders in the tool development process, designers of DSTs are able to develop resources and programs that are better aimed at an intended audience, allowing for verification of the audience's knowledge of the subject, as well as what is desired and needed when using a DST (Breuer et al. 2009). As Bartels et al. (2013) note, any "process to engage agriculture stakeholders around the issue of climate change that fails to systematically understand and incorporate their values, beliefs, identities, goals, and social networks may fail to bring about the desired adaptation and/or mitigation responses." The USDA's Northeast Climate Hub (2016) also finds that for stakeholders to incur the most benefit from DSTs, the information presented must be easily accessible and in a user-friendly format, applicable to time-sensitive and sector-specific management decisions, and of the correct geographic scale.

As important as the development of tools, the dissemination of DSTs is best served when coupled with a strong outreach effort to increase their use and to measure impact (USDA Notheast Climate Hub 2016). The Cooperative Extension model of providing research-based information and training to help stakeholders make more informed decisions is invaluable for outreach and effective use of DSTs. The Cooperative Extension System (that exists in the USA and in several other developed and developing countries) has been found to be an important "boundary organization" that can span the space between research, practice, and the use of trusted existing social networks and social capital to help facilitate local climate change adaptation (Brugger and Crimmins 2014). It is critically important to engage farmers in training and use of new DSTs via extension specialists who can help ascertain the adaptive responses and benefits of DSTs for stakeholders, and this is most useful when there is a clearly defined adaptive response and benefit for decision-making presented from the tool (Fraisse et al. 2006).

In developing agricultural DSTs and climate services, monitoring and evaluation throughout the process is key to discern the utility of what has been created. With agricultural DSTs, informal evaluation through communication with tool developers, extension professionals, and farmers at meetings, conferences, or other forums can provide a means for developers to obtain initial and ongoing feedback. Additionally, formal evaluation through social science research methods provides specific, in-depth insights into how intended audiences view the tools. Formal evaluation can be conducted through rigorous qualitative methods to gather input via real-life, multilevel perspectives, through interviews or through focus group meetings with farmers or other stakeholders. Quantitative research, through online or mailed surveys, can also assess the magnitude and frequency of responses to a series of questions about the needs, barriers to adoption, and usefulness of DSTs. Pre- and Post-Evaluation surveys can be used before and after training sessions to assess the training and tools, and togather input on needed improvements.

The Global Framework for Climate Services, created through the World Meteorological Organization (WMO) provides an "end-to-end" system for the

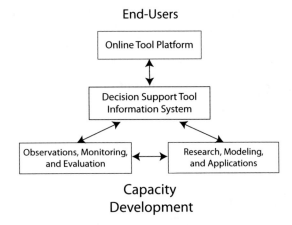

Fig. 22.1 Development process of agricultural DSTs (Adapted from © 2011 World Meteorological Organization)

development of climate services, which is driven by the principles discussed above (WMO 2014). The framework strives for quality data, successful communication and user interaction, targeted monitoring and evaluation, and continuous improvement of climate services. In Fig. 22.1, the framework is modified to specifically describe agricultural decision support tools. The figure graphically represents the connected relationship between the online tool platform, the "behind-the-scenes" decision support tool information system, and how observations, monitoring, evaluation, research, modeling, and applications feed into this feedback system of capacity development and end-user interaction with the tool.

22.3 Stakeholder Access to DSTs

Accessibility is another important factor in the utility and effectiveness of agricultural DSTs. While many tools are created as open and available services by governments and NGOs, the private sector is also playing an increasing role in developing DSTs. Private companies such as IBM and Monsanto are realizing the value of weather and climate information, and have acquired organizations such as the Weather Company and Climate Corps, respectively, in order to capitalize on the market for these services (The Weather Company 2015; Upbin 2013). However, Webber and Donner (2016) caution against the commercialized model for climate services due to the possibility of this limiting the production of useful tools and their application in decision-making as a result of private influence from political priorities and economic austerity. They also note that a commercial model threatens to disadvantage developing countries where the ability to pay for this type of climate information or climate service is limited. Despite these cautions though, privatization of climate services and DSTs will likely continue as observed in the EU region and parts of Asia, including India and China. For instance, in India, many climate services are currently managed by the Indian Meteorological Department (IMD), especially with its recent formation of a Climate Services Division (Rathore and Chattopadhyay 2016). However, as the country pushes to meet goals set by national and state climate action plans, funding gaps are becoming evident, necessitating private investment to implement mitigation and adaptation strategies to meet plan goals (Varma et al. 2015). Given this context, it is worthwhile to briefly discuss the implications of the various systems that are likely to dominate DST development, the benefits and options provided by these systems, and the challenges. The major systems include:

22.3.1 Government-Owned and -Operated Climate Services

Due to the societal relevance of climate change and its impacts and the need for development of innovative open-access products, climate services can be considered viable areas of focus for nonprofits and governments. Thus, government-owned and -operated models will be one of the main structures that will dominate climate services, due to the fact that some governments desire to keep climate related services within their control, on par with other social services they provide. This is increasingly relevant given the need for collection and maintenance of climate data, as this responsibility falls to governments in many countries. For example, in India, climate data is treated with sensitivity and not easily shared with nongovernmental entities. Private organizations and NGOs that collect data independently also cannot release forecasts autonomously, and therefore must cross-check basic forecasts for agriculture with the Indian IMD and Ministry of Earth Sciences. While data quality is preserved in this process, it is also difficult for NGOs and private organizations to guarantee agricultural climate service information on a timely schedule, especially as the desire for information increases among farmers (Ewbank et al. 2015). Therefore, the sustainability of this mechanism for delivery of climate service products by NGOs and private companies should be considered upon analysis of decision-tool development methods.

In another example of the distribution of climate services solely by governments, several levels of administration are often involved. In Europe, the German Climate Service Center (CSC) on the national level, the Joint Programming Initiatives (JPI) on the European level, and the Regional Science Service Centers (RSSC) on the international level (The International Conference on Climate Services 2012) are all involved with the development and distribution of climate services. Since the capacity of governments and international agencies to invest in the development of climate services is a constraint in many cases, DSTs in the agricultural sector will likely remain in the domain of many government entities. In Norway, the government invests and develops climate services and provides the services for free to all users. A good example is the pest and disease-monitoring tool, VIPS, developed with investments from the government and widely used by farmers across the country (NIBIO 2016).

In the USA, climate-related data, research, and information from the National Oceanic and Atmospheric Administration (NOAA) support the private sector and the nation's economy, and allow people to make informed climate-related decisions with tools and resources that help them answer specific questions. Programs like the Regional Climate Centers (RCC) and Regional Integrated Sciences and Assessments have epitomized national climate services.

22.3.2 Privately Owned and Operated DSTs/Climate Services

Private investments in the agricultural sector have increased since the 1980s in production, marketing, and extension, and this increasing trend is also observed in the developing world, where the private sector is beginning to own and operate DSTs/climate services as part of Extension services to farmers. This follows a demand-driven approach, providing farmers with the necessary information, with a focus on increasing productivity. However, farmers' capacity and willingness to pay for these informational tools depends on the reliability and returns received from using the climate information. Private investments can be complementary though when governments do not have the funds to invest on their own, especially in developing countries (Nagothu 2016).

22.3.3 Public–Private Partnerships (PPPs)

A middle-path approach between solely private or public provision of climate services such as DSTs is the public–private partnership approach, where it is possible to bring both private and government interests together to ensure more effective delivery of DSTs to farmers at affordable prices and with a high quality of service. The role of PPPs to provide DSTs/climate services is likely to increase in the future in many countries. One of the reasons attributed is the lack of funding capacity by governments in developing countries to fully support the development of DSTs. Several tools developed by universities using government data and support have since been privatized to allow for continued development and uptake by farmers. One success story along this model is the development of the *Adapt-N* tool by Cornell University, which allows for more efficient application of nitrogen by corn growers based on climate data. The tool has since been privatized by the Agronomic Technology Corporation (Agronomic Technology Corp. 2016), with continued Cornell involvement.

The role of PPPs in the agricultural sector may be crucial given the future projections of climate change and the need for sustainable agricultural development (Nagothu 2016). One example where it is desirable for a PPP is to provide crop insurance to protect farmers from climate change and variability. New models such as weather-based crop insurance indices are emerging where the weather data can be provided by government agencies, and where private insurance companies can complement the tools with the insurance policy and payment. Thus, costs can be shared and farmer interests can be protected to reduce their vulnerability to climate change.

Finally, another line of thought exists that in order to protect the rights of farmers and ensure a fair bargain, climate services should be independent from governmental, business, and political interventions, and should be close to the scientific and academic practices (Brasseur 2015). Regardless of each of these models and the views of different scientists and stakeholders, a close dialogue between different stakeholders in the process is essential irrespective of the approach to providing climate services and tools. It is likely that a combination of approaches (e.g., private, public–private, and government-owned and -operated tools) will continue to emerge in the arena of agricultural DSTs. At a minimum, it is critical that there is adequate funding for free or low-cost government, NGO, or university-developed DSTs that provide research-based information for farmers.

As decision tools and services continue to be developed throughout the world, the factors of quality data, dissemination and communication, evaluation and improvement, and the public and private aspects around these tools need to be taken into account, as well as the supportive frameworks to provide these services. International organizations such as the (WMO) Global Framework for Climate Services and public–private governance mechanisms such as the Global Alliance for Climate Smart Agriculture (GACSA) can support the development and sharing of these climate services.

22.4 Examples of Specific Agriculture and Climate Decision Support Tools Around the Globe

DSTs are often developed with specific regions, commodities, or with specific climate variability, such as El Niño Southern Oscillation (ENSO), in mind. This regional or commodity focus is needed for the success and utility of these tools, as more broad climate forecasts and tools would not provide the specificity required to bring about changes in adaptive capacity at the local or farm level. In order to gain a perspective on the various types, ranges, and applications of DSTs throughout the world, the next section presents a series of examples, beginning with the tools developed by the authors and broadening out to other related and relevant tools. A brief introduction to each tool is included in Table 22.1.

22.4.1 Cornell Climate Smart Farming Tools

The Climate Smart Farming (CSF) Program at the Cornell Institute for Climate Smart Solutions (CICSS) has developed a set of climate and agriculture DSTs for farmers and other stakeholders in the Northeastern USA. The CSF tools combine weather and climate data with agricultural models and are updated on a daily basis to create accurate short-term projections. The CSF Program is also working to incorporate data

22 Agricultural Decision Support Tools

Table 22.1 Climate-based agriculture DSTs discussed in this text, organized by name, region of focus, organization that created the tool, and tool focus

Tool/Website	Region	Organization	Tool focus
Cornell Climate Smart Farming	Northeast (NE) USA	Cornell Institute for Climate Smart Solutions (CICSS)	Decision support for NE-specific farms, using historical climate data, crop information, and climate change projections
VIPS (Plant and Disease Forecasting Service)	Norway	National Institute for Bioeconomy Research (NIBIO)	Decision support and Web-based information and notification of pests and diseases
AgroClimate	Southeast (SE) USA	Southeast Climate Consortium (SECC)	Decision support for SE-specific farms using historical climate data and crop information
Africa RiskView	Africa	World Food Programme	Food security outlooks using climate data, crop parameters, and livelihood information

from downscaled climate change projections into its tools for the Northeast, which will provide a long-term climate context for the data shown in each tool. This will provide context for how climate change may affect certain parameters in the future, helping stakeholders make decisions based on future climate scenarios.

The CSF Program began developing DSTs in 2014, and current tools available include a Growing Degree Day Calculator, Freeze Risk Tools, and a Water Deficit Calculator. In creating these tools, the CSF program collaborated heavily with the Northeast Regional Climate Center (NRCC) at Cornell University-part of the RCC Program administered by NOAA in the USA (DeGaetano et al. 2010). The tools rely on the NRCC-led data through the Applied Climate Information System (ACIS), which is an operational system that provides access to climate data and products to users via Web services (DeGaetano et al. 2014), and is replicated at multiple RCCs throughout the country. While this system is used to drive the CSF decision tools, as well as NRCC-specific tools, it is also the base for climate tools being developed elsewhere, such as the Useful to Usable website from Purdue University in Indiana, and the AgroClimate tools developed for the Southeast USA. In particular, the CSF Tools also utilize daily temperature observations from the National Weather Service (NWS) Cooperative Observer Network, and daily precipitation derived from NWS radar data. These data are interpolated to a 4 km × 4 km grid following the methods of DeGaetano and Belcher (2007), and DeGaetano and Wilks (2009), respectively. This use of interpolated gridded data allows farmers in the region to access accurate information for their farm via the CSF DSTs even if there is not a government-operated or private weather station near their site.

In keeping with a participatory approach to tool development, the CSF tools have been created with a priority placed on farmer needs and utility. The CSF Program prioritized development of certain tools initially based on an assessment of the key agricultural production areas in the Northeastern USA, key vulnerabilities, and input on key information that would help farmers increase climate adaptation. Examples of these assessments include the agriculture chapter of the Climaid Report commissioned by the State of New York, and an assessment commissioned by the USDA NE Climate Hub (Tobin et al. 2015)

Through these assessments, the CSF team determined the critical impacts, vulnerabilities and opportunities of climate change to NE agriculture, including longer growing season, warmer temperatures during the growing season, freeze risk to grapes and apples (a key product in New York and the Northeast), and the potential for short term drought (Rosenzweig et al. 2011; Tobin et al. 2015). Other tools will be prioritized for development to address other key risks in the Northeast, including heat stress in dairy, and waste and water management.

In the development of the CSF tools, members of the project team interacted with research faculty, extension specialists, farmers, and other stakeholders at annual meetings and conferences, focus groups, and through other forums to determine the needs for the CSF DSTs in the Northeast. The CSF team also conducted formal outreach at major agricultural meetings in the Northeast, introducing the CSF DSTs to facilitators, guiding discussions to understand participants' willingness to use these tools, and getting feedback on the tools' utility and design. Cooperative Extension specialists were also consulted, as well as the CSF Farmer Advisory committee, in the design and facilitation of the DSTs, focus group workshops, and surveys (Chatrchyan 2016). Surveys of agricultural producers and agricultural advisers are being conducted to assess farmers' perception of risk related to climate change, and decision-making factors related to agricultural management, with a specific emphasis on weather and climate information and tools.

To provide an example of the CSF tool interface, the first decision tool created for the CSF website—the Growing Degree Day (GDD) Calculator—is illustrated in Fig. 22.2 and further explained (along with the other relevant tools) in the next section. In keeping with the US land grant mission of Cornell University to provide applied research results to stakeholders to improve their lives and businesses, the GDD Calculator and other tools were developed using federal research and foundation dollars and are made available to end-users free-of-charge. The CSF website also includes links to regionally specific tools developed by organizations such as the NRCC and the Network for Environment and Weather Applications (NEWA), national products from NOAA and the National Drought Mitigation Center/ University of Nebraska, and privatized tools such as Adapt-N.

22.4.1.1 CSF Growing Degree Day (GDD) Calculator

As the pilot tool for the CSF website, the GDD calculator became available to the general public in early 2016. GDD measures heat accumulation to help agricultural producers predict when crops (or pests) will reach important developmental stages, and growing degree day calculators are prevalent in the toolboxes of many organizations. However, the CSF GDD tool is unique as it is designed to represent the climate change context for farmers in the Northeastern USA by using a moving 15-year average of the most recent climate data, as opposed to the standard 30-year climate normal. The 15 year average captures the most recent climate change signal (Wilks

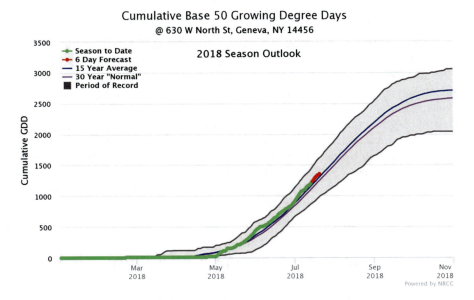

Fig. 22.2 The Climate Smart Farming growing degree day calculator tool output for Geneva, NY during summer 2018. www.climatesmartfarming.org

and Livezey 2013) and allows farmers to better contextualize GDD accumulation given climate change, and potentially improve crop yields based on enhanced information about the best time periods to fertilize, apply pesticides, and harvest crops.

22.4.1.2 CSF Grape and Apple Freeze Risk Tools

One benefit of agricultural DSTs is the ability for developers to tailor them specifically as desired. Grapes and apples are grown throughout the Northeastern USA, and the economic viability of farming in these areas is greatly tied to the fruit crops. With climate change, growing grapes and apples may become even more lucrative due to the opportunities for more varieties given a warmer climate or declines in other regions. But the risk for killing freezes may not disappear as quickly as the temperatures rises, resulting in increased risk for damaging cold snaps, which have been seen (especially in the spring after warmer winters have forced fruit trees out of dormancy) more frequently in the region in the past decade. In order to provide a warning for when freeze risk to grapes and apples may occur, the CSF team developed these tools to provide farmers with a graphical indication of potential freeze risk based on hardiness temperature. The tool graphs hardiness temperature versus observed temperature over a specified date range, and serves as a warning system and barometer for possibly damaging freeze events to multiple crop varieties.

22.4.1.3 CSF Water Deficit Calculator

One of the most recently developed CSF tools, and possibly the most used due to the serious short-term drought farmers experienced in the Northeast in 2016, is the CSF water deficit calculator. This tool estimates soil water content within a crop's effective root zone to inform decision-makers about current and forecasted water deficits. This information is used by farmers and irrigation system managers to determine the optimum frequency and duration of watering that is necessary to avoid plant stress. The tool is also the first CSF tool to incorporate longer term climate information by calculating water deficit probabilities over the next 30 days from historical data (2002–present), with plans to incorporate climate projections from the CMIP5 climate models to provide context for how climate change will affect the potential for water deficits by season and crop type in the future.

22.4.2 Pest and Disease Forecasting Service (VIPS), Norway

VIPS is a web-based notification and information service designed for integrated pest and diseases in cereals, oilseeds, potatoes, vegetables, and fruit crops. VIPS also includes a decision support program for selection of agent and dosage against weeds in corn. VIPS was established in 2001 as a joint project between the then Bioforsk and Norwegian Agricultural Counseling (NLR) under the publicly funded "Action Plan to reduce risks of pesticide." The service is open and free of cost to all users.

While it is well known that the processes of plant growth, pest population dynamics and plant disease epidemiology are all weather driven processes, and thus that crop protection against pests must adapt to climate change, is it less well known that reducing crop losses to pests can also mitigate climate change. The IPCC has reported that agriculture is responsible for over a quarter of total global greenhouse gas emissions (IPCC 2007). Loss of crop yields to plant pests and disease therefore causes unnecessary greenhouse gas emissions and consequently the emissions per unit food produced increases with increasing loss to pests. As a result, a reduction in crop loss to pests will reduce greenhouse gas emissions from food production. In other words, in the field of crop protection, climate change adaptation and mitigation are two sides of the same coin. Despite the significant improvement of pest management practices, there are still significant losses caused by diseases, insect pests and weeds that incur above 30% of the production for crops in general, and 37% for rice.

Development of "monitoring tools and forewarning systems" is thus identified as one of the key strategies for effective pest management. The potential in innovation through forecasting of pest and disease risks is based on the reasoning that farmers' daily adaptation to the day-to-day variability in weather is a short-term analogy to the need for adaptation to long-term changes in climate. While it is uncertain what the exact effects of climate change will actually be in each farming region, it is certain that climate change will affect crop-pest interactions (IPCC 2014). Therefore, any new scientific insights, strategies and tools that can be used to adapt to variability, both within-season and between-seasons, are also relevant for adaptation to future climate change.

Models for pest and disease forecasting tools synthesize scientific knowledge of the biological responses to weather and other environmental conditions of pest and disease organisms formulated into mathematical logics or equations. The models are driven by weather and/or biological data inputs to output predictions about the risk of pest and disease attacks (NIBIO 2016). Models that alert the risk of pest attack or disease outbreak can help advisors and growers decide whether or when it is necessary to use pesticides. Treatment with pesticides only as needed will make plant protection cheaper and reduce adverse environmental impacts compared to traditional spray applications. VIPS currently uses weather data from the Agriculture Meteorological Service (LMT) of Norway consisting of a network of 82 automatic weather stations located in the main agricultural areas, beside weather forecast data from the Norwegian Meteorological Institute and pest damage surveys in the field done by NLR personnel (Brodal et al. 2007).

VIPS is constantly evolving and improving existing models and new models for several pests and diseases. A new version of VIPS was released in 2016 for easier integration of several models, and to include multiple languages—allowing for international cooperation and implementation of VIPS in other countries (NIBIO 2016). Norway is a country with varying topography, which means that climatic conditions change significantly over short distances. By including weather forecasts, the tool will contribute to greater precision in forecasts, particularly for farms located far away from the nearest station. Efforts are being made to test the accuracy of forecasts compared with measured values, and what impact they will have in providing practical warnings.

22.4.3   AgroClimate Website

The AgroClimate website of decision tools was developed by the Southeast Climate Consortium (SECC) in order to provide necessary information and tools on adaptations being used to combat varying climate forecasts in the Southeastern USA (Fraisse et al. 2006). In developing their suite of tools, the SECC invested significant effort to understand the potential benefits and needs of climate forecasts for the main agricultural commodities in the Southeastern USA (Jones et al. 2000). The tools were developed by compiling a database on climate forecasts within the region (Fraisse et al. 2006). In creating the AgroClimate tools, developers established a climatological database by averaging daily weather values from the past 50 years into "monthly values of average minimum temperature, average maximum temperature, and total precipitation." Subsequently, crop models of peanut, tomato, and potato plants were developed in order to adequately assess the risk crops could endure under varying climate forecasts within the region. Finally, the team developed prototypes which were distributed to extension specialists and producers for testing and evaluation (Fraisse et al. 2006). From this, two tools were initially

Table 22.2 Useful survey questions for evaluation of DSTs, developed by AgroClimate (Breuer et al. 2009)

AgroClimate Decision Support Tool Survey Questions
Are producers interested in climate information, especially climate forecasts?
What level of accuracy in forecasts do decision-makers want before using them as decision aids?
What are the management options that producers can adopt in light of climate forecasts?
How should seasonal climate variability forecasts be presented and delivered?
How do potential users evaluate the content and presentation available in AgroClimate?

created–; one of which allowed users to analyze the risk associated with climate forecasts and the other allowing users to analyze and simulate the risk to crop yield based on planting dates and climate forecasts (Fraisse et al. 2006).

After the creation of the initial AgroClimate site, the creators used a series of questions (Table 22.2) in order to determine the effectiveness and usefulness of the suite of tools and to elucidate further fields for development. These questions can be applied to most agricultural decision support tools (Breuer et al. 2009).

Currently, AgroClimate is the main information delivery mechanism in the Southeastern USA and includes multiple tools focusing on climate risk, crop yield risk, crop development, crop disease, and climate information such as drought indices and climate forecasts/outlooks (Farahani et al. 2010). The latest version of the AgroClimate Climatolgy tool is shown in Fig. 22.3, which along with other tools on the site, can be used to compare growing conditions given different ENSO phases.

22.4.4 Other International DSTs

Africa RiskView provides decision support tools for countries in Africa, and is available online and as a downloadable software package. It was developed by the World Food Programme to monitor weather-related food security risk in Africa, with a primary focus on drought. Africa RiskView creates food security outlooks by translating globally available rainfall data, crop parameters, and livelihood information (Hansen and Coffey 2011). The tool also calculates response costs to severe drought and food security events to assist in financial planning and facilitate rapid resource allocation. The tool estimates drought in a region via satellite rainfall estimates to calculate a Water Requirement Satisfaction Index. This allows the user to determine how the season is evolving precipitation-wise, observe potential weather-related impacts on agriculture, and estimate the number of people potentially affected and in need of aid, as well as estimate the potential response costs at the continental, regional, and country levels (Climate and Disaster Risk Solutions 2011; Hansen and Coffey 2011).

Other similar tools such as Uruguay's National Agricultural Information System (SNIA) also inform climate risk management decisions from the farm to the national scale and have been developed through governments and public–private partnerships

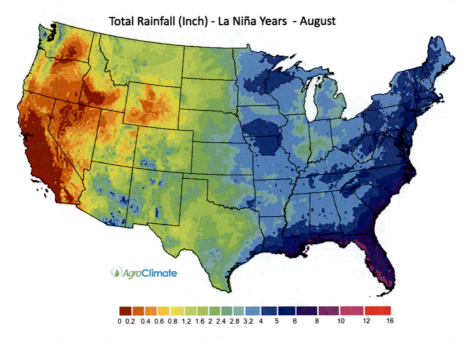

Fig. 22.3 The Southeast climate consortium's AgroClimate climatology tool output for rainfall in august during a La Niña year. www.agroclimate.org

in order to promote the sustainability of agriculture and food security at multiple scales (Hansen and Coffey 2011; World Bank and CIAT 2015). Uruguay's NAIS delivers climate information and DSTs to stakeholders through collection of spatial information and tools developed by the National Institute for Agricultural Research, which partners with the International Research Institute for Climate and Society at Columbia University. The system combines monitoring of weather, vegetation conditions, seasonal climate forecasts, and soil moisture content within an internet-based GIS platform to inform climate risk management decisions. For example, in 2010, the Uruguay Ministry of Agriculture promoted adaptation measures based on the system's outlook for an elevated probability for below-average rainfall by offering low-interest loans and encouraging the use of drought-resistant crop. This outlook was validated months later when official drought emergencies were declared, and farmers who heeded the warning benefited. Due to successes such as this one, Uruguay's SNIA is receiving high levels of support from the Ministry of Agriculture and the World Bank, allowing for continued improvement of agricultural data, outlooks, forecasts, tools, climate risk assessments, and response to increased demand for outputs from this decision support system (Hansen and Coffey 2011).

The majority of the aforementioned DST websites and services in this chapter have been developed in the past 5–10 years, though older decision support systems such as Yield Prophet, which has served Australia since 1994, are also available. This tool provides field-specific soil, crop, and management data that can nearly double yield potentials in water limited systems (Hansen and Coffey 2011; "Yield Prophet," 2017). The tool uses simulated growing conditions from planting to harvest based on a hundred years of weather data to produce a comprehensive crop report that is compatible with computers, smart phones, and tablets. Farmers are assisted in their use of the tool by a not-for-profit agricultural research organization that helps users select appropriate climate forecasts, weather data, and field-specific information. Yield Prophet also has additional features beyond the basic crop report, allowing farmers to access the latest seasonal climate forecasts and impacts on yields, such as updates on yield potential, growth stage, water stress, nutrient stress, frost/freeze risk, etc. And finally, the tool responds to inputs such as fertilizer and irrigation in order to provide simulated scenarios given multiple possible strategies, allowing farmers to test a myriad of possible weather and climate impact adaptation schemes (Hansen and Coffey 2011).

Delivery mechanisms of DSTs throughout the world are also varied. While access to efficient internet networks is rising, it is not as widespread in developing countries. In order to circumvent this, SMS-based text message delivery has been considered in certain regions. A specific tool in Haiti—the Noula Platform—is a good example of mobile phone-based delivery of climate service information for rapid adaptation (Hansen and Coffey 2011; "Noula – Crisis Management Portal – Haiti," 2017). This platform provides information based on local weather forecasts and partners with the country's largest cell service providers so that selected telephone towers can be targeted to ensure timely delivery of messages. The public can also interact with this tool by calling a short code to receive additional information via text and report on emergencies. While not originally designed for agricultural use, this system is prime for adoption by agricultural stakeholders because of its potential use for informing farming communities about pending risks and allowing for possible rapid adaptive management strategy implementation (Hansen and Coffey 2011).

22.5 Synthesis and Conclusions

With the proliferation of climate services and agricultural decision support tools from organizations such as the ones detailed above, it is evident that the spread of applicable climate and weather resources to stakeholders can promote adaptation and reduced risk in the face of climate change, and ultimately promote a more food secure planet. Agricultural producers that can make efficient use of these precision climate and agricultural model outputs should be better able to make more informed decisions from the farm level to the institutional, community, and policy level. However, achieving this decision-making capability requires even further development of climate services and DSTs, and support for integration of these tools into the daily decision-making process alongside other adaptation measures to climate extremes and change (WMO 2014). It is also not enough for modelers and programmers to build an increasing number of online tools and expect them to be used. There are serious questions about access to technology, training, and support that need ongoing focus. Cooperative Extension advisors and commercial farm consultants can play a key role in training and information dissemination if there is adequate support.

Farmers must also understand the value of this climate information and be able to easily access it. One of the largest hurdles for both tool development and extension outreach is adequate capacity and funding, which is a key limiting factor for the scope of many tools being developed, such as the Cornell CSF tools and others. Moving forward, funding agencies that understand the need for more accurate models and tools for agricultural producers need to provide increased funding for their development and maintenance, and to ensure that regional systems of tools are being coordinated and linked. The private sector is increasingly interested in applicable data opportunities. Public–private partnerships, along with foundations, may become increasingly critical to ensure adequate funding and stability for these models and suites of tools, while at the same time ensuring access for all types of users, with varying abilities to obtain and pay for technology. With further research and development at all scales, agricultural decision support tools will play a role in the continued development of sustainable practices and promotion of food security under a changing climate.

The paper is explained in the video titled "Agriculture decision making tool by CUCSS".

Acknowledgements We thank Rick Moore and Brian Belcher for programming the Climate Smart Farming Tools and making these possible in the year, and Savannah Acosta for compiling and analyzing literature on decision support tools. We also thank our collaborators at the Norwegian Institute for Bioeconomy Research. We appreciate the organizers of the Conference on Weather and Climate Decision Tools for Farmers, Ranchers, and Land Managers, at the University of Florida, which led to a great exchange of ideas in preparation for finalizing this manuscript. And finally, we would like to acknowledge our funders: USDA NIFA Federal Capacity Funds (Hatch and Smith Lever funds), the USDA NE Climate Hub (through an Agricultural Research Service, Cooperative Agreement), and insightful funding from the New World Foundation, Local Economies Project. We gratefully acknowledge all support provided for this project.

References

Adger, W. N., Arnell, N. W., & Tompkins, E. L. (2005). Successful adaptation to climate change across scales. *Global Environmental Change, 15*(2), 77–86. https://doi.org/10.1016/j.gloenvcha.2004.12.005.

Agronomic Technology Corp. (2016). *Adapt-N*. Retrieved from http://www.adapt-n.com/

American Meteorological Society. (2015). *Climate services*. Retrieved from https://www.ametsoc.org/ams/index.cfm/about-ams/ams-statements/statements-of-the-ams-in-force/climate-services1/.

Bartels, W.-L., Furman, C. A., Diehl, D. C., Royce, F. S., Dourte, D. R., Ortiz, B. V., et al. (2013). Warming up to climate change: A participatory approach to engaging with agricultural stakeholders in the Southeast US. *Regional Environmental Change, 13*(1), 45–55. https://doi.org/10.1007/s10113-012-0371-9.

Brasseur, G. (2015). *Climate services and the private sector*. Hamburg: Climate Service Center HZG Retrieved from http://www.climate-services.org/wp-content/uploads/2015/05/Brasseur-Private-Sector.pdf.

Breuer, N. E., Cabrera, V. E., Ingram, K. T., Broad, K., & Hildebrand, P. E. (2007). AgClimate: A case study in participatory decision support system development. *Climatic Change, 87*(3–4), 385–403. https://doi.org/10.1007/s10584-007-9323-7.

Breuer, N. E., Fraisse, C. W., & Hildebrand, P. E. (2009). Molding the pipeline into a loop: The participatory process of developing AgroClimate, a decision support system for climate risk reduction in agriculture. *ResearchGate, 3*(1). Retrieved from https://www.researchgate.net/publication/267397537_Molding_the_pipeline_into_a_loop_the_participatory_process_of_developing_AgroClimate_a_decision_support_system_for_climate_risk_reduction_in_agriculture.

Brodal, G., Schiøll, A., Hole, H., Brevig, C., & Rafoss, T. (2007). *VIPS – warning and prognoses of pests and diseases in Norway*. NJF report. Retrieved from http://www.bioforsk.no/ikbViewer/page/publication?p_document_id=33310

Brugger, J., & Crimmins, M. (2014). Designing institutions to support local-level climate change adaptation: Insights from a case study of the U.S. Cooperative Extension System. *Weather, Climate, and Society, 7*(1), 18–38. https://doi.org/10.1175/WCAS-D-13-00036.1.

Chatrchyan, A. (2016). *USDA Project initiation: Project N. NYC-Chatrchyan, Multistate no. NC1179*.

Climate and Disaster Risk Solutions. (2011). *Africa RiskView online*. Retrieved from http://www.un-spider.org/sites/default/files/AfricaRiskViewOnlineNewsletter.pdf.

Cox, P. G. (1996). Some issues in the design of agricultural decision support systems. *Agricultural Systems, 52*(2), 355–381. https://doi.org/10.1016/0308-521X(96)00063-7.

Decker, D. J., Krueger, C. C., Jr, R. A. B., Knuth, B. A., & Richmond, M. E. (1996). From clients to stakeholders: A philosophical shift for fish and wildlife management. *ResearchGate, 1*(1), 70–82. https://doi.org/10.1080/10871209609359053.

DeGaetano, A. T., & Belcher, B. N. (2007). Spatial interpolation of daily maximum and minimum air temperature based on meteorological model analyses and independent observations. *Journal of Applied Meteorology and Climatology, 46*(11), 1981–1992. https://doi.org/10.1175/2007JAMC1536.1.

DeGaetano, A. T., & Wilks, D. S. (2009). Radar-guided interpolation of climatological precipitation data. *International Journal of Climatology, 29*(2), 185–196. https://doi.org/10.1002/joc.1714.

DeGaetano, A. T., Brown, T. J., Hilberg, S. D., Redmond, K., Robbins, K., et al. (2010). Toward regional climate services: The role of NOAA's regional climate centers. *Bulletin of the American Meteorological Society, 91*(12), 1633–1644.

DeGaetano, A. T., Noon, W., & Eggleston, K. L. (2014). Efficient access to climate products using ACIS Web services. *Bulletin of the American Meteorological Society, 96*(2), 173–180. https://doi.org/10.1175/BAMS-D-13-00032.1.

Ewbank, R., Ingham, A., & Schechter, E. (2015). *Developing climate services: experiences in fecast use from Kenya, India and Nicaragua. Programme experience*. London: Christian Aid Retrieved from http://programme.christianaid.org.uk/programme-policy-practice/sites/default/files/2016-03/developing-climate-services-experience-feb-2015.pdf.

Farahani, H., Fraisse, C., Templeton, S., Davis, R., & Khalilian, A. (2010). *Agroclimate tools for South Carolina. Proceedings of the 2010 South Carolina Water Resources Conference*. Columbia, SC: Columbia Metropolitan Convention Center.

Fraisse, C. W., Ingram, K. T., Garcia, Y., Garcia, A., Hoogenboom, G., Hatch, U., Cabrera, V. E., Zierden, D., Breuer, N., Paz, J., & Bellow, J. (2006). AgClimate: A climate forecast information system for agricultural risk management in the southeastern USA [electronic resource]. *Computers and Electronics in Agriculture, 53*(1), 13–27. https://doi.org/10.1016/j.compag.2006.03.002.

Hammer, G. (2000). A general systems approach to applying seasonal climate forecasts. In G. L. Hammer, N. Nicholls, & C. Mitchell (Eds.), *Applications of seasonal climate fore-

casting in agricultural and natural ecosystems (pp. 51–65). Retrieved from http://link. springer.com/chapter/10.1007/978-94-015-9351-9_4). Dordrecht: Springer. https://doi. org/10.1007/978-94-015-9351-9_4.

Hansen, J., & Coffey, K. (2011) *Agro-climate tools for a new climate-smart agriculture.* Retrieved from https://cgspace.cgiar.org/handle/10568/21666

Howden, S. M., Soussana, J.-F., Tubiello, F. N., Chhetri, N., Dunlop, M., & Meinke, H. (2007). Adapting agriculture to climate change. *Proceedings of the National Academy of Sciences, 104*(50), 19691–19696. https://doi.org/10.1073/pnas.0701890104.

IPCC. (2007). *Climate change 2007: Synthesis report. Contributions of working groups I, II, and III to the fourth assessment report of the intergovernmental panel on climate change.* Geneva: IPCC Retrieved from https://www.ipcc.ch/pdf/assessment-report/ar4/syr/ar4_syr.pdf.

IPCC. (2014). *Climate change 2014: Synthesis report. Contribution of working groups I, II and III to the fifth assessment report of the intergovernmental panel on climate change* (p. 151). Geneva: IPCC.

Jones, J. W., Hansen, J. W., Royce, F. S., & Messina, C. D. (2000). Potential benefits of climate forecasting to agriculture. *ResearchGate, 82*(1–3), 169–184. https://doi.org/10.1016/ S0167-8809(00)00225-5.

Meinke, H., & Stone, R. C. (2005). Seasonal and inter-annual climate forecasting: The new tool for increasing preparedness to climate variability and change in agricultural planning and operations. *Climatic Change, 70*(1–2), 221–253. https://doi.org/10.1007/s10584-005-5948-6.

Meinke, H., Nelson, R., Kokic, P., Stone, R., Selvaraju, R., & Baethgen, W. (2006). Actionable climate knowledge: From analysis to synthesis. *Climate Research, 33*(1), 101–110. https://doi. org/10.3354/cr033101.

Nagothu, U. S. (2016). *Climate change and agricultural development: Improving resilience through climate smart agriculture.* Agroecology and Conservation: Routledge.

NIBIO. (2016). *VIPS – Beslutningsstøtte for integrert plantevern - Bioforsk.* Retrieved from http://www. bioforsk.no/ikbViewer/page/prosjekt/tema?p_dimension_id=23995&p_menu_id=24011&p_sub_ id=23996&p_dim2=24003

Noula – Crisis Management Portal – Haiti. (2017). Retrieved from http://www.noula.ht/

Prokopy, L. S., Hart, C. E., Massey, R., Widhalm, M., Klink, J., Andresen, J., et al. (2015). Using a team survey to improve team communication for enhanced delivery of agro-climate decision support tools. *Agricultural Systems, 138*, 31–37. https://doi.org/10.1016/j.agsy.2015.05.002.

Rathore, L. S., & Chattopadhyay, N. (2016). *Weather and climate services for farmers in India.* Geneva: World Meteorological Organization Retrieved from https://public.wmo.int/en/ resources/bulletin/weather-and-climate-services-farmers-india.

Rosenzweig, C., Solecki, W., & DeGaetano, A. (2011). *Responding to climate change in New York State: Synthesis report.* Albany, NY: The New York State Energy Research and Development Authority.

The International Conference on Climate Services. (2012) *Toward a climate service enterprise conference report.* Paper presented at the The Second International Conference on Climate Services. Brussels, Belgium. Retrieved from http://www.climate-services.org/wp-content/ uploads/2015/05/iccs2_report_screen_low_resolution-2.pdf

The Weather Company. (2015). *IBM plans to acquire The Weather Company's product and technology businesses; extends power of Watson to the internet of things.* Retrieved from https:// business.weather.com/news/ibm-plans-to-acquire-the-weather-companys-product-and-tech- nology-businesses-extends-power-of-watson-to-the-internet-of-things

Tobin, D., Janowiak, M., Hollinger, D., Skinner, R., Swanston, D., Steele, R., et al. (2015). *Northeast and Northern Forests Regional Climate Hub assessment of climate change vulnerability and adaptation and mitigation strategies* (p. 65). Davis, CA: Climate Hub USDA.

Upbin, B. (2013). Monsanto buys climate corp for $930 million. *Forbes* Retrieved from http:// www.forbes.com/sites/bruceupbin/2013/10/02/monsanto-buys-climate-corp-for-930-million/.

USDA Notheast Climate Hub. (2016). *Research and extension education needs white paper (draft).* Retrieved from http://www.climatehubs.oce.usda.gov/northeast/regional-assessments/regional

Varma, A., Le-Cornu, E., Madan, P., & Dube, S. (2015). *The role of the private sector to scale up climate finance in India*. New Delhi: Deutsche Gesellschaft für Internationale Zusammenarbeit (GIZ) GmbH Retrieved from https://www.giz.de/en/downloads/giz2015-en-nama-india-private-financial-institutions-climate-finance-final-report.pdf.

Webber, S., & Donner, S. D. (2016). Climate service warnings: cautions about commercializing climate science for adaptation in the developing world. *Wiley Interdisciplinary Reviews: Climate Change*. https://doi.org/10.1002/wcc.424.

Wilks, D. S., & Livezey, R. E. (2013). Performance of alternative "normals" for tracking climate changes, using homogenized and nonhomogenized seasonal U.S. surface temperatures. *Journal of Applied Meteorology and Climatology, 52*(8), 1677–1687. https://doi.org/10.1175/JAMC-D-13-026.1.

World Bank and CIAT. (2015). *Climate-smart agriculture in Uruguay*. Washington, DC: The World Bank Group Retrieved from http://sdwebx.worldbank.org/climateportal/doc/agricultureProfiles/CSA-in-Uruguay.pdf.

World Meteorological Organization. (2014). *Agriculture and food security exemplar to the user interface platform of the global framework for climate services*. Retrieved from https://www.wmo.int/gfcs/sites/default/files/Priority-Areas/Agriculture%20and%20food%20security/GFCS-AGRICULTURE-FOOD-SECURITY-EXEMPLAR-FINAL-14147_en.pdf

Yield Prophet. (2017). Retrieved from http://www.yieldprophet.com.au/yp/WhatIsYP.aspx

Chapter 23
Multilevel Governance for Climate Change Adaptation in Food Supply Chains

Ari Paloviita and Marja Järvelä

Abstract The vulnerability of food supply chains to climate change is higher compared to other industries due to its dependency on climatic conditions, temperature and water supply. As a robust response to the vulnerability of food supply chains, it is essential to find ways of linking the concepts of sustainable development, climate change adaptation and risk governance into one paradigm. The risk governance of food supply chains is conducted by and across both private and public spheres. Hence, in this chapter, we introduce a dual system of governance to match the objectives of climate change adaptation, and discuss the multiplicity and potential integration of both corporate-led private governance and public governance based on the authority of governments and their institutions. The aim of this chapter is to highlight climate change adaptation in relation to the practices of risk governance of the food supply chains within a multilevel framework of private and public policies. It explores the outlook of climate change adaptation in food supply chains, probing the extent to which governance should be framed as an intergovernmental issue, a national/local issue, an upstream supply chain issue or a downstream supply chain issue. The study is carried out by delving into the international adaptation literature with focus on different levels of framing the food supply chain and its adaptation to climate. We conclude that it is important to marry the efficiency of food businesses with the attainment of wider societal objectives such as sustainable development, climate change adaptation and food security, in order to increase resilience of the overall food system.

Keywords Multilevel governance · Climate change adaptation · Food supply chain · Vulnerability · Sustainable development · Resilience

---

Electronic supplementary material: The online version of this chapter (https://doi.org/10.1007/978-3-319-77878-5_23) contains supplementary material, which is available to authorized users.

A. Paloviita (✉) · M. Järvelä
Department of Social Sciences and Philosophy, University of Jyväskylä, Jyväskylä, Finland
e-mail: ari.paloviita@jyu.fi

© Springer Nature Switzerland AG 2019
A. Sarkar et al. (eds.), *Sustainable Solutions for Food Security*,
https://doi.org/10.1007/978-3-319-77878-5_23

23.1 Introduction

Food supply chains are characterized by increasing connectivity and complexity, which affect their vulnerability to unexpected changes. There is growing understanding that climate change will have an increasing impact on food supply chains and food security (Beermann 2011, Paavola et al. 2015; Paloviita and Järvelä 2015). This means that the need for adaptation to climate change impacts becomes increasingly apparent (Dow et al. 2013) and indicates that the adaptive capacity of the food supply chain with regard to its resilience must be addressed (Smith et al. 2003). In fact, climate change is one among a set of interconnected food systems risks which are also influenced by rapid changes in biodiversity, land cover, availability of fresh water, and the nitrogen and phosphorus cycles (Vermeulen et al. 2012). This calls for the development of theoretical and practical frameworks to strengthen the capacity of food supply chains to respond to climate change impacts. This chapter aims to highlight from a network perspective climate change adaptation and the practices of risk governance of the food supply chains within a multilevel framework of private and public policies. Although climate change adaptation has been primarily a matter that is dealt with on a regional and local scale, we argue that food supply chains, governments and intergovernmental agencies have a crucial role in climate change adaptation.

Impacts of climate change, which are expected to be widespread, complex, temporally and geographically variable, and profoundly influenced by social and economic conditions (Vermeulen et al. 2012), can pose major threats to food supply chain resilience. Food supply chains have been typically designed to optimize for cost or customer service (Christopher and Peck 2004), but with climate risk reduction in mind, resilience needs to be highlighted as an additional new objective. As a critical dimension of resilience, adaptive capacity refers to the degree to which the system can build and increase the capacity for learning and adaptation (Klein et al. 2003). Ability to cope with climate change impacts effectively requires that all key activities related to the food supply chain are analyzed in terms of identifying climate risks. After analyzing the risk-related information of all food system activities, the assessment of overall vulnerability of the food supply chain is needed. We argue that effective coping with climate change risks and vulnerabilities calls for a multilevel governance framework, including both public (on a governmental level) and private (on a corporate level) policies, reflecting an essentially dual system of governance. This multilevel governance framework is essential in order to build both food system resilience and food security. In addition, we emphasize the need to address a strategic combination of climate change mitigation and adaptation requirements at the same time (Beermann 2011), since impacts of food systems on climate change and impacts of climate change on food systems are intertwined (Vermeulen et al. 2012).

Consequently, multilevel governance of the food supply chain, including climate risk, seems to be a task of increasing complexity. Analytically, the hazards of the food system can stem either from external impulses (such as climate change) or from internal causes (such as market failure or disconnect in the food supply chain).

One idea to alleviate risk in the food supply chain has been to reduce complexity, which is best manifest in the local food alternative. However, even a food supply chain that pursues the local food framework needs to adapt to external impulses. Hence, Morgan et al. (2006) conclude "…agri-food studies could also benefit from more critical engagement with theories of multilevel governance because far from being a local matter, food supply chain localization will need to draw support from every tier of the multilevel policies that govern our lives today."

Undoubtedly, the scale of the food supply chain has been one of the most pertinent issues of agri-food studies (see for example Thompson and Scoones 2009). One of the main reasons for this is the perceived vulnerability of food supply chains that are increasingly extended across the globe. The counter-tendency has sometimes been identified as re-territorialization of the food system (Battaglini et al. 2015). Consequently, the development of food supplies is shifting towards higher diversity of food supply chains that vary according to local and regional circumstances. Thus, it seems more appropriate to envision the food supply chain as an entity of particular actions, processes and agents rather than something to be presented as a universal system. Accordingly, even the risks and vulnerabilities of the food supply chain can be more coherently identified as embedded in their territorial context rather than validated globally.

The article includes three thematic sections focusing on (1) climate risk identification in food supply chains, (2) multilevel governance represented by public and private policies and (3) building food system resilience and food security. Based on these thematic discussions we propose a framework outlining a multilevel governance of climate change adaptation of the food supply chain (Fig. 23.1), and introduce the conclusion.

23.2 Identifying Climate Risks in Food Supply Chains

There are four essential activities in the food system and in any type of food supply chain: producing (agriculture), processing and packaging (and manufacturing), distributing and retailing and consuming (Ericksen 2008). Although studies of climate change impacts on food systems have focused on production activities, there is an emerging understanding of how climate change will also affect storage, processing, manufacturing, transport, retail and consumption (Vermeulen et al. 2012). Identifying climate change-related risks across all these activities can be part of a vulnerability assessment of the food supply chain, which encompasses the analysis and identification of exposure, sensitivity and adaptive capacity (Smit and Wandel 2006). By identifying risks and vulnerabilities in the food supply chain, corporations can save costs due to proactive risk management and measurements, and develop new products through innovation (Beermann 2011). Svensson (2000), for example, places vulnerability and related concepts, such as risk, uncertainty and reliability within the context of contingency planning, whereas in the research by

Peck (2005) supply chain vulnerability is taken to be related to risk, such that something (e.g. food supply chain) is at risk, vulnerable or likely to be lost or damaged.

Due to the distribution of risks on various activities in the food supply chain, multilevel governance framework for managing risk or vulnerability to climate change becomes essential. Actually, a risk can be attributed to an organization, a supply chain, a region, an economy, the environment or society at large. Simultaneously, the vulnerability of the food supply chain can refer broadly to the climatic conditions that increase the food security risk faced by households or communities as a result of a climate shock (Krishnamurthy et al. 2014) or to the consequences for organizations from extreme weather events (Linnenluecke and Griffiths 2010). Also, climate change can have more long-term effects on vulnerability of the food supply chain (e.g. food cold chain risks, James and James 2010) in ways that increase risk and demand new investment. Furthermore, it has been argued that low-income producers and consumers of food will be more vulnerable to climate change due to their limited ability to invest in adaptive institutions and technologies (Vermeulen et al. 2012; Wheeler and von Braun 2013).

In terms of organizations, Kamalahmadi and Parast (2016) found a new trend in the literature, that focuses on resilience in the context of small and medium-sized enterprises (SMEs), many of them especially vulnerable to disruptions. If we accept the notion that food supply chains are inter-organizational networks embedded within a complex socio-ecological environment and characterized by many uncontrollable drivers, including climate change, we must also accept that there are dependencies and interconnectedness to be addressed at the network level (Kamalahmadi and Parast 2016; Peck 2005). Thus, in adapting the food supply chain to climate change the first priority is in facilitating risk identification and control by the (numerous) private actors in and through multilevel networks of the food supply chain.

Although climate risk may seemingly be isolated to specific regions or industry segments (e.g. typically those most directly reliant on the natural environment, such as agriculture), economic consequences or food security implications can be, in fact, magnified through flow-on effects (Linnenluecke and Griffiths 2010). Given the interdependencies between organizations and their supply chains, risk can be a result of various supply chain disruptions (Christopher and Peck 2004). Moreover, even climate change adaptation policies as implemented by governments may comprise a risk to food supply chains (e.g. strong public support policy for bioenergy, see Wheeler and von Braun 2013; Murphy et al. 2011). Finally, adaptation lead by focal corporations in food supply chains may come as a risk to individual producers, when climate change alters the relative productivity of specific regions, affecting input costs (Porter and Reinhardt 2007). Consequently, the buyers (focal corporations) can easily replace their suppliers or even relocate the hub of a supply chain away from an area that is overly exposed to climate change (West 2014; Paloviita 2015).

Thus, supply chain risk can be placed into five categories relating to (1) process and (2) control risks internal to the firm, (3) demand, (4) supply risks internal to the supply chain network, and (5) environmental risk external to the network (Christopher and Peck 2004). Climate change risk represents the last category, namely a risk external to networks. Nevertheless, the risk may land directly on the focal firm or on those upstream and downstream, hence requiring a broader perspective when it comes to contingency planning. As to the concerns of public interests, the broader perspective needs to be opened towards public policies and the facilitating role they can perform through measures of multilevel governance.

23.3 Multilevel Governance and Coping Risks: Public and Private Policies

Governance arrangements for food supply chain management focus especially on reducing risk and uncertainty, which are usually blind to trade-offs and resulting risks (van Bueren et al. 2014). For example, there are concerns around national and regional mismatches between responsibility for, and vulnerability to, climate change; this means that the governance of integrated adaptation and mitigation policies to achieve food security need mechanisms to avoid unequal distribution of costs and benefits (Vermeulen et al. 2012).

Existing governance activities, such as policy development, reflect objectives, attributes, norms and standards of individual and collective interests (Dow et al. 2013). Multilevel governance is the keyword that is designated to characterize the process of public development and transition in specific fields of individual and collective interests. In general terms, multilevel governance has been defined "as decision-making that is steered not only by public but by private and other interests, and as a process that takes place across multiple geographic scale levels and sectors" (Keskitalo 2010, 4; see also Boland 1999). Some definitions highlight the non-hierarchical networking mode of multilevel governance. For example, Biermann (2007) defines multilevel governance as normally denoting "new forms of regulation that differ from traditional hierarchical state activity and implies some form of self-regulation by societal actors, private–public co-operation in the solving of societal problems, and new forms of multilevel policy." Finally, some other definitions put more emphasis on the regional dynamics of decentralization. For example, regionalization as part of multilevel governance has often been perceived as one aspect of European integration that actually limits to some extent the powers of the nation states. This perspective is interesting when applied to food supply chain risk management that leads to increasingly coordinated action on the European Union (EU) level, yet simultaneously yields space for subnational regional innovation in risk management (see also Hooghe 1996). Nevertheless, nation states are still seen as key actors in the arena of international climate policy as well as in the larger scene of sustainable development.

From the perspective of governance and risk management, the food supply chain is sometimes simply conceived to be something more manageable than the entire food system. According to the FAO report *Climate change and food security: a framework document"* (2008) "the main conceptual difference between a food system and a food supply chain is that the system is holistic, comprising a set of simultaneously interacting processes, whereas the chain is linear, containing a sequence of activities that need to occur for people to obtain food." So, thinking in terms of a linear path makes it easier for policy makers to identify risk, and more specifically, the risk of disrupted processes in the food supply chain. Thinking in terms food supply chain instead of food system, however, should not imply missing the point of the locally or regionally embedded character of risk. In order to avoid short cuts in understanding the circumstances in which the food supply chain is vulnerable to climate change, the identification of risk still needs to differentiate between risks created by internal impulses and those created by external impulses, and set the policies to manage risk accordingly.

Recently, there has been a further attempt to define the politics of public risk management in a more ambitious manner, namely as multilevel *risk* management (e.g. Corfee-Morlot et al. 2011). The leading idea is that, as part of the general framework of climate change adaptation, nation states as well as the subnational and supranational actors should increase their capacities to make a coordinated effort to alleviate climate risk in a coordinated way and on the basis of identified risks. This applies to public food supply chain management as well as other relevant sectors of climate policy. This obligation is, however, quite challenging due to many reasons. Firstly, there is an obvious lack of scientific information on some crucial aspects of risk in food delivery, hence impeding systematic identification of risk. For example, Wheeler and von Braun (2013) argue that globally, there is ample information on climate change and food availability but much less scientific information concerning the climate change impacts on food access, utilization and stability dimensions. However, all these dimensions should be covered in order to achieve a more holistic understanding of food security. Secondly, public government has been mostly institutionalized in a centralized top down mode and therefore it does not give much space for two-way vertical dynamics in adaptation and managing risk. Thirdly, inclusive risk governance would demand effective joint efforts of many public and private stakeholders that seldom find well-functioning channels for long-term cooperation.

In the private sphere, balancing upstream and downstream supply chain risks continues to be challenging as well. This is because purchasing and selling are taken care of by different departments within organizations (Wognum et al. 2011). Moreover, balancing overall food supply chain risks and societal risks is even more challenging because of the dual structure of governance: private and public (Barling 2007). Consequently, coping with risks requires institutional arrangements that facilitate the processes of interaction and coordinate problem solving, focusing especially on the trade-offs between different values and interests (van Bueren et al. 2014). Hence, climate risk can be described as a "wicked problem", which refers to "an approach to understanding the dynamics of a major proposed change with multiple and conflicting inputs and multiple possible outcomes, all of which play over time against, or occasionally with, each other" (Sun and Yang 2016).

Climate change adaptation is often associated with the re-territorialization of risk governance (Keskitalo 2010). This idea fits well with some of the quality demands of the sustainable food supply chain, such as flexibility and reducing complexity. When horizontally networking actors join their efforts, the regional food supply is likely to be best secured (e.g. McAdam et al. 2014). However, this does not suffice for the sustainability of the food supply chain, especially not if we think in terms of some other food security values, such as equity of access to food, food safety and the transparency of the functioning food supply chain and the larger food system. Questions about climate risks to food security or organizational objectives lead to distributional questions and associated governance challenges, depending on whether values and objectives are common and shared, or private to certain actors (e.g. a multinational food corporation). For example, the objectives of climate change adaptation can be associated with livelihoods, profits, environmental stewardship, food security or other multiple objectives (Vermeulen et al. 2012).

At least two important remarks therefore highlight the role of public governance. Firstly, there is a need for supra-regional public risk governance, which implies that in addition to effective horizontal actor-networks, even the vertical linkages of the food supply chain need to be established and consolidated (c.f. Juhola 2010). Secondly, it becomes evident that even if a deliberate focus on the food supply chain's eco-efficiency may advocate a turn towards increasingly regional models of food security and the adaptation to climate change, food supply chain management also implies some effective regulatory measures in the level of the nation state and even beyond. This is because in the international policy arena (IPCC, Paris Agreement, etc.) the adaptation line of discussion continues to accentuate the strong role of nation states even if the initiative and innovation are expected to happen in the local level. Furthermore, intergovernmental organizations, such as FAO, UNEP, OECD and World Bank will continue to have important influence in programming and funding sustainability transformations in agriculture and are increasingly starting to envision for development programs concerning the entire food chain instead of agriculture only (see for example Bagherzadeh et al. 2014; IFAD 2017).

Much of the academic discourse on adaptation policies have so far been focused on the barriers that should be overcome in order to pursue targets of sustainable development and effective risk governance. The empirical research on barriers of implementation concerning effective climate adaptation has revealed, for example, that there is still much fragmentation in the adaptation policies, many problems in coordinating multiple issues of time and scale, as well as major disconnect of tasks, resources and interests. Moreover, local actors often perceive more strongly than actors of higher levels the barriers for their action (Biesbroek et al. 2011; Juhola 2010). These observations in recent climate change adaptation research imply that it is important in food supply chain risk management, to identify and include the barriers of implementation and in particular, how they are embedded in the territories where food is supplied.

23.4 Building Food System Resilience and Food Security

According to Vermeulen et al. (2012) there is no global food system but rather, a set of partially linked food supply chains, varying from local to global supply chains. Hence, a starting point for building food system resilience is to consider food supply chain resilience in particular circumstances. Christopher and Peck (2004) call for supply chain strategies that embody a higher degree of resilience, implying flexibility and agility. Sheffi (2005) suggests that to achieve built-in flexibility, companies should adopt standardized processes, use concurrent instead of sequential processes, and plan to postpone and align procurement strategies with supplier relationships. In addition, building supply chain resilience requires an improved understanding of the network that connects businesses, upstream suppliers and downstream customers, a reconsideration of the supply base strategy, as well as overall design principles for supply chain resilience and a high level of collaborative work across supply chains (Christopher and Peck 2004).

According to Kamalahmadi and Parast (2016), sourcing strategies are the main supply chain resilience strategies that are discussed in the literature. In terms of supply base strategy, the reduction of the supplier base and single sourcing can be dangerous from the resilience perspective, whereas alternative sources of supply can decrease the risk (Christopher and Peck 2004). Sheffi (2005), in turn, suggests focusing on a small group of key suppliers with deep relationships (sharing the risk) or having an extensive supplier network with arm's-length relationships (distributing the risk). Finally, cultural change is an important factor in building resilience, which can be related to corporate culture (Sheffi 2005) or supply chain risk management culture (Christopher and Peck 2004), where the culture of risk management should be extended beyond the boundaries of corporate risk.

Because governments design national climate change adaptation strategies with food security in mind, there is also a need for an improved understanding of the networks that connect food supply chains, governments, households and communities. Hence, collaborative working both across and beyond food supply chains is needed in the multilevel governance of climate risk. For supply chain collaboration, Ponis and Koronis (2012) highlight the importance of information sharing, collaborative work and joint decision-making, whereas Kamalahmadi and Parast (2016) raise inter-firm trust as one of the key prerequisites for collaboration, since lack of trust increases supply chain risks. Thus, climate change adaptation in the food system requires information sharing and joint decision-making between different levels and forms of governance in order to exploit synergies.

In order to better understand the networks that should connect private and public actors, it is important to specify the patterns of food system dynamics with due consideration to their embedded character. From a global perspective, Thompson and Scoones (2009) denote agri-food systems as embedded in complex ecological, economic and social processes. Hence, instead of trusting mainstream narratives of technological change and economic growth, more attention should be paid to the complex system dynamics in order to identify vulnerable parts of the processes.

Thus, the dynamics of agri-food systems contain a multifaceted risk of disruption, making it challenging even for public policy to effectively promote food supply chain resilience.

In fact, many of the public adaptation policies concerning the food supply chain are still in initial phases and should be implemented by joint effort of actors that still need to be motivated to take action (Daniell et al. 2011). Moreover, the institutional frameworks for adaptation may still be insufficient and fragmented (Juhola 2010). Hence, topical questions remain open, such as: who (in addition to food enterprises) are setting the public agenda for increasing resilience of the food supply chain; how big a role should experts take; and what is the role of consumers in establishing food security norms under climate change adaptation (see also Adger et al. 2005; Smit et al. 2000). Many authors underline the importance of public leadership that includes not only regulation for food safety guarantees but also targets of sustainability and food supply chain resilience which is best defined in the territorial context (see for example Ignaciuk 2015).

Nevertheless, we can perhaps speak of a global value community with regard to enhancing food security and the more operative food supply chains. At the least, attributes such as sustainability, eco-efficiency and transparency can be mentioned as concerted qualities associated with increased resilience of the food supply chain (see Smith et al. 2016; Smith 2008). Furthermore, when putting stress not only on the internal dynamics of the food supply chain but also on the larger food system, it is important to emphasize the efficiency not only of delivery but also of the access to food. Targeting improved access in a sustainable manner implies consideration of developing equity of access as part of adaptation strategy both on the global level and the territorial level.

Finally, from the perspective of the public sphere there is a necessity to consider transparency not only in the private field but also within public action. Therefore, issues of leadership and coordination become pertinent (Keskitalo 2010). Considering the urgent character of climate change adaptation it is, however, important to balance between efficiency and democracy of decision-making. Even if participation has been cited as an important aspect of appropriate decision-making, some authors remind that increasing the number of voices in a political process also increases the complexity of decision-making—sometimes in ways that contradict efficiency (see for example Few et al. 2007; Lee et al. 2013). This issue is linked interestingly with the discussion on technological change and epistemic communities (e.g. Keskitalo 2010; Vogel et al. 2007) that dominates the climate adaptation policy scene. Even if adaptation is bound to create a more inclusive concept of public governance in order to be legitimate, its structure will be molded by specific political patterns as established in a great variety of territorial settings involving, in addition to states, many subnational and even supranational actors. It is evident that these territorial settings may include conflicts of interest that intermingle with traditional discord over local issues. However, they may generate even completely new conflict situation concerning the actual adaptation measures to be implemented locally. Reconciliation is important and apparently only a part of it can be efficiently and appropriately treated though legislation and legal system. Therefore the territorially effective networks need to create more flexible measures of negotiation to solve socio-political and socio-economic conflicts that are likely to appear. In the global South the conflicts are often related to natural disaster and the consequent major crises of food supply chain (Harris et al. 2013) whereas in the North the conflict may involve less urgency on human survival and wellbeing but can still cause severe barriers to change because they may concern vital interests on values and livelihood expressed by different stakeholders (e.g. Storbjörk 2007).

Even if equity in food access has been expressed as one of the main targets of the sustainable food system, there is still a long way to secure sufficient access to food for all people (FAO 2014; Juhola and Neset 2016). In fact, there are good reasons to expect even increasing inequality of access to food due to climate change, as it affects various parts of the globe differently. Especially, extreme weather shocks, drought and flood may harm present cultivation in regions where livelihoods of big populations depend directly on food production. Moreover, poverty in general still limits access to proper nourishment. Thus, the reduced capacity of food systems to assure food security to populations vulnerable to hunger and malnutrition is one of the principal concerns related to climate risk (Vermeulen et al. 2012). Yet, climate change adaption is a major concern for all populations, rich and poor. The reason is twofold. Firstly, rich populations and western food consumption patterns tend to accelerate climate change, which leads to greater adaptation needs globally. Secondly, western consumers are increasingly dependent on cheap food produced by suppliers in poor countries, which often are very vulnerable to extreme weather events.

Cheap food is partly a result of an efficient food supply chain, which avoids surplus capacity and inventory. However, additional capacity and inventory can be extremely beneficial in building supply chain resilience, which calls for the re-examination of the efficiency vs. redundancy trade-off (Christopher and Peck 2004). Low-inventory food supply chains that are based on just-in-time mode of delivery are highly vulnerable to weather disruptions (Vermeulen et al. 2012) and thus less resilient. On the other hand, according to Sheffi (2005), increasing redundancy is a very expensive measure because companies must pay for the redundant stock, capacity and workers.

Hence, from the food supply chain perspective, food security remains a complex issue to be addressed in a comprehensive way by analyzing the inner dynamics of both private and public sectors, and the complex interplay between these sectors. It is important to remember that depending on the context, the food security issue is predominantly either an issue of quantity or quality. In both cases, it is useful to approach the food security in the context of multilevel and multisectoral dynamics involving a variety of actor-networks. Consequently, it is important to identify the key actors and their groups because avoiding disruption in the food supply chain is not only a technical matter but, very often, a social and political process. For those processes, mutual social recognition and trust consolidating the social and political process of adaptation to climate change are as important as economic growth targets and technological measures.

23.5 Framework and Its Application in the Finnish Food System

The conceptual framework on systematizing the links between climate risk, governance, climate change adaptation, vulnerability and resilience is presented in Fig. 23.1.

Instead of simply being an expression of a specific theoretical approach, we aim to provide a framework that can stimulate decision-making in the context of climate change adaptation. Hence, this chapter is a comprehensive representation of theoretical and conceptual ideas on how to frame climate risk and adaptation policies in a multilevel governance setting. It is commonly understood that vulnerability is mainly concerned with the inner conditions or the intrinsic characteristics of the food system that make it liable to experience climate risk (see for example Birkmann 2013). Following this understanding, we argue that even so, feedback processes and interventions by multilevel governance need to be addressed when looking for effective measures to reduce vulnerability and climate risk. In general terms, this can be translated into the question of how internally coherent a system has to be in order to become more resilient.

The framework is based on a feedback-loop system, in which climate risk is caused by the overall vulnerability of the food system. As previously explained, climate risk can be a physical, a regulatory or a financial risk, but also a litigation risk or a competitive risk. For example, a company may face a heightened risk of

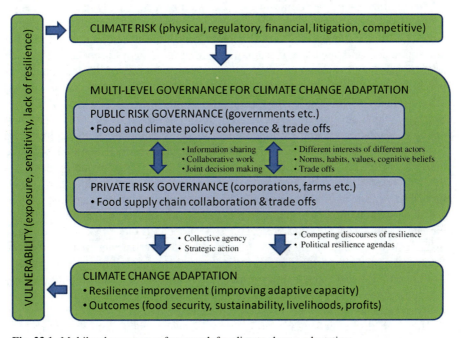

Fig. 23.1 Multilevel governance framework for climate change adaptation

climate litigation and may be exposed to the risk of lawsuits, if it fails to disclose the risks to shareholders and financial regulators. Similarly, if companies do not take measures to reduce climate risks they may be competitively disadvantaged. Identifying direct and indirect climate risks, in turn, is followed by the action of climate change adaptation. In this stage, multilevel (risk) governance for climate change adaptation should be seen as an overall subject of the whole framework. Within this framework, multilevel governance deals with various actors, rules, conventions and mechanisms concerned with risk reduction and climate change adaptation while actors represent different interests of public risk governance (governments) and private risk governance (food supply chains led by food manufacturing and retail industries).

Within public governance, the challenges are associated with the integration of agricultural policies, food policies, climate change mitigation policies and climate change adaptation policies. Policies led by different ministries and governance agencies result in trade-offs, which in turn, require a more coherent policy response to promote food security. Conflicts and trade-offs are inherent within private governance as well. This requires a more collaborative approach in the food supply chain between primary producers, food industries, retailers and consumers to promote food supply chain resilience. Finally, the dual structure of the food system governance, public and private, introduces a broad set of different interests, norms, habits, values and cognitive beliefs. Trade-offs in the multilevel governance can be dealt with by increased information sharing, collaborative work and joint decision-making, and ultimately, by increasing trust.

The application of the framework is illustrated by a case study related to design of multilevel governance of climate change adaptation in the Finnish food system. Although climate change will affect most urgently the developing countries, the development of climate change adaptation is an issue also in European states, such as in Finland (Keskitalo et al. 2012). Finland has been one of the forerunners of public programming in adaptation, as the adaptation policy process in Finland started already in 2002, after the publication of the National Climate and Energy Strategy, which strongly focused on mitigation (Keskitalo et al. 2012). The first national adaptation program was launched in 2005 (Ministry of Agriculture and Forestry 2014) and has been since then further developed in parallel with the national energy and climate strategy. The Ministry of Agriculture and Forestry has the main responsibility of planning the climate change adaptation policy. The main three issues raised in Finnish national public policy with regard to managing climate risk in agriculture are firstly the support to developing species that will best adapt to climate change, secondly the preparedness to flood and other risk in water cycle management and thirdly the issue of controlling invasive species mainly from the perspective of their posing threat on biodiversity. In the following example, we address only the first issue and more specifically the question of how public policy is tackling the issue of adaptation by promoting research and development of crop cultivars in Finland.

However, the implementation of climate change adaptation in the entire food supply chain is still at early stage. Adaptation implies change in the agenda of all actors of the food chain. Looking upstream towards farms, generating change in products often implies changing suppliers, acquiring new skills and other assets. Hence, instead of managing climate risk in one sector (agriculture) only, national public policies should shift to managing climate risk in the overall food supply chain or more broadly in the national food system. This may require that the responsibility of climate change adaptation should be shared with several ministries, not just the Ministry of Agriculture and Forestry, which typically has an agriculture-focused perspective towards the food system. In Finland there is no specific ministry for food, and there are continuous tensions between the Ministry of Agriculture and Forestry and the Ministry of the Environment. In comparison, in the UK there is the Department for Environment, Food and Rural Affairs, which may provide a more appropriate governance structure for policy making in a context of climate change adaptation in the food system.

Moreover, limited practical integration across and between national and local levels was reported when evaluating the actual implementation of the adaptation strategy (Keskitalo et al. 2012), which means that the practical implementation stills needs much coordination and joint efforts by different actors (Juhola 2010). Our framework emphasizes the fact, that climate change adaptation agendas are set not only on national levels, but through multilevel governance linkages between actors within integrated multi-participant groups.

A Finnish example of a multilevel adaptation process that draws upon several different types of actor has been coherently described by Paavola et al. (2015), who focused on the converting the concept of response diversity into a tool of reflection and decision-making in barley breeding and cultivation in Finland. As concerns the social and technological transformations towards promoted resilience in crop cultivation it is important to increase interaction between public and private actors in the specific food chain to reach an effective concept of resilience that is shared and targeted by the various actors. This can be created for example by working in plant breeding towards enhanced response diversity of specific crops to make them more persistent in cases of extreme weather conditions and in circumstances of uncertainty of climate development. The key actors in this adaptation process are agri-food researchers and commercial breeders, but several other potential actors involved in Finnish crop production and agri-food supply were identified, such as farmers, National Emergency Supply Agency and the food industry. In this case, a multilevel adaptation process has included collaboration between researchers and the breeding company as well as discussions between the plant breeding company and farmers. However, it was noted that a commercial firm can be part of a long-term adaptation process, but cannot be solely responsible for it, as demonstrated by data in the study by Paavola et al. (2015, p.50): "what a commercial firm can do is quite restricted".

Hence, other stakeholders should be involved. As Bristow and Healy (2014, 930) note, "the institutions of purposive adaptation in regions go beyond firms and firm-related actors, and also incorporate a variety of other self-organizing institutions of collective agency, notably those of state, governance and community". First, the work of The National Emergency Supply Agency, an organization working under the

Ministry of Economic Affairs and the Employment and which is tasked with planning and measures related to developing and maintaining the security of food supply, is generally associated with reduced vulnerability to crises and weather extremes (Paloviita et al. 2017). In the study of Paavola et al. (2015) plant breeders suggested that the National Emergency Supply Agency should take more responsibility for ensuring resilience with barley breeding in the long-term perspective. Since the National Emergency Supply Agency is maintaining safety stocks for the times of crisis, the redundant stocks in Finland have been at a high level in international comparison. In addition, a strong network around food emergency supply has emerged, including private actors, such as food companies and retailers.

Second, Finnish agriculture is almost exclusively based on family farms, who have very limited economic means for new investment, which makes them vulnerable in implementing any changes in their product sample (Järvelä and Kortetmäki 2015). By tradition the state in Finland has been supporting investment at farms by various means, even through direct financial support. During the last two decades, this tradition has become replaced to a great deal by mandatory and yet negotiated EU policy measures, which has made the governance for change more complex. Due to the complexities of multilevel governance of climate adaptation and risk it seems that the public private interface has still much to accomplish before reaching a state of art of promoting new climate persistent crops, especially in new range of species (Lehtonen 2015). Sharing information is not easy since scientific experiments may still be at early stage and uncertainties are evident as concerns the sustainability of new crop choices. One policy response from public policy sources to this has been a general recommendation of cultivating a variety of crops at one farm instead of relying on monoculture.

Third, public sector can support this process in many ways, for example by selecting measures of direct financial support or by taking a major role in research and development. In Finland, much of this public policy support has been performed through research programmes contributing to the sustainability of the food system and food security. For example, A-LA-CARTE (Assessing limits of adaptation to climate change and opportunities for resilience to be enhanced A-LA-CARTE 2015) was a 4-year collaborative research project of FICCA (The Finnish Research Programme on Climate Change), which concerned agri-food systems as a case study, focusing on food supply of farms and access to food by consumers to manage climate-related risks. In addition, the Finnish Climate Change Panel, which is an independent, interdisciplinary think tank, provides scientific advice for policy making and reinforces interdisciplinary insight in the operation of different sectors. Concerning the adaptation to climate change in the food sector, the panel has examined, e.g. the responsibilities associated with flood risk preventing and alleviating crop damages in agriculture (Juhola et al. 2015).

In an ideal world, climate risk provides windows of opportunity for change and innovation in food system governance (see for example Birkmann 2013, 33). From this perspective, renewal and learning, as well as transformability, should be built-in characteristics of multilevel governance for climate change adaptation (Tàbara et al. 2010). Accordingly, further research on the food supply chain should be critically considered and suggest what governance structures and relationships need to remain and which should change or disappear. Corporations and governments, together with civil society, have a joint opportunity to design and manage these governance structures and relationships in order to improve food system resilience and food security.

23.6 Conclusion

We argue that climate change adaptation is a multilevel governance obligation for a great variety of actors, given the multilevel nature of climate risk governance and its complexities. The role of governments in climate change adaptation should be framed in relation to commercial food supply chains and vice versa. As concerns the willingness of cooperation between actors at different positions of the food chain there is no easy solution for setting comprehensive public/private policies. However, there are increasingly effective networks at subnational and local level that can enhance joint efforts in reforming agriculture and land use. These networks are also increasingly important not only in reconciling between different interest of actors but also in building capacities for change and adaptation, which can potentially increase the effectiveness of coping strategies.

It is important to recognize that new knowledge and ideas of alternative options in crop cultivation do not inevitably lead to their acceptance and legitimate use. For example, in plant breeding one of the development strategies tested is molecular breeding. After some experiments supported by public funds in research and development and parallel opposing civic activity there is presently a quasi-total silence in Finland about the whole issue of GMOs. Hence, even if new networks emerge enhancing capacities for sharing information, collaboration and joint decision-making on multilevel basis, it is not evident that these new constellations can handle expected and unexpected political controversies or open discussion about sensitive issues such as experiments on molecular breeding.

Anyhow, climate change adaptation at national level needs innovation towards multilevel risk governance that implies concerted action between the many public and private actors concerned. Not only risk points in the food chain need to be located and identified but also main priorities, trade-offs, constraints and conflicts need to be acknowledged. For example, unclear divisions of responsibility for climate change adaptation governance can introduce a severe barrier for all actors in the food supply chain and therefore increase the overall vulnerability of the food system. Looking downstream in the food supply chains there is also a growing interest among consumers to adapt their diets to climate risk, even if the new climate friendly choices are still a minority lifestyle in Finland.

The paper is explained in the video titled "Multi-level governance framework for climate change adaptation".

Acknowledgements We would like to thank the Kone Foundation for financing our research project: "Future Food Security in Finland—Identifying and Analyzing Vulnerability Aspects in the Finnish Food System".

References

Adger, W. Neil, Arnell, Nigel W. and Tompkins, Emma L. (2005) Adapting to climate change: perspectives across scales. [in special issue: Adaptation to Climate Change: Perspectives Across Scales] Global Environmental Change Part A, 15 (2), 75-76. https://doi.org/10.1016/j.gloenvcha.2005.03.001.

Assessing limits of adaptation to climate change and opportunities for resilience to be enhanced A-LA-CARTE. (2015). Retrieved August 4, 2017, from http://www.syke.fi/projects/alacarte.

Bagherzadeh, M., Inamura, M., & Jeong, H. (2014). *Food waste along the food chain*. OECD Food, Agriculture and Fisheries Papers, No. 71. Paris: OECD Publishing. Retrieved August 8, 2017, from https://doi.org/10.1787/5jxrcmftzj36-en.

Barling, D. (2007). Food supply chain governance and public health externalities: Upstream policy interventions and the UK state. *Journal of Agricultural and Environmental Ethics, 20*, 285–300.

Battaglini, E., Babović, M., & Bogdanov, N. (2015). Framing resilience in relation to territorialisation. In A. Paloviita & M. Järvelä (Eds.), *Climate change adaptation and food supply chain management* (pp. 119–131). Oxon: Routledge.

Beermann, M. (2011). Linking corporate climate adaptation strategies with resilience thinking. *Journal of Cleaner Production, 19*, 836–842.

Biermann, F. (2007). "Earth system governance" as a cross-cutting theme of global change research. *Global Environmental Change, 17*, 326–337.

Biesbroek, G. R., Termeer, C. J. A. M., Klostermann, J. E. M., & Kabat, P. (2011). Barriers to climate change adaptation in the Netherlands. *Climate Law, 2*, 181–199.

Birkmann, J. (2013). Measuring vulnerability to promote disaster-resilient societies and to enhance adaptation: Discussion of conceptual frameworks and definitions. In J. Birkmann (Ed.), *Measuring vulnerability to natural hazards: Towards disaster resilient societies* (pp. 9–79). Tokyo: United Nations University Press.

Boland, P. (1999). Contested multi-level governance: Merseyside and the European Structural Funds. *European Planning Studies, 7*(5), 647–664.

Bristow, G., & Healy, A. (2014). Regional resilience: An agency perspective. *Regional Studies, 48*(5), 923–935.

Christopher, M., & Peck, H. (2004). Building the resilient supply chain. *International Journal of Logistics Management, 15*(2), 1–13.

Corfee-Morlot, J., Cochran, I., Hallegatte, S., & Teasdale, P.-J. (2011). Multilevel risk governance and urban adaptation policy. *Climatic Change, 104*, 169–197.

Daniell, K. A., Manez Costa, M. A., Ferrand, N., Kingsborough, A. B., Coad, P., & Ribarova, I. S. (2011). Aiding multi-level decision-making processes for climate change mitigation and adaptation. *Regional Environmental Change, 11*, 243–258.

Dow, K., Berkhout, F., & Preston, B. L. (2013). Limits to adaptation to climate change: A risk approach. *Current Opinion in Environmental Sustainability, 5*, 384–391.

Ericksen, P. (2008). Conceptualizing food systems for global environmental change research. *Global Environmental Change, 18*, 234–245.

FAO, IFAD, & WFP. (2014). *The state of food insecurity in the world 2014. Strengthening the enabling environment for food security and nutrition*. Rome: FAO.

Few, R., Brown, K., & Tompkins, E. L. (2007). Public participation and climate change adaptation: Avoiding the illusion of inclusion. *Climate Policy, 7*, 46–59.

Harris, K., Keen, D., & Mitchell, T. (2013). *When disasters and conflicts collide: Improving links between disaster resilience and conflict prevention*. London: Overseas Development Institute.

23 Multilevel Governance for Climate Change Adaptation in Food Supply Chains   495

Hooghe, L. (1996). *Cohesion policy and European integration: Building multi-level governance.* Oxford: Oxford University Press.

IFAD. (2017). *The economic advantage, assessing the value of climate-change actions in agriculture.* Rome: IFAD.

Ignaciuk, A. (2015). *Adapting agriculture to climate change: A role for public policies.* OECD Food, Agriculture and Fisheries Papers, No. 85. Paris: OECD Publishing. Retrieved February 20, 2017, from https://doi.org/10.1787/5js08hwvfnr4-en.

James, S. J., & James, C. (2010). The food cold-chain and climate change. *Food Research International, 43*, 1944–1956.

Järvelä, M., & Kortetmäki, T. (2015). Coping with climate change: Rural livelihoods, vulnerabilities and farm resilience. In A. Paloviita & M. Järvelä (Eds.), *Climate change adaptation and food supply chain management* (pp. 147–157). Oxon: Routledge.

Juhola, S. (2010). Mainstreaming climate change adaptation: The case of Finland. In E. C. H. Keskitalo (Ed.), *Developing adaptation policy and practice in Europe: Multi-level governance of climate change.* Dordrecht: Springer.

Juhola, S., & Neset, T.-S. (2016). Vulnerability to climate change in food systems: Challenges in assessment methodologies. In A. Paloviita & M. Järvelä (Eds.), *Climate change adaptation and food supply chain management* (pp. 57–69). Oxon: Routledge.

Juhola, S., Kokko, K., Ollikainen, M., Peltonen-Sainio, P., Haanpää, S., Salonen, J., & Airaksinen, M. (2015). *Adaptation to climate change: Risks, responsibilities and costs, Abstract.* Retrieved August 4, 2017, from http://www.ilmastopaneeli.fi/uploads/reports/Abstract_Adaptation%20 to%20climate%20change%20-%20risks,%20responsibilities%20and%20costs.pdf.

Kamalahmadi, M., & Parast, M. M. (2016). A review of the literature on the principles of enterprise and supply chain resilience: Major findings and directions for future research. *International Journal of Production Economics, 171*(1), 116–133.

Keskitalo, E. C. H. (2010). *Developing adaptation policy and practice in Europe: Multi-level governance of climate change.* Dordrecht: Springer.

Keskitalo, E. C. H., Westerhoff, L., & Juhola, S. (2012). Agenda-setting on the environment: The development of climate change adaptation as an issue in European states. *Environmental Policy and Governance, 22*, 381–394.

Klein, R. J. T., Nicholls, R. J., & Thomalla, F. (2003). Resilience to natural hazards: How useful is the concept? *Environmental Hazards, 5*, 33–45.

Krishnamurthy, P. K., Lewis, K., & Choularton, R. J. (2014). A methodological framework for rapidly assessing the impacts of climate risk on national-level food security through a vulnerability index. *Global Environmental Change, 25*, 121–132.

Lee, M., Armeni, C., Cendra, J., Chaytor, S., Lock, S., Maslin, M., Redgwell, C., & Rydin, Y. (2013). Public participation and climate change infrastructure. *Journal of Environmental Law, 25*(1), 33–62.

Lehtonen, H. (2015). Evaluating adaptation and production development of Finnish agriculture in climate and global change. *Agricultural and Food Science, 24*, 219–234.

Linnenluecke, M., & Griffiths, A. (2010). Beyond adaptation: Resilience for business in light of climate change and weather extremes. *Business & Society, 49*(3), 477–511.

McAdam, M., McAdam, R., Dunn, A., & McCall, C. (2014). Development of small and medium-sized enterprise horizontal innovation networks: UK agri-food sector study. *International Small Business Journal, 32*(7), 830–853.

Ministry of Agriculture and Forestry. (2014). *Finland's National Climate Change Adaptation Plan 2022.* Government Resolution 20 November 2014, Publications 5b 2014.

Morgan, K., Marsden, T., & Murdoch, J. (2006). Worlds of food: Power, place and provenance in the food chain.Oxford: Oxford University Press

Murphy, R., Woods, J., Black, M., & McManus, M. (2011). Global developments in the competition for land from biofuels. *Food Policy, 36*, S52–S61.

Paavola, S., Himanen, S., Kahiluoto, H., & Miettinen, R. (2015). Making sense of resilience in barley breeding: Towards usability of the concept of response diversity. In A. Paloviita & M. Järvelä (Eds.), *Climate change adaptation and food supply chain management* (pp. 43–54). Oxon: Routledge.

Paloviita, A. (2015). Food processing companies, retailers and climate-resilient supply chain management. In A. Paloviita & M. Järvelä (Eds.), *Climate change adaptation and food supply chain management* (pp. 194–205). Oxon: Routledge.

Paloviita, A., & Järvelä, M. (2015). *Climate change adaptation and food supply chain management* (p. Routledge). Oxon.

Paloviita, A., Kortetmäki, T., Puupponen, A., & Silvasti, T. (2017). Insights into food system exposure, coping capacity and adaptive capacity. *British Food Journal, 119*(12), 2851–2862.

Peck, H. (2005). Drivers of supply chain vulnerability: An integrated framework. *International Journal of Physical Distribution and Logistics Management, 35*, 210–232.

Ponis, S. T., & Koronis, E. (2012). Supply chain resilience: Definition of concept and its formative elements. *The Journal of Applied Business Research, 28*(2), 921–929.

Porter, M. E., & Reinhardt, F. L. (2007, October). A strategic approach to climate. *Harvard Business Review, 85*, 22–26.

Sheffi, Y. (2005). Building a resilient supply chain. *Harvard Business Review, 1*(8), 1–4.

Smit, B., & Wandel, J. (2006). Adaptation, adaptive capacity and vulnerability. *Global Environmental Change, 14*, 282–292.

Smit, B., Burton, I., Klein, R. J., & Wandel, J. (2000). An anatomy of adaptation to climate change and variability climatic change. *Climatic Change, 45*, 223.

Smith, B. G. (2008). Developing sustainable food chains. *Philosophical Transactions of the Royal Society B: Biological Sciences, 363*(1492), 849–861.

Smith, J. B., Klein, R. J. T., & Huq, S. (Eds.). (2003). *Climate change, adaptive capacity and development*. London: Imperial College Press.

Smith, J., Lang, T., Vorley, B., & Barling, D. (2016). Addressing policy challenges for more sustainable local–global food chains: Policy frameworks and possible food "futures". *Sustainability, 8*, 299.

Storbjörk, S. (2007). Governing climate adaptation in the local arena: Challenges of risk management and planning in Sweden. *Local Environment, 12*(5), 457–469.

Sun, J., & Yang, K. (2016). The wicked problem of climate change: A new approach based on social mess and fragmentation. *Sustainability, 8*(1312), 1–14.

Svensson, G. (2000). A conceptual framework for the analysis of vulnerability in supply chains. *International Journal of Physical Distribution and Logistics Management, 30*, 731.

Tàbara, J. D., Dai, X., Jia, G., Mcevoy, D., Neufeldt, H., Serra, A., Werners, S., & West, J. J. (2010). The climate learning ladder. A pragmatic procedure to support climate adaptation. *Environmental Policy and Governance, 20*, 1–11.

Thompson, J., & Scoones, I. (2009). Addressing the dynamics of agri-food systems: An emerging agenda for social science research. *Environmental Science & Policy, 12*(4), 386–397.

van Bueren, E. M., Lammerts van Bueren, E. T., & van der Zijpp, A. (2014). Understanding wicked problems and organized irresponsibility: Challenges for governing the sustainable intensification of chicken meat production. *Current Opinion in Environmental Sustainability, 8*, 1–14.

Vermeulen, S. J., Campbell, B. M., & Ingram, J. S. I. (2012). Climate change and food systems. *Annual Review of Environment and Resources, 37*, 195–222.

Vogel, C., Moser, S. C., Kasperson, R. E., & Dabelko, G. D. (2007). Linking vulnerability, adaptation, and resilience science to practice: Pathways, players, and partnerships. *Global Environmental Change, 17*, 349–364.

West, J. (2014). *The long hedge: Preserving organizational value through climate change adaptation*. Sheffield, UK: Greenleaf Publishing.

Wheeler, T., & von Braun, J. (2013). Climate change impacts on global food security. *Science, 341*, 508–513.

Wognum, P. M., Bremmers, H. J., Trienekens, J. H., van der Vorst, J. G. A. J., & Bloemhof, J. M. (2011). Systems for sustainability and transparency of food supply chains—Current status and challenges. *Advanced Engineering Informatics, 25*, 65–76.

Chapter 24
Climate Change and Food Security in India: Adaptation Strategies and Major Challenges

Atanu Sarkar, Arindam Dasgupta, and Suman Ranjan Sensarma

Abstract India has made rapid strides in improving food production and the country has become not only self-sufficient in food production, but now exports to several other countries as well. However, climate change has emerged as a major threat to India's hard-earned success. Much of India's population depends on climate-sensitive sectors such as agriculture, forestry, and fishing, and thus the livelihoods of hundreds of millions of people are at risk. In fact, the country has already witnessed adverse impacts of climate change on food production, transportation, storage, and distribution. Rising temperatures, erratic rainfall, extreme weather conditions (such as prolonged droughts and floods), changing soil fertility, and new pest infestations are major factors contributing to stagnant agricultural growth. "Climate-smart agriculture" is considered a pragmatic approach to ensuring food security in a changing climate. Adaptation strategies based on the principles of climate-smart agriculture can counter the impacts of climate change, such as the promotion of conservation agriculture, the sustainable management of natural resources and the promotion of climate-smart crops. However, the existing problems of transboundary water conflict, universal insurance of crops, the significant reduction in food wastage needed and the improvement of food distribution are essential to achieving the goals for adaptation. It is also important to note that ready acceptance of "climate-smart agriculture" by farmers cannot be expected, even if the necessary technologies are made accessible to them. Rather, more community-based participatory research is needed to explore socioeconomic and location-specific variables that are influencing farmers' preferences towards the approach.

Keywords Climate change · Food security · India · Conservation agriculture · Climate-smart crops · Food wastage · Public distribution of food · Crop insurance

---

A. Sarkar (✉)
Division of Community Health and Humanities, Faculty of Medicine, Memorial University, St. John's, NL, Canada
e-mail: atanu.sarkar@med.mun.ca

A. Dasgupta
Post-Graduate Department of Geography, Chandernagore College, Hugli, West Bengal, India

S. R. Sensarma
Government Advisory, Infrastructure and Government Services (IGS), KPMG, New Delhi, India

© Springer Nature Switzerland AG 2019
A. Sarkar et al. (eds.), *Sustainable Solutions for Food Security*,
https://doi.org/10.1007/978-3-319-77878-5_24

24.1 Climate Change: Indian Scenarios

For the Indian subcontinent, the Intergovernmental Panel on Climate Change (IPCC) projected a rise in temperature from 2 to 4.7 °C, with the most probable level being around 3.3 °C by the year 2100 (Solomon et al. 2007). Indian Scientists have indicated that across the region temperatures will increase by 3–4 °C towards the end of the twenty-first century (Kumar et al. 2006; Lal and Harasawa 2001). Although this warming has been projected to be widespread across the country, comparatively more changes will be pronounced over northern parts of the country (Kumar et al. 2006). While precipitation has been projected to rise, there will be spatial variations, with some pockets showing an increase and others experiencing a decline in rainfall (Sulochana 2003). Most models have projected an increase in precipitation between 10% and 40% from the baseline period (1961–1990) to the end of the twenty-first century, with the maximum expected increase in rainfall over the north western and central parts of India (Kumar et al. 2006). A trend analysis of India's precipitation from 1901 to 2012 shows a gradual decrease in intensity. During the same period, drought occurrences have increased considerably with an unequal distribution along with a rise in torrential rains (Mishra 2014). The effects of possible changes in the intensity of the monsoons will be significant since a large part of the country receives varying intensities of precipitation during the summer monsoon season (Lal and Harasawa 2001). Almost three-fourths of India's precipitation is concentrated over 4 months of the monsoon season (June–September) (Mishra 2014). Therefore, adequate summer monsoons are essential to the annual precipitation of the entire Indian subcontinent (Solomon et al. 2007). The north-eastern states, which are known for moist weather, are also facing a significant rise in temperature, leading to short-duration torrential rains and promoting stormier and drier seasons. In fact, storm occurrences in India have increased significantly during the past century and rising temperatures may lead to more frequent and stronger storms in the near future. Torrential rains have increased the frequency and intensity of landslide disasters, particularly in the mountainous regions of the Himalayas (Mishra 2014).

India has traditionally been highly vulnerable to natural disasters due to its unique geo-climatic conditions. According to the Annual Disaster Statistical Review (2016), over the last decade, India was one of the top five countries that is most frequently hit by natural disasters (Guha-Sapir et al. 2016). The 2016 report also shows that the total number of people affected by various disasters was the highest (569 million) since 2006, far above the 2006–2015 annual average (224 million). This unprecedented rise was mainly due to the human impact of drought in India, which affected 330 million people in both 2015 and 2016, the highest number of people affected by a natural disaster ever reported (Guha-Sapir et al. 2016). Unsurprisingly, almost 2% of annual GDP is lost due to natural disasters (Mani et al. 2012). Out of 35 total states and union territories, 27 are known to be disaster-prone (Mishra 2014). Around 40 million hectares of agricultural land (almost three times the size of New York State) are prone to floods with 8 million hectares being affected every year (GoI 2002). Flood disasters are the largest cause of economic damage and losses of human lives and livestock (Kumar et al., 2006). The changing volume, timing and pattern of precipitation, and the melting of snow influence streamflow patterns and further intensify the severity of floods (Gosain et al. 2006). India's 8000-km-long coastline that stretches from Gujarat to West Bengal is extremely vulnerable to cyclones (GoI 2006). Lying in the path of tropical hurricanes from the Bay of Bengal, the eastern coast of India is particularly at risk of being damaged by storms and floods (Emanuel 2005; IPCC 2007). The National Institute of Oceanography assessed the impacts of climate change on sea levels and determined that the occurrence of extreme events may change with a projected increase in the occurrence and strength of cyclones in the Bay of Bengal (DEFRA 2005).

Indeed, other non-climatic factors, such as the blocking of natural drainage systems by human encroachment, widespread soil erosion due to deforestation and declining water carrying capacity due to siltation in river beds and other water bodies have further exacerbated the flood risks. The receding of Himalayan glaciers reduces river flows and affects their catchment areas. The increasing frequency of droughts and declining water availability in the channels have resulted in the over-extraction of groundwater and depletion of the water table (Mishra 2014).

Under the circumstances, the present chapter makes a modest attempt to examine the effects of climate change on the food security situation in India and analyze the existing adaptive strategies and their critical overviews.

24.2 Impacts on Food Security

Climate change is a major concern for India because the majority of India's population depends on climate-sensitive sectors such as agriculture, forestry, and fishing for livelihood (Dev and Sharma 2010). The agriculture sector alone represents 23% of India's Gross National Product (GNP) and thus plays a crucial role in the country's development and shall continue to occupy an important place in the national economy. It sustains the livelihood of nearly 70% of the population (Khan et al. 2009). The existing problem of food security in India, if not addressed in time, will become more precarious due to changes in climate, since more than one-third of the population is estimated to be absolutely poor and one half of all children are malnourished in one way or another (Dev and Sharma 2010).

Climate change has adversely affected India's agriculture sector and is an emerging threat to food security. Agriculture yields in India are very sensitive to climate changes such as unnatural high or low precipitation, erratic temperatures and humidity. There is evidence of a rise in the frequency and intensity of extreme weather conditions (flood, drought, heat waves) severely affecting India's food production and food security. Studies show that for every 1 °C rise in temperature, wheat production is reduced by four to five million metric tons. There is extensive evidence on the impacts climate change has on the yield and quality of food crops, beverage crops (tea, coffee), medicinal plants, dairy, and fish (GoI 2018). The current challenges facing Indian food security involve problems impacting the food-producing sector and production, transportation, storage, and access to food. Amid the climatic changes, the biggest challenge for India is to enhancing the current level of food production and maintaining its equitable access across the country to ensure food security in a sustainable manner.

The concept of food security is very complex and multidimensional. Therefore, the issues of food production, distribution, the quality of food, the capability of purchasing, and the sustainability of the entire process must be included. Climate change affects food security in complex ways, as it has been found that it impacts crops, livestock, forestry, fisheries, and aquaculture, and can cause grave social and economic consequences in the form of reduced incomes, eroded livelihoods, trade disruptions, and adverse health impacts (Chakrabarty 2016). However, it is important to note that the net impact of climate change depends not only on the extent of the climatic shock but also the underlying vulnerabilities. According to the Food and Agriculture Organization, both biophysical and social vulnerabilities determine the net impact of climate change on food security (FAO 2016).

If we compare decadal growth pattern of India's population with the growth of production of food grains, it has been observed that during 1981–1991, the decadal population growth was 24%, which declined to 21% during 1991–2001, whereas the growth rate of production and yield have declined for crop groups/crops during the period 1996–2008 compared to the period 1986–1997 (Dev et al. 2010). The growth rate of food grain production (rice, wheat, pulses, and other cereals) declined from 2.9% to 0.93% during the same period (Dev and Sharma 2010). This is undoubtedly a matter of concern and there is a need for considerable enhancement in the production of food grains to match the growth of the population. If the growth patterns of Indian agriculture are considered, it has been observed that the growth of agriculture decelerated from 3.5% between 1981–1982 and 1996–1997 to around 2% between 1997–1998 and 2004–2005 (Chand et al. 2014). Eventually, the very low growth has adverse effects on the farm economy and the livelihood of the farming community and poses a serious threat to national food security. Some initiatives were taken towards the end of the 10th national plan and during the 11th national plan to revive the agriculture sector. Consequently, the growth rate accelerated to 3.75% during 2004–2005 to 2012–2013 (Chand et al. 2014).

However, the country faces major challenges in increasing its food production to the tune of 300 million tons by 2020 in order to feed its ever-growing population, which is likely to reach 1.30 billion by the year 2020. To meet the demand for food from this increasing population, the country's farmers need to produce 50% more grain by 2020. Although the arable land of India has increased from 178 thousand hectares to 224 thousand hectares (26% rise) from 1950 to 2000, during the same period the population has grown from 360 million to 1 billion (almost three times). So rapid population growth along with burgeoning urbanization and industrialization have reduced the per capita availability of arable land from 0.48 hectares in 1950 to 0.15 hectares in 2000 and are likely to further reduce it to 0.08 hectares by 2020 (Mall et al. 2006). Indian agriculture, and thereby India's food production, is highly vulnerable to climate change largely because the sector continues to be highly sensitive to monsoon variability, as 60% of the cropped area is still in the rain-fed zone. Wheat and rice, two crops central to nutrition in India, have been found to be particularly sensitive to climate change (Chakrabarty 2016). Wheat growth in northern India is highly sensitive to temperatures greater than 34 °C (Lobel et al. 2012). The IPCC report of 2007 echoed similar concerns about wheat yield: a 0.5 °C rise in winter temperature is likely to reduce wheat yield by 0.45 tons per hectare in India. Acute water shortage conditions, together with thermal stress, will affect rice productivity even more severely (Metz et al. 2007).

Rising demand for water in agriculture and dwindling and unpredictable river flow due to climate change are causing conflict within the regions. In India, one can witness several cases of water disputes that are inter-state, intra-state and also international (with the neighboring countries Nepal, Bangladesh; and two nuclear-armed countries, i.e., Pakistan, and China). Unfortunately, the condition is escalating due to a lack of pragmatic policies and political will. Several tribunals addressing individual rivers have failed to resolve the disputes and have affected the farmers of the involved states due to a lack of proper decision making at the right time (SANDRP 2016). In addition to interstate and international riparian concerns, there is increasing evidence of local grassroots-level water disputes. There are growing conflicts between pastoralists and different agricultural as well as domestic users, traditional farmers have to compete with investors (such as tourist resorts or factories) and there are examples of villages competing with each other for the same water resources (Taenzler et al. 2011).

There is also a great concern about the impact of climate change on the decline in soil fertility, soil moisture, changes in the water table, rising salinity, the increasing resistance of pests to many pesticides and the degradation of irrigation water quality in various parts of India (Sinha et al. 1998; Mall et al. 2006). Fluctuations in the production of crops in India due to a high dependency on monsoons also has had adverse impacts on the prices of agricultural commodities. As a result, the agricultural market has become unstable and this has given opportunities to middle men to speculate the prices of agricultural commodities. But the farmers suffer as they do not get the justified price due to a lack of government control over the agricultural market (Dev et al. 2010). Insurance for crop loss due to natural disasters is a very important tool for giving farmers relief and economic support.

Climate change is affecting the population's nutritional intake (micronutrients such as vitamins, minerals, and antioxidants), due to negative impacts on fruits and vegetables. One of the most threatening biological responses to terminal heat stress and deprived soil water availability is the loss of production of horticulture crops (Malhotra 2017). The productivities of thermo-sensitive crops such as tea, coffee, cardamom, cocoa, cashew, and black pepper will be directly affected by the rise in temperature (2–3 °C). The increase in night temperatures is going to affect the flowering of mangos and cashews. Floods and summer droughts are likely to affect crop coconut productivity (Rao et al. 2013). Due to poverty and also for cultural reasons, plants are the principal sources of protein in India. However, studies show that rising atmospheric CO_2 can lower the protein content of rice, wheat, barley, and potato. Considering Indian dietary patterns, the study estimated that the average Indian may lose more than 5% of their dietary protein and an additional 53 million people may become at risk. Thus, elevated atmospheric CO_2 may widen the disparity in protein intake within the country, with the people living on plant-based diets being the most vulnerable (Medek et al. 2017).

In India, climate change has brought new patterns of pests and diseases that are affecting all plants and animals and posing new risks for food security, food safety, and human health. Climate change has triggered major changes in geographical distribution and the population dynamics of insect pests, insect-host plant interactions, the activity and abundance of natural enemies, and the efficacy of crop protection technologies, which eventually affect both crop production and food security. Various insect pests currently limited to tropical/subtropical regions will move to temperate regions along with a shift in the areas of cultivation of their host plants. On the other hand, the population of native insect species (some are natural predators of pests) vulnerable to high temperatures in the temperate regions may decrease as a result of global warming. It has also been found that the relative efficacy of pest control measures such as host-plant resistance, natural enemies, bio-pesticides, and synthetic chemicals is likely to change as a result of global warming and climate change (Sharma 2016; Karuppaiah and Sujayanad 2012).

The wastage of food in the stages of production, handling, transportation, storage, and consumption put enormous stress on the economy, environment, and food security. Wasting a kilogram of wheat and rice would mean wasting the 1500 L and 3500 L of water, respectively, that go into their production. According to FAO, every year, almost one-third of food produced globally for human consumption is either lost or wasted. The associated economic, environmental, and social costs of this loss are around $1 trillion, $700 billion, and $900 billion per year, respectively. In India, the cost of food wastage (harvest and post-harvest losses of major agricultural produce) is estimated at around $14 billion per annum at 2014 wholesale prices (Bordoloi 2016; Athar 2018). Poor production of food is not the only reason for food insecurity in India. Farm output has been setting new records in recent years, with increased output from 208 million tons in 2005–2006 to 263 million tons in 2013–2014. India needs 225–230 million tons of food per year, so even when accounting for recent population growth, food production is clearly not the main issue. The most significant but long-ignored factor is that a high proportion of the food that India produces never reaches consumers. $8.3 billion worth of food, or nearly 40% of the total value of annual production, gets wasted and thus India is regarded as the world's biggest waster of food. Apart from perishable food, an estimated 21 million tons of wheat rots or is eaten by insects and rodents as a result of inadequate storage and poor management by government.

Increasing temperatures and unstable, moist weather conditions could result in grain being harvested with more moisture required for stable storage. The lack of drying facilities in these regions creates hazards for food safety and even causes complete crop loses. The absence of modern food distribution chains, too few cold-storage centers and refrigerated trucks, poor transportation facilities, an erratic electricity supply, and a lack of incentives to invest in the sector are believed to be the major reasons for the massive wastage of food (Biswas 2014). About 30–40% of the fruit and vegetables grown in India (40 million tons, amounting to US$13 billion) get wasted annually due to gaps in the cold chain such as poor infrastructure, insufficient cold storage capacity, the unavailability of cold storage in close proximity to farms, and poor transportation infrastructure. This has resulted in instability in prices and farmers cannot get remunerative prices in addition to rural impoverishment and farmers' frustrations (Ali 2004). Even with modern storage technologies, foods get wasted due to poor storage management and a lack of monitoring. The study shows there are 10%, 20%, and 25% post-harvest losses in durables (cereals, pulses, and oilseeds), semi-perishables (potato, onion, sweet potato, tapioca), and products like milk, meat, fish, and eggs, respectively, in modern storage (Maheshwar and Chanakwa 2006). Grain storage is a component of the grain marketing supply chain that evens out fluctuations in the supply of grain from one season, usually the harvest season, to other seasons, and from one year of abundant supply and releasing to lean years. So any losses in storage are bound to disturb the supply chain and, in turn, also affect market prices.

Climate change mostly affects small and marginal farmers, as they depend mostly on rain-fed mono cropping. Their food stocks begin to run out 3 or 4 months after harvest and farm jobs also remain mostly unavailable and by the next monsoon/sowing season, food shortages peak to hunger (Ramachandran 2014). Climate change will also have an adverse impact on the livelihoods of fishers and forest-dependent people, who are already vulnerable and food insecure. The increasing number of weather-related hazards will restrict fishermen from venturing into the sea, which may result in declines in fish catch. In addition, the rise in sea surface temperature will also disturb the marine ecosystem and have negative impacts on fish and other marine resource availability. Forest-dependent people, mainly the tribal population, will face a difficult situation as the yield of forest products will also be affected due to climate change (Chakrabarty 2016). Landless agricultural labourers wholly dependent on agricultural wages are at the highest risk of losing their access to food (Dev 2012). Besides climate change, poverty and unemployment are also an important factor in India, which reduces purchasing power as well as the capacity to access food and other basic necessities of life. Reducing poverty and generating employment among the population are among the biggest challenges facing India, particularly in relation to ensuring food security for the population living below the poverty line.

Climate changes will also have adverse effects on urban food security, especially for those who belong to a low-income group and are living in slums and informal settlements, as they are often exposed to hazards such as floods and landslides. For example, the city of Kolkata (erstwhile Calcutta) is particularly vulnerable to natural disasters, as climate change is likely to intensify the frequent flooding in the Hooghly River during monsoons. The poor inhabitants of Kolkata are most vulnerable as their homes are located in low-lying areas or wetlands that are particularly prone to tidal and storm surges (Dasgupta et al. 2012). Given that food is the single largest expenditure for poor urban households, displacement, loss of livelihood or damage to productive assets due to any extreme weather event will have a direct impact on household food security. The urban poor have also been identified as the group most vulnerable to increases in food prices following production shocks and declines that are projected under future climate change.

24.3 Current Adaptation Strategies and Challenges

Reducing risks to food security from climate change is one of the major challenges of India and adaptation actions to reduce risks are necessary. However, the extreme complexities of India's food insecurity due to existing factors and further exacerbation of the situation by climate change have made developing adaptation strategies very challenging. Poor people in India are more at risk of food insecurity, as they face more severe resource and economic constraints due to the high population growth rate and limited arable land (Kumar 2003). Food insecurity among the poor farming communities due to climate change has been further complicated by the decreasing per capita availability of arable land and is compounded by the challenge of meeting rising food demand due to high population growth (Roul et al. 2015). One of the most important widespread issues is linking food security in India to poverty and sustainability. Low food intake, poor health, low agricultural productivity and low income are perennial forces in the vicious poverty cycle in India. So, the issue is not only the availability of food but the affordability of food for vulnerable populations in adequate quantity and quality. So the question is whether increased food production can meet the needs of the rising population in a sustainable manner. A long-term strategy is needed to reduce the vulnerability of the farming community and sustainably intensify agricultural productivity with minimum degradation of land and natural resources being used (Varadharajan et al. 2013).

The following section will address risk factor-specific adaptation strategies based on the impacts of climate change. It is worth mentioning that several adaptation strategies have already been put in place by various stakeholders. The Indian Council for Agricultural Research (ICAR), the focal organization for India's agriculture research, technology development and transfer of technology, has accorded high priority to understanding the impacts of climate change and developing adaptation and mitigation strategies to meet the challenges posed by climate change to the agricultural system (Srinivasa et al. 2016). In 2011, the ICAR launched a network project called National Innovations in Climate Resilient Agriculture (NICRA, http://www.nicra-icar.in/nicrarevised/) to enhance the resilience of Indian agriculture to climate change and climate vulnerability through strategic research and technology demonstrations. The research on adaptation and mitigation covers crops, livestock, fisheries, and natural resource management. The project consists of four components: (a) enhancing the resilience of Indian agriculture to climatic variability and climate change through strategic research on adaptation and mitigation; (b) validating and demonstrating climate-resilient technologies on farmers' fields; (c) strengthening the capacity of scientists and other stakeholders in climate-resilient agriculture; and (d) drawing policy guidelines for wider-scale adoption of resilience-enhancing technologies and options. The expected outputs are (a) the selection of promising crop genotypes and livestock breeds with greater tolerance to climatic stress; (b) existing best bet practices for climate resilience demonstrated in 150 vulnerable districts; (c) the strengthening of infrastructure at key institutes for climatic change research; and (d) adequately trained scientific manpower to take up

24 Climate Change and Food Security in India: Adaptation Strategies and Major... 505

climate change research in the country and empower farmers to cope with climate variability (ICAR 2018).

The country adopted the National Agroforestry Policy in 2014 to encourage and expand tree plantation in an integrated manner with crops and livestock to improve productivity, and the employment, income, and livelihoods of rural households; to protect and stabilize ecosystems; and to promote resilient cropping and farming systems to minimize risks during extreme climatic events (Srinivasa et al. 2016). The policy also envisages meeting the raw material requirements of wood-based industries, of small timber for the rural and tribal populations, and for reducing the pressure on forests.

As the part of the adaptation to climate change, the Government of India (National Action Plan on Climate Change) emphasized dryland agriculture, risk management, access to information and the use of biotechnology (GoI 2018).

1. Adaptation for dryland agriculture: Sixty percent of India's agricultural land falls under the dryland/rain-fed zone. Thus, the impacts of climate change will significantly affect the nation's food security. To overcome the adverse impacts of climate stress on yield, the current policies have focused on (a) developing drought- and pest-resistant crop varieties; (b) improving the methods of conservation of soil and water; (c) stakeholder consultation and training for farming communities on agroclimatic information and dissemination; and (d) financial support to enable farmers to invest in and adapt relevant climate-smart technologies.
2. Managing risks due to extreme climatic events: The priority areas are: (a) strengthening agriculture and insurance; (b) better GIS-based, web-enabled natural resource mapping and utilization; (c) risk and vulnerability assessment and prediction; and (d) effective risk communication and management based on vulnerability and risk assessment.
3. Knowledge dissemination and translation: The strategies are: (a) developing micro-level databases of natural resources (soil, water), land-use patterns, plant genotypes, and weather patterns; (b) regular, accurate monitoring of natural resources, their degradation and impacts on agricultural production; and (c) the collation and dissemination of customized information to farming communities.
4. The development and utilization of climate-smart biotechnology: The targeted technological solutions are: (a) drought-, flood-, and pest-resistant crops; and (b) genetic engineering to develop more carbon-responsive C-4 crops for increased productivity, the high utilization of CO_2 and sustaining thermal stresses (GoI 2018).

Climate-Smart Agriculture or CSA is considered a pragmatic approach that helps to guide actions needed to transform and reorient agricultural systems to effectively support development and ensure food security in a changing climate. According to FAO, the CSA has three main objectives: (a) sustainably achieving agricultural productivity and incomes; (b) adapting and building resilience to climate change; and (c) reducing and/or removing greenhouse gas emissions, where possible. CSA is not a monolithic approach; rather, it considers the existing diversity of social, economic and environmental contexts including agro-ecological zones and farming systems accordingly. CSA, therefore, identifies and integrates a package of climate-resilient technologies and practices for the management of natural resources at the farm level while considering the linkage between agricultural production and ecosystems services at the landscape level (Likhi 2017). Thus, CSA provides a broad enabling framework to help all stakeholders to identify sustainable agricultural strategies suitable to the local conditions. The following section will address adaptation strategies based on the principles of CSA applicable to Indian conditions (FAO 2015; Shelat 2014).

24.3.1 Conservation Agriculture

Conservation agriculture embodies three main principles: reduced or zero soil disturbance, permanent organic soil cover to reduce soil loss, and crop diversification. According to the FAO, crop diversification with intercropping and rotation improve the nutritional security of farm households and reduce the risk of total crop failure in unfavorable or erratic weather. There is also the potential for gender equity in conservation agriculture; for example, male-driven soil tillage work can be replaced by female-driven weeding work. This is extremely important in the current situation due to a rising feminization of the agriculture workforce due to the outmigration of the male workforce to cities for better opportunities (FAO 2013).

In South Asia, conservation agriculture has emerged as a new paradigm to achieve sustainable agriculture goals. Conservation agriculture is an approach developed to manage farmland for sustainable crop production, while simultaneously preserving soil and water resources. Conservation agriculture largely depends on three major principles: (a) the maintenance of a permanent vegetative cover or mulch on the soil surface; (b) minimal soil disturbance (no/reduced tillage); and (c) a diversified crop rotation. Given the effects of conservation agriculture on soil, water, and economic viability, this management has been widely recommended and adopted with mixed success (Jat et al. 2014).

The Green Revolution, which improved food security in India, ignored a vast swath of rain-fed agro-ecosystems. Unpredictable patterns of rainfall, biotic and abiotic stresses and adherence to traditional farm practices have resulted in poor cropping intensities and low yields (Pradhan et al. 2018). However, for India's rain-fed areas, further local-level adjustments are needed before implementing any broad policy on conservation agriculture (Jat et al. 2014). The tribal population, the most vulnerable sector of India's society, lives in the rain-fed areas located in the central region of the country (also known as the Poverty Belt). Since more than 90% of the nation's tribal people are totally dependent on agriculture for their livelihood and food consumption, poor food production has led them to suffer from chronic malnutrition (Pradhan et al. 2018). Excessive and inappropriate tillage increases soil degradation and erosion, reducing soil productivity and soil organic carbon. Agricultural intensification is needed in the rain-fed regions without further degrading of the natural resource base. It is paramount that the implementation of a new agricultural production system to improve household food and nutritional security as a part of adaptation to climate change be viewed in the context of enhancing farm productivity, environmental quality, and profitability of agriculture (Pradhan et al. 2015).

Reduced and no tillage for planting wheat is gaining increasing acceptance with farmers in South Asia because of reduced land preparation costs. More than 13,500 on-farm trials have been conducted in South Asia to evaluate different resource conservation technologies for rice and wheat. These trials showed that reduced and zero tillage with residue mulch performed better than the conventional till broadcast (Jat et al. 2014). However, the experience in China shows that despite some benefits in crop production, conservation agriculture can also reduce crop production through undesirable effects on soil's physiochemical and biological conditions. According to the researchers, the key limiting factor in the application of conservation agriculture in China is the uncertainty of the long-term effects on soil and crop yields. The effects of conservation agriculture on crop yield increased and decreased with decreasing and increasing annual precipitation, respectively (Zhang et al. 2015). A study in India showed that reduced tillage combined with planned intercropping had higher system productivity and net benefits. The results demonstrated that conservation agriculture could concurrently increase crop yield, diversify crop production and improve soil quality, which eventually would move towards the sustainable intensification of crop production, improving household income and ensuring food security (Pradhan et al. 2018).

24.3.2 Natural Resources Management

Irrigation plays very significant roles in reducing the negative effects of climate change on agricultural productivity, with the biggest reduction being in rain-fed farming systems (Birthal et al. 2014). However, it is important to note that irrigated lands are not insulated from climate change. Changing precipitation has already affected the river systems. Poor rainfall and receding glaciers are major threats to normal water flow. In addition, unsustainable groundwater extraction has made the situation more precarious.

India has 20 river basins and 12 of them are considered major, covering a total catchment area of 2.53 million km^2. Large irrigation potential has been the foundation of India's agricultural growth and past food security. However, the large river irrigations have been significantly affected by climate change by impacting their hydrological cycle. Eventually, these changes have affected the supply of water from inflow from rivers, reservoirs, tanks, ponds and groundwater resources and impacted their ability to replenish water. The projected increase in precipitation variability, which implies longer drought periods, would lead to an increase in irrigation requirements, even if total precipitation during the growing season remains the same. In addition, the irrigation demands could become even greater if rain-fed areas are not able to meet expected food needs. It has been projected that a significant increase in runoff may be witnessed in India. Predictably, as the increased runoff will be mostly in the wet season, the extra water will remain unavailable in the dry season. On the other hand, the extra water in the wet season may increase the frequency and duration of floods. A detailed simulation study on the impact of climate change on water resources in the river systems established that in the near future, while some river systems may have decreased water flow, other rivers may have a high intensity of floods (Kaur et al. 2003). Therefore, adequate water storage infrastructure is needed as a part of adaptation measures. The increased melting and recession of Himalayan glaciers associated with climate change could further alter the run-off (GoI 2013).

Groundwater resources are being rapidly depleted because of unsustainable extraction levels that exceed natural recharge rates. While the global average of the contribution of groundwater in irrigation is about 40%, in India it is expected to be over 50%. Some Indian states have a very high dependency on groundwater for irrigation (60–80%) (Dhawan 2017). Groundwater is crucial, even where crops are irrigated by surface water, particularly when prolonged droughts affect the supply of water. Thus, climate change is likely to increase the demand for groundwater to facilitate irrigation management (Palanisami et al. 2013). Despite the growing scarcity, groundwater irrigation in India remains highly inefficient from a technical point of view. For example, only 3% of India's 8.5 million irrigation well owners used drip or sprinkler irrigation and 88% delivered water to their crops by flooding through open channels. Now, almost 40% of the wells are showing a decline in groundwater level (Dhawan 2017). Groundwater irrigation often encounters competition from urban and industrial demand for water, resulting in conflict in many places. The rapid depletion of groundwater results in an increase in the energy required to pump water, making irrigation more expensive and in turn adding more stress to agriculture (Palanisami et al. 2013).

Studies on watershed development and joint forest management programs in India show that direct human pressure on forests has been considerably high and climate change is making the situation worse. It is important to have integrated knowledge management and effective coordination in the management of the water shed and forest sector and social vulnerability assessment. The research should focus on long-term and multi-sector planning, implementing and interventions for climate change adaptation. Internalization of the valuation of environmental services in the mainstream economy, information sharing, and improvement of governance can ensure sustainability (CCA RAI 2014).

As part of adaptation strategies to climate change, the following steps should be implemented:

(a) Integration of the uses of surface and groundwater needs to be developed.
(b) Reducing wastage, better water-harvesting capacity, and increasing the rate of storage recharge (particularly by community-based micro-storage facilities and aquifer recharge).
(c) Ensuring natural recharge rates are closer to groundwater extraction rates so these reservoirs become more sustainable.
(d) Lining of water transport systems to reduce seepage losses.
(e) Developing the ability to carry over water from one season to the next by proper storage such as in the vadose zone (between the soil surface and water table) above aquifers.

(f) Improved storage structures with sluice modification, sluice management, canal lining, and rotational irrigation with bore well supplementation.

(g) Starting the vigorous evaluation of industrial and sewage wastewater usage to reduce dependence on freshwater supplies.

(h) Better governance through inter-departmental coordination with a proper framework of sustainable water management and optimized water recycling.

(i) Better coordination with the meteorological department for better planning in selecting crops, selecting dates for planting, spacing, and input management.

(j) Promotion of rice intensification, machine transplantation, alternate wetting and drying.

(Pathak et al. 2003; Naresh et al. 2013, 2014).

Before developing macro-level national policy on rain-fed areas, it is important to generate evidence at the micro level, at least at the block (*taluka*, similar to a county) level. Local agriculture institutions and their affiliated extension centres (also known as *Krishi Vigyan Kendra*) can play pivotal roles in generating evidence by (a) analyzing biophysical and socioeconomic impacts and risks of climatic variations on local watershed development programmes; (b) assessing the sensitivity and vulnerability of the watersheds based on the projected climate scenarios; (c) studying the existing maladaptation practices in the watersheds that can exacerbate the effects of climate change; (d) addressing the risks and opportunities (such as the convergence of watershed development with other ongoing development activities) arising due to climate change; and (e) building and strengthening climate change adaptive capacities of communities within a watershed, without foregoing development for the poor. Therefore, research on assessment of the impacts of climate change should encompass soil, water, forests, pastureland, livestock, agriculture practices and the livelihood of the communities.

24.3.3 Climate–Smart Crops

Adaptation strategies for dealing with the impacts of climate change depend not only on climate parameters but also on the existing system's ability to adapt to change. The potential depends on how well the crops adapt to the concomitant environmental stresses due to climate change. Climate-smart horticulture can be a practical solution, but any single specific agricultural technology or practice cannot be universally applied. Rather, it is an approach that requires site-specific (agroecological region) assessments to identify appropriate production technologies. In the event of working out adaptation and mitigation strategies, it will be appropriate to utilize modelling tools for impact analysis for various horticultural crops. As of now, except for potato and coconut, there are no good simulation models for horticultural crops in India. Innovative methods are essential in developing simulation models for important horticultural crops such as mango, grape, apple, orange, citrus, litchi, and guava. The existing climate-smart crops (which are able to cope with high temperature, frost, and excess moisture stress conditions) should be integrated to the specific agro-ecological region. The production systems for each crop should address improved water-use efficiency and adapting to hot and dry conditions. For example, for sowing or planting dates, fertilizer application can be determined by forecasting weather and measuring soil nutrients. Providing irrigation during critical stages of crop growth and the conservation of soil moisture reserves are the most important interventions. Other crop management practices such as mulching with crop residues and plastic mulches to conserve soil moisture, and growing crops on raised beds to protect them from excessive soil moisture due to heavy rain are the most important adaptation strategies (Malhotra 2017).

Rice and wheat are important crops for a large portion of India, being the main sources of food and the agriculture of which is a main source of income. Genome editing, which has been a focal point of research in recent times, is now rightly being targeted towards improving the quality of crops. Research using CRISPR genome editing tools has increased the ability to target and modify crop genes for the development of improved varieties in terms of yield and ability to withstand climate-related stress such as drought, flood, diseases (fungal and pest), high temperature and low humidity. Genome editing is a successful and feasible venture and the latest developments and improvements of CRIPSR tools could further enable researchers to modify more genes in rice and wheat with increased efficiency. Indian scientists have made significant progress in CRISPR technology and it should be used as a part of the development and promotion of climate-smart crops across the country (Mazumdar et al. 2016; Rani et al. 2016; Srivastava et al. 2017). The Philippines-based International Rice Research Institute, which is supported by the Bill and Melinda Gates Foundation, is actively involved in collaborative research on such crops (Temple 2017). CRISPR genome editing tools are also used in fisheries and horticulture. Studies in India showed the recent advances in gene editing techniques in farmed fish (*Labeo rohita* also known as *rohu*) and bananas (*rasthali*) (Chakrapani et al. 2016; Kaur et al. 2018).

24.3.4 Farmers' Prioritization of CSA Technologies

Despite the known benefits of CSA technologies, the current rate of overall adoption by Indian farmers is fairly low. A community-based participatory study in India assessed farmers' preferences and willingness to pay for some selected climate-smart technological options, such as water-smart (rainwater harvesting, drip irrigation, laser land levelling, drainage management, cover crops), energy-smart (zero/energy tillage), nutrient-smart (site-specific integrated nutrient management, green manure, intercropping with legumes), carbon-smart (agro-forestry, fodder management, integrated, pest management, concentrate feeding for livestock), weather-smart (housing for livestock, crop insurance), and knowledge-smart (contingent crop planning, climate-smart crops, seed and fodder banks). The study showed that many socioeconomic and location-specific variables have a significant effect on farmers' preferences towards a particular CSA technology. For example, farmers from areas with low rainfall and high coefficients of variations in annual rainfall preferred risk mitigation technologies such as crop insurance, weather-based crop agro-advisories, and rainwater harvesting. On the other hand, farmers from high rainfall zones had a low preference for rainwater harvesting and crop insurance. Despite the demonstrable benefits of many CSA technologies, the concerned farmers may not be willing to invest in them. Therefore, adaptation policies should highlight the crucial role of providing information about available CSA technologies and making financial resources available to the farmers to adopt various locally relevant CSA technologies. More experienced farmers were more confident about choosing CSA technologies, except certain ones such as rainwater-harvesting technologies. Female farmers usually preferred integrated pest management, weather-based crop agro-advisories, and contingent crop planning compared to male farmers. Preferences for rainwater harvesting and climate-smart housing for climate change adaptation were significantly negative for female farmers. The results also indicated that low-income farmers were more likely to prefer site-specific integrated nutrient management, integrated pest management, and laser land levelling compared to rainwater harvesting, contingent crop planning, and crop insurance (Khatri-Chhetri et al. 2017).

24.3.5 Addressing Fundamental Issues: Conflict Resolution, Crop Insurance, Food Wastage, and the Public Distribution of Food

For conflict resolution with regard to water disputes, it is paramount to pay special attention to conflict components in each context. To avert future climate change-induced conflicts over scarce water resources, different available options should be availed of for peaceful crisis management and conflict resolution. There is a need to enact a new national water policy to combat, mitigate, and adapt to water scarcity situations which may arise out of climate change. If the uncertainties arising out of climate change are to be integrated into water management planning, there is an urgent need for enhancing water storage capacity, and effective technological and financial planning to stop overconsumption and to apply the more sensible use of ground and surface water. The successful operation of a national water policy that is responsive to climate challenges requires strong research-based knowledge and appropriate institutional support at all levels of governance. Since the Indian Constitution reserves the power of the federal government in establishing legislation on the use of interstate rivers and on the adjudication of interstate disputes over water, it is important to use this precious natural resource amicably and to accept shared benefits. To reduce future water stress, all the stakeholders should be part of monitoring and planning on climate change impacts on water resources, improving management capacities, building water management institutions to serve local purposes and reliance on dialogue and confidence building (Taenzler et al. 2011).

Though the Government of India has initiated efforts for providing crop insurance to farmers since independence, a major boost was given in the form of the Comprehensive Crop Insurance Scheme (CCIS) in 1985 during the Seventh Five Year Plan period, which covered the risk in the cultivation of major crops against natural calamities and pests and diseases (Raju and Chand 2009). The National Agricultural Insurance Scheme (NAIS) which is also known as the *Rashtriya Krishi Bima Yojana*, replaced the CCIS in 1999–2000. NAIS operates in all States and Union Territories of India. The percentage of farmers covered under NAIS increased from 9.08% in 2000 to 15.95% in 2007 (Brahmanand et al. 2013). Still, a substantial percentage of farmers remain without crop insurance coverage and, not surprisingly, regular crop failures due to climate change result in the suicides of thousands of Indian farmers. A study in India demonstrates that fluctuations in temperature, particularly during India's agricultural growing season, when heat also lowers crop yields, significantly influence suicide rates. The study concludes by estimating that around 60,000 additional suicides in 30 years (since 1980) can be attributed to warming and the author could not find any evidence of adaptation to reduce the number of suicides (Carleton 2017). Therefore, crop insurance being made available to all the vulnerable farmers is essential to prevent such unfortunate consequences of climate change-driven crop failure.

If the current rate of food wastage continues, no adaptation strategy will be able to achieve the desired food security. Therefore, the adaptation strategy should consider the prevention of food wastage at all stages. The proposed solutions are (a) improved access to low-cost handling and storage technologies (evaporative coolers, metal silos); (b) real-time wireless/mobile-enabled sensors to monitor the storage conditions of perishable food during transportation and data transfer to the clients; (c) the establishment of mega food parks to increase the processing of perishables; (d) the development of innovative and intelligent packaging for perishables; (e) the identification/creation of businesses that buy unwanted food/produce directly from distributors/manufacturers for discounted retail sale; (f) the development and expansion of secondary markets for items with cosmetic damage; (g) the expansion (in number and capacity) of large food grain storage facilities, which are essentially managed by government (Food Corporation of India). Evidence shows that technology is central to addressing food waste, but the ultimate success requires the readiness to change attitudes of the stakeholders along the value chain (Bordoloi 2016; Athar 2018).

The Public Distribution System (PDS) plays a very important role in developing countries to ensure the fair distribution of subsidized food grains (especially rice and wheat) and other essential commodities such as sugar and kerosene to the poor population through a network of fair price shops. In India, this scheme was launched in 1947 and it is still the biggest social scheme for the poor, especially for people living under the poverty line. Although millions of poor citizens of India receive benefits from this scheme, which acts as a food security system, there are still several problems that have prevented this scheme from producing the desired effect on the availability, accessibility and optimum utilization of food grains. The major problem faced by the PDS in India is widespread corruption. A large number of poor and deserving households are left out and a lot of fake cards (which are required for receiving food) are also issued. People often do not get the sanctioned amount of food grains from the fair price shops due to the corruption of the shop owners or ignorance of the beneficiaries. The shop owners are notorious for stealing and overcharging, supplying poor-quality food grains, trading cards for money, and complicated bureaucratic procedures prevent many poor Indians from obtaining such cards. The study shows that more than half of the PDS food grain does not reach the intended people (Kumar et al. 2014). The PDS is the only available system for access to food for the vulnerable population across the country. The benefits of CSA can be achieved by ensuring equity only after revamping the PDS. Better governance, such as effective grassroots monitoring, empowering the local regulatory authority, addressing public grievances, and developing a decentralized functioning system are the keys to the effective functioning of PDS.

24.4 Use of Drama Theory: A Behavioral Decision Model in Food Security, Adaptive Strategies for Public Policy Perspectives

This section gives a view about the application of a behavior model to change people's and government's perspective in terms of adoption and food security. We proposed the potential use of drama theory, a behavioral decision model, to find a win-win situation for the multiple players involved in their dynamic responses. Rationally, any choice is determined in terms of costs and benefits and what maximizes their net benefits. But drama theory captures the "non-rational" aspects of the decision-making process such as crisis, emotion, and self-realization. This approach addresses how players apply rational emotional-pressure on each other to redefine the game prior to it being played (Stubbs et al. 1999).

Drama theory is further developed by Howard et al. (1993), Bennett and Howard (1996) and Howard (1999, 2007). This approach addresses how players (characters) apply rational emotional-pressure on each other to redefine the game prior to it being played (Stubbs et al. 1999). The fundamental difference between game theory and drama theory is that a drama allows for the possibility of the game itself changing even though the environment remains informationally closed; that is, it considers the possibility of endogenous changes arising from interactions within the game itself (Howard 1994). The elements of a frame which corresponds to the game theoretic concept of a game include a set of characters that have options in which characters interact through a series of episodes. A dramatic episode is an interaction between parties in which a set of issues is at stake. It ends when some of the issues are decided so that there is now a new set of issues (Howard 2007). Characters' choices influence the outcome of each particular episode, and also what episodes happen next. Within each episode, the phases are referred to as scene setting, buildup, climax, and denouement, a structure repeated on a large-scale for the drama as a whole. In the buildup stage, characters communicate, with each pressing for a particular position, a scenario they wish to have created and in the climax of the episode, the frames themselves come under pressure. The "moment of truth" occurs in a drama-theoretic model at the end of the buildup and they move on to the climax. It is a point in an interaction where each party has communicated a position and stated intentions it regards as "final."

Howard (1999) developed a technique to analyze conflict and cooperation to solve real-world problems, which is called Confrontation Analysis. This is also called Dilemma Analysis. This is derived from drama theory. This technique uses a card table model to analyze the conflict. A card table model of the moment of truth consists of:

1. A set of characters, each holding a number of cards, for each of the characters has a position. This is a specification of each card (belongs to all the characters) and whether it should be played or not played.
2. Fallback positions/threatened future. (A character's fallback consists of the cards it is threatening to play if its position is not accepted. The character may not intend to carry out its threat; that is, it may be bluffing. In any case, threatened future is a scenario that would result from the implementation of everyone's declared fallback).

Howard (1994) mentioned that there are only six dilemmas the character faces during the confrontation phase. Later, he again modified it as three dilemmas (Drama Theory 2). In Drama Theory 2 (Levy et al. 2009) states that all dilemmas are defined in terms of doubts. It is simpler and also more insightful. Instead of asking "Is this whole future preferred to this one?" it asks "Which option choices in this future are preferred to those in this one?" The metaphor of drama describes the interaction of different characters and how they change their preferences, develop and perceive the new outcomes. When none of the dilemmas exist, then the characters have an agreement that they fully trust each other to carry out the action. To solve the food security problem, close cooperation is needed among government and citizens. The government also needs to change their policy based on public opinion. The effectiveness of ongoing traditional approaches may be limited without the addition of measures to understand how to engage citizens in cooperative behavioral change (Hefny 2012).

A simplified government and farmers interaction is modeled using Drama Theory II to analyze how parties can find mutual ground to collaborate. Figure 24.1 shows a dilemma for adaptation strategies related to giving subsidies to farmers by the government. In this figure, a frame of confrontation has been depicted where farmers and government both are in a rejection dilemma as both the parties are in a situation where there is a possibility of little change of their respective positions. This situation must be avoided as this case would be a lose-lose situation for both parties and not taking adaptive strategies would lead to a bigger cost to the society.

To eliminate the rejection dilemma, parties can choose a new fallback position (Fig. 24.2). Here, communication between both parties plays a big role. Both parties' cooperation has added value. After communicating with each other, the gov-

24 Climate Change and Food Security in India: Adaptation Strategies and Major... 515

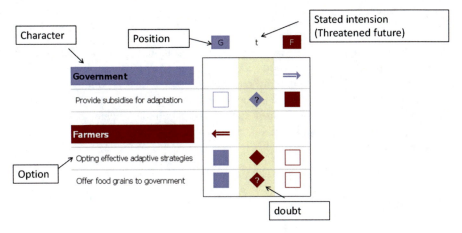

Fig. 24.1 Confrontation card table: government and farmers

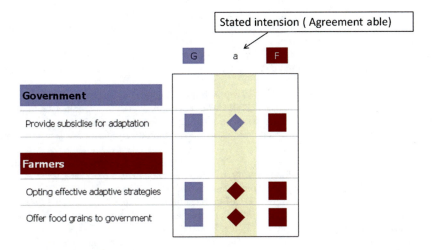

Fig. 24.2 Possible collaboration

ernment comes up with a subsidies option and thus the farmers are able to take adaptive strategies and agree to sell their food grains to the government, which turns the confrontation, and eventually leads to the adopting of different options and preferences by the parties, to a win-win situation. Now, in this new phase of the episode, both parties' positions are compatible with each other's; thus, no dilemma exists. Thus, drama theory leverages how to redefine the parties' options and preferences through continuous communication to find their acceptable positions.

24.5 Conclusion

"Climate-smart agriculture" seems to be a pragmatic approach to counter the impacts of climate change on food security. However, extensive community-based research is required to develop more realistic strategies. Regular dialogues between all the stakeholders are key to the acceptance and successful implementation of the approaches. Lastly, the core issues such as transboundary water conflicts, crop insurance, food wastage and food distribution are essential components to achieve the goals for adaptation.

References

Ali, N. (2004). *Post-harvest technology for employment generation in rural sector of India* (pp. 63–105). India: Central Institute of Agricultural Engineering, Indian Council of Agricultural Research.

Athar, U. (2018). As millions go hungry, India eyes ways to stop wasting $14 billion of food a year. *Reuters*. Retrieved May 5, 2018, from https://in.reuters.com/article/india-food-hunger/as-millions-go-hungry-india-eyes-ways-to-stop-wasting-14-billion-of-food-a-year-idINKBN1EU0UM.

Bennett, P. G., & Howard, N. (1996). Rationality, emotion and preference change: Drama theoretic–models of choice. *European Journal of Operational Research, 92*(3), 603–614.

Birthal, P. S., Negi, D. S., Kumar, S., Agarwal, S., Suresh, A., & Khan, M. T. (2014). How sensitive is Indian agriculture to climate change? *India Journal of Agricultural Economics, 69*(4), 474–487.

Biswas, A. K. (2014). *India must tackle food waste*. World Economic Forum. Retrieved March 5, 2018, from https://www.weforum.org/agenda/2014/08/india-perishable-food-waste-population-growth/.

Bordoloi, B. (2016). Curbing food wastage in a hungry world. *Business Line*. Retrieved March 5, 2018, from https://www.thehindubusinessline.com/opinion/curbing-food-wastage-in-a-hungry-world/article9285737.ece.

Brahmanand, P. S., Kumar, A., Ghosh, S., Roy Chowdhury, S., Singandhupe, R. B., Singh, R., et al. (2013). Challenges to food security in India. *Current Science, 104*, 841–846.

Carleton, T. A. (2017). Crop-damaging temperatures increase suicide rates in India. *Proceedings of National Academy of Sciences, 114*(33), 8746–8751 http://www.pnas.org/content/114/33/8746.

CCA RAI. (2014). Climate proofing of public schemes in India: Selected cases of Watershed Development Programmes and Joint Forest Management Programme. Climate change adaptation in rural areas of India. Deutsche Gesellschaft für Internationale Zusammenarbeit (GIZ) GmbH, Bonn. Retrieved March 5, 2018, from https://www.giz.de/en/downloads/giz2014-en-cca-rai-climate-proofing-india.pdf.

Chakrabarty, M. (2016). Climate change and food security in India. Observer Research Foundation New Delhi-2. *Brief Series, 157*, 1–11.

Chakrapani, V., Patra, S. K., Panda, R. P., Rasal, K. D., Jayasankar, P., & Barman, H. K. (2016). Establishing targeted carp TLR22 gene disruption via homologous recombination using CRISPR/Cas9. *Developmental & Comparative Immunology, 61*, 242–247.

Chand, R. (2014). *From slow down to fast track: Indian agriculture since 1995*. Working Paper National Centre for Agricultural Economics and Policy Research, New Delhi 1–34.

Chand, K., Thakur, S., & Kumar, S. (2014). Climate change and food security in India: Contemporary concern and issues. *International Journal of Development Research, V4*(2), 359–365.

Dasgupta, S., Roy, S., & Sarraf, M. (2012). Urban flooding in a changing climate: Case study of Kolkata, India. *Asian-African Journal of Economics and Econometrics, 12*, 135.

DEFRA. 2005. Climate change impacts on sea level in India. Key Sheet 4. National Institute of Oceanography, Goa, India. In *Investigating the impacts of climate change in India*. Report by Department of Environment, Food and Rural Affairs (DEFRA), UK and Ministry of Environment and Forests (MoEF), Government of India (GoI), 2005. Retrieved March 5, 2018, from http://www.defra.gov.uk/environment/climatechange/internat/devcountry/india2.htm.

Dev, M. S. (2012). *Impact of ten years of MGNREGA: An overview*. Mumbai: Indira Gandhi Institute of Development Research.

Dev, M. S., & Sharma, A. N. (2010). Food security in India: Performances. *Challenges and Policies, Oxfam India Working Paper Series, 7*, 1–46.

Dhawan, V. (2017). *Water and agriculture in India—Background paper for the South Asia expert panel during the Global Forum for Food and Agriculture (GFFA)*. German Asia-Pacific Business Association, German Agri-Business alliance, TERI. Retrieved March 5, 2018, from https://www.oav.de/fileadmin/user_upload/5_Publikationen/5_Studien/170118_Study_Water_ Agriculture_India.pdf.

Emanuel, K. (2005). Increasing destructiveness of tropical cyclones over the past 30 years. *Nature, 436*, 686–688.

FAO. (2013). *Climate-smart agricultural sourcebook*. Rome: Food and Agriculture Organization of the United Nations. Retrieved March 5, 2018, from http://www.fao.org/docrep/018/i3325e/i3325e.pdf.

FAO. (2015). *Climate-smart agriculture: A call for action*. Synthesis of the Asia-Pacific Regional Workshop, Bangkok, Thailand, 18–20 June 2015. Retrieved March 5, 2018, from http://www.fao.org/3/a-i4904e.pdf.

FAO. (2016). Chapter 2: Climate change, agriculture and food security—A closer look at the connections. In *The state of food and agriculture*. FAO, Rome: 22–25.

GoI. (2002). *High Powered Committee (HPC) on disaster management report. HPC Report*. New Delhi: National Center for Disaster Management, Ministry of Agriculture, Govt of India.

GoI. (2006). *Crisis management: From despair to hope. Second Administrative Reforms Commission Third Report*. New Delhi: Govt of India.

GoI. (2013). *Annual report 2013–14*. New Delhi: Central Water Commission, Ministry of water resources, Government of India.

GoI. (2018). *National action plan on climate change*. New Delhi: Prime Minister's Council on Climate Change, Ministry of Environment, Forest, and Climate Change, Government of India. Retrieved March 5, 2018, from http://www.moef.nic.in/ccd-napcc.

Gosain, A. K., Rao, S., & Basuray, D. (2006). Climate change impact assessment on hydrology of Indian river basins. *Current Science, 90*(3), 346–353.

Guha-Sapir, D., Hoyois, P., Wallemacq, P., & Below, R. (2016). *Annual disaster statistical review 2016—Number and trends*. Brussels, Belgium: Centre for Research on the Epidemiology of Disasters (CRED), Institute of Health and Society (IRSS) and Université Catholique de Louvain.

Hefny, M. A. (2012). Changing behavior as a policy tool for enhancing food security. *Water Policy, 14*, 106–120.

Howard, N. (1994). Drama theory and its relation to game theory. *Group Decision and Negotiation, 3*(187–206), 207–235.

Howard, N. (1999). *Confrontation analysis: How to win operations rather than war*. Washington, DC: CCRP, Department of defense.

Howard, N. (2007). *Oedipus decision maker: Theory of drama and conflict resolution*. Retrieved from http://www.dilemmasgalore.com.

Howard, N., Bennett, P. G., Bryant, J. W., & Bradley, M. (1993). Manifesto for a theory of drama and irrational choice. *The Journal of the Operational Research Society, 44*(1), 99–103.

ICAR. (2018). *National innovations in climate resilient agriculture*. Retrieved March 5, 2018, from http://www.nicra-icar.in/nicrarevised/index.php/home1.

IPCC. (2007). Summary for policy makers. In S. D. Solomon, M. Qin, Z. Manning, M. Chen, K. B. Marquis, A. M. Tignor, & H. L. Miller (Eds.), *Climate change 2007: The physical science basis*.

Contribution of Working Group I to the Fourth Assessment Report of the Intergovernmental Panel on Climate Change (pp. 1–18). Cambridge, UK: Cambridge University Press.

Jat, M. L., Singh, B., & Gerard, B. (2014). Chapter Five—Nutrient management and use efficiency in wheat systems of South Asia. *Advances in Agronomy, 125,* 171–259.

Karuppaiah, V., & Sujayanad, G. K. (2012). Impact of climate change on population dynamics of insect pests world. *Journal of Agricultural Sciences, 8*(3), 240–246. Retrieved March 5, 2018, from https://pdfs.semanticscholar.org/443d/b7d239e3cedc5956ba3f970f977c2c75042d.pdf.

Kaur, R., Srinivasan, R., Mishra, K., Dutta, D., Prasad, D., & Bansal, G. (2003). Assessment of a SWAT model for soil and water management in India. *Land Use Water Resource Research, 3,* 1–7.

Kaur, N., Alok, A., Shivani, Kaur, N., Pandey, P., Awasthi, P., & Tiwari, S. (2018). CRISPR/Cas9-mediated efficient editing in *phytoene desaturase* (*PDS*) demonstrates precise manipulation in banana cv. Rasthali genome. *Functional & Integrative Genomics, 18*(1), 89–99.

Khan, S. A., Kumar, S., Hussain, M., & Kalra, N. (2009). Climate change, climate variability and Indian agriculture: Impacts vulnerability and adaptation strategies. In S. N. Singh (Ed.), *Climate change and crops, environmental science and engineering.* Berlin: Springer.

Khatri-Chhetri, A., Aggarwal, P. K., Joshi, P. K., & Vyas, S. (2017). Farmers' prioritization of climate-smart agriculture (CSA) technologies. *Agricultural Systems, 151,* 184–191.

Kumar, M. D. (2003). *Food security and sustainable agriculture in India.* The Water Management Challenge, Working Paper No. 60, International Water Management Institute, Colombo, Sri Lanka. Retrieved from https://ageconsearch.umn.edu/bitstream/92666/2/WOR60.pdf.

Kumar, P., & Pal, K. S. (2014). ICT enabled public distribution system for developing countries. *International Journal for e-Learning Security (IJeLS), 4,* 354–358.

Kumar, K. R., Sahai, A. K., Kumar, K. K., Patwardhan, S. K., Mishra, P. K., et al. (2006). High-resolution climate change scenarios for India for the 21st century. *Current Science, 90,* 334–345.

Lal, M., & Harasawa, H. (2001). Future climate change scenarios for Asia as inferred from selected coupled atmosphere-ocean global climate models. *Journal of Meteorological Society of Japan, 79,* 219–227.

Levy, J. K., Hipel, K. W., & Howard, N. (2009). Advances in drama theory for managing global hazards and disasters, Part 1: Theoretical foundation. *Group Decision and Negotiation, 18*(4), 303–316.

Likhi, A. (2017). *Climate smart agricultural practices in Haryana, India: The way forward & challenges. People. Spaces, Deliberation.* The World Bank. Retrieved March 5, 2018, from https://blogs.worldbank.org/publicsphere/climate-smart-agricultural-practices-haryana-india-way-forward-challenges.

Lobel, B. D., Sibley, A., & Ortiz-Monasterio, I. J. (2012). Extreme heat effects on wheat senescence in India. *Nature Climate Change, 2,* 186–189. https://doi.org/10.1038/NCLIMATE1356.

Maheshwar, C., & Chanakwa, T. S. (2006). Postharvest losses due to gaps in cold chain in India: A solution. *Acta Horticulturae, 712,* 777–784. Retrieved March 5, 2018, from http://www.actahort.org/books/712/712_100.htm.

Malhotra, S. K. (2017). Horticultural crops and climate change: A review. *Indian Journal of Agricultural Sciences, 87*(1), 12–22.

Mall, R. K., Singh, R., Gupta, A., Srinivasan, G., & Rathore, L. S. (2006). Impact of climate change on Indian agriculture: A review. *Climatic Change, 78,* 445–478. https://doi.org/10.1007/s10584-005-9042-x.

Mani, M., Markandya, A., Sagar, A., & Strukova, E. (2012). *An analysis of physical and monetary losses of environmental health and natural resources in India.* Policy Research Working Paper 6219, The World Bank 9.

Mazumdar, S., Quick, W. P., & Bandyopadhyay, A. (2016). CRISPR-Cas9 mediated genome editing in rice, advancements and future possibilities. *Indian Journal of Plant Physiology, 21*(4), 437–445.

Medek, D. E., Schwartz, J., & Myers, S. S. (2017). Estimated effects of future atmospheric CO2 concentrations on protein intake and the risk of protein deficiency by country and region. *Environmental Health Perspectives, 125*(8), 087002. https://doi.org/10.1289/EHP41.

Metz, B., Davidson, O., Bosch, P., Dave, R., & Meyer, L. (2007). *Climate change 2007: Mitigation of climate change*. Working Group III Report of the Fourth Assessment Report of the Intergovernmental Panel on Climate Change, Published For IPCC, Cambridge University Press.

Mishra, A. (2014). An assessment of climate change-natural disaster linkage in Indian context. *Journal of Geology and Geosciences, 3*, 167. https://doi.org/10.4172/2329-6755.1000167.

Naresh, K. S., Aggarwal, P. K., Saxena, R., Rani, S., Jain, S., & Chauhan, N. (2013). An assessment of regional vulnerability of rice to climate change in India. *Climate Change, 118*(3–4), 683–699. https://doi.org/10.1007/s10584-013-0698-3.

Naresh, K. S., Aggarwal, P. K., Swaroopa Rani, D. N., Saxena, R., Chauhan, N., & Jain, S. (2014). Vulnerability of wheat production to climate change in India. *Climate Research, 59*(173–187), 5–187.

Palanisami, K., Kakumanu, K. R., Khanna, M., & Aggarwal, P. K. (2013). Climate change and food security of India: Adaptation strategies for the irrigation sector. *World Agriculture, 3*, 20–26.

Pathak, H., Ladha, J. K., Aggarwal, P. K., Peng, S., Das, S., Singh, Y., et al. (2003). Trends of climatic potential and on farm yields of rice and wheat in the Indo Gangetic Plains. *Field Crops Research, 80*, 224–234.

Pradhan, A., Idol, T., Roul, P. K., Mishra, K. N., Chan, C., Halbrendt, J., et al. (2015). Effect of tillage, intercropping and residue cover on crop productivity, profitability and soil fertility under tribal farming situations of India. In C. Chan & K. Fantle-Lepczyk (Eds.), *Subsistence farming: Case studies From South Asia and beyond* (pp. 77–94). Wallingford, UK: CABI.

Pradhan, A., Chan, C., Roul, P. K., Halbrendt, J., & Sipes, B. (2018). Potential of conservation agriculture (CA) for climate change adaptation and food security under rainfed uplands of India: A transdisciplinary approach. *Agricultural Systems, 163*, 27–35.

Raju, S. S., & Chand, R. (2009). *Problems and progress in Agricultural Insurance in India*. NCAP Policy Brief No. 31. New Delhi: National Centre for Agricultural Economics and Policy Research.

Ramachandran, N. (2014). *Persisting under nutrition in India: Causes, consequences and possible solutions* (pp. 3–27). New Delhi: Springer.

Rani, R., Yadav, P., Barbadikar, K. M., Baliyan, N., Malhotra, E. V., Singh, B. K., et al. (2016). CRISPR/Cas9: A promising way to exploit genetic variation in plants. *Biotechnolnology Letters, 38*(12), 1991–2006.

Rao, S. L. H. V. P., Gopakumar, C. S., & Krishnakumar, K. N. (2013). Impacts of climate change on horticulture across India. In H. Singh, N. Rao, & K. Shivashankar (Eds.), *Climate-resilient horticulture: Adaptation and mitigation strategies*. New Delhi, India: Springer.

Roul, P. K., Pradhan, A., Ray, P., Mishra, K. N., Dash, S. N., & Chan, C. (2015). Influence of maize-based conservation agricultural production systems (CAPS) on crop yield, profit and soil fertility in rainfed uplands of Odisha, India in conservation agriculture. In C. Chan & K. Fantle-Lepczyk (Eds.), *Subsistence farming: Case studies from South Asia and beyond* (pp. 95–108). Wallingford, UK: CABI.

SANDRP. (2016, October 6). *Interstate river water disputes in India: History and status*. South Asia Network on Dams, Rivers and People. Retrieved March 5, 2018, from https://sandrp.in/2016/10/06/inter-state-river-water-disputes-in-india-history-and-status/.

Sharma, H. S. (2016, February 4–5). *Climate change vis-à-vis pest management*. Conference on National Priorities in Plant Health Management, Tirupati, India. Retrieved March 5, 2018, from http://oar.icrisat.org/9333/1/NPPHM-2016_Extended-Summary.pdf.

Shelat, K. N. (2014). *Climate-smart agriculture, the way forward—The Indian perspectives*. National Council for Climate Change, Sustainable Development and Public Leadership (NCCSD), Ahmedabad and Central Research Institute for Dryland Agriculture (CRIDA), Hyderabad. Retrieved March 5, 2018, from http://www.nicra-icar.in/nicrarevised/images/publications/Tbu_CSA_Book.pdf.

Sinha, S. K., Singh, G. B., & Rai, M. (1998). *Decline in crop productivity in Haryana and Punjab: Myth or reality?* (p. 89). New Delhi: Indian Council of Agricultural Research.

Solomon, S., Qin, D., Manning, M., Alley, R. B., Berntsen, T., et al. (2007). *Climate change 2007: The physical science basis*. Contribution of Working Group I to the Fourth Assessment Report of the Intergovernmental Panel on Climate Change, Cambridge University Press, Cambridge, UK and New York, USA (pp. 19–71).

Srinivasa, R., Gopinath, K. A., Prasad, J. V. N. S., & Prasannakumar, A. K. S. (2016). Sustainable food security in tropical India: Concept, process, technologies, institutions, and impacts. In D. L. Sparks (Ed.), *Advances in agronomy* (Vol. 140, pp. 201–217). Amsterdam: Elsevier.

Srivastava, D., Shamim, M., Kumar, M., Mishra, A., Pandey, P., Kumar, D., et al. (2017). Current status of conventional and molecular interventions for blast resistance in rice. *Rice Science, 24*(6), 299–321.

Stubbs, L., Tait, A., & Howard, N. (1999). *How to model a confrontation—Computer support for drama theory.* Proceedings of the Command and Control Research and Technology Symposium, Naval War College, Newport.

Sulochana, G. (2003). The Indian monsoon and its variability. *Annual Review of Earth and Planetary Sciences, 31*, 429–467.

Taenzler, D., Ruettinger, L., Ziegenhagen, K., & Murthy, G. (2011). *Water, crisis and climate change in India: A policy brief.* The Initiative for Peacebuilding—Early Warning Analysis to Action (IfP-EW). Retrieved March 5, 2018, from https://www.adelphi.de/en/system/files/mediathek/bilder/2011_water_crisis_and_climate_change_in_india_a_policy_brief.pdf.

Temple, J. (2017, May 4). Reinventing source: Rice for a world transformed by climate change. *MIT Technology Review.* Retrieved March 5, 2018, from https://www.technologyreview.com/s/604213/reinventing-rice-for-a-world-transformed-by-climate-change/.

Varadharajan, K. S., Thomas, T., & Kurpad, A. V. (2013). Poverty and the state of nutrition in India. *Asia Pacific Journal of Clinical Nutrition, 22*, 326–339.

Zhang, W., Zheng, C., Song, Z., Deng, A., & He, Z. (2015). Farming systems in China: Innovations for sustainable crop production. In V. Sadras & D. Calderini (Eds.), *Crop physiology. Applications for genetic improvement and agronomy* (2nd ed., pp. 43–64). San Diego, CA: Academic Press.

Part V
Conclusion

Chapter 25
Conclusion

Atanu Sarkar and Gary W. vanLoon

In recent years, we have witnessed far reaching changes in climate around the world and the impact on local weather is already beginning to have an impact on food security. In this context, actions directed toward adaptation are believed to be feasible and necessary to ensure risk reduction. The suggested adaptation actions needed to sustain food security in the changing climate can take many forms, some of which are very complex. Most focus in large part on production. Emerging research, however, suggests that there is a need to consider broader perspectives of food security as we move into an era of changing climate (Campbell et al., 2016).

This volume, *Sustainable Solutions for Food Security: Combating Climate Change by Adaptation,* consists of three major focus areas: (a) climate-smart crops, adaptive breeding and genomics, (b) natural resource and landscape management, and (c) strategies for access to food, technology, knowledge and equity, and risk governance.

Rising temperature, flood, and drought are the known climatic factors for reduced productivity or even the complete loss of crops. Each variety of crops has a certain threshold temperature, where, if exceeded, yield losses occur. At high temperature there can be cellular injury and impaired reproductive development resulting in growth inhibition. Pollen is highly sensitive to elevated temperatures resulting in impaired fertilization and abortion of fruits/seeds. At the same time, in varying degrees plants do have a natural tolerance to a high temperature through physiological and morphological adaptations. Nevertheless, survival of plants under extreme

---

A. Sarkar (✉)
Division of Community Health and Humanities, Faculty of Medicine, Memorial University,
St. John's, NL, Canada
e-mail: atanu.sarkar@med.mun.ca

G. W. vanLoon
Department of Chemistry and School of Environmental Studies, Queen's University,
Kingston, ON, Canada
e-mail: gary.vanloon@chem.queensu.ca

© Springer Nature Switzerland AG 2019
A. Sarkar et al. (eds.), *Sustainable Solutions for Food Security*,
https://doi.org/10.1007/978-3-319-77878-5_25

environments by adaptive and acclimitization mechanisms can be severely compromised resulting in reduced yield. Therefore, identification and promotion and selective breeding of heat-tolerant crops and adjustment to cropping geographies especially in the projected climate change hot spots offer the ways to ameliorate most of the immediate challenges of food insecurity. There are similar needs for varieties of crops that are resistant to other stresses such as drought or flooding that result from changing climate patterns.

Maize, wheat, and rice are the major grains consumed across the world and fortunately there are significant genetic variations for tolerance against heat and drought/ submergence available in gene banks. In order to ensure constant development of new breeds of climate-smart staple crops, particularly making them accessible to the vulnerable countries in sub-Saharan Africa, South Asia, and Central America, adequate funding is required for research on new breeding techniques, training for breeding expertise, and capacity building of small to medium-sized seed companies.

The recent advances in genomics provide a unique potential to adapt newly established crop species to changing climate and enabling improved and consistent high yields. For example, installation of a C_4 photosynthetic pathway into major crops like rice could potentially increase yields by half, double the water-use efficiency, and reduce nitrogen fertilizer utilization by two-third. Scientists discovered that some natural plants with a C_4 photosynthetic pathway have high photosynthetic efficiency leading to significant reductions in requirement of water and nitrogen. The existing varieties of crops like rice and wheat have less efficient C_3 photosynthesis that causes more water losses, and needs large amount of nitrogen for synthesis of protein in order to maximize their photosynthetic rate.

While novel technologies have succeeded in developing some climate-resilient staple crops, initiatives are also underway to promote diversification of crops including a focus on underutilized, neglected, and minor crops. The underutilized plant species are often indigenous and part of food culture of the local people who grow them. Undoubtedly, the promotion of these crops would increase genetic, species and ecosystem biodiversity by safeguarding crop vulnerability to climate change, pests and diseases and eventually the initiatives would provide highly nutritious food from diverse sources. In fact, people are increasingly showing interest in such crop diversification in many parts of the world. For example, Bambara groundnut, a species indigenous to various parts of Africa and Southeast Asia has high growth potential in drought-prone areas and is known to have significant nutritional properties. Currently a number of researchers are involved in developing improved agronomic practices directed toward increased yield and growth promotion.

Amid threatened food security due to global climate change, efficient management of natural resources is the key to sustainable intensification of food production. A global bio-economic model shows that until the middle of the century, more land will be converted to croplands and a gain of productivity will continue due to uses of high yield seeds and other technological advances. But, if current level of greenhouse gas emission continues, the negative impacts of climate change on food and agriculture will increase greatly in the latter half of the century. Therefore, effective and efficient management of natural resources will play pivotal roles to slow down the impacts with the assumption that greenhouse gas emission continues to rise. Henceforth, in well irrigated and in rain-fed systems, efficient use of every drop of available green water is necessary to avoid drought stress situations. Irrigation too must be developed in ways that eliminate overuse through timely application and use of water-saving technologies like drip irrigation. For efficient management of water and soil nutrients, particularly in developing countries, several approaches such as conservation agriculture, direct seeding, alternate wetting and drying, and site specific nutrient management are proven to be effective and feasible. But, there are few support systems to disseminate such knowledge and demonstrate to the most vulnerable farmers and the communities facing the most extreme climate change-related challenges with least adaptive capacity. Therefore, cooperation among local community based organizations and national and international institutions is warranted.

In the resource constraint countries, low-cost integrated water and land management, for instance the 'Smart-valley' approach in West Africa is found to be an effective adaptation strategy for major food crop like rice. In such approach the farmers build and maintain those low-cost water control structures entirely on their own. Despite its effectiveness, less than one-tenth of the potential land in sub-Saharan Africa is under this approach, indicating a need for more proactive and innovative approaches to promote its application over wider areas. The "System of Rice Intensification" or SRI methods integrates the management of rice plants, soil, water, and nutrients and has been shown (a) to enhancethe growth of the plants' root systems, and (b) to improve abundance and diversity of the soil biota. In turn these methods help the farmers to grow more productive and robust rice plants by producing more complete expression of the plants' genetic potentials. Thus the SRI is a classic example of climate-resilient rice production, based on the principles and true applications of principles of agroecology. The method, originally developed in Madagascar, is now benefitting over 20 million farmers over 50 countries.

In the globalized world, private participation is a key in adaptation strategy development and implementation. However, there is a need for balanced participation among the members of the community-public-private partnership, particularly in pricing of any services. For example, use of wastewater is a very effective mean in restoration of dwindling wetlands, particularly during frequent drought like conditions. Nevertheless, economic valuation of wastewater when used in the context of environmental restoration projects is imprecise and ambiguous, resulting in potential conflicts between private decision and institutional variables. Therefore, an appropriate mechanism is necessary for obtaining a meaningful environmental valuation of wastewater. Although drip irrigation is a tested technology, particularly in dry regions, it is yet to be deployed effectively on a global scale. Studies suggest that social, economic, technological and environmental factors defining the prices and availability of water should be considered together to enable determining effective policies on irrigation technology uptake.

Similar to land-based agriculture, fisheries and aquaculture are under serious threat by the impending impacts of climate change on aquatic ecosystems. The impacts include rising temperature, acidification of ocean water, extreme weather conditions and climatic events, and sea-level rise. Fisheries and aquaculture, contribute significant sources of food, nutrition, income, and livelihoods for hundreds of millions of people around the world, and require urgent attention to adapt to the emerging risks. The attention must be directed toward inland fisheries as well as fish supply in open oceans. A study of shrimp aquaculture in Sri Lanka shows that strategies like better pond management practices including site selection, its construction, preparation, maintenance, selection of post-larvae for stocking, bottom sediment management and disease management together with reducing pollution, and conservation of sensitive ecosystems are effective means for shrimp culture to adapt against climate stressors.

Beverages like tea or coffee are generally less discussed in the discourse of climate change and food security, but they can be impacted in significant ways by even small changes in local climates. The quality of tea leaves is very sensitive to temperature variability, extreme weather condition like drought, heavy rain, and frost events. Interestingly, a study in Japan's oldest and most famous green tea-producing region shows how the "terroir" concept utilized in French wine industry can help in adaptation against adverse climate stresses. In the "terroir" concept the wine grape growers use traditional knowledge of natural environmental elements and the agriculture practice elements such as soil types, and other exposures in the microclimatic conditions.

According to the United Nations population division, in 2009, for the first time in human history the global urban population surpassed the rural population (United Nations n.d.). Hence, urban agriculture must also be considered as an emerging opportunity for addressing food insecurity in the context of climate change. Intuitively, cities are considered (vis-à-vis rural realm) densely populated lands with very little available space and urbanization is considered as an assault on nature. In the urban agriculture discourse the focus is how to contribute to supplying the locally grown food in a sustainable manner. A study in Barcelona (Spain) shows the landscape mosaic around the city or rural space with urban relations. The study concludes by stating that resilience of urban agriculture could be ensured by appropriate design along with landscape architecture, infrastructure planning, civil engineering, urban ecology, economics, and the arts. It is important to note that the conditions of towns and cities in the developing countries are often extremely precarious due to out-of-control inward migration (particularly from rural to urban), lack of resources to cope with unprecedented population growth, poor city planning, lack of security, and growing conflicts and no efforts to improve rural livelihoods in the foreseeable future. Therefore, in the existing critical conditions a western model of urban agriculture may not be appropriate, and hence more household-based innovative approaches must be considered as viable alternatives.

Technological solutions alone cannot bring any perceivable change in adaptation actions; it is essential that the stakeholders such as communities, government officials, technocrats, and scientists work in tandem. Furthermore, innovation should not be limited to lab scientists and technocrats; rather bottom-up community-driven innovation and implementation should be integral to achieve the goals for adaptations. In fact, there is ample evidence of failures of climate change adaptation projects due to ineffective community participation. Less than effective outcomes can also be attributed when projects exhibit an elitist, authoritarian approach that can intimidate poor, marginalized, and vulnerable local communities. A study in West Africa shows that stakeholders see a participatory environment as the most critical feature to ensure effective community participation. The participatory environment should include a three-step process—community entry, setting good objectives, and plan implementation. Participation of communities also requires certain situational conditions or sociopolitical environment that influences the execution of participatory programs. On the other hand, several implementing agencies have attributed failures in apparently worthwhile initiatives to the community's own internal problem, illiteracy, poor knowledge and skills, ignorance, and their lack of motivation. Researchers have learned from the communities that for successful climate change adaptation action, there should be clear objectives with realistic goals along with long-term development perspectives. It is an undeniable fact that illiteracy and ignorance of the community members and economic uncertainty are frequently responsible for their inclination toward short term, material gains. It is the funding agencies' or implementing bodies' responsibility to debug the potential pitfalls prevailing within the communities; these pitfalls may be the community's lack of skills and knowledge or unequal and unfair participation by the community or marginalized groups.

Traditional ecological knowledge and emotionally bonded moral values of a subsistence community in semiarid West Timor, Indonesia play significant roles in maintaining food sovereignty without overuse of the local resources. However, the generations-old knowledge is not fully capable of sustaining the communities in today's world due to the traditional diet, deficient in nutrient values and the produce that lacks market demand. Moreover, climate change has impeded the sustainability of traditional practices. Hence, the authors conclude by inferring the importance of strong collaborations between technology and local knowledge to the development process. Undoubtedly, access to local traditional food for Arctic indigenous communities has been severely threatened by climate change. But warming climate also offers them opportunities to grow vegetables and fruits that were previously beyond their distributional range. An initiative to promote agroforestry (combining woody perennials with crops) in subarctic Ontario (Canada) has been found to be very effective for horticulture. Interestingly, the modern agroforestry initiative has not replaced long-standing practices but rather it has been harmonized with the traditional harvesting of game bird by utilizing the by-products to enrich the soil with nutrients.

Food insecurity due to climate change is not just an exclusive problem of the resource constraint countries. Farming communities living in the hinterland of Australia, the driest continent in the world, are facing major challenges to remain viable under constant threat of climate change. The outside world assumes Australia's market and export-oriented agriculture as providing a strong guarantee for securing food security for its own population and beyond. But the authors argue that the smallholder subsistence and market-oriented farmers living in remote rural areas of Australia are facing an existential crisis due to extreme climate change-related challenges without being able to adapt. In addition to that, the existing export driven policy compels the already over-stretched farming communities to remain competitive and eventually become more vulnerable due to unsustainable use of natural resources, like water. Finally the authors criticize the current pace of dissemination of knowledge and practices of climate-smart agricultural practices, not able to catch up with the urgent need.

In the era of information overload, it is a challenge for the farmers to take the best decisions regarding farm practices. Historically, agriculture extension personnel from local agriculture institutions or government agencies have been playing central roles in disseminating information to the farmers. Uncertainties arising out of the variable influences of climate change have made the decision making more perplexing. Therefore, more precise tools are warranted in order to give accurate forecasting in local microclimate areas that will benefit individual farmers. Cornell University (USA) has developed a website and climate-based agricultural decision support tools that aim to assist the farmers, particularly for the farms located far away from the nearest climate station, to make better informed decisions in the face of increasing climate uncertainty. Based on the experience of promoting this and similar tools, the authors have opined that greater participation of other stakeholders is needed. This participation is needed for further development of climate services and to improve utilization of such decision support tools, appropriating in diverse locations, integration of these tools into the daily decision-making process alongside other adaptation measures to climate extremes, and indeed constant flow of funding to support the initiative.

Addressing the vulnerability of food supply chains due to climate change and the required adaptation actions are the key to counter food insecurity. A study in Finland shows that to adapt any such disruption of food supply chains, it is essential to find ways of linking the concepts of sustainable development and risk governance into one paradigm. The authors have stated that the risk governance related to food supply chains ought to be practiced within a multilevel framework of private and public policies. For example, when a publicly supported research develops a genetically modified climate-resilient crop that provides a higher shelf life, this may be very promising for trade. However, the new technology might not get acceptance within the policy discourse due to opposition from civil society organizations and eruption of unexpected political controversies. Therefore, climate change adaptation with regard to food supply chain at national level needs novel approaches toward multilevel risk-governance that implies concerted action between the many public and private actors concerned.

Lastly, the analysis of India's current policy on sustaining food security amid changing climate shows the complexities of the issues. "Climate-smart agriculture" is considered as a pragmatic approach to continue increasing food production to meet demand for rising population. However, the policy makers cannot ignore the fundamental problems related to growing tensions due to transboundary water conflict, ensuring financial security for the vulnerable farming communities by providing insurance coverage for crops, realizing a significant reduction in post-harvest food wastage, and achieving efficiency in the existing food distribution for the poorer segments of the society. It is unrealistic to expect the farmers to readily accept the technologies related to "climate-smart agriculture" even if they are made accessible. Therefore, more community-based research is needed to explore socio-economic and location-specific variables influencing farmers' preferences.

Since adaptations are intricate and context specific, it is a challenge to cover every nuance of actions in a single volume. Albeit very important, we have limited here the contributions of those that are based on theory and/or are descriptive studies. Rather, the focus has been for the contributors to share their respective empirical studies. Food policy experts have emphasized the importance of conducting action-oriented research and sharing the results across the world for rapid and effective implementation of adaptation plans (Campbell et al. 2016). This volume provides a well-integrated collection of chapters representing diverse examples within this new area of research. Furthermore, the authors' diverse perspectives along with their individual expertise areas have synergized the multiple dimensions of the subject. The majority of the chapters are based on novel approaches within the emerging field and hence the volume will remain relevant for future research scholars and policy makers.

The volume is contributed by the authors from all the continents giving genuinely diverse perspectives of the adaptations. Thus the readers will find the volume to be comprehensive, systematic, and as one providing an in-depth analysis of individual adaptation actions.

References

Campbell, B. M., Vermeulen, S. J., Aggarwal, P. K., Corner-Dolloff, C., Girvetz, E., Loboguerrero, A. M., Ramirez-Villegas, J., Rosenstock, T., Sebastian, L., Thornton, P. K., & Wollenberg, E. (2016). Reducing risks to food security from climate change. *Global Food Security, 11*, 34–43.

United Nations. (n.d.) *Urban and Rural Areas 2009*. UN Population Division. http://www.un.org/en/development/desa/population/publications/urbanization/urban-rural.shtml

Index

A
Abiotic stresses, 61, 112
Abscisic acid (ABA) levels, 52
ACC oxydase, 52
Active soil aeration, 233, 247
Adaptation
  agroecosystem, 13
  Australia
    chemical sprays, 446
    exports, 445
    fruit industry, 445
    irrigation infrastructure, 445
    management actions, climate
      change, 445
    markets, 445
    minimum irrigation application, 446
    reforms, market-based, 446
    technological responses, 446
  capacity building component, 13
  climate change, 13, 347
  comprehensive and holistic scientific
    research, 13
  crop insurance, 17
  DSTs, 474
  implementation of innovative strategies, 15
  India
    climate change, 504
    climate-smart biotechnology, 505
    climate-smart crops, 509, 510
    conflict resolution, 511
    conservation agriculture
      (*see* Conservation agriculture)
    crop insurance, 512
    CSA, 505, 510, 511
    dryland agriculture, 505
    food insecurity, 504

food wastage, 512
knowledge dissemination and
  translation, 505
managing risks, 505
National Agroforestry Policy, 505
natural resources (*see* Natural resources
  management)
PDS, 512, 513
vulnerability, 504
laboratory, policy makers and farmers, 16
local climates, 13
morphological (*see* Morphological
  adaptations)
and risk prediction, 16, 17
scale, 14, 15
socio-economic environment, 13
strategies, 5, 7
to warmer climates (*see* Warmer climates)
water management, 17, 18
Adaptive crop management
  CGIAR CRP, 192
  CSA, 192
  nutrient use efficiency
    (*see* Nutrient use efficiency)
  warmer climate, 201
  WUE (*see* Water use efficiency (WUE))
Adoption
  crop diversification (*see* Crop
    diversification)
  diversified food systems, 128
  Green Revolution agricultural
    techniques, 127
Adoption approaches, drip irrigation, 272
  See also Drip irrigation technology
African Orphan Crops Consortium, 161
AGRHYMET data base, 174

© Springer Nature Switzerland AG 2019
A. Sarkar et al. (eds.), *Sustainable Solutions for Food Security*,
https://doi.org/10.1007/978-3-319-77878-5

Index

Agricultural production areas, 5
Agricultural Research Council (ARC) of
    South Africa, 133
Agricultural systems, 130
Agriculture
    DST (*see* Decision support tools (DST))
    workforce, feminization, 12
AgroClimate website, 471, 472
Agroecological principles, irrigated rice
    active soil aeration, 233
    beneficial interactions, 232
    plant density, 232
    soil conditions, 233
    transplanting, 232
    water applications, 233
Agro-ecosystem, 4
Agroforestry, 129, 135, 137
    challenges and opportunities, 413
    community garden, 418
    definition, 413
    ecosystems services, 413
    in import substitution, 418–423
    nutrients, 423
    potatoes, in Fort Albany, 418
    and Sharing-the-Harvest, 413
    in subarctic CA, 425–428
Alcohol dehydrogenase 1 (*ADH1*), 55
Alternate wetting and drying (AWD), 72,
    194, 195
Alternative splicing (AS) mechanism, 59, 60
Amblyopyrum muticum, 57
Amino acids, 157
1-Aminocyclopropane-1-carboxylic acid
    (ACC) synthase, 52
Analytic framework, drip irrigation technology
    categories, 274
    diffusion models
        description, 274
        external influence, 275
        general form, 274
        internal influence, 274
        mix influence, 275
    duration models, 276
    innovativeness, logit and probit models,
        275–276
Ancestral territories, AtoinMeto, 387–390
Anthesis-silking interval (ASI), 71
Antioxidant, 38, 39
Aquaculture
    as capture fisheries, 288
    description, 288
    food and nutritional security (*see* Food and
        nutritional security)

    shrimp (*see* Shrimp aquaculture)
    world capture fisheries and aquaculture
        production, 288
Aquatic animals, 7
Arabidopsis thaliana plants, 56
Arid zones, 136
Artificial microRNA (amiRNAs), 119
Ascorbate peroxidase (APX), 38, 39
AtoinMeto
    Adat laws, 388–390
    ancestral territories, 387–390
    changing climate, 391
    clan-based land and water management, 388
    clan regime, 382–383
    clan system, 388, 389
    climatic hazards, 383
    emotional bonding, 388
    food (corn) production process, 390
    food and economic access, 392–393
    food availability, 390
    food system, in community, 388
    government's programs and infrastructure
        developments, 402
    grounded solutions, to food security,
        402, 403
    pillars and bottom-up process, food
        security, 392, 394
    public transport, 391, 392
    rainfall, 396, 397, 402
    utilization of food, 392, 393 (*see also* West
        Timor, AtoinMeto)
Australia
    adaptation (*see* Adaptation)
    climate change, 449
    competition, 450
    dairy producers, 449
    economic vulnerability, food producers, 450
    environmental changes, 438, 439
    export products, 449
    food justice, 450
    food poverty, 451
    food producers, 451
    food security
        agricultural regions, 441
        anthropogenic climate change, 440
        drought experiences, 440
        ecological consequences, 440
        export products, 440
        GDP, 440
        global trade liberalisation
            agenda, 440
        resilience, 440
        United Nations, 442

Index

FoodShare coordinator, 452
government policies, 452
landscape of productivity
  dairy industry, 443
  farmers, 444
  industrial-scale production, 442
  liberalisation initiatives, 443
  milk processing companies, 444
  nutritional vulnerability, 444
  regional food supply chain, 443
  Southwest Victoria, 442
  vignette, 443
  vulnerability and food insecurity,
    443, 444
  water regulation, 442
  welfare net, 444
lock-ins (*see* Lock-in, irrigation)
markets and climate awareness, 449
NGO, 452
resilience, 438, 439, 441, 446, 449
SER, 451, 452
social and economic costs, 451
State and Federal Government Water
  Policy, 453–454
supply chains, 449
vulnerability, 450
Auxin, 39

B

Bambara groundnut, 135, 140
beneficial characteristics, 157
constraints
  advantages, 161
  agricultural and consumer
    communities, 160
  agricultural sustainability, 159
  American market, 161
  biofortification, 160
  biological limitations, 159
  boiling, 158
  breeding companies, 161
  cookability, 159
  cowpea, 161
  Domestication Syndrome, 160
  EU Framework 6 Programme
    project, 159
  export crop, 161
  FAO Agroecological Mapping
    Project, 159
  farmer-based selection, 160
  global production levels, underutilised
    crops, 159
  inferior scale, 159
  niche markets, 160
  orphan crops, 160
  seed traits, 159
  social media, 161
  social perception, 160
  genetic improvement, 161
  growth habit, 151
  in morphology, 151
  nutritional qualities, 151
  research, 152
  resilience traits, 153–156
  and underutilised crops, 152
Barren land conversion, 185
Benin and Togo
  map, 215
  maximum temperature index, 216, 217
  minimum temperature index,
    216, 217
  precipitation index, 215, 216
Better management practices (BMP's),
  303, 307
  aquaculture extension services, 307
  description, 302
  management adaptations, 303
  management interventions, 303
  mangroves, 305
  shrimp farm, in mangrove area,
    305, 306
  site selection, 302
  technical interventions, 303, 305
Biofortification, 160
Biological nitrification inhibition (BNI), 77
Biomass, 247
Breeding
  genomic technologies, 101, 102
  germplasm, 100
  GS, 98, 99
  MAB, 97, 98
  underutilised crop, 163
Bundle sheath (BS) cells, 113, 114, 116,
  118–120

C

Canopy temperature, 34, 35
Capture fisheries, 288, 292, 299
Carbon dioxide enrichment effect, 134
Carbon markets, 15
Catalase (CAT), 38, 39
Cellular metabolic process, 37
CERES-rice, 140
CERES-wheat model, 136, 140

Index

C₄ evolution
- atmospheric CO_2 partial pressure, 117
- genetic and anatomical precondition, 116
- organelles, 116
- phases, 116, 117
- photorespiration, 116, 117

Chemical fertilizers, 233, 235, 237, 247, 248
Chickpea, 96
Chlorophyll, 36, 37
CIMMYT, 69–71, 73, 75, 77, 79, 80
Cis-element ABRE, 58
Climate adaptive breeding
- carbon dioxide
  - maize, 78, 79
  - rice, 78, 79
  - wheat, 78, 79
- drought
  - maize, 68–71
  - rice, 72
  - wheat, 71
- ESA, 80
- feeding, 67
- flooding and salinity
  - maize, 75, 76
  - rice, 75, 76
  - wheat, 75, 76
- funding of public sector, 79
- heat
  - maize, 73, 74
  - rice, 75
  - wheat, 74, 75
- IITA, 80
- nutrient use efficiency
  - maize, 76–78
  - rice, 76–78
  - wheat, 76–78
- principal, 79
- seed systems, 80
- temperature extremes, 67

Climate change
- adaptations, 18
- adverse effects, 16
- on agricultural production, 9
- agriculture, 212
- agroecologically informed practices, 230
- categorization, 13
- context, 15
- disasters and exacerbate extreme events, 460
- economic consequence, 9
- FAO, 12
- and food insecurity, 4, 7, 9, 12, 212
- food production, 8
- and food security, 12

- genetic resources, 7
- genomic (*see* Genomic technologies)
- GHGs emissions, 245
- India (*see* India)
- information and agricultural development, 460
- IPCC, 460
- livestock production, 7
- mitigation, 13
- negative impact of pests and weeds, 6
- risks, 11
- SDG-13, 13
- smart-valley approach (*see* Smart-valley approach)
- snow geese breeding, in subarctic Canada, 411, 414, 415
- SRI-grown rice phenotypes, 240
- storm damage, 242

Climate change adaptation (CCA), 301, 307,
- action of, 490
- appropriate decision-making, 487
- community-based programs (*see* Community-based CCA programs)
- Finnish food system, 490
- food security, 487
- framework, 484
- governments function, 493
- implementation, 490
- information sharing and joint decision-making, 486
- innovation, 493
- limitation, 357
- multi-level governance, 481
- national and local levels, 491
- objectives, 485
- policies, 482, 490
- regional and local scale, 480
- research, 485
- responsibility, 491, 493
- re-territorialization, risk governance, 485
- and risk reduction, 490
- stakeholders perspectives, 356, 357
- strategies, 486

Climate change and food security
- adaptation actions, 526
- climate resilient staple crops, 524
- community or marginalized groups, 527
- C_3 photosynthesis, 524
- C_4 photosynthetic pathway, 524
- farm practices, 528
- fisheries and aquaculture, 526
- flood and drought, 523
- impact, 523, 525

irrigation, 525
low-cost integrated water and land management, 525
maize, wheat and rice, 524
private participation, 525
quality of tea leaves, 526
rising temperature, 523
SRI methods, 525
survival of plants, 523
uncertainty, 528
urban agriculture, 526
vulnerability, food supply chains, 528
Climate change, on shrimp culture, 295
BMPs (*see* Better management practices (BMPs))
in coastal planning and ecosystem restoration, 302
cyclones, 298
droughts, 298
elements, 296
farmer conceptions, 302
floods, 299
growing season, 297
and health, 300–301
National Climate Change Policy, 301–302
precipitation, 296–297
salinity, 296, 297
SLR, 299, 300
temperature, 297
weather condition and climatic events, 298
Climate resilient agriculture, 134
Climate resilient crops
Bambara groundnut (*see* Bambara groundnut)
Climate services, *see* Decision support tools (DSTs)
Climate smart agriculture (CSA), 505, 510, *see* Adaptive crop management
Climate smart crops
concomitant environmental stresses, 509
CRISPR genome editing tools, 510
horticultural, 509
production systems, 510
rice and wheat, 510
Climate smart farming (CSF)
development, 468
farmer needs and utility, 467
GDD calculator, 468
grapes and apples freeze risk, 469
impacts, 468
NRCC-led data, 467
NWS radar data, 467
Program, 466, 467

tools, 466
water deficit calculator, 470
Website, 468
Climate-related shocks, 3
Climate-resilience, 235
Climate-resilient crops
C_4 photosynthesis, 115, 116
Climate-sensitive approach, 12
Climatic change, Uji tea cultivation, 316
Climatic risks, 9
Coast Conservation and Coastal Resources Management Department (CC&CRMD), 302
Commercialization models, 16
Commodities prices, 8
Community-based CCA programs
community entry and village governance, 364–367
content analysis method, 363
good objective, setting up
ability, 367
challenges and characteristics, 368–369
factor affecting objectives, 367, 369, 370
features, 367
measures, 370, 371
plan implementation, 371–372
poverty situation, 367
guidelines, for interview, 363
interview, phases, 362
participatory condition, 363
qualitative research methods, 362
sectors and institutions, 361
stakeholders' perception, community participation, 362
steps, community entry, 364, 365
vulnerability, 360–361
Wa West district, in Ghana, 360–361
Community greenhouses, 411
Community participation
agencies, implementation, 356
CCA programs (*see* Community-based CCA programs)
control of stakeholders, 359
frameworks, in academic domains, 357
outcomes
evaluation measures, 373
hire evaluator, 372
survey, 372
process conditions and outcome, 364
project implementation, 356
stakeholders perspectives, in CCA, 356–357
tools and methods, 359

536

Index

Community participation (*cont.*)
  yardsticks, 357
    abstract criteria, 358
    no fixed criteria for participation, 358
    participatory discourse, 358
    privileged stakeholders, 358
Composting
  carbon and N, 423
  Cree smoke geese, 423
  description, 423
  factors, 423
  meat producers, in CA, 423
  metals/metalloids, 425, 427
  organochlorines, 425, 426
  in subarctic CA, 425–428
  waterfowl byproducts, 424
Conservation agriculture (CA), 197, 199, 200
  crop diversification, 506
  development, crop yield, 507
  farmland management, 506
  Green Revolution, 506
  principles, 506
  soil and water resources, 506
  wheat, 507
Contingent valuation method (CVM), 256, 260, 262
Cooler canopy, 34, 35
Cowpea, 161
C_4 photosynthesis
  advantages, 114–116
  anatomy
    bioinformatics approaches, 121
    in C_3 plants, 121
    genes controlling vein spacing, 120
    installing, 120
    Kranz, 120, 121
    M and BS cells, 120
    measurement of leaf vein density, 120
    plasmodesmatal connectivity, 120
    rice and *Setaria*, 121
  biochemistry engineering
    amiRNAs, 119
    gdch lines, 119
    genetic modifications, 118
    leaf anatomy and physiology, 118
    maize transporters, 118
    malate, 118, 119
    metabolic transporters, 118
    NADP-ME, 118, 119
    OAA, 118, 119
    PEP, 118
    3-PGA, 119
    photorespiration repositioning, 118
    pyruvate, 119

climate-resilient crops, 115, 116
evolution, 116, 117
Crassulacean acid metabolism (CAM), 32
Crop diversification
  agriculture, 126
  climate resilient agriculture, 134
  crop performance, 135, 136
  domestic crops, 131
  farming systems, 141
  food and nutritional security
    ARC of South Africa, 133
    availability, 132
    consumer trends, 131
    consumption of staples, 133
    FAO/WHO, 132
    food provision, 132
    global affluence, 131
    grain yield, 132
    micronutrient deficiencies, 132
    Moringa, 133
    prevalence of malnutrition, 133
    rural households, 133
    stability, 132
    trickle-down impact, 133
    underutilised crops, 132
  food insecurity, 126
  food security, 126
  global population, 126
  global strategy, 143
  the Green Revolution, 127
  hydrological cycle, 135, 136
  investment and financial support, 142
  livelihood protection, 138, 139
  match climates, 127, 128
  national policies, 126
  pathogen suppression, 137, 138
  pests, 137, 138
  poverty alleviation, 138, 139
  resilience and nutritional security
    adoption, 128
    farming households, 128
    functionality, 130, 131
    off-farm diversification, 128
    on-farm diversification, 129
    SDG, 128
    spatial/temporal process, 128
    structural levels, 129, 130
  resources and knowledge systems, 139
  specialisation and intensification, 126
  technology, 140
  vast populations, 131
  WUE in maize, wheat and rice, 195, 196
Crop improvement, 92, 95, 97, 99, 101, 102
Crop insurance, 17, 510–512, 516

Crop management, 41, 42
  adaptive (*see* Adaptive crop management)
Crop production
  in Central Eurasia, 6
  Scandinavia, 6
Crop rotation, 129
CSA technologies, 510, 511
Cultivated cacti, 32
Cultural landscape
  and cities, 332
  contemporary, 340
  land mosaics, 339, 340, 342–347
  social and economic relationships, 333
  urban agriculture, 333
Cyclones/hurricanes, 7

D
De Bruijn graphs (DBG), 93
Decision Support System for Agrotechnology
      Transfer (DSSAT) crop modeling
      software, 175
Decision support tools (DSTs)
  accessibility, stakeholder
      developing countries, 463
      government, 464
      PPPs, 465, 466
      private companies, 463
      private investments, 465
      utility and effectiveness, 463
  adaptation and risk reduction, 474
  Africa RiskView, 472
  agricultural development, 462
  agroclimate climatology, 473
  AgroClimate website, 471, 472
  analytical phase, 461
  audiences, 462
  Australia, 473
  climate change, 461
  Climate Smart Farming Website, 469
  communication, 462
  Cooperative Extension System, 462
  CSF (*see* Climate smart farming (CSF))
  definition, climate services, 460
  delivery mechanisms, 474
  design, 461
  development process, 463
  dissemination, 462
  extreme weather events and shifting
      climate, 461
  integration, 474
  public-private partnerships, 475
  quantitative research, 462
  regional/commodity, 466

  seasonal weather forecasting, 461
  stakeholders, 461, 462
  Uruguay's NAIS, 473
  user-friendly format, 462
  VIPS, 470, 471
  Water Requirement Satisfaction Index, 472
  website, 460
  WMO, 462
  Yield Prophet, 474
Dehydroascorbate reductase, 38
Delta agriculture, in Baix Llobregat, 338
DeNovoMAGIC, 93
Desalinated water pricing, wetlands
  ecosystem services, 257, 258
  empirical valuation methodology
      calculation, 263–265
      opportunity costs, 263
      price policy, 263
      recreational costs, 263
      transport costs, 263
  externalities, 257
  human and monetary resources, 258
  market structure, 266
  private market *vs.* environmentally efficient
      prices, 265
  proposal, 258
  valuation methods (*see* Environmental
      valuation methods, wetlands)
  value transfer method, 258
  water services, 257
Desert locust, 6
Differentially methylated loci (DML), 59
Diffusion
  drip irrigation, 272, 274 (*see also* Drip
      irrigation technology)
  Smart-valley approach, 221–223
Direct seeded rice (DSR), 196, 197
Disease
  resistance, 289
  white spot, 290
Domestic crops, 131
Domestic rice production, 127
Domestication Syndrome, 160, 163
Doubled haploid (DH), 70
Drama theory, 513–515
Drip irrigation technology
  adoption of diffusion, in Spain
      Cartagena community, 279
      diffusion rates, 279
      economic aspects, 282
      environmental factors, 282
      farm characteristics, 282
      inter-firms diffusion, 279
      probability, 282

538   Index

Drip irrigation technology (*cont.*)
    variables, 280
    water availability, 282
  adoption of innovation, 271, 277
    economic characteristics, 277–278
    environment factors, 278
    farm characteristics, 278
    farmer characteristics, 277
    innovation characteristics, 278
  analytic analysis (*see* Analytic framework,
    drip irrigation technology)
  approaches, of adoption and diffusion,
    272, 273
  definition, 270
  innovation–decision–diffusion process, 271
  limitations, 270
  OECD countries, 271
Drought avoidance, 153
Drought escape, 154
Drought resistance, SRI-grown plants,
  241, 242
Drought stress, 6, 7
Drought tolerance, 154
Drought tolerant crops
  ABA accumulation, 58
  abiotic stress tolerance, 61
  abiotic stresses, 58
  AREB/ABFs, 58
  AREB1, 58
  AS mechanism, 59, 60
  cis-element ABRE, 58
  DML, 59
  DNA marker applications, 61
  DRA1, 59
  DREB/CBF, 58
  epigenetic modifications, 59
  food security, 62
  gene transcription, 59
  genetic variability, 60, 62
  HAT, 59
  miRNAs, 61
  MITEs, 59
  mitigation of stress effects, 58
  molecular networks, 58
  pre-breeding, 61
  reprogramming of gene expression, 61
  rice varieties, 59
  TE activity, 59, 60
  transgenic maize, 61
  water available in soil, 58
  water stress conditions, 59
  ZmNAC111, 59
The Drought Tolerant Maize for Africa
  (DTMA) project, 70

E
East Asia Pacific (EAP), 175
Ecosystem services, 129
  benefits, 257
  definition, 257
  desalinated water pricing, 257
  valuation estimation, 258
  valuation techniques, 257
  value transfer method, 262
  water filtration services, 261
Energy-consuming reactions, 113
Energy transition, 346
Environment Impact Assessment (EIA), 303, 305
Environmental valuation methods, wetlands
  applications of, 256, 259
  categories, 256
  CVM, 262
  dynamic behavior, 262
  externalities, costs and benefits, 259
  hedonic approach, 261
  HP method, 261
  peri-urban park, in Catalonia, 260
  revealed preferences methods, 256
  TCM and CVM, 260
  transfer method, 262
  travel cost technique, 259, 260
  water filtration services, 261
Epigenetic
  drought-induced changes, 59
  modifications, 54, 55, 59
  regulation, 54, 58, 59
  stress, 55
  TF accessibility, 54
Error-prone repair mechanisms, 99
Escape strategy, 52
Ethylene responsive factors (ERFs), 52–54
EU Framework 6 Programme project, 159
Extreme temperatures, 5, 6

F
FAO Agroecological Mapping Project, 159
FAOSTAT agricultural database, 174
FAOSTAT harvested area, 183
Farmer-based management choices, 14
Farming
  community, 8, 16, 130, 141
  DSTs (*see* Decision support tools (DST))
  systems, 141
Feed demand, 178, 179
Field-based screening methods, 37
Field trials, 162
Fisheries, 288, 291
  See also Aquaculture

Index 539

Flood tolerant crops
   ADH1 and *PDC1*, 55
   agricultural yield, 51
   Amblyopyrum muticum, 57
   Arabidopsis thaliana plants, 56
   C_2H_4 accumulation, 52
   development, 58
   energy cost, 52
   environmental condition, 52
   epigenetic changes, 55
   ERFs, 52–54
   escape strategy, 52
   flooding conditions, 52
   food security, 51
   FR13A, 53
   GA, 52
   gene expression, 55
   gene pools, 56
   hormone, 52
   hydrological events, 52
   maize (*Zea mays*) plants, 55
   miRNAs, 55, 56
   plant DNA methylation patterns, 55
   quiescence strategy, 53
   stress resilient crops, 51
   SUB1 locus, 53
   TFs, 53–55
   Triticum aestivum, 57
   VIN-3, 55
   water excess, 52
   waterlogging, 52
   Zea maiz (milho), 57
Flooding, 28
Fluorophore, 93
Food and Agriculture Organization
   (FAO), 12
Food and nutritional security
   aquaculture production systems, 294
   capture fisheries, 292, 293
   fish, 292–293
   income, 292
   nutritional value and health benefits, 292
   omega-3 (n-3) fatty acids, 293
   shrimp aquaculture systems, classification,
      294, 295
Food consumption score (FCS), 219
Food demand, 176, 177, 183, 187
Food grains, 3
Food insecurity, 410, 413
   and climate change, 4
   genetic resources, 7, 8
   mechanisms, 12
Food production
   and consumption, 176–178

   on global land use
      barren land conversion, 185
      cropland and mosaic distribution,
         185–187
      distribution, 183, 184
      FAOSTAT harvested area, 183
      GDP per capita, 183
      HadGEM, 183, 184
      IMPACT model, 183
      IPSL, 183
      land cover types, 183, 184
      MODIS data, 185
      multinomial logit, 183
      population growth, 184
      urban areas, 185
Food production unit (FPU), 172
Food security
   agroforestry, 418–423
   AtoinMeto, 384, 387–388
      (*see also* AtoinMeto)
   Australia (*see* Australia)
   crop diversification
      (*see* Crop diversification)
   definition, 410
   destabilized, 5
   development, food production, 524
   Ethiopia, 438
   global, 91
   greenhouses, 411
   India (*see* India)
   innovative intervention, 413
   and insecurity, 410
   interventions, in northern CA, 411 (*see
      also* Subarctic Ontario (ON))
   locally based, 417
   NTT provincial government, 398–401
   SRI management
      higher yields, 238
      microbes, 238
      milling outturn, 238
      morpho-physiological changes, 237
      production and net income, costs of, 239
Food sovereignty, 131, 141
   AtoinMeto, 384, 392
Food supply chain
   adaptation, climate change (*see* Climate
      change adaptation)
   adaptive capacity, 480
   agricultural policies, 490
   application, 490
   characteristics, 480
   climate risk
      categories, 482
      conditions, 482

540   Index

Food supply chain (*cont.*)
  demand, new investment, 482
  distribution, 482
  exposure, sensitivity and adaptive capacity, 481
  focal corporations, 482
  long-term effects, 482
  organizations, 482
  production activities, 481
  regions/industry, 482
  small and medium-sized enterprises, 482
  vulnerability, 481
 consumers, 493
 coping, 480
 corporations and governments, 493
 crop cultivation, 493
 design, 480
 direct and indirect climate risks, 490
 EU policy measures, 492
 financial support, 492
 Finland, 492
 framework, 489
 impacts of climate change, 480
 multi-level governance (*see* multi-level governance, food supply chain)
 resilience (*see* Resilience)
 stakeholders, 491
Food system, 12, 342, 345, 346
Food wastage, India, 502, 512, 516
Former Soviet Union (FSU), 176
Fossil fuels, 76
FR13A, 53

G
GDP per capita, 183
Gene expression
 ABA dependent, 58
 epigenetic modifications, 54
 miRNAs, 55
 reprogramming, 61
Genetic manipulation, 41
Genetic material, 126
Genetic resources, 7, 8
Genetic variability, 60, 62
Genome editing, 510
 advantages, 100
 applications, 101
 CRISPR/Cas, 99–101
 engineered nucleases, 99
 error-prone repair mechanisms, 99
 pest resistance, 100
 precise insertion/deletion of DNA, 99
 virus resistance, 100

Genome wide association analysis (GWAA), 69
Genome-wide association studies (GWAS), 96
Genome-wide genotyping
 application, 96
 chickpea, 96
 costs, 96, 97
 GBS, 96
 GWAS, 96
 heat stress, 96
 high-density genotyping arrays, 96
 QTL, 96, 97
 RRS, 96
 salinization, 97
 SNPs, 96, 97
 temperature stress, 96
 WGR, 96, 97
Genome-wide selection (GWS), 69
Genomic breeding, 161
Genomic selection (GS), 69, 98, 99
Genomic sequencing technologies
 long read sequencing, 93
 optical mapping, 95
 SGS, 92, 93
Genomic technologies
 in crop breeding, 101, 102
 genome-wide genotyping, 96
 molecular markers, 92 (*see* Molecular markers, crop breeding)
 pangenome analysis, 95
 sequencing (*see* Genomic sequencing technologies)
 SNPs, 92
Genomics tools, 69
Genomics-assisted breeding, 164
Genotyping by sequencing (GBS), 96
Germplasm banks, 56
Ghana
 administrative units, Upper West Region, 360
 CCA programs, 361
 climate change, 361
 district's climatic dynamics, 360
 temperature, 360, 361
 Wa West district, 360
Gibberellins, 39
Giberellic acid (GA), 52
Global Action Plan for Agricultural Diversification (GAPAD), 143
Global affluence, 131
Global bioeconomic model
 AGRHYMET data base, 174
 agricultural area, 178–180
 agricultural productivity, 172
 climate change, 180, 181
 cultivated land, 181

Index 541

FAO database, 172
FAOSTAT agricultural database, 174
food consumption and production, 176–178
food insecurity, 172
food production (*see* Food production)
FPU, 172
GCMs, 172
household income, 172
IMPACT, 172, 173, 175, 176, 180–183
LandSHIFT, 174
livestock, 172
MAgPIE, 174
malnutrition, 172, 182
MODIS data, 174
NLSD, 174
population, 182
satellite data, 173
SDG2, 182
socioeconomic and climate projections, 175, 176
socioeconomic factors, 174
yields, 181
Global circulation models (GCMs), 172
Global climate models (GCMs)
   DSSAT crop modeling software, 175
   GHG emissions, 175
   HadGem2-ES, 175
   IPCC, 175
   IPSL-CM5A-LR, 175
   RCPs, 175
   SSPs, 175, 176
Global warming, 7
Glutathione, 38
Glutathione reductase (GR), 38, 39
Glutathione S-transferase, 38
Google Earth image, 222
The Green Revolution, 8, 10, 126, 127, 181, 235, 236, 249
Green tea, 312, 315
Greenhouse gas (GHG) emissions, 76, 175, 230, 239, 240, 245
Greenhouses, 411
Gross Domestic Product (GDP), 215
Groundwater irrigation, 508
Growing Degree Day (GDD) Calculator, 468
Guaiacol peroxidase, 38

H
Haber–Bosch process, 76
HadGEM, 183, 184
HadGem2-ES model, 175
Hard-to-cook, 158

Harvested areas, 179
Healthy foods, 410
Heat-induced cellular injury, 32
Heat resistant crops, 32
Heat shock proteins (HSPs), 36, 39
Heat stress, 7
Heat-stress tolerance
   breeding, 40
   crop management, 41, 42
   genetic manipulation, 41
Heat tolerance
   food production, 28
   global air temperature, 28
   global warming, 28
   heat resistant crops, 32
   heat shock/stress, 28
   high temperature, 28–31
   morphological (*see* Morphological adaptations)
   oxidative damage and plant defence mechanism, 38
   physiological mechanisms (*see* Physiological mechanisms)
   stress (*see* Heat-stress tolerance)
   summer crops, 31, 32
   winter crops, 29, 31
Heat Tolerant Maize for Asia (HTMA) Project, 73
Heat transcription factors (HSF), 39
Heat waves, 28
Hedonic Prices (HP) method, 256, 261
High-density genotyping arrays, 96
High-throughput precision phenotyping, 70
HiRise, 93
Histone acetyltransferases (*HAT*) genes, 59
Hormones, 38, 39
Hydrological cycle, 135, 136

I
ICAR, *see* The Indian Council for Agricultural Research (ICAR)
Import substitution
   agroforestry, 418–423
   definition, 417
   food localization, 417–418
   subarctic agroforestry intervention, 418, 419
Improved crop phenotypes, 233
Improved Maize for African Soils (IMAS) Project, 77
India
   adaptation strategies and challenges (*see* Adaptation)

India (*cont.*)
 climate change
  Annual Disaster Statistical Review, 498
  cyclones, 499
  drought, 498
  intensity of monsoons, 498
  IPCC, 498
  melting of snow, 499
  non-climatic factors, 499
  north-eastern states, 498
  rainfall decline, 498
  temperatures, twenty-first century, 498
 drama theory, 513–515
 impacts, food security
  agriculture yields, 499
  biophysical and social vulnerabilities, 500
  cold chain, 502
  complex and multidimensional, 500
  crops, 501
  demand, water, 501
  food prices, 503
  food wastage, 502
  GNP, 499
  hazards, 502
  malnourished children, 499
  moist weather conditions, 502
  nutritional intake, 501
  pests and diseases, 502
  population and food production, 500
  small and marginal farmers, 503
  soil fertility, 501
  technology storage, 503
  urban, 503
 implementation, 516
The Indian Council for Agricultural Research (ICAR), 504, 505
Indigenous Canadians, 410
Indonesia, 527
Inferior scale, 159
Inland fishery production, 7
Inland valleys, 212
Innovations, drip irrigation, 271, 272
 See also Drip irrigation technology
Integrated Nutrient Management (INM) Division, 203
Integrated pest management (IPM), 242, 247, 249
Intellectual property protection, 16
Intensive canal system, 10
Intergovernmental Panel on Climate Change (IPCC), 4, 115, 175, 498
International Institute of Tropical Agriculture (IITA), 80

International Model for Policy Analysis of Agricultural Commodities and Trade (IMPACT), 172, 173, 175, 176, 180–183
International Plant Nutrition Institute (IPNI), 202, 203
International Rice Research Institute (IRRI), 72
International wheat improvement network (IWIN), 71
IPSL-CM5A-LR model, 175
Irrigated rice production, 230, 241

K
Kranz anatomy, 120, 121

L
Labour supply, 247, 248
Land mosaic
 in Barcelona metropolitan area, 346
 to cultural landscapes, 342–347
 economic and social shifts, 340
 environmental and economic values, 340
 growing expectations, 347
 heavy grey engineering techniques, 341
 mosaics loss, at Metropolitan Region of Barcelona, 343
 from urban agriculture, 336–338
 urban plan for Barcelona, 341
Land planning, 341–342, 344
Landscape architecture, 333, 347
Landscape design, 332
LandSHIFT model, 174
Latin America and the Caribbean (LAC), 176
Leaf structural modifications, 33, 34
Leaf water status, 37
Ligation-based approaches, 92
Livelihood protection, 138, 139
Livestock production, 7
Local average treatment effect (LATE), 218, 219
Lock-in, irrigation
 dairies, 446
 drought, 447, 448
 dry climate and marginal land, 447
 feedlot systems, 447
 governance, corporations, 447
 market and technological drivers, 448
 private companies, 448
 research, 446
 SER, 448
 short-term cash flow, 447
 social welfare, 448

Index 543

transactions, 447
water markets, 447
Long read sequencing
genome assemblies, 93
short read synthesis, 94
single-molecule based, 94
Long-term drought, 28
Low-income countries, 16
Lund-Potsdam-Jena managed Land (LPJmL)
gridded dynamic vegetation
model, 174

M

Macroeconomic level, 8
MAgPIE model, 174
Maize, 55
climate adaptive breeding
carbon dioxide, 78, 79
drought, 68–71
flooding and salinity, 75, 76
heat, 73, 74
nutrient use efficiency, 76–78
nutrient use efficiency
NE, 202, 203
SSNM, 201, 202
production zones, 5
WUE
AWD, 194, 195
CA, 199, 200
crop diversification, 195, 196
DSR, 196, 197
irrigated systems, 192–194
rain-fed systems, 197–199
Maize-growing areas, 5
Malate, 118, 119
Malnutrition, 12
Malnutrition *vs.* under-nutrition, 157
Marker-assisted breeding (MAB), 69–70, 97, 98
Marker-assisted recurrent selection (MARS), 70
Marker-assisted selection (MAS), 61, 76
Mean temperatures, 5
Mesophyll cells, 114, 116–120
Metabolic transporters, 118
"Metropolitan Ecological Matrix", 334, 348, 349
Methane (CH_4) emissions, 245
Micronutrient deficiencies, 132
Micronutrients *vs.* calories, 157
MicroRNAs (miRNAs), 55, 56, 58, 61
Middle East and North Africa (MEN), 176
MinHash Alignment Process, 94
Miniature inverted-repeat transposable
elements (MITEs), 59
MinIon platform, 94

Mitigation
climate change, 13
local and global stresses, 13
scientific/technical innovation, 13
Modern Green Revolution, 127
MODIS data, 174, 185
Molecular markers, 92, 161
crop breeding
GS, 98, 99
MAB, 97, 98
Monoculture farming systems, 141
Moringa, 133
Morphological adaptations
canopy temperature, 34, 35
high temperature-induced yield losses,
33, 34
leaf structural modifications, 33, 34
plant responses, 33
reproductive development, 35, 36
root system architecture, 35
Multi-level governance, food supply chain
adaptation, 484, 485
climate change, 492
climate risk, 480, 489
design, 490
development, food supplies, 481
eco-efficiency, 485
external impulses, 481
FAO report, 483
framework, 480, 482, 484
hazards, 480
identification, risk, 484
intergovernmental organizations, 485
obligation, 493
policies adaptation, 485
private sphere, 484
public development and transition, 483
reduction, risk and uncertainty, 483
risk management, 483
risk reduction, 490
self-regulation, 483
supra-regional public risk governance, 485
trade-offs, 490
vulnerability, 481
Multiple advanced generation inter-cross
(MAGIC) populations, 164
Multiple cropping systems, 130

N

NADP-dependent malic enzyme (NADP-ME),
118, 119
National Climate Change Policy, 301–302
National Land Survey Dataset (NLSD), 174

544   Index

Natural resources management
  adaptation strategy, 508
  biophysical and socio-economic
    impacts, 509
  forest management programs, 508
  groundwater, 508
  reduction, climate change effects, 507
  river irrigations, 507
New Breeding Techniques (NBTs), 71
Next generation sequencing (NGS), 161
Niche markets, 160
Nitrogen fertilizers, 77
Nitrogen-use efficiency (NUE), 114, 115, 122
Nitrous oxide (N_2O) emissions, 245
No climate change (NoCC), 178
Non-photorespiratory processes, 38
Normalised difference vegetation index
  (NDVI), 37
Nutrient cycling, 129
Nutrient expert (NE), 202, 203
Nutrient use efficiency
  in maize, wheat and rice
    NE, 202, 203
    SSNM, 201, 202
Nutrition, 3
Nutritional security, 157

O

Off-farm diversification, 128
Omega-3 (n-3) fatty acids, 293
On-farm diversification, 129
Optical mapping, 95
Organochlorines, 418, 425, 426
Orphan crops, 160
Overheating, 32
Overlap-layout-consensus (OLC), 93
Oxaloacetic acid (OAA), 118
Oxford Nanopore, 94
Oxidative damage and plant defence
  mechanism
    antioxidant, 38, 39
    hormones, 38, 39
    HSPs, 39
    ROS, 38
Oyster-tecture proposal, 348–350

P

PacBio's sequencing platform, 94
Paddy fields, 230, 233, 240, 242–243, 245
Pangenome analysis, 95
Pathogen attacks, 7
Pathogen suppression, 137, 138

Pests, 137, 138
  control, 129
  and disease forecasting service (VIPS),
    Norway, 470, 471
  outbreaks, 7
  resistance, 100
Phosphoenolpyruvate (PEP), 118
Phosphoenolpyruvate carboxylase (PEPCase),
  114, 116, 118, 119
Phosphoglycolate, 113
Photorespiration, 38, 113–118
Photosynthesis, 38
  BS cell, 113, 114
  C_3, 113
  C_4 (see C_4 photosynthesis)
  phosphoglycolate, 113
  photorespiration, 113, 114
  Rubisco, 113
  types, 114
Physiological mechanisms
  leaf chlorophyll, 36, 37
  leaf water status, 37
  photorespiration, 38
  photosynthesis, 38
Plant breeding programs, 8
Plant defence mechanism
  and oxidative damage, 38
Plasmodesmatal connectivity, 120
PMT-P program (Supplementary Food –
  Recovery Program), 395, 396
Policy issues, 127
Poor cookability, 158
Poverty alleviation, 138, 139
PromethION, 94
Public Distribution System (PDS), India, 512, 513
Public goods, 257
*Public Intellectual Property Resource for
  Agriculture*, 16
Public-private partnership (PPPs) approach,
  465, 466
Public–private–community partnerships, 16
Pyruvate decarboxylase 1 (*PDC1*), 55

Q

QTL mapping, 61
Quantitative trait loci (QTL), 34, 40, 41, 61,
  96, 97
Quiescence strategy, 53

R

Radial oxygen loss (ROL), 52
Radiation use efficiency (RUE), 112, 114

Index

Rainfall distribution, 6
Rain-fed systems, 11, 197–199
Rainwater-fed pond, 11
Reactive oxygen species (ROS), 28, 38
Reduced representation sequencing (RRS), 96
Reduction, in GHG emissions, 240, 245
Relationships, rural with urban agriculture,
    334, 336–338, 342, 347
Representative concentration pathways
    (RCPs), 175
Reproductive development, 35, 36
Resilience
  food supply chains
    agri-food systems, 486
    cheap food, 488
    climate change adaptation, 486
    conflicts of interest, 487
    cultural changes, 486
    economic growth, 488
    equity in food access, 488
    flexibility and agility, 486
    food consumption patterns, 488
    global value community, 487
    low-inventory, 488
    multi-level governance, climate risk, 486
    poverty, 488
    private and public sectors, 488
    public adaptation policies, 487
    reconciliation, 487
    redundant stock, capacity and
      workers, 488
    strategies, 486
    technological change, 487
    transparency, 487
Revealed preferences methods, 256
Ribulose bisphosphate carboxylase oxygenase
    (Rubisco), 113, 114, 118, 119
Rice, 52
  C_4 photosynthesis (*see* C_4 photosynthesis)
  climate adaptive breeding
    carbon dioxide, 78, 79
    flooding and salinity, 75, 76
    heat, 75
    nutrient use efficiency, 76–78
  consumes, 212
  consumption, 212
  cultivation, 213
  farmers, 221, 223
  farming, 218
  intensification, 212
  nutrient use efficiency
    NE, 202, 203
    SSNM, 201, 202
  production, 112–113, 212, 213

WUE
  AWD, 194, 195
  CA, 199, 200
  crop diversification, 195, 196
  DSR, 196, 197
  irrigated systems, 192–194
  rain-fed systems, 197–199
  yield, income and food security,
    223–225
Risk
  communication, 16
  DST (*see* Decision support tools
    (DST))
Roche/Life 454, 93
Rogers' diffusion theory, 221–223
Root system architecture, 35
Rural space
  as chance, for cities adaptation, 348
  citizens and food, 347
  land planning, 342
  transport and energy exchange, 345
  vs. urban space, 333, 334

S

Salicylic acid, 39
Satellite data, 173
Sawah, Market Access and Rice Technologies
    for Inland Valleys (SMART-IV)
    project, 213–215, 222
Sea level rises (SLR), 7, 299, 300
Second generation sequencing (SGS),
    92, 93
Second green revolution, 112
Semi-arid regions of southern India, 8, 9
Sencha cultivation, 315
Shared socio economic pathways (SSPs),
    175, 176
Short read synthesised long read
    sequencing, 94
Shrimp aquaculture
  Chinese white shrimp, 290
  climate change impacts (*see* Climate
    change, on shrimp culture)
  Indian white shrimp, 290
  industry, in Sri Lanka, 291
  intensive shrimp farming, 289
  marine shrimp farming, 289
  Pacific white shrimp, 290
  production and trends, 290, 291
  species, 289, 290
  traditional shrimp culture, 289
Shrimp culture systems, 294, 295
Single-molecule based long read sequencing, 94

546 Index

Single-nucleotide polymorphisms (SNPs)
  application, 92
  arrays, 96, 99
  chips, 96
  cost, 96
  crop diversity and phenotypic variation, 92
  datasets, 97
  densities, 96
  DNA sequencing, 92
  in plant populations, 96
  in recombinant populations, 97
  stress-related genes, 97
  wheat spike ethylene, 96
Site Specific Nutrient Management (SSNM),
    201, 202
Small-scale fisheries, 7
Smart-valley approach
  adoption, 213, 219, 220
  agricultural practices, 213
  agriculture, 212
  climate change, 212
  data analysis, 218, 219
  description, 213, 214
  diffusion, 221–223
  dissemination, 213
  impact, 223–225
  rice cultivation, 213
  rice production, 213
  rice yield, income and food security,
    223–225
  sampling, data collection and processing,
    217, 218
  SMART-IV project, 213
  soil and water management technology, 213
  study area, 215, 216
  water control, 212
SmidgION, 94
Snorkel1 (SK1), 52
Snorkel1 (SK2), 52
Snowball approach, 217, 218
Social media, 161
Socioeconomic factors, 174
Soil and solar radiation, 126
South Asia (SAS), 175
Spain, drip irrigation, 271, 279
  See also Drip irrigation technology
Spatial scale, 14
Stakeholders perspectives, in CCA, 356–357
Standard rice management, 237
Stated preferences methods, 256, 261
Stay-green genotypes, 37
Strategic Environment Assessment (SEA),
    303, 305
Stress resilient crops, 51

SUB1 (Submergence 1) locus, 53
Subarctic Ontario (ON)
  climate-change projections, 414, 415
  community greenhouses, 411
  crops variety, in community and home/yard
    gardens, 419–422
  cultural context, 414
  food insecurity, 410
  Fort Albany, physical description, 414
  geographic location, 412, 413
  innovative food security intervention, 413
  potential nutrient content, locally-sourced
    materials, 423, 424
  protocol, Cree traditional food system,
    429, 430
  school-based nutritional educational
    programs, 410
  sharing-the-harvest
    intervention, 417
    lesser snow goose management,
    416–417
    potential lead contamination, 416
    traditional food intake, 415–416
  snow geese breeding, 411
Sub-Saharan Africa (SSA), 4, 127, 175, 212
Succulents, 32
Summer cereal crop, 32
Summer crops, 31, 32
Superoxide dismutase (SOD), 38, 39
Sustainable development, 483, 485
Sustainable development goals (SDGs), 12,
    13, 128
Sustainable livelihoods, AtoinMeto, 383, 385,
    388, 402
Synthesis-based approaches, 92
System of Crop Intensification (SCI), 234
System of Rice Intensification (SRI)
  agroecological principles (see
    Agroecological principles, irrigated
    rice)
  as climate-smart agriculture (CSA), 230
  to climate stress
    colder temperatures, 242
    drought resistance, 241, 242
    pests and diseases, 244, 245
    shorter growing season, 244
    storm damage, 242
    water productivity, 240
    water saving, 240, 241
    water use efficiency (WUE), 241
  comparison research, 236
  constraints and limits, 245, 246
    access to biomass, 247
    labour supply and motivation, 247, 248

Index 547

pest and disease control, 247
water control, 246
weed control, 247
critiques, 236
to food security (*see* Food security, SRI
management)
GHGs emissions, 245
Green Revolution, 236
growing environment, plants, 237
methods and effects, 230
morphological and physiological
effects, 237
new rice management, 237
phenotypical difference, 230, 231
principles and practices, 235
researchers' evaluations, 235
root systems, 231
soil-plant microbiome, 236
synergistic effects, 234
System of Rice Intensification/SRI methods, 525
System of Wheat Intensifications (SWI)
methods, 241

T
Temporal scale, 15
Terroir concept, Uji Tea cultivation
agriculture practices elements
application, 325
bio-climatic indicators, 326
cultivation interventions, 325
cultivation practices, 324
fertilizer and soil maintenance, 325
leaf covering, 325
micro-climate condition, 325
defined, 321
implementation, 321, 322
natural environmental elements
agriculture plants, 322
air temperature, 322–323
clay soil, 323
climatic factor, 322
cultivars, 324
factors, 321
humidity factor, 323
samidori cultivar, 324
soil factor, 323
tea plantations, 324
tea plants, 324
topographical factor, 323, 324
Thermo-tolerance mechanism, 32
Tocopherols, 38
Transcription activator-like effector nucleases
(TALENs), 99

Transcription factors (TFs), 55
abiotic stresses, 53, 54
accessibility, 54
AP2/EREBP, 54
epigenetic regulation, 54
genetic transformation, 54
plant tolerance, 53
transcriptional regulation, 53, 54
Transgenic, 61
Transposable element (TE) activity, 59, 60
Travel cost method, 256, 258–260, 265
Tribal community, 384
Triticum aestivum, 57
Tropical forests, 7

U
Uji tea cultivation
adaptation methods, 320
average mean air temperature, 317, 318
climatic conditions
air temperature, 317
changes, 316, 317
extreme weather events, 319
observed changes, harvesting
period, 317
temperature, 316
climatic factors, 319
covering materials, 313, 314
definition, Uji Tea, 315, 316
location, Uji Area, 312, 315
natural environmental characteristics, 316
social surveys, 319
tea growers, 313
tea leaf plucking under shade, 314, 315
terroir concept, 321 (*see also* Terroir
concept, Uji Tea cultivation)
traditional, 312, 313
types of tea, 315
vs. winegrape cultivation, in France's, 320
UK Department for International Development
(DIFID), 70
Underutilised crops, 162
application, 162
and Bambara groundnut (*see* Bambara
groundnut)
diversification (*see* Crop diversification)
global production levels, 159
international research centre, 152
planning, 163
United Nations food security, 442
United States Agency for International
Development (USAID), 70
Unsaturated fats, 157

548

Index

Urban agriculture
  agrarian park, 338, 339
  agro-forestry mosaic, in Barcelona, 347
  benefits, 332
  challenges, 336
  cities and urban areas, 336
  delta agriculture, in Baix Llobregat, 338
  description, 333
  green/blue infrastructure, 348
  to land mosaics, 336–338
  outdoor activity, 335
  planning, 342
  population growth, 336
  and rural relationship, 334
  vs. rural space, 333
  space and time, 332
  technical innovation, 334
  Yokohari's types, 337, 338
Urban space
  citizens and food, 347
  land planning, 342
  transport and energy exchange, 345
  vs. rural space, 333, 334
Uruguay's National Agricultural Information
    System (SNIA), 472, 473

V
Value transfer method, 258, 259,
    262, 263
Vernalization insensitive 3 (*VIN-3*), 55
Virus resistance, 100
Virus-transmitting vectors, 137
Vulnerability, 127, 130, 131, 138
  Australia (*see* Australia)
  food supply chain (*see* Food supply
    chain)

W
Warmer zones, 137
Water
  availability, 270, 282
  drip irrigation, 270
  economic and social activities, 270
  effectiveness, in agriculture, 270
  excess, 52
  management, 17, 18
  reservoir, 10
  reuse, 255
  scarcity, 136
  stress, 68

  sustainable use, 270
  traditional irrigation systems, 270
Water deficit calculator, CSF, 470
Water Efficient Maize for Africa
    (WEMA), 70
Water management, *see* Smart-valley
    approach
Water use efficiency (WUE), 112, 114, 115
  in maize, wheat and rice
    AWD, 194, 195
    CA, 199, 200
    crop diversification, 195, 196
    DSR, 196, 197
    irrigated systems, 192–194
    rain-fed systems, 197–199
Waterfowl harvesting, 414, 416, 417, 431
Waterlogging, 52
Weather extremes, 112
Weather-dependent activity, 134
West Africa, *see* Smart-valley approach
West Timor, AtoinMeto
  communities, in West Timor, 385
  description, 386
  food availability, 390
  in Kupang and TTS Regencies, 386, 387
  land area, 386
  local knowledge, 382, 383, 386, 389
  nutrition security, in East Nusa Tenggara
    (NTT), 393, 395
  semi-arid region, 383, 402
  subsistence community, 383–384
  sustainable management, water sources,
    389, 398
Wetlands
  in agricultural landscapes, 258
  desalinated water, 255
  recreational activities, 256
  restoration, 255
  valuation methods, 256 (*see also*
    Environmental valuation methods,
    wetlands)
  water reuse, 255
Wheat
  climate adaptive breeding
    carbon dioxide, 78, 79
    drought, 71
    flooding and salinity, 75, 76
    heat, 74, 75
    nutrient use efficiency, 76–78
  nutrient use efficiency
    NE, 202, 203
    SSNM, 201, 202

Index 549

WUE
    AWD, 194, 195
    CA, 199, 200
    crop diversification, 195, 196
    DSR, 196, 197
    irrigated systems, 192–194
    rain-fed systems, 197–199
Whole-genome resequencing (WGR), 96, 97

Winter crops, 29, 31
World Food Programme (WFP), 219

Z
Zea maiz (milho), 57
Zero-mode wave guides, 94
Zinc finger nucleases (ZFNs), 99